Horizontal Gene Transfer

Horizontal Gene Transfer

Edited by

Michael Syvanen

Department of Medical Microbiology and Immunology
University of California
Davis
California
USA

and

Clarence I. Kado

Davis Crown Gall Group
University of California
Davis
California
USA

CHAPMAN & HALL
London · Weinheim · New York · Tokyo · Melbourne · Madras

Published by
Chapman & Hall, an imprint of Thomson Science, 2–6 Boundary
Row, London SE1 8HN, UK

Thomson Science, 2–6 Boundary Row, London SE1 8HN, UK

Thomson Science, 115 Fifth Avenue, New York, NY 10003, USA

Thomson Science, Suite 750, 400 Market Street, Philadelphia, PA 19106, USA

Thomson Science, Pappelallee 3, 69469 Weinheim, Germany

First edition 1998

© 1998 Chapman & Hall

Thomson Science is a division of International Thomson Publishing

Typeset in 10/12pt Palatino by Cambrian Typesetters, Frimley, Surrey

Printed in Great Britain by The University Press, Cambridge

ISBN 0 412 79310 5

A catalogue record for this book is available from the British Library

Library of Congress Catalog Card Number: 98–70514

♾ Printed on acid-free text paper, manufactured in accordance with ANSI/NISO Z39.48–1992 (Permanence of Paper).

Contents

Contributors

Adkins, Ronald M.
Human Genetics Center, SPH
University of Texas Health
Sciences Center at Houston
PO Box 20334
Houston, TX 77225
USA

Anthony, Karen
Department of Biological Sciences
University of Alberta
Edmonton T6G 2EP
Canada

Avancini, Rita
Departamento de Parasitologia
I.B.-C.P. 6109
Universidade Estadual de
Campinas
13083-970 Campinas (SP)
Brazil

Barreiro, Virginia
Department of Genetics
University of Stockholm
S-10791 Stockholm
Sweden

Berry, William B.N.
Department of Geology and
Geophysics
University of California
Berkeley, CA 94720
USA

Bessen, Debra
Department of Epidemiology
and Public Health
Yale University
60 College Street
P.O. Box 3333
Newhaven, CT 06510–8039
USA

Blum, Stephanie
Genetik
Fachbereich Biologie
Carl-von-Ossietzky Universität
Oldenburg
Postfach 2503
D-26111 Oldenburg
Germany

Boussy, Ian A.
Department of Biology
Loyola University of Chicago
6525 N. Sheridan Rd
Chicago, IL 60626
USA

Briles, David
Department of Microbiology
BBRB 658
University of Alabama,
Birmingham
845 19th Street South
Birmingham, AL 35294
USA

Burmester, A.
Lehrstuhl für Allgemeine,
Microbiologie & Mikrobengenetik
Friedrich-Schiller-Universität Jena
Neugasse 24, D-07743
Germany

Calendar, Richard
Department of Molecular and Cell
Biology
University of California
Berkeley, CA 94720
USA

Carlson, Karin
Department of Microbiology
Box 581
University of Uppsala
Biomedical Center
S-75123 Uppsala
Sweden

Christie, Gail
Department of Microbiology and
Immunology
Virginia Commonwealth
University
Richmond, VA 00000
USA

Clark, Jonathan B.
Department of Ecology and
Evolutionary Biology
310 Biosciences West
University of Arizona
Tucson, AZ 85721
USA

Cooper, Andrew J.
Department of Microbiology
407 So. Goodwin
University of Illinois
Urbana, IL 61801
USA

Courvalin, Patrice
Unité des Agents Antibactériens
CNRS EP J0058
Institut Pasteur
25 rue du Docteur Roux
75724 Paris Cedex 15
France

Cryderman, Diane
Department of Biochemistry
4–711 Bowen Science Bldg
University of Iowa
Iowa City, IA 52242-1109
USA

Czempinski, K.
Lehrstuhl für Allgemeine,
Microbiologie & Mikrobengenetik
Friedrich-Schiller-Universität, Jena
Neugasse 24, D-07743
Germany

Davenport, Laura
Department of Molecular Biology
Vanderbilt University
Nashville, TN 00000
USA

Day, Martin
School of Pure and Applied
Biology
University of Wales
PO Box 915
Cardiff CF1 3TL
UK

Doolittle, Russell F.
Center for Molecular Genetics
University of California, San Diego
La Jolla, CA 92093
USA

Fabry, Stefan
Lehrstuhl für Genetik
Universität Regensburg
D-93040 Regensburg
Germany

Fekete, Richard
Department of Biological Sciences
University of Alberta
Edmonton T6G 2EP
Canada

Frantsve, Julie
Department of Biology
Loyola University of Chicago
6525, N. Sheridan Rd
Chicago, IL 60626
USA
and
220 Perkins Hall
Harvard University
5 Oxford St
Cambridge, MA 02138
USA

Frost, Laura
Department of Biological Sciences
University of Alberta,
Edmonton
Alberta T6G 2E9
Canada

Fründt, Corinne
The Friedrich Miescher Institute
PO Box 2543
CH-4002 Basel
Switzerland

Gelvin, Stanton B.
Department of Biological Sciences
Purdue University
West Lafayette, IN 47907-1392
USA

Gogarten, Peter J.
Department of Molecular and Cell
Biology
Universitiy of Connecticut
Storrs, CT 06269-3044
USA

Goussard, Sylvie
Unité des Agents Antibactériens
CNRS EP J0058
Institut Pasteur
25 rue du Docteur Roux
75724 Paris Cedex 15
France

Grillot-Courvalin, Catherine
Unité des Agents Antibactériens
CNRS EP J0058
Institut Pasteur
25 rue du Docteur Roux
75724 Paris Cedex 15
France

Graupner, Stefan
Genetik
Fachbereich Biologie
Carl-von-Ossietzky Universität
Oldenburg
Postfach 2503
D-26111 Oldenburg
Germany

Haggård-Ljungquist, Elisabeth
Department of Genetics
University of Stockholm
S-10691 Stockholm
Sweden

Hall, Ruth M.
CSIRO Division of Biomolecular
Engineering
PO Box 184
North Ryde 2113, NSW
Australia

Hartman, Hyman
IASB
880 Spruce Street
Berkeley, CA 94707
USA

Heinemann, Jack A.
Department of Plant and Microbial
Sciences
University of Canterbury
PB 4800
Christchurch
New Zealand

Hilario, Elena
Department of Molecular and Cell
Biology
University of Connecticut
Storrs, CT 06269-3044
USA

Hollingshead, Susan K.
Department of Microbiology
BBRB 654
University of Alabama,
Birmingham
845 19th Street South
Birmingham, AL 35294
USA

Ichikawa, Takanari
Max Planck Institut für
Züechtungsforschung
50829 Cologne
Germany

Itoh, Masanobu
Department of Biology
Loyola University of Chicago
6525 N. Sheridan Rd
Chicago, IL 60626
USA
and
Department of Applied Biology
Kyoto Institute of Technology
Matsugasaki
Sakyo-ku, Kyoto 606
Japan

Kado, Clarence I.
Davis Crown Gall Group
University of California
Davis, CA 95616
USA

Kidwell, Margaret G.
Department of Ecology and
Evolutionary Biology
310 Biosciences West
University of Arizona
Tucson, AZ 85721
USA

Kirk, David L.
Department of Biology
Washington University
St. Louis, MO 63130
USA

Klimke, William A.
Department of Biological Sciences
University of Alberta
Edmonton T6G 2E9
Canada

Kobayashi, Michiyoshi
Institute of Applied Biochemistry
University of Tsukuba
Tsukuba, Ibaraki 305
Japan

Krassilov, Valentin A.
Palaeontological Institute
Profsoiuznaya 123
Moscow, 117647
Russia

Lake, James A.
Molecular Biology Institute and
MCD Biology
University of California
Los Angeles, CA 90095
USA

Lampe, David J.
Department of Entomology
University of Illinois at Urbana-
Champaign
505 So. Goodwin
Urbana, IL 61801
USA

Lawrence, Jeffrey
Dept. of Biological Sciences
215 Capp Hall
University of Pittsburgh
Pittsburgh, PA15260
USA

Li, Wen-Hsiung
Human Genetics Center
University of Texas Health Science
Center at Houston
PO Box 20334
Houston, TX 77225
USA

Lorenz, Michael G.
Genetik
Fachbereich Biologie
Carl-von-Ossietzky Universität
Oldenburg
Postfach 2503
D-26111 Oldenburg
Germany

Manchak, Jan
Department of Biological Sciences
University of Alberta, Edmonton
Alberta T6G 2E9
Canada

Martin, William, F.
Institüt für Genetik
Technische Universität
Braunschweig
Spielmannstr. 7
D-38023 Braunschweig
Germany

McKane, Melissa
Department of Biological Sciences
138 Biology Bldg
University of Iowa
Iowa City, IA 52242-1324
USA

McWeeny, Kerri
Genomics and Molecular Biology
Dept. L/MD
Abbott Laboratories
100 Abbott Park Road
Abbott Park IL 60064

Meacham, Christopher A.
Jepson Herbarium
1001 Valley Life Sciences Building
University of California
Berkeley, CA 94720
USA

Meins, Frederick, Jr
Friedrich Miescher Institute
PO Box 2543
CH-4002 Basel
Switzerland

Meyer, Alain D.
Institut des Sciences
V'tales, CNRS
91198 Gif sur Yuette
France

Meyer, Birte
Genetik
Fachbereich Biologie
Carl-von-Ossietzky Universität
Oldenburg
Postfach 2503
D-26111 Oldenburg
Germany

Milkman, Roger
Department of Biological Sciences
138 Biology Bldg
University of Iowa
Iowa City, IA 52242-1324
USA

Miller, Robert V.
Department of Microbiology and
Molecular Genetics
307 LSE
Oklahoma State University
Stillwater, OK 74078
USA

Mosig, Gisela
Department of Molecular Biology
Vanderbilt University
Nashville, TN 37203
USA

Myung, Heejoon
Department of Molecular and Cell
Biology
University of California
Berkeley, CA 94720
USA

Nam, Jaesung
Department of Biological Sciences
Purdue University
West Lafayette, IN 47907-1392
USA

Odegrip, Richard
Department of Genetics
University of Stockholm
S-10691 Stockholm
Sweden

Olendzenski, Lorraine
Department of Molecular and Cell
Biology
University of Connecticut
Storrs, CT 06269-3044
USA

Plummer, Dahlia
Department of Biology
City College of the City University
of New York
Convent Avenue at 138th Street
New York, NY 10031
USA

Purugganan, Michael D.
Box 7614, Department of Genetics
North Carolina State University
Raleigh, NC 27695
USA

Raleigh, Elisabeth
New England Biolabs
32 Tozer Road
Beverly, MA 01915
USA

Ripp, Steven
Department of Microbiology and
Molecular Genetics
307 LSE
Oklahoma State University
Stillwater, OK 74078
USA

Rivera, Maria
Molecular Biology Institute and
MCD Biology
University of California,
Los Angeles
Los Angeles, CA 90095
USA

Robertson, Hugh M.
Department of Entomology
320 Morrill Hall
University of Illinois at Urbana-
Champaign
505 So. Goodwin
Urbana, IL 61801
USA

Roth, John
Department of Biology
University of Utah
Salt Lake City, UT 84112
USA

Salyers, Abigail A.
Department of Microbiology
University of Illinois
601 So. Goodwin
Urbana, IL 61801USA

Schmitt, Rudiger
Lehrstuhl für Genetik
Universität Regensburg
D-93040 Regensburg
Germany

Shoemaker, Nadja B.
Department of Microbiology
University of Illinois
407 So. Goodwin
Urbana, IL 61801
USA

Shoun, Hirofumi
Institute of Applied Biochemistry
University of Tsukuba
Tsukuba, Ibaraki 305
Japan

Sikorski, Johannes
Genetik
Fachbereich Biologie
Universität Oldenburg
Postfach 2503
D-26111 Oldenburg
Germany

Simmons, Gail M.
Department of Biology
City College of the City University
of New York
Convent Ave. at 138th Street
New York, NY 10031
USA

Simon, Alex
Department of Biology
City College of the City University
of New York
Convent Ave. at 138th Street
New York, NY 10031
USA

Soto-Adames, Felipe N.
Department of Entomology
University of Illinois at Urbana-
Champaign
505 So. Goodwin
Urbana, IL 61801
USA

Stark, Klaus
Lehrstuhl für Genetik
Universität Regensburg
D-93040 Regensburg
Germany

Syvanen, Michael
Department of Medical
Microbiology and Immunology
University of California
Davis, CA 95616
USA

Takaya, Naoki
Institute of Applied Biochemistry
University of Tsukuba
Tsukuba, Ibaraki 305
Japan

Wackernagel, Wilfried
Genetik
Fachbereich Biologie
Universität Oldenburg
Postfach 2503
D-26111 Oldenburg
Germany

Walden, Kimberly K.O.
Department of Entomology
University of Illinois at Urbana-
Champaign
505 So. Goodwin
Urbana, IL 61801
USA

Williamson, Donald I.
Port Erin Marine Laboratory
University of Liverpool
Isle of Man, IM9 6JA
UK

Wittstock, Marcus
Genetik
Fachbereich Biologie
Carl-von-Ossietzky Universität
Oldenburg
Postfach 2503
D-26111 Oldenburg
Germany

Wöstemeyer, Joh.
Lehrstuhl für Allgemeine,
Microbiologie & Mikrobengenetik
Friedrich-Schiller-Universität, Jena
Neugasse 24
D-07743 Jena
Germany

Wöstemeyer, A.
Lehrstuhl für Allgemeine,
Microbiologie & Mikrobengenetik
Friedrich-Schiller-Universität, Jena
Neugasse 24
D-07743 Jena
Germany

Yu, Sidney
Department of Molecular and
Cell Biology
University of California
Berkeley, CA 94720
USA

Plate 1 (*left to right*): (a) Chris Meacham, Hyman Hartman, John Roth, Roger Milkman; (b) Abigail Salyers, Ved Malik; (c) Catherine Pujol, Patrice Courvalin, Catherine Grillot, Hugh Robertson; (d) Russell Doolittle, Wen-Hsuing Li; (e) Jack Heinemann, Erh-Min Lai; (f) Audience of a video seminar given by Don Williamson; (g) Michael Lorenz, Wilfried Wackernagel, Robert Miller, Michael Syvanen; (h) Cynthia Liebert, Juan Lopez-Pila, Gerald Liddel, Michael Puruggnan, Bill Martin.

(a)

(b)

(c)

(d)

(e)

(f)

(g)

(h)

Plate 2 (*left to right*): (a) View from dining commons of conference centre; (b) sail boats for use by participants; (c) Laura Frost, Gerald Liddel, Richard Calendar; (d) DeLill Nasser, Rosemary Redfield, Hugh Robertson; (e) Michael Syvanen, Susan Hollingshead; (f) Stanton Gelvin, Lan-Ying Lee, Phillip Harriman, Gerald Liddel, DeLill Nasser; (g) Grace Richter, Laura Frost; (h) Jack Heinemann, Lori Daane.

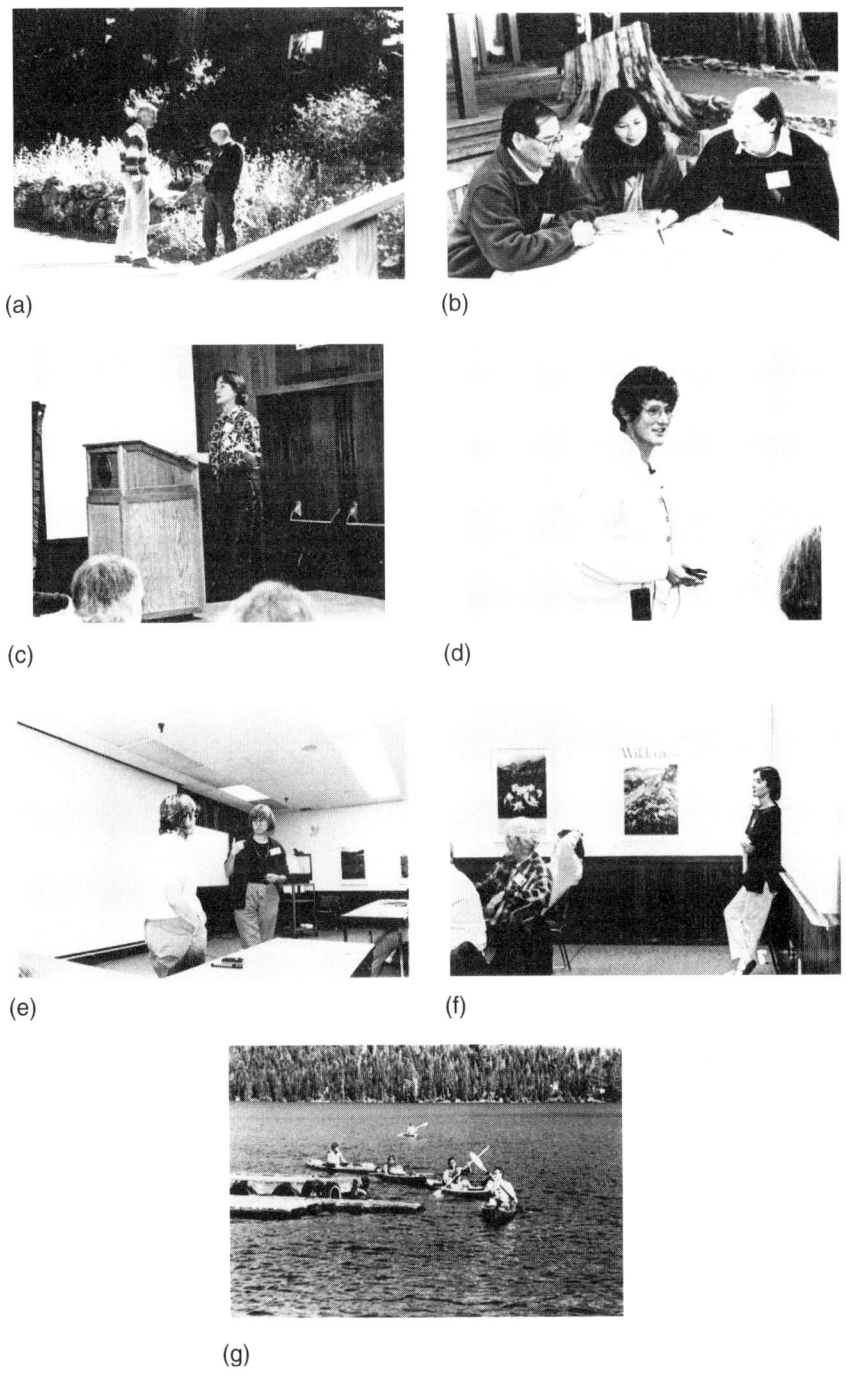

Plate 3 (*left to right*): (a) Wilfried Wackernagel, Jan Zadoks; (b) Clarence Kado, Lan-Ying Lee, Ruth Hall; (c) Abigail Salyers; (d) Laura Frost; (e) Susan Hollingshead, Anne Summers; (f) Abigail Salyers and audience; (g) Canoeing on Fallen Leaf Lake

Preface

The horizontal transmission of nucleic acids is thoroughly documented, especially with viruses and other host–parasite relationships. With the increasing body of information deposited in the sequence databases, DNA transfer between distinct groups of organisms and between kingdoms is becoming apparent. In this book, we include examples of gene transfer involving all of life's kingdoms. This necessitated bringing together a highly disparate group of active scientists to discuss and review the current understanding of horizontal gene transfer. This brings into a single volume such disparate subjects as movement of DNA into bacterial cells, mobile genetic elements in metazoans, plant and animal infection mechanisms, molecular evolution, and macroevolutionary theory. An underlying interest in this subject concerns the safety of releasing genetically engineered organisms into the environment.

The motivation for bringing people from different areas together at the Fallen Leaf Lake Conference in September, 1996, was to assess a single hypothesis. It has been conjectured that horizontal gene transfer plays a significant role in evolution. If this is true then three conditions must be realized. First, can genes, or more specifically DNA, move from one species to an unrelated one? Thus, a section of this book is devoted to the subject of transfer mechanisms, a phenomenon well documented in bacteria. The discovery of plasmid transfer laid the basis for our current thinking about horizontal gene transfer. Transfer mechanisms also include microorganisms that transfer DNA to eukaryotic cells; hence a number of papers are given on *Agrobacterium*, on bacterial transfer to yeast, on actinomycetes, and on eukaryotic parasites that inject their genomes into their host's cells.

Second, what is the evidence that genes transfer in nature? The primary evidence supporting evolutionarily significant horizontal transfers involves phylogenetic reasoning. This is an area where the evidence is accumulating in the gene and protein sequence data bases. Two problems are repeatedly encountered: defining the topology of a gene tree and estimating divergence times following molecular clock assumptions. There are several contributions discussing results obtained from phylogenetic analysis and problems associated with this approach.

The third question raised by the central hypothesis is that if the mechanisms exist and events can be documented, does horizontal gene transfer actually play any significant evolutionary role? Or, is a theory that incorporates migrant DNA useful in explaining more general biological phenomena? To this end, there are chapters that directly address macroevolutionary patterns and trends.

We gratefully acknowledge and are indebted to the following sponsors who enabled us to bring together the contributors to this volume.

- The School of Medicine, The Division of Biological Sciences and The College of Agricultural and Environmental Sciences of the University of California, Davis
- The National Science Foundation (grant number 9618669)
- The United States Department of Agriculture (grant number USDA APHIS 96-1100-0073)
- The University of California Systemwide Biotechnology and Education Program
- The National Institutes of Health (grant number PHS AI41259-01).

We are most grateful for additional support from the Sorvall Company, and Blackwell Scientific Publishers. The Fallen Leaf Lake Conference on Horizontal Gene Transfer (Fallen Leaf Lake, California, September 12–15, 1996) was chaired by Michael Syvanen. We acknowledge the help of Anne Summers, Michael Lorentz, Robert Miller, Laura Frost, Margaret Kidwell and Russell Doolittle, as members of the international organizing committee. Clarence Kado was the director of the conference.

The chapters in this volume were peer reviewed prior to their acceptance and we are most grateful for the help of the following reviewers: Michael Lorenz, Margaret Kidwell, Robert Miller, Michael Hart, David Demezas, Moses Vijayakumar, Lorain van Waasbergen, Damon Lisch, Jonathan Clark, Joana Silva, Patrick O'Grady, Holly Wichman, Martin Wojcieshowski, Ray Wu, Jack Meeks, Laura Frost, Stanton B. Gelvin, Anne Britt, Christopher Schardl, Allan Campbell and Rüdiger Schmidt.

Michael Syvanen
Clarence Kado

Introduction

1

Michael Syvanen and Clarence I. Kado

In 1959, a series of papers appeared showing that Hfr strains of *Escherichia coli* could transfer genetic information to certain mutant strains of *Salmonella typhimurium* (Baron *et al.*, 1959; Miyake and Demerec, 1959). In that same year, Tomochiro Akiba and Kunitaro Ochiai (Ochiai *et al.*, 1959; Akiba *et al.*, 1960) discovered infectious multiple-antibiotic resistant plasmids in pathogenic bacteria. This discovery led directly to the finding that these resistance plasmids moved between different bacterial species (Mitsuhashi, 1977), thereby demonstrating that horizontal gene transfer was a natural process among wild-type bacteria. The Akiba and Ochiai results changed the way we think about the epidemiology of antibiotic resistant bacteria. Studies on the agents responsible for these horizontal transfers resulted in many of the findings that were employed, with surprising ease, in the creation of the first transgenic organisms in the 1970s.

Papers probing the theoretical implications of horizontal gene transfer did not begin to appear until the 1970s. Some of these early conjectures cite horizontal gene transfer as a mechanism to explain widespread convergence and homoplasy in plant evolution (Krassilov, 1977) and plant systematics (Went, 1971), in animal evolution (Anderson, 1970; Reanney, 1976) and in speciation (Hartman, 1977). Other speculative ventures included explanations for the biological unities (Syvanen, 1984) and for the widespread occurrence of parallelisms in the fossil record (Krassilov, 1977; Erwin and Valentine, 1984; Reanney, 1976; Syvanen, 1984).

By the mid 1980s, numerous mechanisms for natural horizontal gene transfer were firmly established, not only for bacteria but also for metazoans. Horizontal gene transfer had become an increasingly plausible explanation for a number of puzzling taxonomic and paleontological phenomena. However, there remained a paucity of observations giving direct support to these speculations. With the rapid increase in the nucleic acid database over the past decade, this situation is changing, and this book covers some of these more recent developments.

Today, researchers in many unrelated areas are making observations related to horizontal gene transfer, leading to the unusual breadth of topics incorporated in this volume. This book is not an attempt at a comprehensive survey of horizontal gene transfer; it focuses primarily on material from active research areas.

Part One deals with mechanisms of gene transfer. This section is dominated by research into bacterial plasmids – a natural emphasis, given the large number of plasmids that will stimulate conjugal transfer of DNA from bacteria to an extremely broad range of other organisms, including other bacteria, yeasts and fungi and plants. The chapters by Heineman (Chapter 2) and Kado (Chapter 6) deal with the evolution of conjugal plasmids. These are the genetic units that have evolved a flexibility that enables them to survive in hosts during vertical evolution, as well as to adapt to new hosts after horizontal transfer. A sampling of the diversity of mechanisms controlling the mobility of plasmids and their associated genes is presented in the chapters by Klimke *et al.* (Chapter 3), Salyers *et al.*, (Chapter 4) and Hall (Chapter 5). Three chapters (7, 8 and 9) are devoted to *Agrobacterium tumefaciens* and emphasize the importance of this plant pathogen as a model for horizontal DNA transfer. In addition, there are two chapters of non-plasmid mediated transfer into other eukaryotic cells; these mechanisms include endocytosis mediated by mammalian cell transfer (Courvalin *et al.*, Chapter 9) and a fungus-to-fungus endoparasitism (Wöstemeyer *et al.*, Chapter 10).

Those who have written about horizontal gene transfer have usually assumed that viruses would be major vectors (Syvanen, 1987). This assumption is based on the fact that many viruses have broad host ranges which allow them to infect unrelated species. Additionally, viruses can carry specific polynucleotides from their previous host in their virions and can incorporate host genes into their genomes. Little new information on this subject is available and because this book emphasizes active research areas, we have omitted a discussion of metazoan or plant viruses as vectors of transmission.

The abundance of transfer mechanisms among bacteria would make it appear that horizontal DNA transfer would be easily observed in nature. However, for genes encoded on bacterial chromosomes, as opposed to those on plasmids, actually documenting the occurrence of horizontal gene transfer in nature has only been accomplished in recent years. Indeed, as recently as 1987, results on the distribution of enzyme electromorphs, as interpreted using mathematical population genetic theory, led to the conclusion that there was no horizontal gene movement even among demes of *Escherichia coli*. Soon after these negative results were announced, reports of naturally occurring horizontal transfer began to appear with increasing frequency. Part Two deals with this subject from two different perspectives. The first is the question of the efficiency and operation of known transfer mechanisms in nature – namely, under what natural conditions can phage transduction (Miller and Ripp, Chapter 14) and free DNA transformation (Lorenz *et al.*, Chapter 11; Day, Chapter 12; Wackernagel *et al.*, Chapter 13) be detected? The involvement of conjugal transfer in the spread of antibiotic resistance genes on plasmids has been thoroughly documented over the past decades (Levy and Miller, 1989). Horizontal transfer of chromosomal genes also occurs, as is documented by a group of chapters using comparative sequence analyses of genes from contemporary populations of bacteria. Milkman *et al.* (Chapter 17) convincingly show that exchange has been occurring among the *E. coli* demes, even though each deme maintains its own genetic identity. Lawrence and Roth (Chapter 16) compare the genomes of *E. coli* and *S. typhimurium* and, using a clever deductive argument, estimate the frequency at which entire clusters move into (and are lost from) their chromosomes. Chromosomally encoded antibiotic resistance genes in pathogenic *Neisseria* and *Streptococcus* have been observed with mosaic patterns; i.e. short regions within individual genes appear to have been derived from different species in the same genera (Spratt *et al.*, 1992). Hollingshead (Chapter 15) shows that genes for some of the outer membrane proteins in pathogenic streptococci also display mosaic patterns. Calendar *et al.* (Chapter 18) contribute another piece of evidence that the double-stranded DNA bacteriophage from enteric bacteria also have mosaic chromosomes. This chapter provides additional support for the cassette theory of bacteriophage evolution that was outlined by Campbell and Botstein (1983).

Part Three covers the most thoroughly documented example of horizontal gene transfer among metazoans – namely, the mobile genetic elements found in the drosophilids. Since the movement of *P*-factors into natural fly populations occurred in recent years, we have been able to observe its lateral movement and, further, because laboratory strains of *D. melanogaster* have been continuously maintained during this period, natural isolates today can be compared with ancestral strains. Kidwell (Chapter 19) analyzes the sequence of *P*-factors from a great variety of drosophilid species in an analysis designed to trace *P*-factor movement. There is a large number of mobile elements in *Drosophila*; a number of their phylogenetic trees are conspicuously incongruent with the corresponding species tree. Robertson *et al.* (Chapter 20) illustrate this phenomenon using the case of the *mariner* transposon, as do Simmons *et al.* (Chapter 21) with reference to the retroposon, *hobo*. Purugganan (Chapter 22) describes an interesting mobile element, called *waxy*, that acts like a mobile intron. He shows that *waxy* insertions into transcribed genes are processed out of the resulting mRNA transcripts.

The chapters in Part Four provide a number of examples of transkingdom gene transfers and a discussion of the criteria for identifying such transfers. These examples were revealed in the course of molecular evolutionary analysis of protein sequences. Attention was focused on cases in which specific gene trees were found to be incongruent with the underlying species trees. Doolittle (Chapter 23) gives an overview and reviews some past reports with a particular emphasis on what we recognize today as erroneous claims. This history has helped to establish a more rigorous criterion for identifying gene transfers that occur between remotely related organisms, especially when these transfer events are ancient. Chapters 24–26 discuss a number of specific transkingdom transfers involving eukaryotic recipients.

Chapters 27–30 deal with very ancient branching patterns. Unexpectedly to us at the time we assembled these manuscripts, we found that these chapters were relevant to theories of the origin of the eukaryotic cell. These chapters provide additional empirical support for previously purely speculative theories of the endosymbiotic origin of the eukaryotic nucleus. Because no chapter in this volume discusses possible implications of these exciting findings, we will deal with them separately below.

Finally, Part Five contains articles on macroevolutionary trends. Krassilov (Chapter 31) brings his arguments on the polyphyly of angiosperm origins up to date, and Berry and Hartman (Chapter 32) document the parallelisms in the graptolite fossil record. Independently of molecular evidence, Williamson has argued that species from different invertebrate phyla have hybridized on a number of occasions and that this accounts for some remarkable parallelisms seen between larval forms from different phyla. His ideas were originally published as a monograph (Williamson, 1992) and his current essay (Chapter 33) is written with criticisms of his earlier work in mind. Finally, Meacham and Hartman (Chapter 34) describe a technique for dealing with homoplasy encountered during phylogenetic analysis independent of any biological hypotheses. This could be a tool for deciding among competing hypotheses for the causes of given homoplasy.

1.1   ORIGINS OF THE MODERN EUKARYOTIC CELL

The earliest mention of horizontal gene transfer can be traced to Merechowsky's (1905) suggestion that the eukaryotic mitochondria and chloroplast originated when bacteria invaded the eukaryotic cell and were subsequently incorporated by it (an idea reintroduced later by others: Stanier, 1970; Margulis, 1971). This hypothesis was given major support when it was shown that the 16S RNAs from the chloroplast and mitochondria were more closely related to cyanobacteria and purple bacteria, respectively, than to any other major assemblages (Woese and Fox, 1977). A number of proteins found in these organelles are today no longer encoded in the plastid genomes, but appear to be related to those from bacteria. Martin (Chapter 28) examines the molecular evolution for five enzymes found in both the eukaryotic cytosol and the organelle. His results imply that, besides the major endosymbiosis events, multiple transfers of genes from a variety of other bacterial donors also occurred. That is, if the chloroplast, for example, is the result of an endosymbiotic event, it must have been accompanied by multiple other bacteria-to-eukaryote gene transfers in order to account for today's organellar enzymes.

The relationship between archaebacteria and eukaryotes is not as simple as is indicated by the ribosomal RNA trees. Olendzenski and Gogarten (Chapter 27) examine the relationship of a number of metabolic genes among archaebacteria, bacteria and eukaryotes. For many genes, it is clear that the two bacterial groups are much more closely related to each other than either is to the eukaryotes. They find a number of genes that appear to group Gram-positive bacteria and the archaebacteria to the exclusion of other bacteria and the eukaryotes. That is, there is a major incongruity between the ribosomal RNA tree (which, incidentally, is supported by the trees of other genes whose products are involved in transcription, translation and chromosome integrity) and trees based on these metabolic enzymes. The complete sequence of the genome of *Methanococcus* (Bult *et al.*, 1996) also supports the notion that many of its genes have closer affinity with eubacteria than with eukaryotes. Lake and Rivera's paper (Chapter 30) is easiest to reconcile with others in the field by positing that if the eukaryotic translational apparatus came from the archaebacteria, then it came from multiple archaebacterial sources.

Doolittle *et al.* (1996) have determined the time of last common ancestor of bacteria and eukaryotes based on protein molecular clock estimations. Doolittle arrived at the surprising result that the last common ancestor for eukaryotes and bacteria lived only 2 billion years ago – considerably later than the 4 billion years ago attributed to the first bacterial fossils. This result has caused controversy. In this volume Adkins and Li (Chapter 29) present evidence supporting Doolittle's result and help to clear up some confusions surrounding this issue. They have dated, again using the molecular clock, the time of internal gene duplication events that give rise to three modern genes. They conclude that these events occurred 3.2–3.8 billion years ago. This shows that the molecular data is consistent with the first appearance of life at 4 billion years ago and reconfirms that most orthologous genes point to a time of a last common ancestor 2 billion years ago.

The 2 billion years estimate is not easy to reconcile with tree-like patterns of descent. In particular, the hypothesis that the shape of the 16S RNA tree reflects the phylogeny of the species involved is difficult to reconcile with such a short coalescence time. In addition, these results imply that diversification of all prokaryotes occurred less than 2 billion years ago, long after prokaryotic life had been established. Some of these conflicts are resolved if we accept 2 billion years not as the time of last common ancestor between eukaryotes and bacteria, but, as H. Hartman suggested during a discussion of these data, as a time of major horizontal gene transfer events.

We can imagine the following scenario (Fig. 1.1). Two billion years ago, the three lineages are already formed: Bacteria, Archaea and a very primitive eukaryote which Hartman (1984) calls the Kronocyte. The Kronocyte receives a major influx of genes from Archaebacteria. Because the entire translational, transcriptional and replication functions of Archaebacteria appear to be closely related to those in eukaryotes, it is possible the modern nucleus is a remnant of that archaebacterial cell, as was originally suggested by Mereschovsky (1905). At about this time, there would also have been a major movement of genes from bacteria into both eukaryotes and archaebacteria, possibly making up a quarter of the genes found in both assemblages. Golding and Gupta (1995) have noted many of these same incongruities. They have chosen horizontal gene transfer to explain these phenomena but postulate that an

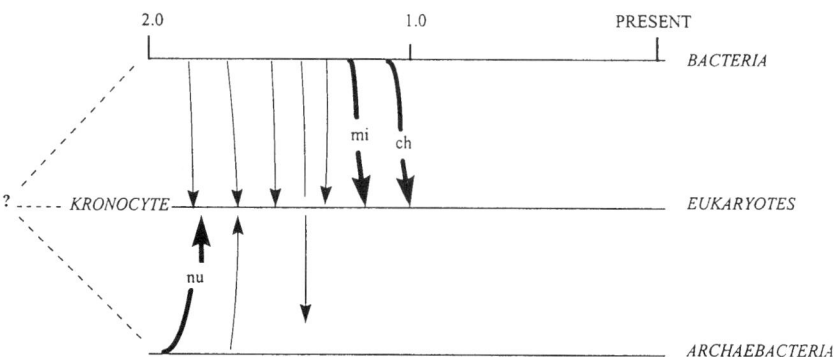

Fig. 1.1 Horizontal transfers and the emergence of the modern eukaryote. Between 1 and 2 billion years ago, major horizontal gene transfer events occurred between the kronocyte and bacteria. The modern eukaryotic cell arose from at least three endosymbiotic events (thick arrows) where the modern nucleus (nu) came from an archaebacterium, and the chloroplast (ch) and mitochondria (mi) came from true bacteria. Multiple other transfer events must have also occurred to account for today's distribution of genes.

endosymbiotic event between an archaebacterium and a Gram-negative bacterium gave rise to the eukaryotic cell; however, their model does not account for the origin of unique eukaryotic characters such as the nucleolus and cytoskeletal proteins.

Two billion years ago happens to have been a very active period in the geochemical record. It has been suggested, based on the relative amounts of Fe^{2+} and Fe^{3+} in surface sediments, that this is when the atmosphere converted from anaerobic to aerobic (Holland, 1994). In addition, based on the C^{13}/C^{12} ratios of inorganic carbonates, a major change in biological carbon activity apparently took place at this time (Karhu and Holland, 1996). Such a major upheaval is unlikely to have occurred without many extinctions, and it would also create powerful selective pressure for new combinations of traits.

If this picture is correct, then the question of the relationship among the three kingdoms is an open one. Determining that relationship will require identifying those genes in the modern eukaryote which are derived from the ancestral kronocyte for which there are bacterial orthologs.

REFERENCES

Akiba, T., Koyama, K., Ishiki, Y. *et al.* (1960) On the mechanism of the development of multiple-drug-resistant clones of *Shigella*. *Jap. J. Microbiol.* **4**: 219.

Anderson, N.G. (1970) Evolutionary significance of virus infection. *Nature* **227**: 1346–1347.

Baron, L.S., Carey, W.F. and Spilman W.M. (1959) Genetic recombination between *Escherichia coli* and *Salmonella typhinurium*. *Proc. Natl Acad. Sci. USA* **45**, 976–982.

Bult, C.J., White, O., Olsen, G.J. *et al.* (1996) Complete genome sequence of the methanogenic archaeon, *Methanococcus jannaschii*. *Science* **273**: 1058–1073.

Campbell, A. and Botstein, D. (1983) Evolution of the lambdoid phages, in *Lambda II*, (eds R.W. Hendrix, J.W. Roberts, F.W. Stahl and R.A. Weisberg), CSH Press, Cold Spring Harbor, NY.

Doolittle, R.F., Feng, D.F., Tsang, S. *et al.* (1996) Determining divergence times of the major kingdoms of living organisms with a protein clock. *Science* **271**: 470-477.

Erwin, D.H. and Valentine, J.W. (1984) Hopeful monsters, transposons and metazoan radiation. *Proc. Natl Acad. Sci. USA* **81**: 5482–5483.

Golding, G.B. and Gupta, R.S. (1995) Protein-based phylogenies support a chimeric origin for the eukaryotic genome. *Molec. Biol. Evol.* **12**: 1–6.

Hartman, H. (1977) Speculation on viruses, cells and evolution. *Evol. Theory* **3**: 159–163.

Hartman, H. (1984) Origin of the eukaryotic cell. *Speculations in Science and Technology* **7**: 77–81.

Holland, H.D. (1994) Early proteozoic atmospheric change, in *Early Life on Earth*, (ed. S. Bengtson), Columbia University Press, New York, pp. 237–244.

Karhu, J.A. and Holland, H.D. (1996) Carbon isotopes and rise of atmospheric oxygen. *Geology* **24**: 867–879.

Krassilov, V.A. (1977) The origin of angiosperms. *Bot. Revs* **43**: 143–176.

Levy, S.B. and Miller, R.V. (eds) (1989) *Gene Transfer in the Environment*, McGraw-Hill, New York.

Margulis, L. (1971) Symbiosis and evolution. *Sci. Amer.* **225**: 48–57.

Mereschowsky, C. (1905) Uber Natur und Ursprung der Chromatophoren in Pflanzenteilen. *Bio. Zentrabl.* **25**: 593–635.

Mitsuhashi, XX (ed.) (1977) *R Factor: Drug Resistant Plasmids*, University Park Press, Baltimore/London/Tokyo, 315 pp.

Miyake, T. and Demerec, M. (1959) *Salmonella–Escherichia* hybrids. *Nature (London)* **183**, 1586–1588.

Ochiai, K., Yamanaka, T., Kimura, K. and Sawada, O. (1959) Inheritance of drug resistance (and its transfer) between *Shigella* and *E. coli* strains. *Nihon Iji Shimpo* **1861**: 34.

Ochman, H. and Selander, R.K. (1984) Evidence for clonal population structure in *Escherichia coli. Proc. Natl Acad. Sci. USA* **81**: 198–201.

Reanney, D. (1976) Extrachromosomal elements as possible agents of adaptation and development. *Bact. Revs* **40**: 552–590.

Spratt, B.G., Bowler, L.D., Zhang, Q.Y. *et al.* (1992) Role of interspecies transfer of chromosomal genes in the evolution of penicillin resistance in pathogenic and commensal *Neiserria* species. *J. Molec. Evol.* **34**: 115–125.

Stanier, R. (1970) Organization and control in prokaryotic and eukaryotic cells, in *20th Symposium of the Society for General Microbiology*, Cambridge University Press, pp. 1–38.

Syvanen, M. (1984) Cross-species gene transfer: implications for a new theory of evolution. *J. Theor. Biol.* **112**: 333–343.

Syvanen, M. (1987) Molecular clocks and evolutionary relationships: possible distortions due to horizontal gene flow. *J. Molec. Evol.* **26**: 15–23.

Went, F.W. (1971) Parallel evolution. *Taxonomy* **20**: 197–226.

Williamson, D.I. (1992) *Larvae and evolution: toward a new zoology.* Chapman & Hall, New York.

Woese, C.R. and Fox, G.E. (1977) Phylogenetic structure of the prokaryotic domain: the primary kindgoms. *Proc. Natl Acad. Sci. USA* **74**: 5088–5090.

PART ONE
Mechanisms of Transfer into Bacteria and Eukaryotes

Looking sideways at the evolution of replicons

2

Jack A. Heinemann

SUMMARY

Horizontally mobile elements (HMEs) may be distinguished from chromosomes because the two types of replicon are subject to divergent evolutionary forces when conditions affect plasmid and chromosomal reproduction differently. The relevant distinction between plasmids and chromosomes is not size but the ability of plasmids to be inherited horizontally – that is, by neighbors – and to be passed vertically to daughters. Plasmid autonomy from cellular evolution is demonstrated by plasmids exchanging between prokaryotes and eukaryotes and from dead to living cells. Our studies of transkingdom conjugations between *Escherichia coli* and *Saccharomyces cerevisiae* now involve integrative plasmids. We find that *oriT*, more than homology, may guide recombination. Dissection of the recombination pathways that dictate the terms under which DNA of prokaryotic origin might survive in eukaryotes is under way. Sex between dead cells is introduced as a mechanism for the evolution of antimicrobial resistance and plasmid addiction systems. Antimicrobial agents discriminate between the physiological processes that are necessary for chromosome and non-chromosomal replicon reproduction. Whereas antimicrobials prevent the reproduction and survival of pathogens, they often fail to inhibit the reproduction and survival of HMEs. The result is, first, that antimicrobials protect horizontal while simultaneously punishing vertical replication; and second, that genes that adapt pathogens to changing environments, such as resistance and virulence/symbiosis determinants, accumulate on HMEs when they reproduce faster than chromosomes. New experiments reveal the influence that antimicrobials could have on replicon organization and thus the mechanism behind the rapid spread of resistant pathogens.

2.1   INTRODUCTION

Molecular biology's eclipse of bacterial genetics has sloppied certain concepts. The uncritical shorthand is that genes are DNA (or RNA) and replicons are chromosomes or plasmids whose size, not biology, is the distinguishing character (discussed in Eberhard, 1989; Heinemann, 1993; Krawiec and Riley, 1990). The problem with distinguishing replicons by nonbiological criteria is illustrated by the recent debate on adaptive mutagenesis, a putative means for traits to be acquired at higher frequencies by bacteria when they are rewarded for demonstrating the mutant trait than when they are not (Lenski and Mittler, 1993). Much of the phenomenon can be attributed not to a change in the Luria and Delbrück (1943) paradigm that mutations arise in individuals independently of their selective value, but to an untested assumption that plasmids replicate like chromosomes. In fact, the genes of interest were on plasmids that replicated under conditions that chromosomes could not (Galitsky and Roth, 1995; Radicella et al., 1995). Ignoring the biological relevance of different replicon types can lead to conceptual and practical problems. For example, assuming that plasmids evolve like chromosomes or behave for the benefit of their hosts may confound our attempts to design lasting antimicrobial agents, to interrupt the cycle of pathogen evolution and to predict the fate of genes released into the environment in the form of genetically modified organisms.

A biologically relevant characteristic of replicons is their reproductive strategy, as these strategies could be discriminated by evolution if they were different physiologically. For example, horizontally mobile elements (HMEs), can be inherited horizontally, from neighbors, and then be passed vertically to daughters, whereas chromosomes tend to specialize in vertical reproduction (Fig. 2.1). Transposons, plasmids and viruses, collectively found in all cellular life forms, are distinguished from individual genes and complex gene clusters such as operons and chromosomes, not because of size differences but because their respective reproductive options are differentially successful in different environments (Heinemann, 1993; Heinemann and Ankenbauer, 1993b, submitted; Heinemann et al., 1996).

These concepts are best illustrated by the biology of plasmids found in bacteria. This chapter will focus particularly on bacterial 'conjugative' plasmids, known for their ability to reproduce both vertically and horizontally (Heinemann, 1991). Note that this special focus on plasmids is not meant to imply that these same processes are not operating in the evolution of viruses and transposons in other types of organisms.

Plasmid autonomy from cellular evolution is demonstrated by three phenomena: plasmid exchange between prokaryotes and eukaryotes (promiscuity); plasmid exchange between dead and living cells; and

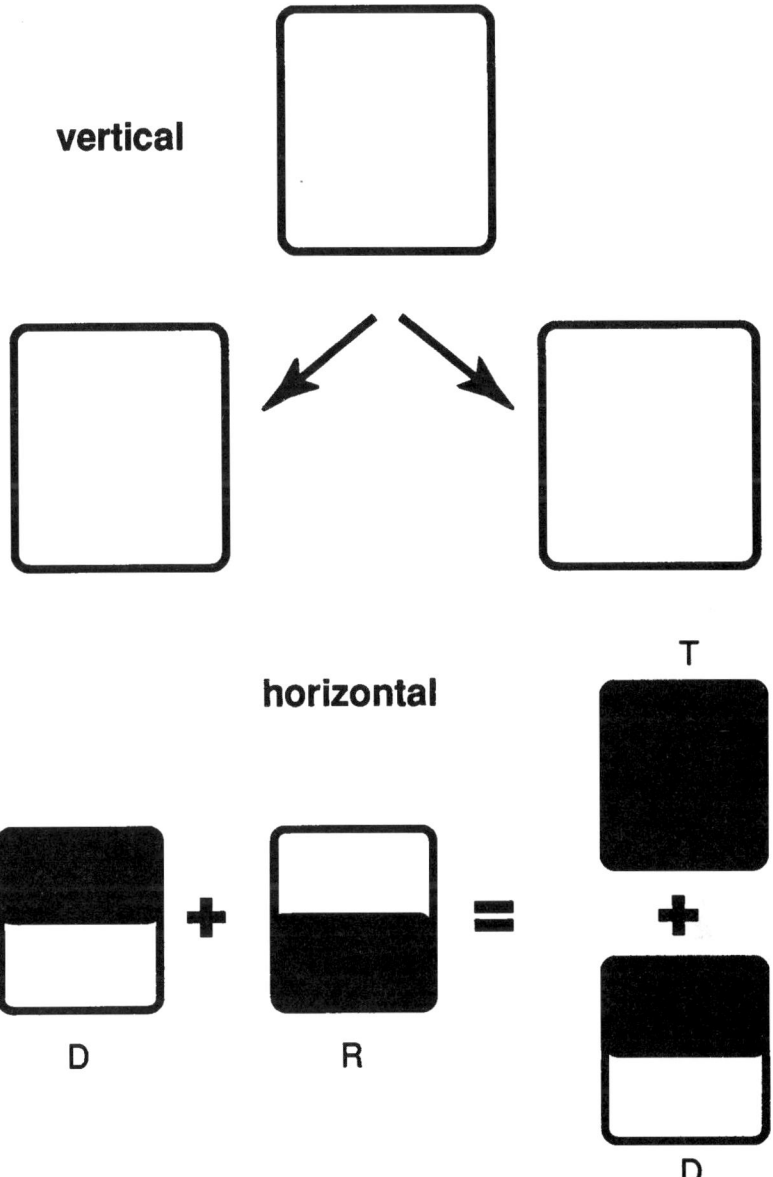

Fig. 2.1 Strategies of reproduction. Vertical: cells and chromosomes (represented by squares) increase in number through clonal division, faithfully perpetuating a single genotype. Horizontal: conjugative plasmids, phage, conjugative transposons and some genes (through transformation) reproduce with chromosomes (vertically) or by transfer to neighbors (horizontally). Horizontal transfer can result in the preservation of one parental genotype (D) and the creation of a recombinant type (T) by contributing to the genome of the recipient cell (R). Thus, under some conditions HMEs may continue to reproduce and evolve when cells cannot.

the ability of plasmids to kill their hosts. Examples of each of these phenomena will be discussed.

2.2   LOVE OF BEASTS

Plasmids reproduce by horizontal transfer to organisms of at least three different biological kingdoms – a microbiological 'bestiality' of sorts (Heinemann, 1991; Mazodier and Davies, 1991). The apparent promiscuity of plasmids violates teleological preconceptions about their function. In assuming that plasmids encoded sexual attributes of prokaryotes, the unexpressed expectation was that the potential for genes to transfer among bacteria, and from bacteria to eukaryotes, would be limited in the same way that recombination between multicellular creatures of different species was limited. When genes were seen to exchange between prokaryotes of different genera, 'bacterial sex' was viewed as either altruistic (donation of beneficial genes, such as antibiotic or phage resistance, to potential competitors) or as understandable as an example of selfish plasmid replication. Still, that plasmids engineered their own transfer to eukaryotes was a surprise undoubtedly because of the connotations in the sexual terminology ascribed to the recombination phenomenon (called conjugation). The barriers to plasmid transmission have been routinely overestimated (Heinemann *et al.*, 1996). Perhaps the most important barrier to inheritance is not transfer, but the ability to achieve replication (Heinemann, 1991). However, even the dedicated *Escherichia coli* replicon pBR322 is one base change at multiple sites from replicating stably in a eukaryotic yeast (Kipling and Kearsey, 1990).

The ability to spread infectiously allows plasmids to reproduce under conditions that chromosomes do not (e.g. Heinemann and Ankenbauer, 1993a, b; Heinemann *et al.*, 1996; Peters and Benson, 1995). Plasmids can move through populations and between adjacent populations, so their fate may be distinguished from a chromosome's both physiologically and geographically. Since HMEs like plasmids can evolve independently of their hosts and their chromosomes, predictions of plasmid biology and of the function of genes ferried by plasmids cannot be extrapolated from the behavior of chromosomes.

The idea that plasmids encode their means to move from prokaryotes to eukaryotes by a mechanism analogous to that sufficient to account for their transfer between closely related bacteria was born in the observation of DNA intermediates in *Agrobacterium tumefaciens* that resembled those expected (but never demonstrated) to exist in bacteria during conjugation (Stachel *et al.*, 1986). Until 1989, *A. tumefaciens*-mediated tumorigenesis in plants was the only known example of a transkingdom DNA exchange. This exchange resulted in the recovery of a piece of the bacterial Ti plasmid from plant chromosomes (Yadav *et al.*, 1980). The idea became truly credible when Buchanan-Wollaston *et al.* (1987) created a hybrid DNA delivery system using the genetic components encoding the inter-prokaryotic conjugation machinery and the Ti pathogenic delivery system. Shortly after, we found that the delivery of DNA to eukaryotes was a general property of conjugative plasmids and not a special property evolved for exclusively pathogenic relationships (Heinemann and Sprague, 1989). An unique activity for Ti and *A. tumefaciens*-encoded DNA transfer gene products is nevertheless still sought by some to justify the routine compromise of the prokaryotic–eukaryotic barrier (e.g. Herrera-Estrella *et al.*, 1990; Howard *et al.*, 1992; Shurvinton *et al.*, 1992).

Conjugation between bacteria (and also, as far as we can tell, between bacteria and eukaryotes) requires the activity of three types of genes (Heinemann, 1991). The first type, called *tra*, acts in *trans* and mediates the requisite cell–cell interactions and the biochemistry of DNA transfer. The second type, called *mob*, also acts in *trans* and is dedicated to plasmid-specific DNA metabolism. The third type, for which there is only one gene, is *oriT*, a *cis*-acting origin at which a strand- and site-specific nick in the plasmid is introduced just prior to strand transfer to a recipient cell. The *oriT* type is epistatic to *mob* because *oriT* is required to initiate DNA metabolism. Mobilization of the plasmid results in the separation of the two strands into each of two cells and doubling of the plasmid number if each template founds a replicon. The diversity of species-specific replication requirements limits the effective host range (although not transfer range) of plasmids unless they can also integrate into an existing replicon, like the T-DNA of the Ti plasmid.

Our studies on transkingdom conjugations between *Escherichia coli* and *Saccharomyces cerevisiae* (Heinemann, 1991; Heinemann and Sprague, 1989; Sprague, 1991) now involve plasmids that cannot replicate (Singh and Heinemann, 1997). Such plasmids, called YIp (yeast integrating plasmid), must recombine with an existing replicon to be inherited. We seek to find the rules obeyed by such recombinations, ultimately by monitoring the involvement of various yeast recombination pathways in interkingdom DNA transmission.

YIp transfer has been difficult to detect using the protocols developed for replicating plasmids (Heinemann and Sprague, 1990). We detected transmission only when a plasmid homologous to the YIp vector was already present in the yeast recipient, providing a target for integration (Heinemann and Sprague, 1990; Singh and Heinemann, 1997). This may have been due to the copy number of the target plasmid (approximately 30/cell) increasing the number of homologous sequences available for recombination (increasing target size) or the presence of a particular sequence within the target plasmid (site specificity). Since yeast homologous recombination is directed by the sequences near the ends of broken molecules, we previously speculated that the target plasmid was the preferred recombination substrate because it contained sequences homologous to *oriT* which potentially biased subsequent recombinational events (Heinemann and Sprague, 1990).

A new protocol has been developed for the relatively efficient transmission of YI plasmids (Singh and Heinemann, 1997). Preliminary evidence suggests that conjugatively transferred molecules prefer to recombine with molecules of bacterial origin (i.e. those that retain *oriT*) over those of yeast origin, even when the target size should favor recombination between yeast sequences. Yeast recipients, containing a chromosomally integrated pBR322, mated with bacterial donors of pBR322-based YIp vectors, carrying both the *URA3* and *LEU2* genes, and produced yeast transconjugants with the YIp markers integrated at the 'pBR322 locus' about twice as frequently as at the *URA3* and *LEU2* loci combined (Fig. 2.2). The pBR322 locus was favored even though *URA3* and *LEU2* comprised two-thirds of the sequences of the YIp vector used in this study.

Integration events at the pBR322/*his3* locus were detected by their unusual genetic linkage pattern. The locus is bracketed by two different and nonfunctional *his3* alleles separated by the plasmid pBR322 and yeast *TRP1* gene (Fig. 2.2). The His⁻ phenotype corrects at a low but easily detected frequency through recombination between the heteroduplications; some fraction of these revertants concomitantly excise the intervening DNA (Klein, 1988). Thus, some His⁺ recombinants are Trp⁻ by virtue of a *TRP1* deletion. His⁺ revertants of Leu⁺/Ura⁺ interkingdom recombinants should yield a subset of cells where *TRP1*, *LEU2* and *URA3*, singly or in combination, disappear coordinately if the YIp is integrated into the pBR322 locus. Interkingdom recombinants that always yield Leu⁺ (Ura⁺) His⁺ derivatives may or may not have integrations elsewhere. This class of recombinants will require a physical characterization to identify the integration site unambiguously. We are now pursuing this class of recombinants using a combination of random spore analysis, chromosome tagging and Southern blotting to confirm the linkage data.

2.3   LOVE OF THE DEAD

Modern antimicrobials are forces of natural selection so precise that they select cellular replicons rather than cells. Antimicrobial agents discriminate between the physiological processes that are necessary for chromosomal and nonchromosomal replicon reproduction (Heinemann and Ankenbauer, submitted). Whereas antimicrobials prevent the reproduction of pathogens, they often fail to inhibit the reproduction and survival of nonchromosomal replicons. The result is that antimicrobials protect

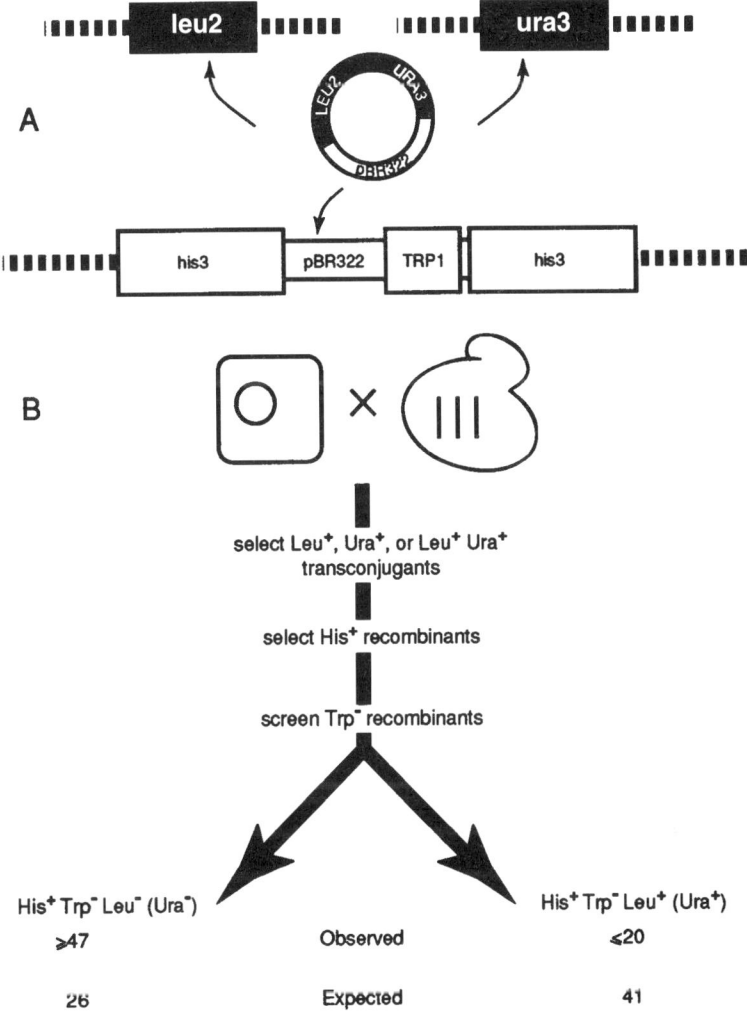

Fig. 2.2 Determining plasmid integration sites with a genetic assay. (A) Potential homologous integration sites in separate *S. cerevisiae* chromosomes (strain 329-6c; Klein, 1988) for plasmids transferred from bacteria by conjugation. (B) Approximately 12-hour-old donor bacteria were transferred to a lawn of freshly cultured yeast spread on medium that permits only the growth of transconjugants (i.e. lacking uracil, leucine or both), by replica plating. To date, 67 independent transconjugants have been subcultured to purity then grown to saturation in the appropriate liquid medium. Dilutions of these latter cultures were spread on minimal medium lacking histidine to select for potential pop-out events at the *his3* locus. Hundreds of His+ colonies were transferred to various other media by replica plating to determine their genotypes. The proportion of transconjugants ($\geq$ 47/67) that simultaneously converted to a His+ Leu$^-$/Ura$^-$ phenotype was significantly different ($P \leq 0.001$) from the proportion expected (26/67) had recombination been simply a function of the length of homologous regions (in kb).

horizontal replication as they select against vertical replication. Antimicrobial agents do not prevent conjugative plasmids in dead cells from transferring back to live cells or from entering dead cells and retransferring to live cells, thereby resurrecting genes from dead cells.

William Hayes (1952) first demonstrated this principle when he used the antibiotic streptomycin to show that plasmids moved in only one direction during bacterial sex. Mixed cultures of dead and living cells produced recombinants in the presence of the antibiotic. Hayes's experiment was an early warning that an antibiotic could prevent reproduction without preventing the flow of genes between the microbiologically 'dead' and living (Heinemann and Ankenbauer, 1993a, b; Heinemann *et al.*, 1996). Some agents allow a plasmid to enter and convert a cell into a vector for the plasmid even though the entire process occurs after the cell could be considered inviable (Heinemann, 1993).

The differences between chromosomes and HMEs, illustrated by their reproductive sensitivities to antimicrobial agents, have led to a biased distribution of genes among replicons; resistance (and virulence and horizontal transfer) determinants usually accumulate first on plasmids, not chromosomes (Chapter 6). HMEs may even be ferrying the resistance determinants which are exclusively observed as chromosomal genotypes.

Antimicrobials create dead vectors that can disperse resistance genes. Dead cells are both stealthy vectors, because we cannot detect them using conventional microbiological culturing, and potent vectors, because of the transfer range of plasmids.

Insusceptibility of plasmid biology to cell death may even potentiate resistance transfer during combination and serial drug treatments (Fig. 2.3). Whereas antimicrobials that disrupt cell integrity or drain membrane potential often stop horizontal transmission, sensitivity to these drugs usually requires the cell to express genes or divide. So when the cell is first violated by a drug that only inhibits the latter, its ability to donate genes is protected from the effects of the former (Heinemann, 1993). In essence, some antibiotic combinations might create ecological niches that enhance the reproductive potential of HMEs by creating vectors that are immune to other environmental assaults of either anthropogenic (e.g. synthetic drugs) or natural origin (e.g. antibiotics).

Separate studies have demonstrated that antimicrobials can increase mutation and interspecies gene transfer, thereby predisposing microbes to inherit the means to resist our medicines (Heinemann and Ankenbauer, submitted). A bias in gene distribution between HMEs and chromosomes permits the physiological effects of antimicrobials to bias the evolution of resistance elements.

Interestingly, not just antimicrobial agents but also some plasmids kill cells. Conjugative plasmids can carry at least one pair of genes known as

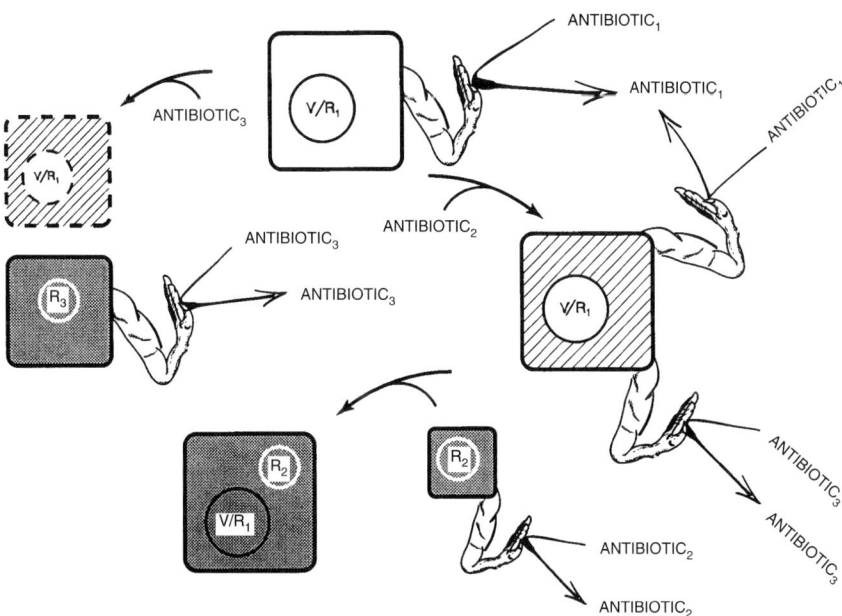

Fig. 2.3 Antibiotic-induced plasmid vectors. Some antimicrobial agents prevent both chromosome and HME reproduction while others only inhibit the former (Heinemann, 1993; Heinemann and Ankenbauer, 1993b). Depending on the order in which a pathogenic organism encounters a series of assaults, the evolution of the HME it contains could differ radically from the chromosome and cell itself. Illustrated is the particular case where a cell ($\square$) made pathogenic by a resistance (R_1) plasmid ($\bigcirc$) encoded virulence determinant (V) is 'killed' ($\boxtimes$) by application of either one of two alternative antibiotics to which the pathogen is genotypically sensitive. Treatment with antibiotic$_3$ renders the cell incapable of replicating the chromosome and the plasmid (dashed lines). Treatment with any of the class of antibiotics (reviewed in Heinemann, 1993) represented by antibiotic$_2$ only effectively prevents vertical reproduction and makes horizontal transfer phenotypically resistant to the class of antibiotics represented by antibiotic$_3$. Reinoculation of the patient with organisms from the environmental reservoir ($\blacksquare$), which usually must already be resistant to the medicating antibiotic (2 or 3), could result in the transfer of both the virulence determinant and resistance to antibiotic$_1$ from organisms first treated with antibiotic$_2$ but not antibiotic$_3$. This and other scenarios (reviewed in Heinemann and Ankenbauer, submitted) describe plausible Darwinian mechanisms to explain the unexpectedly high coincidence of multiple resistance and virulence traits.

psk (*p*ost *s*egregational *k*illing) that effectively addict a cell to the plasmid. The *psk* pairs encode a stable toxin and a labile antidote; mis-segregation of a plasmid away from one daughter during cell division results in an accumulation of toxin relative to antidote and, ultimately, death of the plasmidless cell. Because monocultures of cells addicted to their plasmids are composed almost exclusively of cells with the plasmid (due to death of wayward daughters during the growth of the culture), *psk* has been 'canonized' as a plasmid-encoded mechanism for maintaining the plasmid during vertical reproduction (Gerdes *et al.*, 1986).

We found it curious that HMEs kill wayward daughters rather than re-enter them. In competition with plasmidless cells, the slower growing plasmid-bearing population would almost certainly be displaced. In contrast, efficient horizontal transfer could simultaneously reform wayward daughters and maintain population growth rates. Has addiction been misinterpreted as a strategy more relevant to chromosomes than to HMEs?

Tim Cooper (a PhD student in my laboratory) has begun to test the predictions of the vertical stability model. He has already contrasted the isogenic *psk* plasmids pRK2526 (Add$^+$, *tra*$^+$) and pRK21526 (Add$^-$, *tra*$^+$). The RP4-derived plasmids differ by a deletion of the *parDE* operon (all

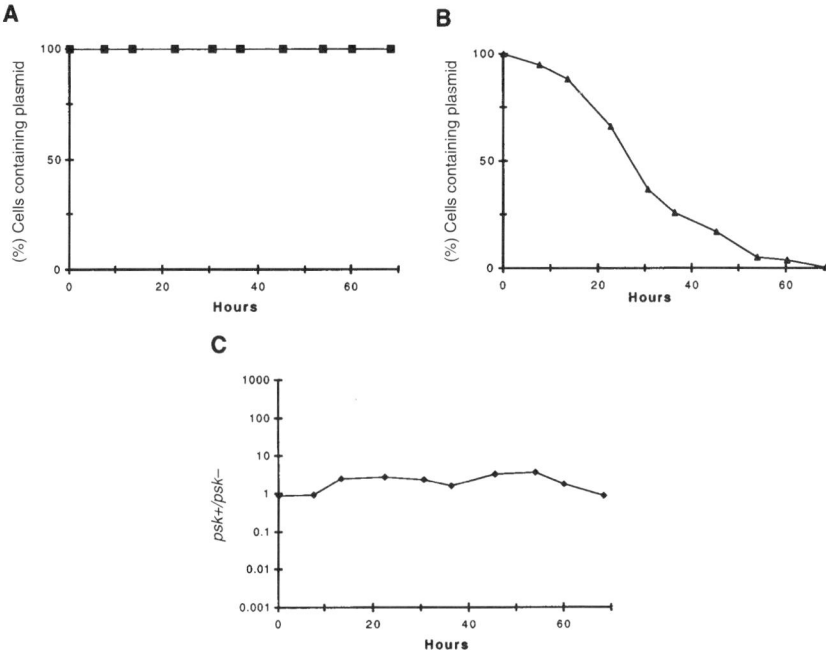

Fig. 2.4 Effects of *psk* on vertical reproduction. (A) Cells addicted to their plasmid grow as monocultures of uniformly plasmid-containing cells. (B) Cells carrying Add$^-$ plasmids grow as cellular monocultures primarily lacking the plasmid. (C) The number of (vertical) plasmid generations in each of the monocultures is the same regardless of addiction. Plasmid generations were inferred from the generations of the subset of cells occupied by the respective plasmids at specified sampling times.

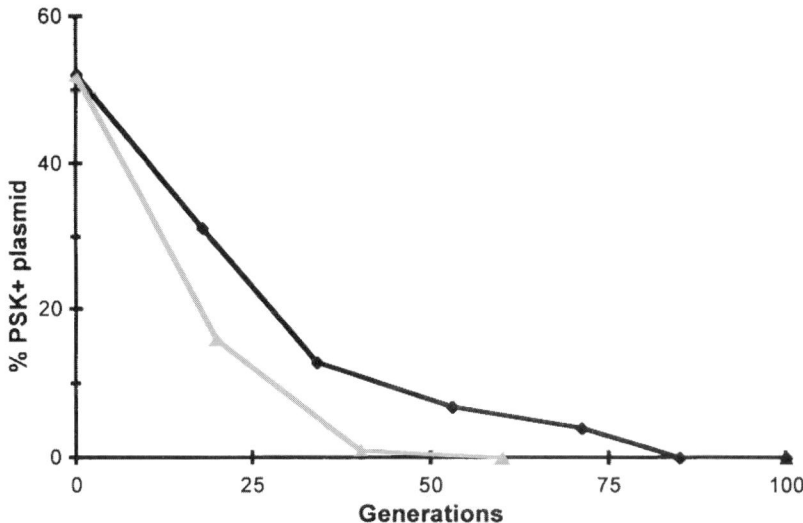

Fig. 2.5 Effects of *psk* on HME competition. When cells addicted to their plasmids are grown in direct competition with cells carrying isogenic but Add⁻ plasmids (again under conditions of limited horizontal transfer), the ratio of cell types and plasmid types favors the cells and the plasmids that preserve wayward daughters, whether grown under conditions of perpetual exponential growth (top, black line) or alternating stationary phases (bottom, gray line).

plasmids from Sia *et al.*, 1995). When essentially no horizontal transfer was permitted, addicted (A) cells (Fig. 2.4A) grown in monoculture were more uniformly occupied by plasmids (Add⁺) than were monocultures of cells (B) carrying the same plasmid lacking the toxin (Add⁻) gene (Fig. 2.4B). However, if A and B cells were grown together, B would quickly displace A from the population and the Add⁻ plasmid would displace the Add⁻ plasmid even though the majority of cells in the mixed culture had no plasmid at all (Fig. 2.5). Even under monoculture conditions, the seeming success of Add⁺ plasmids was illusory because Add⁺ and Add⁻ plasmids reproduced at the same frequency (i.e. mis-segregated just as often) even though a greater proportion of cells in addicted monocultures had plasmids (Fig. 2.4C). None of these observations is expected by the vertical stability model.

Nevertheless, each is a prediction of the competition model. This model hypothesizes that *psk* efficiently secures the plasmid niche. In the case of plasmids, the niche is the cell and the threat to the niche is offered by other HMEs (plasmids, phage) that could inhibit reproduction of the incumbent HME. The competition model would expect that *psk* is an advantage during horizontal, not vertical, reproduction and so Cooper

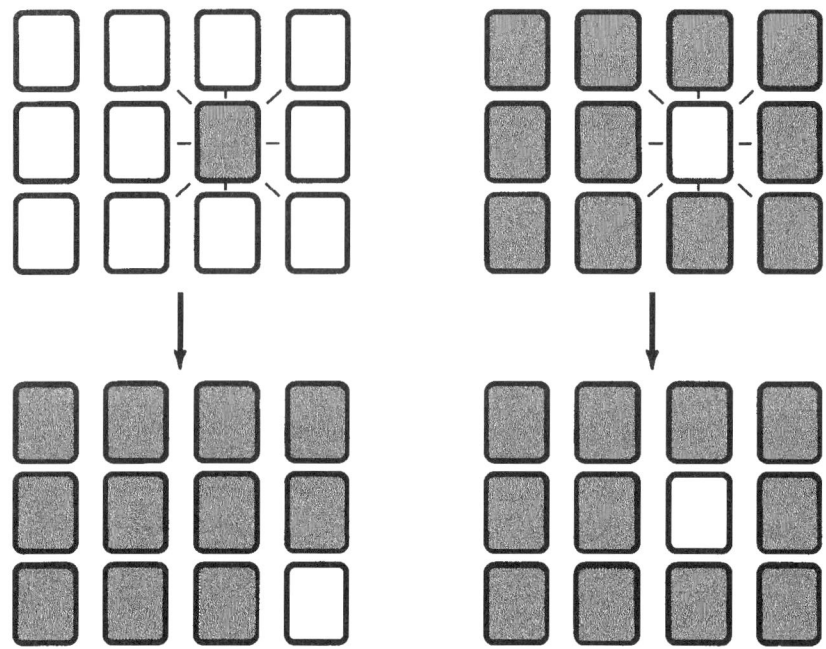

Fig. 2.6 Effect of *psk* on horizontal reproduction and HME competition. Under conditions permitting horizontal transfer but limited vertical generations (left side of figure), the *tra*⁺ plasmid (pRK2526) that addicted its host (■) displaced the much larger population of isogenic, replication incompatible and Add⁻ plasmid (pRK21575, *tra*⁻; □). Addicted cells (■) maintained their plasmids (pRK21558, *tra*⁻) despite challenge from a minority Add⁻ (pRK21526, *tra*⁺) or other Add⁺ (□) plasmid (pRK2526, *tra*⁺) (right side of figure).

tested the effect of *psk* on horizontal reproduction. Preliminary results are consistent with expectations of the competition model. A minority of addicted donors converted a majority population of Add⁻ containing recipients into other addicted donors. A minority of Add⁻ or Add⁺ donors remained a minority, however, when mixed with addicted recipients (Fig. 2.6).

2.4 CONCLUSION

Inheritance of resistance plasmids is the most frequent mechanism by which pathogens become resistant to antimicrobial agents (Amabile-Cuevas *et al.*, 1995). HMEs have continued to evolve despite human use of agents that severely disrupt the reproductive potential of chromosomes (cells). The effects of these agents are but illustrative of the fallacy that subcellular replicons are evolving as components of the cellular genome. HMEs are subject to different evolutionary rules because they depend on different biochemical pathways to reproduce and these pathways are differentially successful. We propose that the different effects of antimicrobials on the reproductive success of HMEs and cells has contributed to the near universality of microbial resistance. Of course, the broad host range of some HMEs was also necessary for the evolution of HMEs that had to reproduce by transferring to species not subject or not susceptible to the particular agents we have employed.

ACKNOWLEDGMENTS

I thank G. Richter, T. Cooper and an anonymous referee with a flair for communicating constructive criticism for review of the manuscript. Tim and I are grateful to D. Figurski for advice, strains and plasmids. Some of the work described was supported by a grant from the University of Canterbury (to J.A.H.) and a University MSc scholarship (to T.C.).

REFERENCES

Amabile-Cuevas, C.F., Cardenas-Garcia, M. and Ludgar, M. (1995) Antibiotic resistance. *American Scientist* **83**: 320–329.

Buchanan-Wollaston, V., Passiatore, J.E. and Cannon, F. (1987) The *mob* and *oriT* mobilization functions of a bacterial plasmid promote its transfer to plants. *Nature* **328**: 172–175.

Eberhard, W.G. (1989) Why do bacterial plasmids carry some genes and not others? *Plasmid* **21**: 167–174.

Galitsky, T. and Roth, J.R. (1995) Evidence that F plasmid transfer replication underlies apparent adaptive mutation. *Science* **268**: 421–423.

Gerdes, K., Rasmussen, P.B. and Molin, S. (1986) Unique type of plasmid maintenance function: postsegregational killing of plasmid-free cells. *Proc. Natl Acad. Sci. USA* **83**: 3116–3120.

Hayes, W. (1952) Recombination in Bact. coli K 12: unidirectional transfer of genetic material. *Nature* **169**: 118–119.

Heinemann, J.A. (1993) Transfer of antibiotic resistances: a novel target for intervention. *APUA Newsletter* **11**: 1, 5–6.

Heinemann, J.A. (1991) Genetics of gene transfer between species. *Trends Genet.* **7**: 181–185.

Heinemann, J.A. and Ankenbauer, R.G. (1993a) Retrotransfer in *E. coli*: conjugation: bidirectional exchange or *de novo* mating. *J. Bacteriol.* **175**: 583–588.

Heinemann, J.A. and Ankenbauer, R.G. (1993b) Retrotransfer of IncP plasmid R751 from *Escherichia coli* maxicells: evidence for the genetic sufficiency of self-transferable plasmids for bacterial conjugation. *Mol. Microbiol.* **10**: 57–62.

Heinemann, J.A. and Ankenbauer, R.G. Submitted: Antibiotics and their influence on resistance evolution. *Microbiol. Rev.*

Heinemann, J.A., Scott, H.E. and Williams, M. (1996) Doing the conjugative two-step: evidence for recipient autonomy in retrotransfer. *Genetics* **143**: 1425–1435.

Heinemann, J.A. and Sprague, G.F. Jr (1989) Bacterial conjugative plasmids mobilize DNA transfer between bacteria and yeast. *Nature* **340**: 205–209.

Heinemann, J.A. and Sprague, G.F. Jr (1990) Transmission of plasmid DNA to yeast by conjugation with bacteria. *Methods Enzymol.* **194**: 187–195.

Herrera-Estrella, A., van Montagu, M. and Wang, K. (1990) A bacterial peptide acting as a plant nuclear targeting signal: the amino acid-terminal portion of *Agrobacterium* VirD2 protein directs a β-galactosidase fusion protein into tobacco nuclei. *Proc. Natl Acad. Sci.* **87**: 9534–9537.

Howard, E.A., Zupan, J.R., Citovsky, V. and Zambryski, P.C. (1992) The VirD2 protein of *A. tumefaciens* contains a C-terminal bipartite nuclear localization signal: implications for nuclear uptake of DNA in plant cells. *Cell* **68**: 109–118.

Kipling, D. and Kearsey, S.E. (1990) Reversion of autonomously replicating sequence mutations in *Saccharomyces cerevisiae*: creation of a eukaryotic replication origin with prokaryotic vector DNA. *Mol. Cell. Biol.* **10**: 265–272.

Klein, H.L. (1988) Different types of recombination events are controlled by the RAD1 and RAD52 genes of *Saccharomyces cerevisiae*. *Genetics* **120**: 367–377.

Krawiec, S. and Riley, M. (1990) Organization of the bacterial chromosome. *Microbiol. Revs* **54**: 502–539.

Lenski, R.E. and Mittler, J.E. (1993) The directed mutation controversy and neo-Darwinism. *Science* **259**: 188–194.

Luria, S.E. and Delbrück, M. (1943) Mutations of bacteria from virus sensitivity to virus resistance. *Genetics* **28**: 491–511.

Mazodier, P. and Davies, J. (1991) Gene transfer between distantly related bacteria. *Ann. Rev. Genet.* **25**: 147–171.

Peters, J.E. and Benson, S.A. (1995) Redundant transfer of F′ plasmids occurs between *Escherichia coli* cells during nonlethal selection. *J. Bacteriol.* **177**: 847–850.

Radicella, J.P., Park, P.U. and Fox, M. (1995) Adaptive mutation in *Escherichia coli*: a role for conjugation. *Science* **268**: 418–420.

Shurvinton, C.E., Hodges, L. and Ream, W. (1992). A nuclear localization signal and the C-terminal omega sequence in the *Agrobacterium tumefaciens* VirD2 endonuclease are important for tumor formation. *Proc. Natl Acad. Sci.* **89**: 11837–11841.

Sia, E.A., Roberts, R.C., Easter, C. *et al.* (1995) Different relative importances of the par operons and the effect of conjugal transfer on the maintenance of intact promiscuous plasmid rk2. *J. Bacteriol.* **177**: 2789–2797.

Singh, K.K. and Heinemann, J.A. (1997) Yeast plasmids. In *Methods in Molecular Biology, Vol. 62: Recombinant Gene Expression Protocols*, (ed. R. Tuan), Humana Press, Totowa, NJ.

Sprague, G.F. Jr (1991) Genetic exchange between Kingdoms. *Curr. Opin. Genet. Devel.* **1**: 530–533.

Stachel, S.E., Timmerman, B. and Zambryski, P. (1986) Generation of single-stranded T-DNA molecules during the initial stages of T-DNA transfer from *Agrobacterium tumefaciens* to plant cells. *Nature* **322**: 706–712.

Yadav, N.S., Postle, K., Saiki, R.K. *et al.* (1980) T-DNA of a crown gall teratoma is covalently joined to host plant DNA. *Nature* **287**: 458–461.

Plasmid specificity and interaction: the similarities and differences between the transfer regions of two compatible plasmids, F and R100-1

3

William A. Klimke, Richard Fekete, Jan Manchak, Karen Anthony and Laura S. Frost

SUMMARY

Bacteria often carry either self-transmissible or mobilizable elements such as plasmids in natural environments. In keeping with their 'selfish' nature, these plasmids can influence the ability of a second co-resident plasmid to replicate (incompatibility) or to transfer (fertility inhibition). In addition, closely related conjugative plasmids may share certain transfer gene products or not, depending on whether that transfer protein defines an aspect of specificity of transfer for that plasmid.

We have compared two compatible plasmids, F (IncFI) and R100-1 (IncFII), which have highly similar transfer regions as determined by DNA sequence analysis. In general, the genes that encode the proteins that initiate DNA transfer during conjugation are specific for the homologous plasmid; however, the exceptions suggest that plasmids continually evolve to optimize their transfer potential. The proteins involved in pilus assembly and mating pair formation (recipient cell recognition, surface exclusion, phage sensitivity, pilus retraction) appear to be interchangeable but minor sequence changes can result in a change of specificity (surface exclusion) or function (phage sensitivity, recipient cell recognition) which allows the two plasmids to coexist in the same cell without competing for the same receptors on a potential recipient cell. Thus two compatible plasmids exhibit behavior that resembles the children's game 'paper, scissors, stone': one plasmid will dominate the other and will tolerate the expression of the other plasmid's transfer region only so long as it can be used to optimize transfer of the dominant plasmid. This chapter compares the known sequence of the transfer regions of F and R100-1 and briefly describes the consequences of sequence differences in terms of mating pair formation, pilus function, surface exclusion and gene regulation; it also looks at the transfer efficiency of chimeric plasmids containing portions of the F and R100-1 *oriT* region fused together.

3.1   INTRODUCTION

The F plasmid is a naturally derepressed conjugative plasmid which expresses about 1–2 F-pili/cell and transfers at high efficiency among the enteric bacteria. A closely related plasmid, R100, and a derepressed mutant, R100-1, were isolated in Japan (reviewed in Meynell *et al.*, 1968) and gained notoriety because they carried multiple antibiotic resistance markers. The F transfer region has been completely sequenced (Frost *et al.*, 1994) and a detailed restriction map for the closely related plasmid NRI is available (Womble and Rownd, 1988). A large portion of the *tra* region of R100-1 has also been sequenced: *geneX* (Fee and Dempsey, 1988); *orf169* in F (Cram *et al.*, 1984); *oriT-finP* (Fee and Dempsey, 1986; McIntire and Dempsey, 1987); *traJ* (Inamoto *et al.*, 1988); *traY* (Finlay *et al.*, 1986); *traA* (Frost *et al.*, 1985); *traLEKBP* (Anthony *et al.*, 1996); *traT* (Ogata *et al.*, 1982); *traD-finO* (Yoshioka *et al.*, 1990). We have recently completed the sequence of the *trbD-traF* portion of R100-1 up to the *SacI* site in mid-*trbA*, leaving the *trbA-traS* sequence (6.8 kb) to be completed.

The F plasmid expresses the F-pilus which is the attachment site for the RNA phages R17 and Qβ as well as the filamentous phages in the Ff group including f1, M13 and fd. It is also thought to be responsible for recognizing suitable recipient cells as well as other donor cells (surface exclusion) via a specialized structure at the pilus tip (Anthony *et al.*, 1994). The F-pilus is composed of pilin (7.2 kDa), arranged with five-fold rotational symmetry to form a filament of 1–2 μm in length. The F-pilin subunit has an acetylated N-terminal alanine which forms part of the major epitope in the protein. This epitope is not found on the sides of the pilus (Frost *et al.*, 1985) nor, surprisingly, at the pilus tip and consequently does not seem to have a role in recipient cell recognition, f1 phage attachment or surface exclusion. Two other regions of the F-pilin protein are involved in RNA phage attachment: the C-terminal domain and a small region near the N-terminus (aa 15–22), which are presumably exposed on the sides of the F-pilus (Frost and Paranchych, 1988). The R100-1 pilus, which principally varies in sequence near the N-terminus, is highly resistant to all three types of phage. It fails to attach R17 or QΒ because of sequence differences in the pilin subunit and, although it attaches f1 phage, further steps in the infectious process are blocked (Willetts and Maule, 1986).

Unlike R100-1, the F conjugation system requires the outer membrane protein OmpA as well as an intact inner core of the lipopolysaccharide in the recipient cell (Anthony *et al.*, 1994). The mature F and R100-1 TraT proteins, which partially determine the specificity of the surface exclusion reaction (along with TraS), differ by only one amino acid (gly→ala) (Finlay and Paranchych, 1986). Experiments in which clones expressing either the F or R100-1 pilin proteins were used to complement a mutation in F *traA* (pilin), suggested that an F-pilus tip protein did indeed exist (Anthony *et al.*, 1994) since the requirement for OmpA and the inner core of the LPS remained and the surface exclusion system was F-like in its specificity.

While the mechanisms for controlling transcription of the major (33 kb) operon of F and R100-1 are superficially the same, the sequence differences among the regulatory elements as well as the patterns of transcription differ substantially and demonstrate that each plasmid has developed specificity in this regard (Dempsey, 1994). One exception is *finO* which encodes a protein essential for antisense control of the positive regulatory protein TraJ. In F, the *finO* gene is disrupted by an IS3 element, leading to constitutive expression of TraJ and derepressed levels of transfer (Cheah and Skurray, 1986; Yoshioka *et al.*, 1987). R100 encodes a functional FinO, which is not plasmid specific, and which interacts with the FinP antisense RNA of F to inhibit *traJ* mRNA translation (van Biesen and Frost, 1994). This fertility inhibition ensures that R100, which is similarly repressed, will have approximately the same success in spreading through a population as F and negates the positive aspects of the *finO* mutation in F.

The protein complex which is responsible for the actual transfer of the DNA (reviewed in Lanka and Wilkins, 1995) during conjugation contains TraI, the relaxase/helicase; TraD, an ATP-dependent transport protein; TraY, which optimizes TraI binding to *oriT* in preparation for transfer; and TraM, which is transcribed from a monocistronic operon immediately downstream from *oriT* and which binds to multiple sites in the *oriT* region. The function of TraM is unknown, but it is thought to be involved in processing the signal that a mating pair has formed and transfer should begin (Kingsman and Willetts, 1978). The mechanism and requirements of the site-directed phosphotransfer (nicking) reaction have been worked out in considerable detail for both F and R100. The specificity of these proteins for their homologous *oriT*s has been previously documented

(Finlay *et al.*, 1986; Willetts and Maule, 1986); however, no correlation between sequence differences in *oriT* and plasmid specificity has been demonstrated beyond computer homologies.

The sequence differences between F and R100 TraY and TraM as well as their respective binding sites at *oriT* and, in the case of TraY, at the Py promoter have been documented for F (Frost *et al.*, 1994) and R100-1 (Inamoto and Ohtsubo, 1990). While no mutations in F *traY* were available until recently (Maneewannakul *et al.*, 1996), mutations in F *traM* were not complementable by R100. The sequence differences between F and R100 TraI and TraD have been documented by Yoshioka *et al.* (1990). While the R100 TraD protein has a string of eight PQQ tripeptides compared with a single PQQ motif in F TraD, R100 *traD* can complement *traD* mutations in F. The large TraI proteins from both plasmids differ in only a few amino acids, but it has not been clear whether F and R100-1 TraI are specific for their homologous *oriT*s since both require TraY for maximal relaxase activity (Inamoto *et al.*, 1991; Nelson *et al.*, 1995) The nick sites in these plasmids differ by two base pairs, suggesting that TraI has sequence specificity (see Fig. 3.2) (Frost *et al.*, 1994). However, this specificity is difficult to separate from the overlying specificity of the TraY proteins which bind near the nick site and attract TraI to *oriT*.

3.2   RESULTS

3.2.1   INSERTIONS AND DELETIONS IN THE F AND R100
TRANSFER REGIONS

The R100-1 plasmid (obtained from Neil Willetts, Edinburgh, Scotland) was used to clone the transfer region into constructs based on the restriction map supplied by Womble and Rownd (1988). The *traE* gene was found to contain an IS2 element (Anthony *et al.*, 1996) and another IS2 element was found further downstream in the R100-1 *tra* operon (data not shown). Consequently, a different isolate of R100-1 was obtained from David Bradley, St John's, Newfoundland and was used to sequence the '*traK-trbA*' portion of the *tra* operon. This part of the *tra* operon was cloned as a 12 kb *Eco*RI fragment ('*traE-traN*') which was subcloned as two *Sal*I-*Eco*RI fragments of 8 and 4 kb with the *Sal*I site near the end of *traC* (conserved in F at nt 9925; Frost *et al.*, 1994). The remainder of R100-1 *traN* was sequenced from a PCR product generated using primers before and after the F *traN* gene (nt 12948–12971 and nt 14837–14920; Frost *et al.*, 1994). The '*traN-trbA*' sequence was obtained from an *Eco*RI-*Sac*I fragment with the *Sac*I (*Sst*I) site being conserved in F at nt 15951 (Frost *et al.*, 1994). The sequence was obtained using USB Sequenase and single-stranded templates generated from clones containing progressively larger deletions in the *tra* region constructed using the Erase-a-

Base kit (Promega). Alternatively, double-stranded DNA was sequenced with oligonucleotide primers using USB Sequenase. If the sequence between F and R100-1 was significantly different, oligonucleotide primers were synthesized and the sequence of the opposite strand was determined.

There is one deletion and there are several insertions in the R100-1 sequence as compared with F (Fig. 3.1). The deletion of *trbH* in R100-1 was noted by Yoshioka *et al.* (1990). This gene is not involved in F transfer since *trbH* mutations are silent with respect to piliation and transfer (Maneewannakul and Ippen-Ihler, 1993). Yoshioka *et al.* (1990) also noted the presence of several additional open reading frames (ORFs) in R100-1. Two ORFs, *orfA* and *orfB*, are found beyond *finO* in R100-1 and F with sequence homology rapidly decreasing within *orfA* . Since these ORFs have no known function in transfer they will not be discussed further. *OrfC*, which is also known as *orf286* in plasmid R6-5, a plasmid closely related to R100-1 (Cram *et al.*, 1991), is responsible for increased expression of *finO* mRNA, possibly by increasing its stability, which leads to tighter repression of transfer in R100-like plasmids (van Biesen and Frost, 1992). *OrfD* is known as *traX* in F and is responsible for acetylating the N-terminus of the mature pilin subunit prior to assembly (Moore *et al.*, 1993). *OrfE*, which is neatly inserted between *traT* and *traD* in R100, has no known function (Yoshioka *et al.*, 1990).

We have found two new sites of insertion in R100-1: one site, located between *traR* and *traC* in F, contains four possible ORFs (*orfG1, G2, H* and *I*) while the second site (*orfF*) is between *traU* and *trbC* in F. The cluster of four ORFs contains an almost perfect tandem repeat encoding two small proteins named OrfG1 and OrfG2. The functions of these proteins or of OrfF, H and I are unknown nor has any homology to proteins in the databanks been found (BLASTP program).

3.2.2   CORRELATION BETWEEN SEQUENCE AND CHANGES IN FUNCTION

By and large the differences in sequence between F and R100-1 *tra* genes were very small with most nucleotide changes involving the third position of codons and therefore not altering protein sequence. The accession number for the sequence is AF005044. Table 3.1 lists the transfer proteins in F and R100-1 and gives the percentage similarity and identity for each one as well as the proposed function. One exception is the *traB,P* genes of F and R100-1. The change in sequence at the end of *traB* and beginning of *traP* was significant but did not affect protein function (Anthony *et al.*, 1996). Another large-scale change in sequence occurred in *traN*, which encodes an outer membrane protein involved in mating pair stabilization (Maneewannakul *et al.*, 1992). This change involved a large central

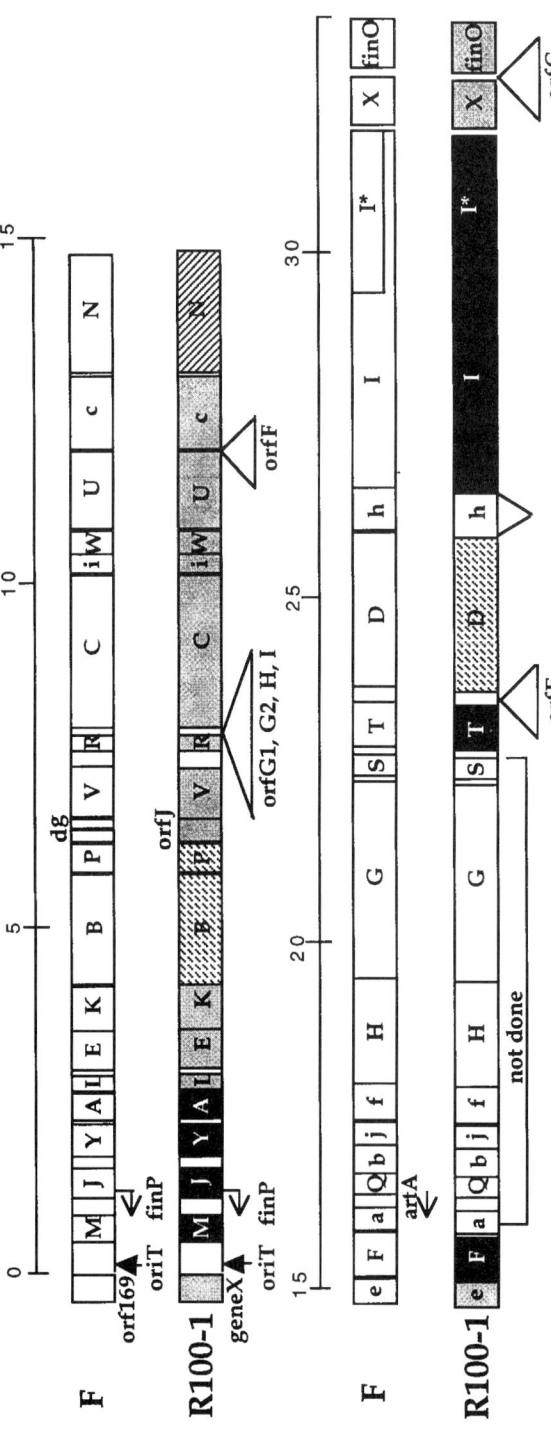

Fig. 3.1 A comparison of the maps for the F and R100-1 transfer regions. The sequence of the F transfer region is supplied in Frost *et al.*, 1994 (accession number U01159). The R100-1 transfer gene products are shaded according to change in specificity/function. ■ (TraM, TraJ, TraY), difference in sequence which corresponds with a difference in specificity and a lack of trans-complementation between F and R100-1. ▨ proteins (TraB, TraP, TraD) which have sizeable sequence differences but no change in specificity and can completely trans-complement each other. ▨ proteins (TraA, TraF, TraT, TraI) that have small changes in sequence but a discernible differences in specificity/function. ☐, none to small changes in sequence that do not affect specificity or function. ▨, (TraN) a protein with a large change in sequence which alters protein specificity but presumably not function. Insertions or deletions are shown below the line for R100-1. The bar above the F transfer region designates the length of the sequence (in kb) from the BglII site near the beginning of the *orf169 (geneX)* gene.

portion of the TraN protein (~200 amino acids) and was shown to be associated with the requirement by F (but not R100-1) for OmpA in the recipient cell. A small difference (three amino acids) in R100-1 TraF correlated with a loss of f1 infectivity. When a mutation in F *traF* was complemented with a clone expressing R100-1 TraF, the resulting pili were still capable of attaching f1 phage but were resistant to further infection, presumably because of a defect in pilus retraction. These results will be discussed in full elsewhere.

3.2.3   THE USE OF CHIMERIC PLASMIDS TO STUDY PLASMID SPECIFICITY

The crucial events in plasmid transfer from the plasmid's point of view are the recognition of the 'correct' *oriT* which is then nicked by TraI, followed by transport of this DNA through the conjugation pore. There are several candidates for the transfer protein (TraD, M, Y) responsible for recognizing the 'correct' *oriT* and ensuring that it is positioned near the transferosome (the pilus assembly/DNA transport machinery). F TraY binds to a major site (sbyA) near *oriT* which is between an integration host factor binding site (IHF-A) and a TraM binding site (sbmC). There is a minor binding site for F TraY (sbyC) which overlaps IHF-A (Fig. 3.2) (Luo *et al.*, 1994). TraM binds to three sites near *oriT*; two (sbmA and sbmB) coincide with the autoregulated promoters for *traM* and the third (sbmC) is involved in transfer (Di Laurenzio *et al.*, 1992; Penfold *et al.*, 1996). The role of TraY appears to be two-fold (at least): it increases the binding and nicking of *oriT* by TraI (Nelson *et al.*, 1995) and maximizes *traM* transcription (Penfold *et al.*, 1996). The mechanism for positioning the resulting relaxosome at the base of the transferosome is not known. TraD is thought to be located at this position and perhaps interacts with the relaxosome in some way.

The organization of the *oriT* region of R100-1 is very similar to that of F with conserved positioning of the nick site as well as TraY and TraM binding sites (Fig. 3.2) (Inamoto *et al.*, 1991; Abo *et al.*, 1991; Abo *et al.*, 1993). It is known that R100-1 *traM* cannot complement a *traM* mutation in F (Willetts and Maule, 1986). However, it is not clear whether this specificity results from the inability of TraM to bind other than its cognate DNA or whether it must interact specifically with other proteins to affect its role in transfer.

In this study two chimeric plasmids were constructed and their ability to complement the *traM* mutation in pOX38-MK3 (Penfold *et al.*, 1996) or be mobilized by pOX38-Km, pOX38-MK3 or R100-1 was tested using a mating efficiency assay described in Penfold *et al.*, (1996) (Table 3.2). pNY300 is a 1082 bp BglII fragment containing *oriT-traM* of F cloned into the BamHI site in pUC18 (Di Laurenzio *et al.*, 1992). pRF105 is a 1049 bp

Table 3.1 Comparison of the gene products of the R100-1/F transfer regions

Gene product R100-1/F	Number aa residues R100-1/F	% Similarity/ identity R100-1/F	Known/proposed function	Reference[a]
Gene X/Orf169	169/169	96.5/94.1	Transglycosylase	Bayer et al., 1995
TraM	127/127	95.3/89.0	Nucleosome at oriT; connects relaxosome to transferosome	
TraJ	223/229	47.5/22.6	Positive regulator of tra operon	
TraY	75/131	58.5/38.6[b]	Positive regulator, relaxosome formation F N- and C-TraY domains homologous to R100-1 TraY	
TraA	119/121	92.4/89.1	Pilin subunit	
TraL	91/91	100/98.9	Pilus assembly	
TraE	188/188	95.2/93.6	Pilus assembly	
TraK	242/242	99.2/98.8	Pilus assembly	
TraB	486/475	97.9/97.3	Pilus assembly	
TraP	195/196	88.7/82.1	Pilus assembly	
OrfJ/TrbD-TrbG[c]	106/65+83	83.1/78.5 (D) 49.4/24.7 (G)	F TrbD and TrbG are fused in R100-1; no known function	
TraV	171/171	97.7/97.1	Lipoprotein; pilus assembly	Doran et al., 1994
TraR	79/73	97.3/97.3	C4-type zinc finger	Doran et al., 1994
OrfG1	138	71.5/61.3	No F equivalent; no known function	
OrfG2	149	(G1/G2)	OrfG1 is similar to OrfG2	
OrfH	73		No F equivalent; no known function	
OrfI	115		No F equivalent; no known function	
TraC	876/875	99.3/98.8	Pilus assembly/outgrowth; ATP/GTP binding motif	

	aa	similarity/identity	Function	Reference
TrbI	128/128	93.8/93.8	No known function	
TraW	210/210	100/99.5	Pilus assembly	
TraU	330/330	99.1/98.5	Pilus assembly	
OrfF	103		No F equivalent; no known function	
TrbC	212/212	93.4/91.5	Pilus assembly; homology to thioredoxin-binding proteins but not within active site	This work
TraN	617/602	82.3/73.6	F mating pair stabilization via OmpA in recipient cell	This work
TrbE	63/55	95.3/93.0	No known function	
TraF	247/247	99.2/98.4	Pilus assembly; affects f1 infectivity	This work
TrbA-TraS[d]			Not done for R100-1	
TraT	243/244	99.6/99.2	Lipoprotein; mediates surface exclusion	
OrfE	245		No F equivalent; no known function	
TraD	738/717	97.6/96.6	DNA transport; pilus assembly; MS2/R17 phage infection; 2 ATP/GTP binding motifs	
TrbH	239		No R100-1 equivalent; no known function	
TraI	1756/1756	97.5/96.0	Relaxase; helicase; 2 ATP/GTP binding motifs	
OrfD/TrX	248/248	93.9/90.3	Acetylates N-terminus of pilin; possible lipoprotein	
OrfC	286		No F equivalent; increases FinO levels	
FinO	186/186	96.8/96.2	Stabilizes antisense RNA FinP; decreases TraJ; increases tra repression	

[a] Only references not found in Frost et al. (1994) are given.

[b] F TraY has homologous N- and C-terminal domains resulting from a gene duplication event. The similarity/identity values are an average for R100-1 TraY (aa 1-75) compared to the N-terminal domain of F TraY (aa 1-75) as well as the C-terminal domain (aa 56-131).

[c] The OrfJ protein resembles a fusion of F TrbD and TrbG. The similarity/identity values are given for F TrbD (D) versus OrfJ (aa 1-65) and F TrbG (G) versus OrfJ (aa 24-83).

[d] The trbA-TraS region of R100-1 has not been completed. Please refer to Frost et al. (1994) for a description of these gene products in F.

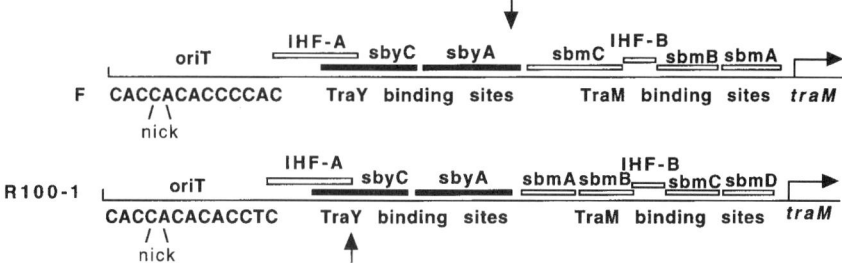

Fig. 3.2 The sequence of the *oriT-traM* promoter region of the plasmids F and R100-1. The position of the IHF-A and IHF-B binding sites, the two binding sites for TraY (sbyA and C), the binding sites for TraM and the start of the *traM* gene are shown for F (Frost *et al.*, 1994) and R100-1 (Abo and Ohtsubo, 1993). The nick site for the TraI relaxase activity is shown by / \. Note the two base difference near the nick site in F and R100-1. The positions of the DraI sites in F and R100-1 that were important in constructing pRF315 and pRF206 (Table 3.2) are indicated as vertical arrows.

BamHI fragment from R100-1 cloned into pUC18. The BamHI site upstream of the R100-1 *oriT* corresponds exactly to the BglII site in F; the sequence of both clones is almost identical (except as noted at *oriT*) until the start of sbyC (Fig. 3.2). pRF315 and pRF206 were constructed by ligating together restriction fragments or polymerase chain reaction (PCR) products from pNY300 or pRF105 such that pRF315 contained the F *oriT* and TraY binding region ligated to the R100-1 TraM binding sites and *traM* gene, while pRF206 has the R100-1 TraY and TraM binding sites and *traM* gene ligated to the F *oriT* region (Fig. 3.2; Table 3.2). It is interesting to note that the only difference between pRF206 and pRF105 is the two-base difference at *oriT* (Fig. 3.2).

Each of these four plasmids, along with pUC18 (negative control), was introduced into *E. coli* cells carrying pOX38-Km, pOX38-MK3 (Penfold *et al.*, 1996) or R100-1. Mating efficiency data for the transfer of the conjugative plasmids as well as mobilization of the test plasmids was carried out for each strain (Table 3.2). The mobilization assays and pOX38-MK3 complementation data suggest the following.

1. Only the F *traM* gene could complement the mutation in pOX38-MK3 resulting in its transfer to *E. coli* ED24 (Spc^R), the recipient cell strain. Thus in the case of single copy plasmids it is important that TraM binds to the DNA that is transferred.
2. All the chimeric plasmids except pRF105 were mobilized by pOX38-Km, suggesting that either F TraY can bind the R100-1 sbyA(C) sites efficiently enough to promote transfer or the requirement for TraY binding is overridden with multicopy, mobilizable plasmids.

Table 3.2 Mobilization and complementation efficiency of F/R100-1 chimeric plasmids in the presence of pOX38-Km, pOX38-MK3 and R100-1

Plasmid/E. coli DH5α[a]	oriT	sbyA	sbmABC	TraM	Mobilization by pOX38-Km (per 100 donor cells)	Mobilization by pOX38-MK3 (per 100 donor cells)	Mobilization by R100-1 (per 100 donor cells)	Ability to complement pOX38MK3
pUC18	–	–	–	–	–	–	–	–
pNY300	F	F	F	F	250	270 (+)	0.08	+++
pRF315	F	F	R100-1	R100-1	140	44 (+)	0.014	–
pRF206	F	R100-1	R100-1	R100-1	200	10 (+/–)	21 (+)	–
pRF105	R100-1	R100-1	R100-1	R100-1	0	0	10 (+)	–

[a] The donor cell contained pOX38-Km, pOX38-MK3 or R100-1 in *E. coli* DH5α plus one of the chimeric plasmids listed in column 1. The recipient cell was *E. coli* ED24 (Spc[R]).
– indicates negative.

3. Mobilization by pOX38-MK3 was efficient for pNY300 while the efficiency of mobilization was reduced for pRF315 and pRF206, suggesting that the R100-1 TraM is less effective than F TraM at promoting transfer of F *oriT*-containing plasmids.

4. R100-1 could mobilize all the plasmids to some extent but only pRF105 and pRF206 were mobilized efficiently. This suggests that the R100-1 TraI protein can nick at both the F and R100-1 *oriT* sequence; however, the R100-1 TraY protein binds the sbyA site in R100-1 more efficiently than that in F.

Preliminary results using an agarose gel to separate the nicked species from other forms of DNA (e.g. supercoiled) suggested that all the plasmids that were mobilized at high efficiency, with the exception of pRF206 in the presence of pOX38-Km, had the nicked species (data not shown). pRF206 was probably nicked at low levels and these nicked species might then be preferentially transferred.

3.3   DISCUSSION

Although the F and R100-1 transfer regions seem virtually identical except for a few noted exceptions, small changes in sequence can have important consequences. The most dramatic example is TraT in which a single amino acid substitution changes the specificity of the surface exclusion system. Comparison of the sequence differences between transfer proteins encoded by F and R100-1 such as TraA (pilin), TraF and TraN, which seem to affect phage sensitivity and mating pair formation, might shed light on the function of these proteins. Interestingly, these two plasmids seem to recognize different moieties on the surface of the recipient cell and do not compete for these receptors. While the specificity of the F and R100-1 TraM proteins for their cognate DNA has long been appreciated, the similarity between the two proteins is remarkably high. There are 12 amino acid differences between F and R100-1 TraM (126 amino acids in total) which change the specificity of these proteins completely. Anti-F TraM antiserum fails to react with R100-1 TraM and both proteins bind to very different sequences in their cognate *oriT* regions. This is typical of the DNA binding proteins (TraY, M, J) that provide specificity during transfer.

The ability of R100-1 TraI (but not F TraI) to nick both F and R100-1 *oriT* sequences is offset by the specificity of the R100-1 TraY protein (but not F TraY) for its own binding sites. Thus in a situation where a cell contains both plasmids (F and the repressed wild-type plasmid R100), an interesting scenario would develop whereby dominance of one plasmid over the other would ensue. In this case, the F plasmid would be repressed by the FinO protein expressed by R100 which is also contributing to its own repression. However, the F plasmid would nevertheless have an increased chance for transfer because of the promiscuity of the R100 TraI protein which can nick both F and R100-1 *oriT*. Similarly, multicopy plasmids have an increased chance for mobilization by competing with the conjugative plasmids for available binding sites at the base of the transferosome. The results given here suggest that the 'successful' plasmid is the one that is relaxed at *oriT* and then finds its way to this transfer site via protein–protein interactions. Since TraM appears to define the specificity of plasmid transfer, we predict that it should interact with a protein at the base of the transferosome. The most likely candidate is TraD since it is the only transfer protein that is situated in the inner membrane, which is essential for DNA transport. These results also suggest that plasmids that are often in contact with one another in nature will evolve in such a way as to maximize their own transfer and diminish as much as possible the transfer of other competing plasmids. Certainly fertility inhibition of F by R100 is a clear demonstration of this, but the results given here suggest that there is a much more subtle interplay between plasmids, involving varying degrees of specificity of transfer proteins for DNA and for interactions with other transfer proteins. The specificity of these interactions can be affected by properties such as plasmid copy number, which might be an important factor in plasmid dissemination as well as its usual role in plasmid maintenance.

REFERENCES

Abo, T. and Ohtsubo, E. (1993) Repression of the *traM* gene of plasmid R100 by its own product and integration host factor at one of the two promoters. *J. Bacteriol.* **175**: 4466–4474.

Abo, T., Inamoto, S. and Ohtsubo, E. (1991) Specific DNA binding of the TraM protein to the *oriT* region of plasmid R100. *J. Bacteriol.* **173**: 6347–6354.

Anthony, K.G., Sherburne, C., Sherburne, R. and Frost, L.S. (1994) The role of the pilus in recipient cell recognition during bacterial conjugation mediated by F-like plasmids. *Mol. Microbiol.* **13**: 939–953.

Anthony, K.G., Kathir, P., Moore, D. *et al.* (1996) Analysis of the *traLEKBP* sequence and the TraP protein from three F-like plasmids: F, R100-1, and ColB2. *J. Bacteriol.* **178**: 3194–3200.

Bayer, M., Eferl, R., Zellnig, G. *et al.* (1995) Gene *19* of plasmid R1 is required for both efficient conjugative DNA transfer and bacteriophage R17 infection. *J. Bacteriol.* **177**: 4279–4288.

Cheah, K.-C. and Skurray, R.A. (1986) The F plasmid carries an *IS3* insertion within *finO*. *J. Gen. Microbiol.* **132**: 3269–3275.

Cram, D.S., Ray, A., O'Gorman, L. and Skurray, R. (1984) Transcriptional analysis of the leading region in F plasmid DNA transfer. *Plasmid* **11**: 221–233.

Cram, D.S., Loh, S.M., Cheah, K.-C. and Skurray, R.A. (1991) Sequence and conservation of genes at the distal end of the transfer region on plasmids F and R6-5. *Gene* **104**: 85–90.

Dempsey, W.B. (1994) Regulation of R100 conjugation requires *traM* in *cis* to *traJ*. *Mol. Microbiol.* **13**: 987–1000.

Di Laurenzio, L., Frost, L.S. and Paranchych, W. (1992) The TraM protein of the conjugative plasmid F binds to the origin of transfer of the F and ColE1 plasmids. *Mol. Microbiol.* **6**: 2951–2959.

Doran, T.J., Loh, S.M., Firth, N. and Skurray, R.A. (1994) Molecular analysis of the F plasmid *traVr* region: *traV* encodes a lipoprotein. *J. Bacteriol.* **176**: 4182–4186.

Fee, B.E. and Dempsey, W.B. (1986) Cloning, mapping, and sequencing of plasmid R100 *traM* and *finP* genes. *J. Bacteriol.* **167**: 336–345.

Fee, B.E. and Dempsey, W.B. (1988) Nucleotide sequence of *geneX* of antibiotic resistance plasmid R100. *Nucleic Acids Res.* **16**: 4726.

Finlay, B.B. and Paranchych, W. (1986) Nucleotide sequence of the surface exclusion genes *traS* and *traT* from the IncF$_0$*lac* plasmid pED208. *J. Bacteriol.* **166**: 713–721.

Finlay, B.B., Frost, L.S. and Paranchych, W. (1986) Origin of transfer of IncF plasmids and nucleotide sequences of the type II *oriT*, *traM*, and *traY* alleles from ColB4-K98 and the type IV *traY* allele from R100-1. *J. Bacteriol.* **168**: 132–139.

Frost, L.S. and Paranchych, W. (1988) DNA sequence analysis of point mutations in *traA*, the F pilin gene, reveal two domains involved in F-specific bacteriophage attachment. *Mol. Gen. Genet.* **213**: 134–139.

Frost, L.S., Finlay, B.B., Opgenorth, A. *et al.* (1985) Characterization and sequence analysis of pilin from F-like plasmids. *J. Bacteriol.* **164**: 1238–1247.

Frost, L.S., Ippen-Ihler, K. and Skurray, R.S. (1994) Analysis of the sequence and gene products of the transfer region of the F sex factor. *Microbiol. Reviews* **58**: 162–210.

Inamoto, S. and Ohtsubo, E. (1990) Specific binding of the TraY protein to *oriT* and the promoter region for the *traY* gene of plasmid R100. *J. Biol. Chem.* **265**: 6461–6466.

Inamoto, S., Yoshioka, Y. and Ohtsubo, E. (1988) Identification and characterization of the products from the *traJ* and *traY* genes of plasmid R100. *J. Bacteriol.* **170**: 2749–2757.

Inamoto, S., Yoshioka, Y. and Ohtsubo, E. (1991) Site- and strand-specific nicking *in vitro* at *oriT* by the TraY-TraI endonuclease of plasmid R100. *J. Biol. Chem.* **266**: 10086–10092.

Kingsman, A. and Willetts, N. (1978) The requirements for conjugal DNA synthesis in the donor strain during F*lac* transfer. *J. Mol. Biol.* **122**: 287–300.

Lanka, E. and Wilkins, B.M. (1995) DNA processing reactions in bacterial conjugation. *Ann. Rev. Biochem.* **64**: 141–169.

Luo, Y., Gao, Q. and Deonier, R.C. (1994) Mutational and physical analysis of F plasmid *traY* protein binding to *oriT*. *Mol. Microbiol.* **11**: 459–469.

Maneewannakul, K. and Ippen-Ihler, K. (1993) Construction and analysis of F plasmid *traR, trbJ,* and *trbH* mutants. *J. Bacteriol.* **175**: 1528–1531.

Maneewannakul, S., Kathir, P. and Ippen-Ihler, K. (1992) Characterization of the F plasmid mating aggregation gene *traN* and of the new F transfer region locus *trbE*. *J. Mol. Biol.* **225**: 299–311.

Maneewannakul, S., Kathir, P. and Ippen-Ihler, K. (1996) Construction of derivatives of the F plasmid pOX-*tra715*: characterization of *traY* and *traD* mutants that can be complemented *in trans*. *Mol. Microbiol.* **22**: 197–205.

McIntire, S.A. and Dempsey, W.B.(1987) *oriT* sequence of the antibiotic plasmid R100. *J. Bacteriol.* **169**: 3829–3832.

Meynell, E., Meynell, G.G. and Datta, N. (1968) Phylogenetic relationships of drug-resistance factors and other transmissible bacterial plasmids. *Bacteriol. Reviews* **32**: 55–83.

Moore, D., Hamilton, C.M., Maneewannakul, K. *et al.* (1993) The *Escherischia coli* K-12 F plasmid gene *traX* is required for acetylation of F pilin. *J. Bacteriol.* **175**: 1375–1383.

Nelson, W.C., Howard, M.T., Sherman, J.A. and Matson, S.W. (1995) The *traY* gene product and integration host factor stimulate *Escherichia coli* DNA helicase I-catalyzed nicking at the F plasmid *oriT*. *J. Biol. Chem.* **270**: 28374–28380.

Ogata, R.T., Winters, C. and Levin, R.P. (1982) Nucleotide sequence analysis of the complement resistance gene from plasmid R100. *J. Bacteriol.* **151**: 819–827.

Penfold, S.S., Simon, J. and Frost, L.S. (1996) Regulation of the expression of the *traM* gene of the F sex factor of *Escherichia coli*. *Mol. Microbiol.* **20**: 549–558.

van Biesen, T. and Frost, L.S. (1992) Differential levels of fertility inhibition among F-like plasmids are related to the cellular concentration of *finO* mRNA. *Mol. Microbiol.* **6**: 771–780.

van Biesen, T. and Frost, L.S. (1994) The FinO protein of IncF plasmids binds FinP antisense RNA and its target, *traJ* mRNA, and promotes duplex formation. *Mol. Microbiol.* **14**: 427–436.

Willetts, N. and Maule, J. (1986) Specificities of IncF plasmid conjugation genes. *Genet. Res. Camb.* **47**: 1–11.

Womble, D.D. and Rownd, R.H. (1988) Genetic and physical map of the IncFII antibiotic resistance plasmids. *Microbiol. Reviews* **52**: 433–451.

Yoshioka, Y., Ohtsubo, H. and Ohtsubo, E. (1987) Repressor gene *finO* in plasmids R100 and F: constitutive transfer of plasmid F is caused by insertion of IS3 into F *finO*. *J. Bacteriol.* **169**: 619–623.

Yoshioka, Y., Fujita, Y. and Ohtsubo, E. (1990) Nucleotide sequence of the promoter-distal region of the *tra* operon of plasmid R100, including *traI* (DNA helicase I) and *traD* genes. *J. Mol. Biol.* **214**: 39–53.

Lateral broad host range gene transfer in nature: how and how much? 4

Abigail A. Salyers, Andrew J. Cooper and Nadja B. Shoemaker

SUMMARY

Bacteria can exchange DNA by conjugation, transformation and transduction. The only one of these mechanisms known to be capable of mediating very broad host range transfers is conjugation. Recently a new kind of conjugal gene transfer element has been discovered: conjugative transposons. These elements are contributing significantly to the spread of antibiotic resistance genes in the Gram-positive bacteria and in the *Bacteroides* group of Gram-negative bacteria, and one has now been found in a species of proteobacteria. The properties of these elements suggest that they are related not only to plasmids but also to bacteriophages. In the case of the *Bacteroides* conjugative transposons, transfer of the element is stimulated by the antibiotic tetracycline, a characteristic that suggests why they have spread so extensively within this genus during the past 30 years. Evidence that extensive broad host range gene transfers occur in nature is reviewed. Conjugative transposons and plasmids are clearly responsible for at least some of these events, but the possibility that many occurred by other modes of horizontal transfer cannot be ruled out.

4.1 HOW DO BROAD HOST RANGE LATERAL GENE TRANSFERS OCCUR?

Some of the most interesting types of gene transfer events are those that occur between bacteria from different genera or phylogenetic groups. A growing body of evidence, which is reviewed in a later section of this chapter, supports the hypothesis that broad host range transfers do occur in nature and may occur fairly frequently. Moreover, the genes being transferred are being expressed in their new hosts. This raises questions about the contribution of broad host range gene transfer events to the evolution of modern bacteria. And, if such transfers occur commonly in nature, how is it possible that distinct species still exist? Broad host range gene transfers also have practical importance because their existence suggests that there is no limit to how far a gene can spread once it enters the bacterial gene pool.

There are three well-established mechanisms of lateral gene transfer: phage transduction, transformation and conjugation. It is generally assumed that conjugation, the direct transfer of DNA from the cytoplasm of a donor to the cytoplasm of a recipient, is the mechanism responsible for very broad host range gene transfers, whereas transformation and transduction are generally regarded as mechanisms of narrow host range gene transfer. The reason for thinking that transduction mediates DNA transfers primarily between closely related bacteria is that most of the bacteriophage receptors so far described have been limited to a single bacterial species. This notion could, however, be an artifact of the small number of phages and hosts that have been studied to date. Moreover, phage DNA released from lysing bacterial cells might be taken up by other bacteria via transformation, thus obviating the necessity for a surface receptor. Another factor that limits the range of productive transduction events is that generalized transduction transfers linear segments of DNA, which must be integrated into the recipient's genome by homologous recombination. Such DNA segments would be lost in hosts whose DNA failed to have sufficient homology with the incoming segments to allow them to be incorporated. Of course, this restriction does not apply to genes carried on the bacteriophage itself or to plasmids packaged and transferred by the phage.

Most studies of natural transformation done to date have tended to support the notion that productive gene transfers mediated by natural transformation occur mainly between strains of the same species. For example, natural transformation systems, in contrast to chemical transformation and electroporation, take up linear DNA segments far more efficiently than plasmids. The reason is that DNA taken up by natural transformation is cut into segments and one strand is removed before the DNA enters the recipient's cytoplasm. To fix these linear segments in its genome, the recipient would have to integrate them via homologous recombination, unless the DNA segment contained a transposable element. As more studies are done of transduction and natural transformation in a wider variety of bacteria, assumptions about the limited range of transfers mediated by these processes may well change.

Although there is some uncertainty about the capacity of natural transformation systems and phages to mediate gene transfers between distantly related bacteria, it is well-established that conjugation systems can mediate such transfers. For this reason, this chapter will focus on conjugation and elements that are transferred by conjugation.

4.2   MOVING BEYOND THE PLASMID PARADIGM

For a long time, plasmids were the only elements known to be transferred by conjugation. We now know, however, that there is at least one other type of conjugal element: conjugative transposons (Clewell and Flannagan, 1993; Salyers *et al.*, 1995b). Some examples of conjugative transposons and other transmissible integrated elements are given in Table 4.1. There is not yet an established definition for what consitutes a conjugative transposon, and many disparate types of elements have been included in this category. Yet, although the different elements listed in Table 4.1 differ greatly in size and DNA sequence, they all share a few common features (Fig. 4.1). Conjugative transposons are normally integrated into the bacterial genome. To transfer, they first excise to form a covalently closed circular intermediate, which is then transferred by conjugation to the recipient (Li *et al.*, 1995; Scott *et al.*, 1994). In the recipient, and presumably also in the donor, the circular intermediate integrates once again into the genome. Transfer of conjugative transposons differs from Hfr transfer because the plasmid integrated in an Hfr strain does not excise prior to transfer but instead transfers part of itself plus some adjacent chromosomal DNA to the recipient. Moreover, the recipients of Hfr-mediated transfer events are not able to retransfer the DNA they receive, because they receive only part of the element responsible for the transfer event. Recipients of conjugative transposons become fully transfer-proficient, as long as the transfer genes are expressed in the recipient. Finally, Hfr-mediated gene transfers are restricted to closely related bacteria because homologous recombination is required to integrate the transferred DNA. Conjugative transposons carry their own integrase genes and can thus integrate into the genome of any host in which the integrase is expressed. The wide range of hosts into which known conjugative transposons can integrate suggests that the integrase genes are indeed widely expressed in many genetic backgrounds.

Conjugative transposons are also different from plasmids that are capable of integrating into the chromosome because they carry an insertion sequence and could excise by homologous recombination between the insertion sequences that flank the integrated form. For one thing, the conjugative transposon integrases appear to have a different mechanism from that of transposases (Bedzyk *et al.*, 1992; Clewell and Flannagan, 1993; Scott, 1992). For another, there is no target site duplication when

Table 4.1 Examples of conjugative transposons and other transmissible integrated elements

Element	Transfer capability	Size	Resistance genes	References
Gram-positive conjugative transposons				
Tn916	Self-transmissible	18–25 kbp	tetM	Clewell et al., 1993; Scott, 1993
Tn5253	Self-transmissible	>60 kbp	tetM, cat	Shoemaker et al., 1980
Vibrio cholerae conjugative transposon				
SXT	Self-transmissible	62 kbp	Resistance to Smx, Tm	Waldor et al., 1996
Bacteroides conjugative transposons				
Tc^r Em^r DOT	Self-transmissible	70 kbp	tetQ, ermF	Stevens et al., 1993
Tc^r Em^r 7853	Self-transmissible	70 kbp	tetQ, ermG	Cooper et al., 1996; Nikolich et al., 1994
XBU4422	Self-transmissible	60 kbp	None known	Shoemaker and Salyers 1990
NBU1	Mobilized by conjugative transposons	11 kbp	None known	Shoemaker et al., 1993, 1996a, b
Tn4555	Mobilized by conjugative transposons	12.5 kbp	cfxA	Smith, 1993

Cm, chloramphenicol; Str, streptomycin; Smx sulfamethoxazole; Tm, trimethoprim.

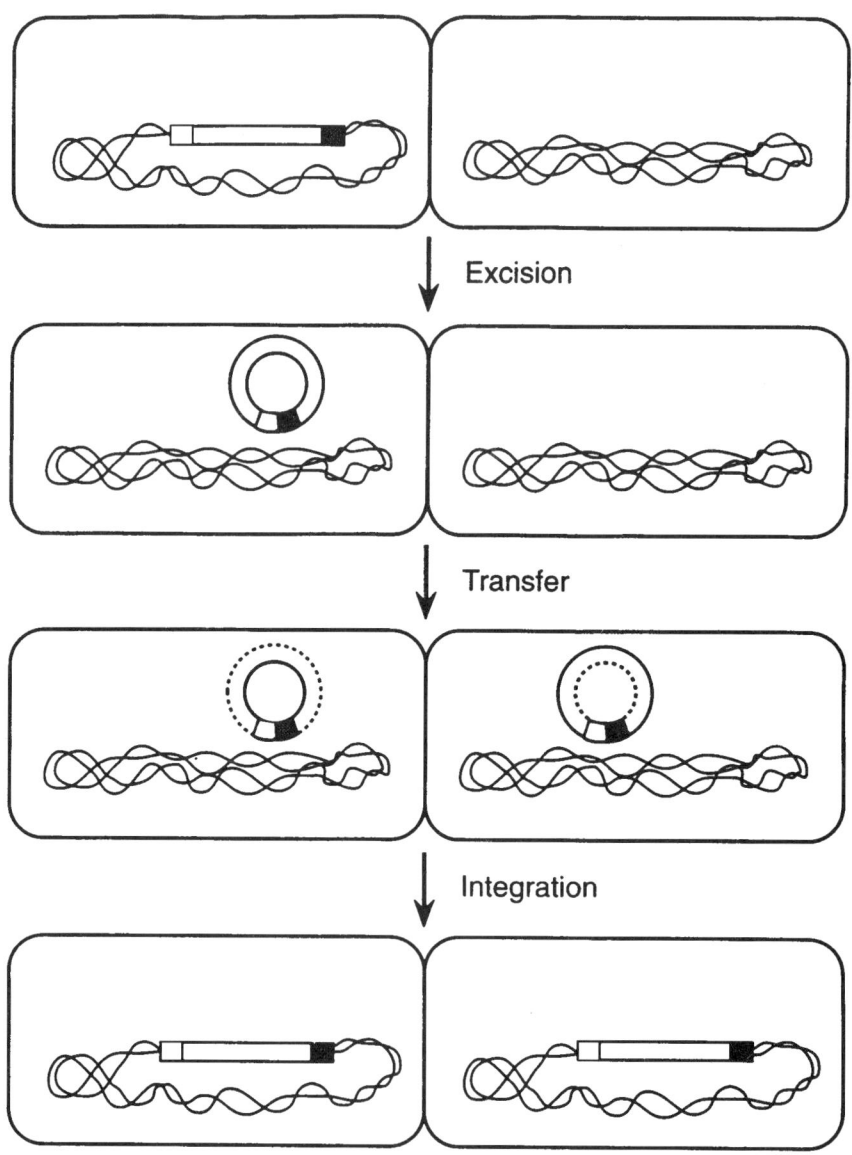

Fig. 4.1 Features shared by most conjugative transposons (Scott, 1992; Salyers *et al.*, 1995b). The conjugative transposon is normally integrated in the donor's genome. To transfer, it first excises by a precise or nearly precise excision mechanism that restores the integrity of the donor molecule and produces a covalently closed circular intermediate. The circular intermediate is transferred similarly to a plasmid; that is, a single-stranded copy of the circle is transferred to the recipient, then made double stranded in the recipient. The circular form then integrates into the chromosome.

the conjugative transposon integrates. In fact, the two ends of a conjugative transposon have no extended regions of identity to each other, and excision appears to be independent of homologous recombination. Conjugative transposons also differ in a number of ways from the gene cassettes and integrons that have contributed to the evolution of plasmids that carry multiple antibiotic resistance genes (Hall and Collis, 1995). The gene cassettes that integrate into integrons have a circular intermediate, but unlike the conjugative transposons they do not carry their own integrase genes nor do they have the capacity to transfer themselves by conjugation. They become transmissible only if they integrate into a self-transmissible element such as a plasmid. The one point of similarity is that the integrase genes of the integrons into which the gene cassettes integrate are members of the lambda integrase family, just as are the integrases of conjugative transposons. Conjugative transposons have some similarities to a family of integrated plasmids that have been found in actinomycetes (e.g. Brown *et al.*, 1990). The main difference is that the conjugative transposons' circular intermediates have not been found to replicate in any host so far tested, unless the act of conjugation itself is considered to be a form of replication, whereas the integrating actinomycete plasmids do replicate as plasmids in some hosts.

An interesting feature of the Gram-positive conjugative transposons is that their integrases are members of the lambdoid phage integrase family, although the mechanism of integration proposed for them appears different from that of phage lambda (Clewell and Flannagan, 1993; Poyant-Salmeron *et al.*, 1990). That is, there is not a specific attachment site sequence on the chromosome that is identical to the *att* site on the phage. The integrases of the integrating actinomycete plasmids are also members of the lambda integrase family, but these plasmids have large *att* sites (Katz *et al.*, 1991). The integrases of the *Bacteroides* conjugative transposons have not yet been characterized, so it is not clear if they too will fall into this gene family, but a type of integrating *Bacteroides* element called NBU1, which is mobilized by the conjugative transposons, does have a lambdoid phage type of integrase gene (Shoemaker *et al.*, 1996a).

Conjugative transposons differ considerably in their integration specificity. The *Bacteroides* conjugative transposons are relatively site-specific, with 5–8 integration sites per chromosome (Bedzyk *et al.*, 1992). By contrast, the type of Gram-positive conjugative transposon exemplified by Tn916 integrates almost randomly in most hosts (Clewell and Flannagan, 1993; Clewell *et al.*, 1995). Conjugative transposons tend to have very broad host ranges. Tn916 can transfer between Gram-positive and Gram-negative bacteria (Clewell *et al.*, 1995). The *Bacteroides* conjugative transposons can transfer co-resident plasmids to *E. coli*, a different phylum of Gram-negative bacteria (Salyers *et al.*, 1995b). Recent work in our laboratory suggests that the *Bacteroides* conjugative transposons can also transfer themselves to *E. coli*, although integration is much less efficient in *E. coli* than integration in *Bacteroides* and does not occur via the ends of the conjugative transposon (unpublished results).

4.3   COOPERATION, NOT COMPETITION: INTERACTIONS BETWEEN PLASMIDS AND CONJUGATIVE TRANSPOSONS

Plasmids and conventional transposons (e.g. Tn3, Tn*10*) have often been described as selfish DNA segments, and this description is true insofar as plasmids and transposons often exclude others of their kind from entering or setting up residence in a host that already harbors one. The conjugative transposons, by contrast, are remarkably sociable. More than one conjugative transposon can enter the same strain. In fact, we have found a clinical isolate of *Bacteroides fragilis* that carried two conjugative transposons of the same type (Shoemaker and Salyers, 1990). Conjugative transposons also interact with plasmids and other integrated elements. Both the Gram-positive and *Bacteroides* conjugative transposons can mobilize plasmids in *trans*. The *Bacteroides* conjugative transposons also act in *trans* on unlinked integrated elements called NBUs to trigger their excision and circularization (Shoemaker *et al.*, 1993). The circular form of the NBU is then mobilized by the conjugative transposons to a recipient, where the NBU integrates into the recipient genome.

An interesting example of an interaction between conjugative transposons and plasmids, which illustrates how they can broaden each other's host range, involves the IncP plasmid R751 and the *Bacteroides* conjugative transposon Tc^r ERL (Shoemaker and Salyers, 1987). R751 can transfer itself from *E. coli* to *Bacteroides* but does not replicate in *Bacteroides*. A derivative of R751 that carries the *Bacteroides* conventional transposon Tn*4351* can transfer to and integrate into the chromosome of *Bacteroides* recipients, but it cannot excise and transfer itself back out of *Bacteroides*. A strain of *Bacteroides* that carries Tc^r ERL as well as the integrated form of R751 was capable of transferring R751 back to *E. coli* recipients, where R751 was recovered as a plasmid. Similarly, the IncP plasmids R751 and RP4 can mobilize the circular form of the *Bacteroides* integrated element, NBU1, from *E. coli* to *Bacteroides*. NBU1 is capable of integrating in *E. coli* as well as in *Bacteroides* spp., but its integration is less site-specific in *E. coli* than in *Bacteroides* (Shoemaker *et al.*, 1996b).

4.4   REGULATION OF TRANSFER FUNCTIONS BY TETRACYCLINE

An unusual feature of the *Bacteroides* conjugative transposons is that their transfer is stimulated 1000-fold to 10 000-fold by low concentrations of the antibiotic tetracycline (Salyers *et al.*, 1995a,b). Much less transfer has also been reported for Tn*925*, a member of the Tn*916* family of conjugative transposons (Salyers *et al.*, 1995b), but this observation has been controversial and the mechanism is unknown. In the case of the *Bacteroides* conjugative transposons, all derivatives of tetracycline have a stimulatory effect on transfer, as does the non-toxic tetracycline analog, autoclaved chlorotetracycline. Tetracycline stimulation of transfer is mediated by a complex regulatory system (Salyers *et al.*, 1995a; Stevens *et al.*, 1993), but precisely how this system senses tetracycline and how the regulatory gene products control expression of transfer functions remains to be established.

Tetracycline stimulation of element transfer suggests a possible explanation for the widespread distribution of the tetracycline-resistant *tetQ*, which has occurred in colonic *Bacteroides* strains over the past three decades. Prior to 1970, tetracycline-resistant strains of *Bacteroides* clinical isolates were uncommon, whereas today nearly all *Bacteroides* clinical isolates are tetracycline resistant, and all of the resistant strains so far tested carry *tetQ* (Speer *et al.*, 1993). In all of the *Bacteroides* clinical isolates, *tetQ* has proved to be carried on a conjugative transposon. Thus, conjugative transposons appear to have been responsible for the rather dramatic increase in tetracycline resistance among *Bacteroides* clinical isolates. It was during this same period, between 1970 and the present, that low-dose, long-term tetracycline therapy for dermatological problems such as acne became popular. Whatever the reason for the dramatic rise in tetracycline-resistant *Bacteroides* clinical isolates, the fact that an antibiotic can stimulate resistance gene transfer as well as select for maintenance of strains that acquire the resistance gene raises questions about the assumption by many people that sub-therapeutic doses of antibiotics, such as those used in the treatment of acne, have no effect on the spread of antibiotic resistance genes.

4.5   HOW MUCH? EVIDENCE FOR WIDESPREAD GENE TRANSFER IN NATURE

Transfer of resistance genes between bacteria of different phylogenetic groups can be achieved readily in a laboratory setting, but these transfers are done under optimal conditions with high concentrations of donors and recipients. In the real world environment, conditions may not be as optimal for transfer. Also, one might expect that bacteria normally found in different sites (e.g. soil and the intestine) might not have much opportunity to exchange DNA, and if they do acquire DNA from a distantly related bacterium they may not be able to express the new gene. Nonetheless, there is a growing body of evidence to support the contention that very broad host range horizontal gene transfers occur frequently in nature, and that in many cases the newly acquired genes are being expressed. The evidence consists of finding resistance genes that are nearly identical at the DNA sequence level (> 95% identical) in distantly related bacteria (Figs 4.2 and 4.3). The evidence summarized in Fig. 4.2 supports the contention that lateral transfers between Gram-negative and Gram-positive bacteria do occur in nature. Most of the species listed in Fig. 4.2 are permanent or transient colonizers of the human body, but there is one interesting exception: *tetQ* has been found in *Prevotella ruminicola*, one of the numerically predominant species in the intestines and rumen of livestock animals, as well as in *Bacteroides*, which is a numerically predominant genus of bacteria found in the human colon (Nikolich *et al.*, 1994a). Finding nearly identical genes in these two genera suggests that bacteria normally found in different sites can exchange DNA, presumably during transient colonization of the animal intestine by human-associated bacteria or vice versa. On the *Bacteroides* conjugative transposons, *tetQ* is usually located upstream of *rteA*, a gene thought to be part of the tetracycline-sensing regulatory system (Salyers *et al.*, 1995a; Stevens *et al.*, 1993). In the *Prevotella* strains that carry *tetQ*, only *tetQ* itself plus a small aminoterminal fragment of

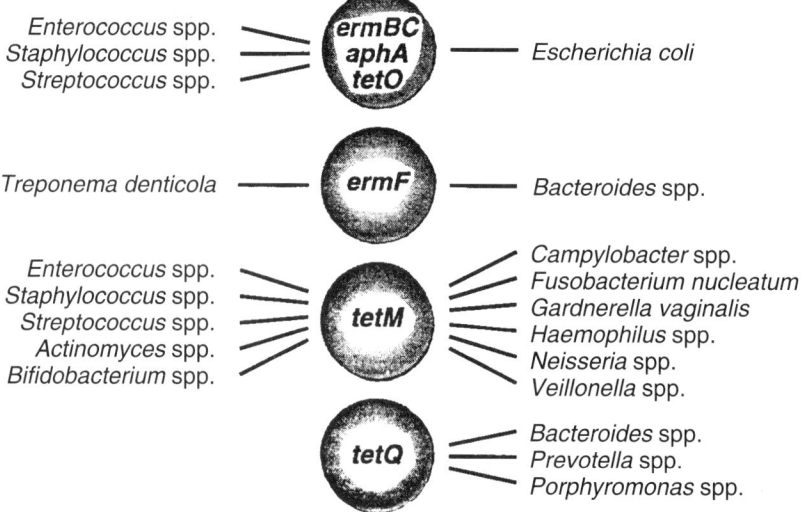

Fig. 4.2 Summary of evidence that lateral gene transfers occur in nature between bacteria from different phylogenetic groups (Salyers, 1995; Roberts *et al.*, 1996). The circles in the middle column are the genes that have been found in natural isolates of the bacterial genera listed in the columns at either side. In all cases, the genes found in the different genera of bacteria are at least 95% identical at the DNA sequence level.

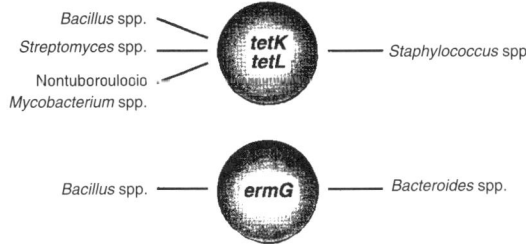

Fig. 4.3 Summary of evidence that lateral gene transfers occur in nature between soil and intestinal bacteria (Salyers, 1995; Cooper *et al.*, 1996). The arrangement of the figure follows the same convention as in Fig. 4.2. The data on *tetK* and *tetL* were obtained using high stringency hybridization rather than sequence comparison. The *ermG* alleles in *Bacillus* and *Bacteroides* spp. were more than 99% identical at the DNA sequence level (Cooper *et al.*, 1996).

rteA was found (Nikolich *et al.*, 1994a). None of the *Prevotella tetQ* alleles has been shown to be on a conjugative transposon, although one was found on a transmissible plasmid. These observations suggest two conclusions. First, the direction of transfer appears to have been from human bacteria to animal bacteria. Second, the transfer was not simply the movement of a *Bacteroides* conjugative transposon into the animal isolates. We know that at least some *Bacteroides* conjugative transposons can transfer themselves intact into a *P. ruminicola* strain in laboratory matings (Shoemaker *et al.*, 1992). Either the conjugative transposons are more unstable in certain *P. ruminicola* strains than in the one used for the laboratory experiments, or the transfer was not a simple conjugal transfer event.

It is interesting to note that *tetM*, another gene that has undergone very wide dissemination among diverse species of bacteria, is almost always associated with conjugative transposons (Clewell and Flannagan, 1993; Clewell *et al.*, 1995; Salyers, 1995). Most of the papers published on the dissemination of *tetM* have relied on DNA hybridization rather than sequence information, but as more *tetM* sequences from diverse organisms begin to accumulate in the databases, it is evident that these genes share at least 98% identity to each other. As already mentioned, *tetQ* alleles in *Bacteroides* strains are associated almost exclusively with conjugative transposons. Thus conjugative transposons appear to have contributed significantly to the spread of these elements in some cases even though they are not clearly linked to the transfer of *tetQ* between human and animal bacteria. Another gene that is frequently carried by conjugative transposons is *ermF*. Other resistance genes shown in Figs 4.2 and 4.3 are generally associated with plasmids rather than conjugative transposons, but since conjugative transposons can mobilize plasmids, it is possible that even in these cases the gene transfer events are being driven by conjugative transposons.

The evidence summarized in Fig. 4.3 suggests that transfer occurs between soil and intestinal bacteria. The *ermG* data are particularly convincing because the gene found in *Bacillus* was over 99% identical to the one found in *Bacteroides* spp. (Cooper *et al.*, 1996). The *ermG* gene in *Bacteroides* was located on a conjugative transposon, but it is not clear whether this was true of the *ermG* gene from *Bacillus*.

An important point to note is that the genes summarized in Figs 4.2 and 4.3 were found because the strains that carried them were resistant to one or more antibiotics. That is, the genes are being expressed in many different hosts. This could be because the promoters themselves have unusually broad host ranges, or can mutate readily to be expressed in new hosts. Alternatively, insertion sequence elements can activate expression of a gene by inserting into the promoter region. Whatever the mechanism, it is clear that some genes can be expressed and function effectively in a wide variety of hosts. This brings us finally to a question that has not yet been addressed: how typical are antibiotic resistance genes as indicators of horizontal transfer? Clearly, the fact that resistance genes tend to be found on self-transmissible elements makes them more likely to be transferred than chromosomal genes located outside such elements. Also, antibiotic resistance genes may have evolved to have promoter regions that are expressed in many hosts, whereas an ordinary housekeeping gene might not be expressed so readily if it moved into a distantly related host. Given these caveats, it is probably best to consider antibiotic resistance genes as indicators of the upper limit of gene transfer. As the number of genes in the databases increases, examples of housekeeping genes that have been transferred between different genera or phylogenetic groups may be found, and this could provide some idea of the lower limit of transmissibility.

ACKNOWLEDGMENTS

This work was supported by grant AI 22383 from the US National Institutes of Health.

REFERENCES

Bedzyk, L.A., Shoemaker, N.B., Young, K.E. and Salyers, A.A. (1992) Insertion and integration of *Bacteroides* conjugative chromosomal elements. *J. Bacteriol.* **174**: 166–172.

Brown, D.P., Idler, K.B. and Katz, L. (1990) Characteristics of the genetic elements required for site-specific integration of plasmid SE211 in *Saccharopolyspora erythaea*. *J. Bacteriol.* **172**: 1877–1888.

Clewell, D.B. and Flannagan, S.E. (1993) The conjugative transposons of gram-positive bacteria, in *Bacterial Conjugation*, (ed. D.B. Clewell), Plenum Press, New York, pp. 369–393.

Clewell, D.B., Flannagan, S.E. and Jaworski, D.D. (1995) Unconstrained bacterial promiscuity: the Tn916-Tn1545 family of conjugative transposons. *Trends Microbiol.* **3:** 229–236.

Cooper, A.J., Shoemaker, N.B. and Salyers, A.A. (1996) The erythromycin resistance gene from the *Bacteroides* conjugative transposon Tc^{r}Em^r 7853 is nearly identical to *erm*G from *Bacillus sphaericus. Antimicrob. Agents and Chemother.* **40**: 506–508.

Hall, R.M. and Collis, C.M. (1995) Mobile gene cassettes and integrons: capture and spread of genes by site-specific recombination. *Mol. Microbiol.* **15**: 593–600.

Katz, L., Brown, D.P. and Donadio, S. (1991) Site-specific recombination in *Escherichia coli* between the *att* sites of plasmid pSE211 from *Saccharopolyspora erythraea. Mol. Gen. Genet.* **227**: 155–159.

Li, L.-Y., Shoemaker, N.B. and Salyers, A.A. (1995) Localization and characterization of the transfer region of a *Bacteroides* conjugative transposon and regulation of transfer genes. *J. Bacteriol.* **177**: 4992–4999.

Nikolich, M.P., Hong, G., Shoemaker, N.B. and Salyers, A. A. (1994a) Evidence for the horizontal transfer of *tet*Q between bacteria that normally colonize humans and bacteria that normally colonize livestock. *Appl. Environ. Microbiol.* **60**: 3255–3260.

Nikolich, M.P., Shoemaker, N.B. and Salyers, A.A. (1994b) Characterization of a new type of *Bacteroides* conjugative transposon, Tc^{r}Em^r 7853. *J. Bacteriol.* **176**: 6606–6612.

Poyart-Salmeron, C., Trieu-Suot, P., Carlier, C. and Courvalin, P. (1990) The integration–excision system of the conjugative transposon Tn1545 is structurally and functionally related to those of lambdoid phages. *Mol. Microbiol.* **4**: 1513–1521.

Roberts, M.C., Chung, W.O. and Roe, D.E. (1996) Characterization of tetracycline and erythromycin resistance determinants in *Treponema denticola. Antimicrob. Agents and Chemother.* **40**: 1690–1694.

Salyers, A.A. (1995) *Antibiotic Resistance Transfer in the Mammalian Intestinal Tract: Implication for Human Health, Food Safety, and Biotechnology*, Springer-Verlag, New York.

Salyers, A.A., Shoemaker, N.B. and Li, L.-Y. (1995a) In the driver's seat: the *Bacteroides* conjugative transposons and the elements they mobilize. *J. Bacteriol.* **177**: 5727–5731.

Salyers, A.A., Shoemaker, N.B., Li, L.-Y. and Stevens, A.M. (1995b) Conjugative transposons: an unusual and diverse set of integrated gene transfer elements. *Microbiol. Rev.* **59**: 579–590.

Scott, J.R. (1992) Sex and the single circle: conjugative transposition. *J. Bacteriol.* **174**: 6005–6010.

Scott, J.R. (1993) Conjugative transposons, in *Bacillus subtilis and Other Gram Positive Bacteria*, (eds A. Sonnenshein *et al.*), American Society for Microbiology, Washington DC, pp. 597–614.

Scott, J.R., Bringel, F., Marra, D. *et al.* (1994) Conjugative transposition of Tn916; preferred targets and evidence for conjugative transfer of a single strand and for a double-stranded circular intermediate. *Mol. Microbiol.* **11**: 1099–1108.

Shoemaker, N.B. and Salyers, A.A. (1987) Facilitated transfer of IncPβ R751 derivatives from the chromosome of *Bacteroides uniformis* to *Escherichia coli* recipients by a conjugative *Bacteroides* tetracycline resistance element. *J. Bacteriol.* **169**: 3160–3167.

Shoemaker, N.B. and Salyers, A.A. (1990) A cryptic 65-kilobase pair transposon-like element isolated from *Bacteroides uniformis* has homology with *Bacteroides* conjugal tetracycline resistance elements. *J. Bacteriol.* **172**: 1694–1702.

Shoemaker, N.B., Smith, M.D. and Guild, W.R. (1980) Dnase resistant transfer of chromosomal *cat* and *tet* insertions by filter mating in pneumococcus. *Plasmid* **3**: 80–87.

Shoemaker, N.B., Wang, G.R. and Salyers, A.A. (1992) Evidence for natural transfer of a tetracycline resistance gene between bacteria from the human colon and bacteria from the bovine rumen. *Appl. Environ. Microbiol.* **58**: 1313–1320.

Shoemaker, N.B., Wang, G.R., Stevens, A.M. and Salyers, A.A. (1993) Excision, transfer, and integration of NBU1, a mobilizable site selective insertion element. *J. Bacteriol.* **175**: 6578–6587.

Shoemaker, N.B., Wang, G.R. and Salyers, A.A. (1996a) The *Bacteroides* mobilizable insertion element, NBU1, integrates into the 3' end of a Leu-tRNA gene and has an integrase that is a member of the lambda integrase family. *J. Bacteriol.* **178**: 3594–3600.

Shoemaker, N.B., Wang, G.R. and Salyers, A.A. (1996b) NBU1, a mobilizable site-specific integrated element from *Bacteroides* spp. can be integrated non-specifically in *E. coli*. *J. Bacteriol.* **178**: 3601–3607.

Smith, C.J. (1993) Identification of a circular intermediate in the transfor of Tn4555, a mobilizable transposon from *Bacteroides* spp. *J. Bacteriol.* **175**: 2683–2691.

Speer, B.S., Shoemaker, N.B. and Salyers, A.A. (1993) Bacterial resistance to tetracycline: mechanisms, transfer, and clinical significance. *Clin. Microbiol. Rev.* **5**: 387–399.

Stevens, A.M., Shoemaker, N.B. and Salyers, A.A. (1993) Tetracycline regulation of genes on *Bacteroides* conjugative transposons. *J. Bacteriol.* **175**: 6134–6141.

Waldor, M.K., Tschape, H. and Mekalanos, J.J. (1996) A new type of conjugative transposon encodes resistance to sulfamethoxazole, trimethoprim, and streptomycin in *Vibrio cholerae* O139. *J. Bacteriol.* **178**: 4157–4165.

The role of gene cassettes and integrons in the horizontal transfer of genes in Gram-negative bacteria

5

Ruth M. Hall

SUMMARY

Over 40 different gene cassettes have been found to date and each consists of a single gene (commonly an antibiotic resistance gene) and a recombination site (59-base element) located downstream of the gene. These cassettes are mobilized by site-specific recombination, and are normally found integrated at a further specific recombination site which is found in a larger structure called an integron. The integron also encodes the recombinase, a member of the λ integrase family, that catalyzes both integration and excision of cassettes. Three different integrase genes (*intI*) have now been identified and these define three distinct integron families (class 1, 2 and 3). Though the amino acid sequences of the integrases are only 40–60% identical, the same cassettes have been found in integrons from different classes, indicating that all integrases recognize the cassette-associated 59-base elements.

Integrons of classes 1 and 2 are themselves mobile elements. The integrons most commonly found in clinical antibiotic resistant strains belong to class 1 and have been detected in many Gram-negative species, particularly Enterobacteriaceae and Pseudomonads. Class 1 integrons are highly evolved and most of them are transposition-defective derivatives of a parental transposon configuration represented by Tn*402*. That these transposition-defective integrons can move is evidenced by their many locations on plasmids, within transposons and on bacterial chromosomes. Class 2 includes Tn7 and close relatives. The ability of integrons to transpose into promiscuous plasmids, either directly or as part of another transposon, is likely to be an important factor in facilitating their spread across species boundaries.

5.1   INTRODUCTION

The acquisition of antibiotic resistance genes by bacteria and their subsequent spread to other bacterial species is a major problem worldwide in all areas – medicine, animal husbandry, horticulture and aquaculture – where antibiotics are used to control bacterial pathogens. That mixing of the gene pools from these different areas can occur is known because identical genes have been found in more than one gene pool, and this further contributes to the resistance problem.

Multiply antibiotic-resistant bacteria were first identified in *Shigella* in the 1950s. Subsequently, it was shown that the resistance genes could be transferred together to other Gram-negative bacteria either by transduction or when cells came into direct contact, and conjugative plasmids carrying several resistance genes (resistance transfer factors) were identified (reviewed in Watanabe, 1963; Davies, 1995). Since that time, many antibiotic resistance genes have been found to reside on plasmids and many distinct types of plasmids have been identified and studied. Amongst these are plasmids that can transfer from cell to cell by conjugation, and others that do not themselves carry conjugation functions but are able to utilize the conjugative-transfer apparatus of another plasmid in order to achieve their own transfer to a new host. Some of these plasmids have a broad host range and can not only move into many bacterial species but also survive in many different hosts. Though plasmids are probably the simplest and most common vehicles for the transfer of antibiotic resistance genes from one cell to another, either within the same species or across species boundaries, genes can also enter new hosts by transduction or transformation. With these latter routes, and also for plasmids that are transferred into but not maintained in a new host, the stable acquisition of new genes by the recipient organism depends on their ability to be incorporated into the genome of the recipient. Likewise, for a gene to be transferred horizontally by a plasmid, it must first become incorporated into the plasmid genome. Thus, intergenome transfer of genetic material is an important factor in horizontal gene transfer.

Stable incorporation of a gene into a new genome can be achieved by a variety of mechanisms. Homologous recombination is effective only if homologous DNA is present in the genome of the new host, and other known recombinational mechanisms, such as transposition and site-specific recombination, which need little or no homology, appear to have played a major role in the inter-genome transfer of genes, particularly antibiotic-resistance genes. Many translocatable genetic elements (mobile DNA) have been identified and these can be classified broadly by the mechanism they use to achieve incorporation into a new genome. Two major mechanisms are known: transposition (IS, transposons, retroposons and retroviruses) and site-specific recombination (integrating phage, integrating plasmids and conjugative transposons).

Recently, a new class of mobile element that moves by site-specific recombination, but consists of only a single gene and a specific recombination site, has been discovered in studies of the acquired antibiotic resistance genes found in Gram-negative bacteria. These mobile elements, known as gene cassettes, differ from other mobile elements that utilize site-specific recombination for movement in several respects, the most important of which is that they do not carry the determinant for the site-specific recombinase that is needed to catalyze their movement. This determinant resides in a companion element, known as an integron, and gene cassettes are normally found incorporated within an integron. The cassette–integron system functions as a natural cloning and expression system that, because the number of cassettes that can be integrated into an integron is not restricted, is able to create arrays of genes of enormous diversity. Integrons and gene cassettes have been found in many different Gram-negative bacteria (Table 5.1), indicating that they have been very successful in achieving horizontal transfer. The system has been able to spread so effectively because many integrons are themselves translocatable elements – either transposons or transposition-defective derivatives of transposons. Here, the basic features of gene cassettes and integrons and the mechanisms by which they move are outlined. More detailed descriptions are found in recent reviews (Hall and Collis, 1995; Recchia and Hall, 1995b).

Table 5.1 Known species range of class 1 integrons

Escherichia coli	*Serratia marcescens*
Shigella flexneri, Shigella sonnei	*Citrobacter diversus*
Salmonella typhimurium,	*Proteus mirabilis*
* Salmonella oranienberg*	*Acinetobacter baumanii*
Morganella morganii	*Pseudomonas aeruginosa,*
Klebsiella pneumoniae,	* Pseudomonas fluorescens*
* Klebsiella aerogenes*	*Xanthomonas*
Enterobacter aerogenes,	*Yersinia pestis*
* Enterobacter agglomerans*	*Alcaligenes denitrificans*

5.2 MOBILE GENE CASSETTES

Over 40 gene cassettes have been identified to date, though the sequences of some of them are incomplete, and it is likely that many more will be found in the future. The features of known cassettes have recently been compiled (Recchia and Hall, 1995b), and all except one comprise a single gene with a specific recombination site, known as a 59-base element, that is located downstream of the gene (Fig. 5.1A). Although the genes are mostly antibiotic-resistance genes, they encode a variety of unrelated enzymes including β-lactamases, dihydrofolate reductases, aminoglycoside, chloramphenicol and streptothricin acetyl-transferases, aminoglycoside adenylyltransferases and inner membrane proteins that actively export chloramphenicol or quaternary ammonium compounds such as those used as antiseptics and disinfectants (Table 5.2). It is thus unlikely that the nature of the gene affects whether it becomes part of a cassette, and the prevalence of antibiotic-resistance genes almost certainly arises from the bias in the source of the genes examined to date, i.e. from clinically resistant organisms. The vast majority of cassettes do not include a promoter (Hall *et al.*, 1991; Recchia and Hall, 1995b) and expression of the genes depends on a promoter supplied by the integron (section 5.3).

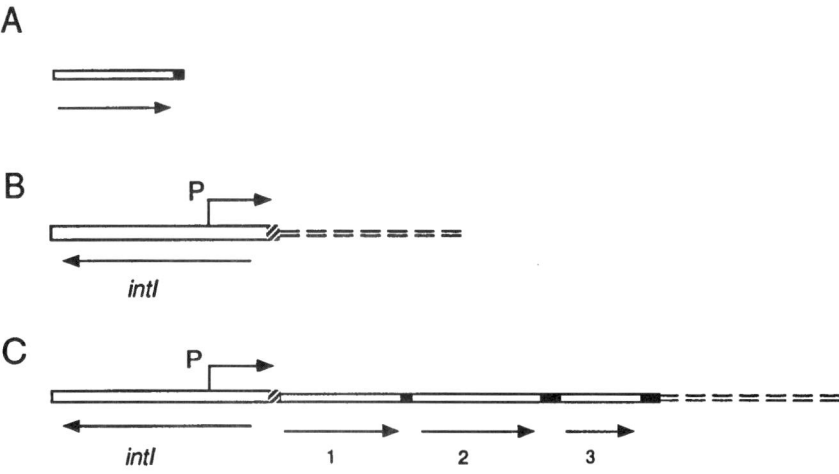

Fig. 5.1 Generalized structures of gene cassettes and integrons. (A) A linearized gene cassette showing the position of the gene (arrow) and the 59-base element (■). (B) An integron showing the position of the *intI* gene (arrow), the *attI* site (▨) and the promoter used for transcription of the cassete-associated genes (P). (C) An integron containing two integrated cassettes.

Table 5.2 Cassette-encoded genes[a]

Protein product	Class	Genes
β-Lactamase	A	*blaP1, blaP2, blaP3*
	B	*bla$_{IMP}$*
	D	*oxa1, oxa2, oxa3, oxa5, oxa7, oxa9, oxa10*
Dihydrofolate reductase	A	*dfrA1, dfrA5, dfrA7, dfrA12, dfrA13, dfrA14*
	B	*dfrB1, dfrB2, dfrB3*
Chloramphenicol acetyltransferase	B	*catB2, catB3, catB5*
Chloramphenicol transporter		*cmlA*
Aminoglycoside (3') adenylyltransferase		*aadA1, aadA2*
Aminoglycoside (2') adenylyltransferase		*aadB*
Aminoglycoside (6') acetyltransferase	1	*aacA1*
	2	*aacA4, aacA(IIa), aacA(IIb)*
	3	*aacA7, aacA (orfB)*
Aminoglycoside (3') acetyltransferase		*aacC1, aacC(Ib)*
Streptothricin acetyltransferase		*sat2 (sat1)*
Quaternary ammonium compound exporter		*qacE*
Open reading frame		orfA, orfC, orfD, orfE, orfF

[a] Genes that differ in sequence by 1% or less are not listed.

Each gene is associated with a unique 59-base element recombination site that confers mobility because it is recognized by the integrases that are encoded by integrons. The 59-base elements vary in both length and sequence, but share common features. They are generally made up of a pair of imperfect inverted repeats and conform to consensus sequences for over 20 bp at their outer ends (Hall *et al.*, 1991; Collis and Hall, 1992b). However, elements with quite different lengths and sequences have been shown to be active recombination sites that are recognized by the IntI1 integrase (Martinez and de la Cruz, 1990; Hall *et al.*, 1991; Collis and Hall, 1992a,b; Collis *et al.*, 1993; Bunny *et al.*, 1995), indicating that they are indeed members of the same family.

Cassettes can exist in two forms. The first is the linear integrated form in which they are found in integrons. The second is a free, closed-circular form that is generated by excision of a cassette from an integron (Collis and Hall, 1992b) and can integrate into an integron receptor site (Collis *et al.*, 1993). Because the circular form is unable to replicate, it cannot be stably maintained *in vivo*. Nonetheless, free circular cassettes are likely to play an important role in the insertion of new genes into another integron present in the same cell, and because of their small size may also be able to survive cell lysis and be transformed into a new host thus assisting the horizontal spread of resistance genes.

5.3   INTEGRONS

Three distinct classes of integrons, each of which encodes a distinct integrase, have been identified to date. Class 1 integrons are most commonly isolated from resistant clinical strains, and have been found in many different species. Class 2 includes transposon Tn7 and close relatives (Hall *et al.*, 1991; Recchia and Hall, 1995b). The third class was found only recently and so far includes only one representative (Arakawa *et al.*, 1995). Members of all three classes share the features that are characteristic of integrons. They encode a site-specific recombinase (IntI) that is a member of the λ integrase family (Ouellette and Roy, 1987), include an integrase-specific recombination site (*attI*) and also contain a promoter to express the cassette-associated genes (Fig. 5.1B).

The same cassettes have been found in each class of integron, suggesting that the cassette-associated 59-base element sites are recognized by all three integrases, IntI1, 2 and 3. This is somewhat surprising granted that the level of amino acid identity between these three integrases ranges from 40 to 60% (Hall and Vockler, 1987; Arakawa *et al.*, 1995). The three *attI* recombination sites cannot be aligned to generate a consensus and are not obviously members of the 59-base element family. The extent of *attI1* has been examined experimentally and more than 34 but no more than 64 bp of the integron sequence together with at least 6 bp of adjacent DNA are required for maximal site activity (Recchia *et al.*, 1994).

Because the vast majority of cassettes do not include a promoter (Hall *et al.*, 1991; Recchia and Hall, 1995b), expression of the genes they encode depends on a promoter present in the integron (Fig. 5.1C). The activity of the P_{ant} promoter in class 1 integrons has been demonstrated experimentally (Stokes and Hall, 1989; Collis and Hall, 1995), and several versions of this promoter that have different strengths have been found in wild isolates (Lévesque *et al.*, 1994; Collis and Hall, 1995; Bunny *et al.*, 1995). The genes in a multicassette array are thus expressed as an operon, and the level of expression is highest for the gene in the promoter-proximal cassette (Collis and Hall, 1995). The presence of promoters in integrons belonging to class 2 and 3 is currently inferred from the fact that resistance genes in the integrated cassettes are expressed.

5.4   CASSETTE INTEGRATION AND EXCISION

Movement of gene cassettes catalyzed by the IntI1 integrase has been demonstrated experimentally. Both integration and excision of gene cassettes involves a single site-specific recombination event (Collis and Hall, 1992a; Collis *et al.*, 1993). Deletion generates a free, covalently closed circular gene cassette (Collis and Hall, 1992b) and an integron that has lost one gene cassette (Collis and Hall, 1992a). Integration of free circular cassettes normally occurs at the *attI1* site in the recipient integron (Collis *et al.*, 1993). However, IntI-mediated cointegration and resolution events can lead to the reassortment of cassettes when more than one is present (Hall and Collis, 1995) so that the order of cassettes does not necessarily reflect the order of uptake involved in creating the final array.

Using cointegration assays, the IntI1 integrase has been shown to catalyze recombination between two 59-base elements and between a 59-base element and *attI1* (Martinez and de la Cruz, 1990; Hall *et al.*, 1991; Recchia *et al.*, 1994) and recently recombination between two *attI1* sites has also been detected (Recchia, 1996). However, in the absence of other constraints, recombination between *attI1* and a 59-base element is most efficient (G.D. Recchia and R.M. Hall, unpublished observation).

Although integrons are clearly the preferred location for gene cassettes, one example of a cassette integrated at a non-specific site has also been documented (Recchia and Hall, 1995a). In this case, the *aadB* gene cassette (the *aadB* gene confers resistance to gentamicin, tobramicin and kanamycin) is integrated at a secondary site in the IncQ plasmid RSF1010, which has a very broad host range. The fact that IntI1-mediated recombination between a 59-base element and secondary sites can occur has been demonstrated experimentally (Francia *et al.*, 1993; Recchia *et al.*, 1994). This reaction permits gene cassettes to be integrated at many different locations, but expression of the cassette-encoded gene depends on the presence of a suitably oriented promoter in the recipient genome. Integration of gene cassettes at secondary sites is not reversible because the cassette-associated recombination site is generally inactivated (Recchia and Hall, 1995a) and the cassette is not flanked by a second IntI-specific recombination site. Integration of a cassette at a secondary site thus leads to the stable incorporation of a new gene into a pre-existing genome.

5.5   CLASS 1 INTEGRONS ARE TRANSPOSONS OR DEFECTIVE TRANSPOSON DERIVATIVES

The efficient spread of cassette-associated resistance genes to many different species depends on the spread of integrons. Class 1 integrons are most widely distributed and are known to be present in many different locations, on plasmids, within transposons and on the bacterial chromosome. While the multiple locations suggested that integrons are themselves translocatable, early studies did not reveal the characteristic features of transposons (Stokes and Hall, 1989). Recently, the complete sequence of Tn402, which is a class 1 integron, has revealed four potential transposition genes (Rådström et al., 1994) and closely related genes found in Tn5063 have been shown to be required for transposition (Kholodii et al., 1995). Tn402 is bounded by inverted repeats of 25 bp. However, most of the class 1 integrons found in clinical isolates over the last 40 years contain a sulphonamide resistance gene, sul1, that is part of a conserved segment (3'-conserved segment) that is not found in Tn402. These integrons are now known to be transposition-defective transposon derivatives (Hall et al., 1994; Brown et al., 1996; H.J. Brown, H.W. Stokes and R.M. Hall, unpublished observations), and their structures fall into two quite distinct groups (Brown et al., 1996; H. Brown, H.J. Stokes and R.M. Hall, unpublished observations). The groups are characterized by the presence of two different IS elements, IS1326 and IS6100, and the absence of some or all of the transposition genes. Loss of the transposition genes appears to be due to deletions adjacent to the IS that are presumably caused by the IS. However, most of the class 1 integrons that have been examined in detail retain intact outer ends (Hall et al., 1994). As a consequence, they are presumably able to move if the transposition proteins are supplied in trans. Indeed, this appears to have occurred as members of both groups of class 1 integrons have been found at several independent locations (Hall et al., 1994; Brown et al., 1996).

5.6   THE 3'-CONSERVED SEGMENT AND EVOLUTION OF CLASS 1 INTEGRONS

Because both groups of transposition-defective class 1 integrons contain the 3'-conserved segment, they must have diverged from one another after the the the 3'-conserved segment was acquired by an ancestral transposon related to Tn402. The 3'-conserved segment is located downstream of the integrated cassettes and contains both the sul1 gene and a truncated copy of the qacE gene that confers resistance to quaternary ammonium compounds including disinfectants and antiseptics (Sundström et al., 1988; Stokes and Hall, 1989; Paulsen et al., 1993). The complete qacE gene is found in Tn402 and has recently been shown to be part of a functional cassette (Recchia, 1996). This raises the possibility that the 3'-conserved segment consists of a series of gene cassettes that have become fused by loss of their 59-base element recombination sites.

Integrons containing the 3'-conserved segment are found in plasmids (pSa and NR1) from the Shigella strains that were isolated in Japan in the

1950s and studied by Watanabe (1963). Thus, evolution of this segment is likely to have occurred either prior to the introduction of antibiotics for therapeutic use, or during the preceding 10 years when sulphonamides were used therapeutically.

ACKNOWLEDGMENTS

I thank my collaborator, Hatch Stokes, and all those who have contributed to our work on this system over the last 10 years.

REFERENCES

Arakawa, Y., Murakami, M., Suzuki, K. *et al.* (1995) A novel integron-like element carrying the metallo β-lactamase gene bla_{IMP}. *Antimicrob. Agents Chemother.* **39**: 1612–1615.

Brown, H.J., Stokes, H.W. and Hall, R.M. (1996) The integrons In0, In2 and In5 are defective transposon derivatives. *J. Bacteriol.* **178**: 4429–4437.

Bunny, K.L., Hall, R.M. and Stokes, H.W. (1995) New mobile gene cassettes containing an aminoglycoside resistance gene, *aacA7*, and a chloramphenicol resistance gene, *catB3*, in an integron in pBWH301. *Antimicrob. Agents Chemother.* **39**: 686–693.

Collis, C.M. and Hall, R.M. (1992a) Site-specific deletion and rearrangement of integron insert genes catalysed by the integron DNA integrase. *J. Bacteriol.* **174**: 1574–1585.

Collis, C.M. and Hall, R.M. (1992b) Gene cassettes from the insert region of integrons are excised as covalently closed circles. *Mol. Microbiol.* **6**: 2875-2885.

Collis, C.M. and Hall, R.M. (1995) Expression of antibiotic resistance genes in the integrated cassettes of integrons. *Antimicrob. Agents Chemother.* **39**: 155–162.

Collis, C.M., Grammaticopoulos, G., Briton, J. *et al.* (1993) Site-specific insertion of gene cassettes into integrons. *Mol. Microbiol.* **9**: 41–52.

Davies, J. (1995) Vicious circles: looking back on resistance plasmids. *Genetics* **139**: 1465–1468.

Francia, M.V., de la Cruz, F. and García Lobo, M. (1993) Secondary sites for integration mediated by the Tn*21* integrase. *Mol. Microbiol.* **10**: 823–828.

Hall, R.M. and Collis, C.M. (1995) Mobile gene cassettes and integrons: capture and spread of genes by site-specific recombination. *Mol. Microbiol.* **15**: 593–600.

Hall, R.M. and Vockler, C. (1987) The region of the IncN plasmid R46 coding for resistance to β-lactam antibiotics, streptomycin/spectinomycin and sulphonamides is closely related to antibiotic resistance segments found in IncW plasmids and in Tn*21*-like transposons. *Nucleic Acids Res.* **15**: 7491–7501.

Hall, R.M., Brookes, D.E. and Stokes, H.W. (1991) Site-specific insertion of genes into integrons: role of the 59-base element and determination of the recombination cross-over point. *Mol. Microbiol.* **5**: 1941–1959.

Hall, R.M., Brown, H.J., Brookes, D.E. and Stokes, H.W. (1994) Integrons found in different locations have identical 5' ends but variable 3' ends. *J. Bacteriol.* **176**: 6286–6294.

Kholodii, G.Y., Mindlin, S.Z., Bass, I.A. *et al.* 1995) Four genes, two ends, and a *res* region are involved in transposition of Tn*5053*: a paradigm for a novel family of transposons carrying either a *mer* operon or an integron. *Mol. Microbiol.* **17**: 1189–1200.

Lévesque, C., Brassard, S., Lapointe, J. and Roy, P.H. (1994) Diversity and relative strength of tandem promoters for the antibiotic-resistance genes of several integrons. *Gene* **142**: 49–54.

Martinez, E. and de la Cruz, F. (1990) Genetic elements involved in Tn*21* site-specific integration, a novel mechanism for the dissemination of antibiotic resistance genes. *EMBO J.* **9**: 1275–1281.

Ouellette, M. and Roy, P.H. (1987) Homology of ORFs from Tn*2603* and from R46 to site-specific recombinases. *Nucleic Acids Res.* **15**: 10055.

Paulsen. I.T., Littlejohn, T.G., Rådström, P. *et al.* (1993) The 3'-conserved segment of integrons contains a gene associated with multidrug resistance to antiseptics and disinfectants. *Antimicrob. Agents Chemother.* **37**: 761–768.

Rådström, P., Sköld, O., Swedberg, G. *et al.* (1994) Transposon Tn*5090* of plasmid R751, which carries an integron, is related to Tn7, Mu, and the retroelements. *J. Bacteriol.* **176**: 3257–3268.

Recchia, G.D. (1996) Mobile gene cassettes and integrons: evolutionary and recombinational studies. PhD Thesis, Macquarie University.

Recchia, G.D. and Hall, R.M. (1995a) Plasmid evolution by acquisition of mobile gene cassettes: plasmid pIE723 contains the *aadB* gene cassette precisely inserted at a secondary site in the IncQ plasmid RSF1010. *Mol. Microbiol.* **15**: 179–187.

Recchia, G.D. and Hall, R.M. (1995b) Mobile gene cassettes: a new class of mobile element. *Microbiol.* **141**: 3015–3027.

Recchia, G.D., Stokes, H.W. and Hall, R.M. (1994) Characterisation of specific and secondary recombination sites recognised by the integron DNA integrase. *Nucleic Acids Res.* **22**: 2071–2078.

Stokes, H.W. and Hall, R.M. (1989) A novel family of potentially mobile DNA elements encoding site-specific gene-integration functions: integrons. *Mol. Microbiol.* **3**: 1669–1683.

Sundström, L., Rådström, P., Swedberg, G. and Sköld, O. (1988) Site-specific recombination promotes linkage between trimethoprim- and sulfonamide resistance genes. Sequence characterization of *dhfrV* and *sulI* and a recombination active locus of Tn*21. Mol. Gen. Genet.* **213**: 191–201.

Watanabe, T. (1963) Infective heredity of multiple drug resistance bacteria. *Bacteriol. Rev.* **27**: 87–115.

Evolution of the selfish Ti plasmid of *Agrobacterium tumefaciens* promoting horizontal gene transfer

6

Clarence I. Kado

SUMMARY

Studies on the origin and evolution of plasmids may provide valuable insights on the promiscuous nature of DNA. The first examples of the selfish nature of nucleic acids are exemplified by primordial oligoribonucleotides which evolved into primitive replicons. The propagation pattern of these molecules was probably similar to that of the current viral RNA ribozymes, which have recently been shown to possess both RNA synthesizing and template-mediated polymerizing capabilities (Ekland and Bartel, 1996). The parasitic nature of nucleic acids is depicted by satellite nucleic acid molecules associated with viruses. The satellites of adenovirus and tobacco ringspot virus serve as examples. Comparative analysis of the replication origins of virions and plasmids shows them to be highly conserved, originating from the simplest autocatalytic protein-free replicon and developing into highly complex and evolved plasmids, replicating by a rolling circle mechanism assisted by proteins. The selfish nature of plasmids is depicted by their ability to engineer their host genetically so that the host cell is best able to cope and survive in a hostile environment. This attribute leads to the perpetuation of plasmids. Sequestering of genes by plasmids occurs when the environmental conditions negatively affect the host. The sequestering mechanism is fundamental and forms the outreach mechanisms to generate and propagate macromolecules of increasing size. The complexity of plasmids increases with the addition of new genes such as those that allow the host to occupy a specific environment normally inhospitable to the host cell.

The 200 kb Ti plasmid resident in *Agrobacterium tumefaciens* represents one of the most advanced plasmid type. This plasmid confers on this organism the unique ability to transmit a specific segment of the Ti plasmid – the T-DNA – and incorporate that DNA into plant cells and into their genome. The introduced T-DNA contains genes solely expressed in the plant cell, converting it into a hospitable environment in the form of tumor (crown gall) and in the synthesis of specific compounds called opines. These opines are specifically used by *A. tumefaciens* through opine catabolic genes also located on the Ti plasmid. Thus, the Ti plasmid has set up an environment for self-perpetuation by genetically engineering the host bacterium to survive in an environment with enormous competitive advantages. Other plasmids of similar size to the Ti plasmid, or larger, are likely to confer specialized features on their respective host cells.

6.1   INTRODUCTION

The self-preservation of nucleic acid molecules is recognized by examining the traits of viroids, viruses and plasmids. These molecules represent the simplest forms of biologically active nucleic acids. The satellite viruses observed with adenoviruses, known as satellite associated adenovirus (SAV) (Mayor *et al.*, 1969), and with tobacco ringspot virus, known as satellite tobacco ringspot virus (STobRV) (Schneider and White, 1976), represent nucleic acids that have parasitized the active virion by sharing the replication and coat protein synthesizing system of the virus. The satellite viral RNAs contain no open reading frames that are not translated into biologically active proteins. The linear multimers produced by rolling circle replication are processed into linear monomers by their self-contained ribozyme and the monomers are circularized much like plasmids (Tol *et al.*, 1991). Experimental supplementation of additional satellite RNA molecules to the inoculum causes a proportional decrease in disease severity and increase in the population of satellite RNA (Bruening, 1990). This effect illustrates saturation of parasitism by the satellite RNA, where more of its molecules are produced than that of the virus.

Interestingly, the conserved element of all self-replicating molecules is the origin of nucleic acid replication (*oriV*) (Table 6.1). Thus, the simplest plasmids contained only a primitive *oriV* and no open reading frames to encode proteins, and were quite small (less than 1 kb). The driving force in sustaining and protecting themselves (e.g. against nucleases) has yielded plasmids mainly in the form of covalently closed circular molecules (although there are several linear viruses and plasmids capped

Table 6.1 Conservation of the origin of nucleic acid replication

Plasmid/virion	Classification	Replication origin/iteron	References
satTRSV RNA	nepovirus	$N_{214}UGAAGN_5$	Tol et al., 1991
lambda	lambdoid phage	$N_9TGAAGN_3$	Scherer, 1978; Grosschedl and Hobon, 1979
pSa	incW plasmid	$N_5TGAAATN_6$	Okumura and Kado, 1992
R6K	incX plasmid	$N_5TGAGAGN$	Stalker et al., 1982
RK2	incP plasmid	$N_7TGAGGGN_4$	Stalker et al., 1982
RSF1010	incQ plasmid	$N_2TGACAGN_{12}$	Line et al., 1987
mini-F	incF plasmid	$N_1TGAGGGN_{16}$	Murotsu et al., 1981; Tolun and Helinski, 1982

with protein, e.g. bacteriophage PRD1 and plasmids of *Streptomyces*, and *Borrelia*) that can adapt relatively quickly to retain the survivability of the host in which they reside. This comes about by sequestering particular genes that confer traits on the host organism for its survival and thus, in turn, for the survival of the plasmid itself.

The Ti plasmid harbored in virulent strains of *Agrobacterium tumefaciens* epitomizes the high level of sophistication that can evolve among plasmids. This 200 kb macromolecule contains all of the information necessary to process and transmit a specific set of plasmid genes to eukaryotic cells (in the natural system, plants serve as the recipient). This process of horizontal gene transfer represents a natural form of genetic engineering conferred by plasmid genes that have encumbered the host bacterial cell to carry out the plant transformation process. The resulting effect of this phenomenon is the formation of non-self-limiting overgrowths comprising transformed plant cells that now produce one or more metabolites (opines) that are specifically used by *Agrobacterium* and not by competing organisms. The genes involved in consuming the opines are also contained on the Ti plasmid (Petit *et al.*, 1970), and details of the octopine catabolic pathway have been deduced (Cho *et al.*, 1996). Opines such as octopine (arginine–pyruvate conjugate) and agrocinopine (glucose, fructose, arabinose phosphate) (Ellis and Murphy, 1981) also promote conjugative transfer of the Ti plasmid to recipient *Agrobacterium* strains that are usually free of the Ti plasmid. Thus, the Ti plasmid has the essential genes for processing and transmitting oncogenes and opine

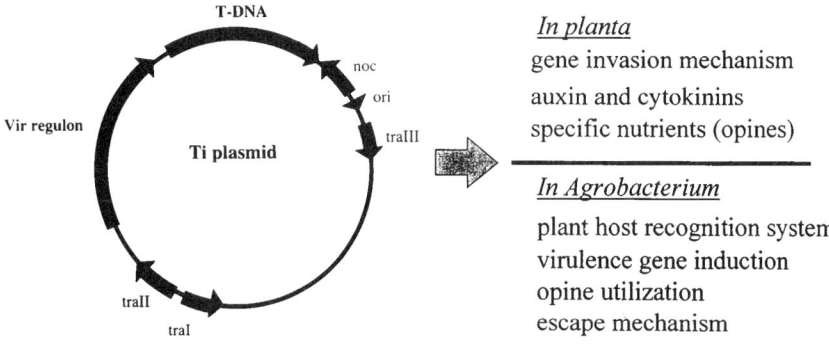

Fig. 6.1 Outreach properties of the *Agrobacterium* Ti plasmid. The genes of the Ti plasmid confer on *Agrobacterium* host specificity, recognition of lignin precursors elaborated by the plant host leading to the activation of virulence (*vir*) genes clustered in the Vir regulon on the Ti plasmid, the utilization of specific opines such as nopaline (*noc*), and conjugative transfer (*tra*). The latter genes (*traI, II, III*) involved in conjugative transfer are important for vertical dissemination (escape mechanism) of the Ti plasmid to ensure that plasmid perpetuation is established in the *Agrobacterium* host. Specific properties conferred on the plant cell and on the *Agrobacterium* cell are listed on the right of the arrow.

synthesizing genes to plants, and possesses catabolic genes for utilizing the opines and plasmid transfer genes. The Ti plasmid clearly benefits from sustaining the vitality of the host bacterium by creating a continuous supply of one or more specific metabolites and thereby provides *Agrobacterium* the necessary competitive advantage over competing microorganisms (Fig. 6.1).

6.2   ORGANIZATION OF THE PROMISCUOUS DNA TRANSFER SYSTEM

The Ti plasmid contains the genetic information in a 30 kb sector (*vir* regulon) that encodes the processing and transfer activities of a second Ti plasmid sector known as the T-DNA (Fig. 6.1). The T-DNA is flanked by sequences represented by 25 bp direct repeats (Fig. 6.2). These repeats are highly conserved among characterized broad host range plasmids and are targets for a specific nicking reaction requiring VirD1 and VirD2 proteins (Waters and Guiney, 1993). For T-DNA processing, the right and left borders are nicked by the VirD2 protein in the presence of magnesium ions and this protein covalently attaches to the 5'-end of the cleaved oligonucleotide as shown by *in vitro* experiments (Pansegrau *et al.*, 1993). Since a single-stranded substrate is required for this reaction, it is thought that a relaxation complex is required to initiate T-DNA transfer (Filichkin and Gelvin, 1993).

pTiC58	T- borders	ΛTATATCCTG:C/TC
RP4/RK2	oriT	ACCTATCCTG:CC
R751	oriT	ACACATCCTG:CC
R64	oriT	GCACATCCTG:TC
pC194	oriT	TCTTATCTTG:AT
F	oriT	GTTTTCGTGG:TGT

Fig. 6.2 Border sequences recognized by the nicking enzymes of broad and narrow host-range plasmids. The sequence similarities between the border sequence of the T-DNA and those of the plasmids listed (Waters and Guiney, 1993) indirectly support the notion that the T-DNA is transferred by a promiscuous conjugative transfer mechanism involving a conjugative pilus (Kado, 1994a).

The *vir* regulon comprises six essential operons containing 24 genes (Rogowsky *et al.*, 1990). These genes are normally not expressed maximally in the *Agrobacterium* cell, but become fully expressed when the bacterial cell encounters a wounded plant cell. Signals in the form of precursors of the lignin biosynthetic pathway (Kado, 1991) are recognized by *Agrobacterium* through interactions between a specific membrane spanning histidine kinase (VirA) (Doty *et al.*, 1996) that phosphorylates VirG, which in turn promotes transcription of the remaining *vir* genes. VirA and VirG are members of the two-component signal transducing superfamily of phospho-relay proteins (Albright *et al.*, 1989; Burbulys *et al.*, 1991; Parkinson and Kofoid, 1992; Winans, 1992; Hoch and Silhavy, 1995), and would suggest that the genes encoding VirA and VirG were sequestered through horizontal gene transfer.

6.3 A PROMISCUOUS CONJUGATIVE PILUS ENCODED BY TI PLASMID GENES

The molecular mechanism of the *Agrobacterium*-mediated gene transfer system was long suspected to mimic a bacterial-type conjugative system (Tempé *et al.*, 1977; Stachel and Zambryski, 1986), but the molecular mechanism of gene transfer has remained elusive. Research focused on the genes responsible for virulence and T-DNA transfer is a logical approach towards gaining important insights on the gene transfer mechanism. Comparative sequence analysis of the *virB* operon of the *vir* regulon was found to be closely similar in both nucleotide and amino acid sequence to the sequences of operons of other plasmids known to encode the conjugative pilus (Kado, 1993, 1994a, b; Shirasu and Kado, 1993). Also, the size and genetic arrangement of the genes of the pilus operon were nearly identical. The strongest evidence for the pilus encoded by the *virB* operon is shown by electron microscope studies of an *A. tumefaciens* strain that was freed of flagella (Chesnokova *et al.*, 1996). The 'bald' strain NT1REB (Chesnokova *et al.*, 1997) containing pUCD2614, a plasmid bearing only the *vir* regulon, showed the presence of a pilus (Fig. 6.3). This confirms our previous studies showing pilus-like structures elaborated by a wild-type *A. tumefaciens* strain (Kado, 1994a). Also Fullner *et al.* (1996) demonstrated that *A. tumefaciens* containing RSF1010 can mediate pilus formation that depends on *virA*, *virG*, *virB* and *virD* genes. Based on these studies, it seems apparent that the pilus encoded and assembled by *virB* genes operates in a highly orderly manner conferring a highly promiscuous feature on *A. tumefaciens*.

6.4   HORIZONTAL GENE TRANSFER BY *AGROBACTERIUM*

The powerful promiscuous DNA transfer system encoded by the Ti plasmid has been harnessed to deliver many foreign genes into plants. This was apparent owing to the broad host range of *A. tumefaciens* in causing the crown gall disease in members of 93 different families of plants (DeCleene and De Ley, 1976). As described above, the T-DNA processing mechanism is initiated at the borders of the T-DNA. Hence, the borders serve as specific sites (*oriT*) for the processing reaction initiated and terminated by a nick at the respective border. Any DNA placed between the borders will be transferred from *A. tumefaciens* to plants. Thus, a large number of foreign genes have been transferred and incorporated into the chromosome of many different plants. The mechanism of DNA transfer is therefore clearly promiscuous.

Besides plant cells as recipients of the T-DNA, the promiscuity of the Ti plasmid system is exemplified by T-DNA transfer from *A. tumefaciens* to *E. coli* (Sprinzl and Geider, 1988), to the yeasts *Schizosaccharomyces pombe* (Sikorski *et al.*, 1990; R. Sikorski and P. Heiter, 1990, personal communication) and *Saccharomyces cerevisiae* (Bundock *et al.*, 1995), and to *Streptomyces lividans* (Kelly and Kado, 1997). Such horizontal gene transfer also occurs between *E. coli* harboring plasmid F and *S. cerevisiae* (Heinemann and Sprague, 1989), and *Streptomyces lividans* (Giebelhaus *et*

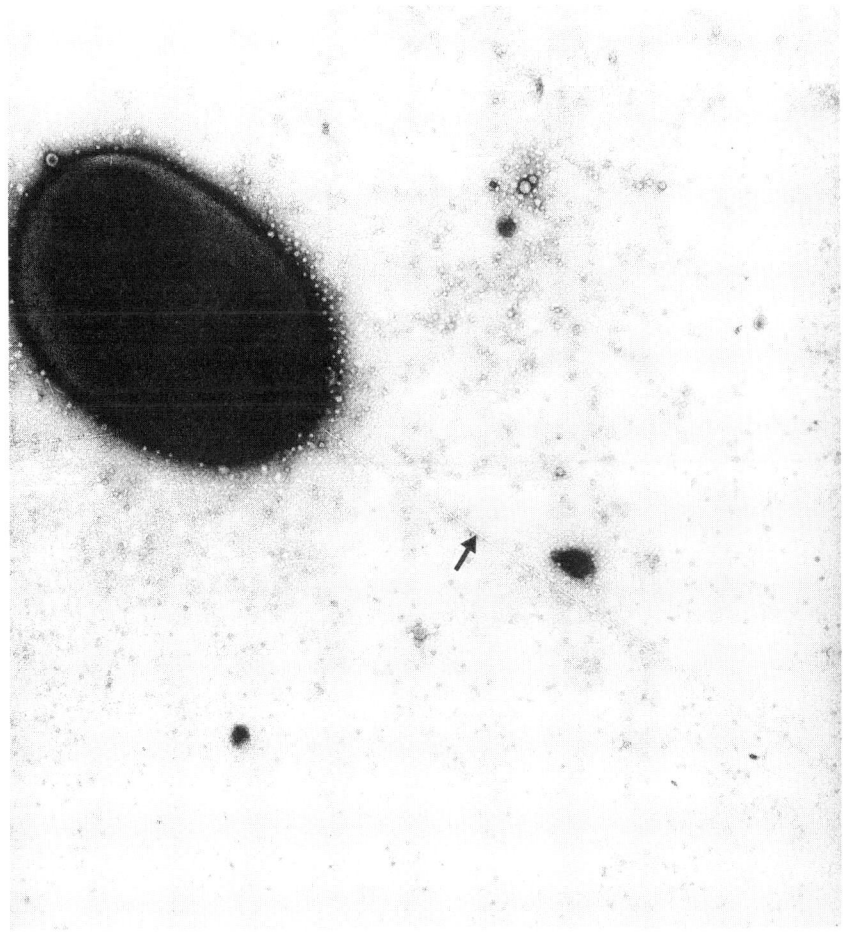

Fig. 6.3 VirB-specific pilus (indicated by the arrow) originating from a bald strain of *A. tumefaciens* NT1REB containing pUCD2614 and induced with acetosyringone. The polar-originating pilus extends beyond the cell length. Electron micrograph (magnification 40 000 ×) taken by Ehr-Min Lai.

al., 1996). For T-DNA transfer between *A. tumefaciens* and *S. lividans*, we have used real-time observations of their interaction by laser equipped confocal microscopy (Fig. 6.4). *A. tumefaciens* is clearly seen hitched to the mycelial hyphae of *Streptomyces* (Kelly and Kado, 1997). This hitching is only seen with *A. tumefaciens* containing pUCD2614 (Rogowsky *et al.*, 1990) induced with acetosyringone. Plasmid pUCD2614 does not contain the plasmid conjugative transfer genes of the TraI, II and III regions (Fig. 6.1). Such displays strongly suggest that the promiscuous conjugative system of

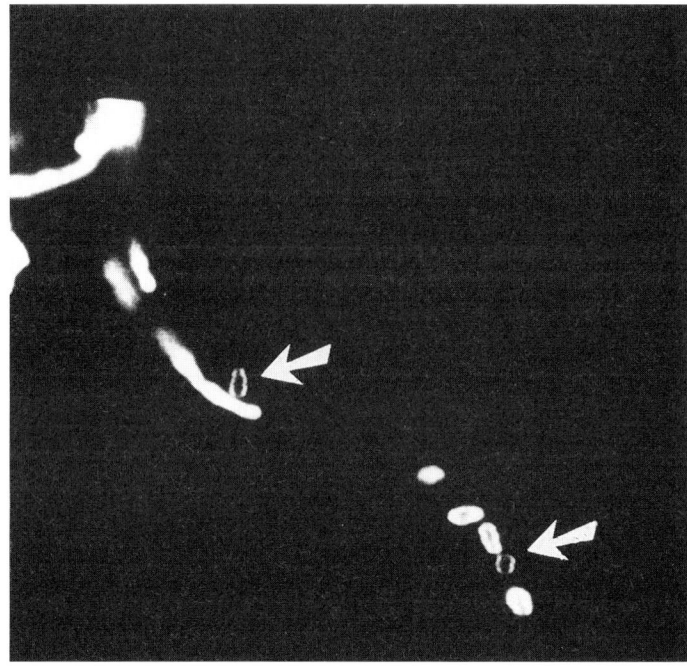

Fig. 6.4 *Agrobacterium*-mediated T-DNA transfer to *Streptomyces lividans* visualized by confocal laser microscopy. Arrows show interaction between *Agrobacterium* and the hyphal tips of *Streptomyces*. Confocal micrograph taken by Brian Kelly.

A. tumefaciens is non-specific in seeking potential recipient cells as candidates for deposition of the T-DNA which may or may not culminate in its integration into the chromosome of the recipient. The level of sophistication of the T-DNA is apparent by the promoters of its genes. These promoters are recognized by the plant transcriptional system to generate the products essential for tumorigenesis, enzymes which catalyze the formation of growth hormones (auxin and cytokinins) and opines (e.g. nopaline, a condensation product of pyruvate and α-ketoglutarate). This represents another example of the selfish property of plasmids to perpetuate themselves by generating an environment favorable for the host's own survival. During the evolution of the Ti plasmid, the auxin and cytokinin genes were apparently sequestered from an ancestral higher cell and may represent an example of horizontal gene transfer. The absence of antibiotic resistance genes on the Ti plasmid is somewhat surprising since *A. tumefaciens* can be recovered from the soil near plant roots where antibiotic-producing organisms such as *Streptomyces* species are often present.

6.5   CONCLUSIONS AND IMPLICATIONS

Promiscuity of stable nucleic acids originated from outside pressures to survive in environments that are normally hostile to molecular perpetuation. Oligonucleotide replication is a fundamental component of molecular perpetuation. Molecular perpetuation is promoted by molecular propagation through the establishment of a niche that is favorable to the nucleic acid molecule, i.e. plasmids, viroids and virions, followed by outreach mechanisms to sequester genes which further confer environmental coping properties on the host in which the plasmid resides. This outreach mechanism provides the added advantage to plasmids. The Ti plasmid of *A. tumefaciens* is a classical representative of a highly developed promiscuous and selfish DNA molecule that is capable of transferring genes horizontally in several different types of higher organisms, in order to promote the survival value of the *Agrobacterium* cell which, in turn, ensures plasmid survival. The Ti plasmid is also equipped to transfer vertically by conventional conjugation. Recent comparative amino acid sequence analyses of the predicted Tra and Trb proteins of octopine Ti plasmids resemble the Tra/Trb proteins of the IncPα plasmid RP4 (Alt-Mörbe *et al.*, 1996). Similarly, some of the Tra proteins of nopaline Ti plasmid C58 show homologies to Tra proteins of RP4 and F, and Mob proteins of RSF1010 (Farrand *et al.*, 1996). Also similar analyses between VirB proteins and those of IncN plasmid pKM101 revealed close similarities (Pohlman *et al.*, 1994). Owing to striking resemblance between Tra/Trb proteins from different plasmids, the genetic outreach system must have operated to evolve the Ti plasmid. Based on the resemblance, it has been hypothesized that the conjugative transfer genes may have originated from at least three separate plasmid sources (Alt-Mörbe *et al.*, 1996). This hypothesis can be modified to include genetic sources other than plasmids; for example, genetic elements from higher cells. The adaptation of the vertical conjugative system by *Agrobacterium* for the transfer of genes horizontally epitomizes the power of genetic outreach. The *Agrobacterium*-mediated interkingdom transfer of genes is a classical example of horizontal gene transfer occurring in nature. It should be noted here that such transfer mechanisms also provide opportunities for genes to be transferred from higher cells into *Agrobacterium*. Thus, both forward and retrotransfer of genes can potentially occur in nature.

ACKNOWLEDGMENTS

I am indebted to Olga Chesnokova, Imran Khan, Brian Kelly and Erh-Min Lai for their unpublished data and to the Davis Crown Gall Group for their capable assistance in developing this concept. This work was supported by US Public Health Services research grant GM45550 from the National Institutes of Health.

REFERENCES

Albright, L.M., Huala, E. and Ausubel, F.M. (1989. Prokaryotic signal transduction mediated by sensor and regulator protein pairs. *Annu. Rev. Genet.* **23**: 311–336.

Alt-Mörbe, J., Stryker, J.L., Fuqua, C. *et al.* (1996) The conjugal transfer system of *Agrobacterium tumefaciens* octopine-type Ti plasmids is closely related to the transfer system of an IncP plasmid and distantly related to Ti plasmid *vir* genes. *J. Bacteriol.* **178**: 4248–4257.

Bruening, G. (1990) Replication of satellite RNA of tobacco ringspot virus. *Seminars in Virology* **1**: 127–134.

Bundock, P., den Dulk-Ras, A., Beijersbergen, A. and Hooykaas, P.J.J. (1995) Trans-kingdom T-DNA transfer *Agrobacterium tumefaciens* to *Saccharomyces cerevisiae*. *EMBO J.* **14**: 3206–3214.

Burbulys, D., Trach, K.A. and Hoch, J.A. (1991) The initiation of sporulation in *Bacillus subtilis* is controlled by a multicomponent phosphorelay. *Cell* **64**: 545–552.

Chesnokova, O., Coutinho, J.B., Khan, I.H. Khan *et al.* (1997) Characterization of flagella genes of *Agrobacterium tumefaciens* and the effect of a bald strain on virulence. *Mol. Microbiol.* **23**: 579–590.

Cho, K., Fuqua, C., Martin, B.S. and Winans, S.C. (1996) Identification of *Agrobacterium tumefaciens* genes that direct the complete catabolism of octopine. *J. Bacteriol.* **178**: 1872–1880.

DeCleene, M. and DeLey, J. (1976) The host range of crown gall. *Bot. Rev.* **42**: 389–466.

Doty, S.L., Yu, M.C., Lundin, J.I. *et al.* (1996) Mutational analysis of the input domain of the VirA protein of *Agrobacterium tumefaciens*. *J. Bacteriol.* 178: 961–970.

Ekland, E.H. and Bartel, D.P. (1996) RNA-catalysed RNA polymerization using nucleoside triphosphates. *Nature* **382**: 373–376.

Ellis, J.G. and Murphy, P.J. (1981) Four new opines from crown gall tumours – their detection and properties. *Mol. Gen. Genet.* **181**: 36–43.

Farrand, S.K., Hwang, I. and Cook, D.M. (1996) The *tra* region of the nopaline-type Ti plasmid is a chimera with elements related to the transfer systems of RSF1010, RP4, and F. *J. Bacteriol.* **178**: 4233–4277.

Filichkin, S.A. and Gelvin, S.B. (1993) Formation of a putative relaxation intermediate during T-DNA processing directed by the *Agrobacterium tumefaciens* VirD1, D2 endonuclease. *Mol. Microbiol.* **8**: 915–926.

Fullner, K.J., Lara, J.L. and Nester, E.W. (1996) Pilus assembly by *Agrobacterium* T-DNA transfer genes. *Science* **273**: 1107–1109.

Giebelhaus, L.A., Frost, L., Lanka, E. *et al.* (1996) The Tra2 core of the IncP plasmid RP4 is required for intergeneric mating between *Escherichia coli* and *Streptomyces lividans*. *J. Bacteriol.* **178**: 6378–6381.

Grosschedl, R. and Hobom, G. (1979) DNA sequences and structural homologies of the replication origins of lambdoid bacteriophages. *Nature* **277**: 621–627.

Heinemann, J.A. and Sprague, G.F. Jr (1989) Bacterial conjugative plasmids mobilize DNA transfer between bacteria and yeast. *Nature* **340**: 205–209.

Hoch, J.A., and Silhavy, T.J. (1995) *Two-component Signal Transduction*, ASM Press, Washington, DC.

Kado, C.I. (1991) Molecular mechanisms of crown gall tumorigenesis. *Critical Revs Plant Sci.* **10**: 1–31.

Kado, C.I. (1993) *Agrobacterium* mediated transfer and stable incorporation of foreign genes in plants, in *Bacterial Conjugation* (cd. D.B. Clewell), Plenum Press, New York, pp. 243–254.

Kado, C.I. (1994a) T-DNA transfer to plants is mediated by pilus-like apparatus encoded by the Ti plasmid *virB* operon. *Adv. Plant Biotechnol.* **4**: 23–36.

Kado, C.I. (1994b) Promiscuous DNA transfer system of *Agrobacterium tumefaciens*: role of the *virB* operon in sex pilus assembly and synthesis. *Mol. Microbiol.* **12**: 17–22.

Kelly, B., and Kado, C.I. (1997) Promiscuous gene transfer of *Agrobacterium tumefaciens* extends to the actinomycete *Streptomyces lividans*. *Amer. Soc. Microbiol. 97th Gen. Mtg (Abstr.)*, p. 417.

Lin, L.S., Kim, Y.J. and Meyer, R.J. (1987) The 20 bp directly repeated DNA sequence of broad host range plasmid R1162 exerts incompatibility *in vivo* and inhibits R1162 DNA replication *in vitro*. *Mol. Gen. Genetics* **208**: 390–397.

Mayor, H.D., Torikai, K., Melnick, J.L. and Mandel, M. (1969) Plus and minus single-stranded DNA separately encapsidated in adeno-associated satellite virions. *Science* **166**: 1280–1282.

Murotsu, T., Matsubara, K., Sugisaki, H. and Takanami, M. (1981) Nine unique repeating sequences in a region essential for replication and incompatibility of the mini-F plasmid. *Gene* **15**: 257–271.

Okumura, M. and Kado, C.I. (1992) The region essential for efficient autonomous replication of pSa in *Escherichia coli*. *Mol. Gen. Genet.* **235**: 55–63.

Pansegrau, W., Schoumacher, F., Hohn, B. and Lanka, E. (1993) Site-specific cleavage and joining of single-stranded DNA by VirD2 protein of *Agrobacterium tumefaciens* Ti plasmids: analogy to bacterial conjugation. *Proc. Natl Acad. Sci. USA* **90**: 11538–11542.

Parkinson, J.S. and Kofoid, E.C. (1992) Communication modules in bacterial signaling proteins. *Annu. Rev. Genet.* **26**: 71–112.

Petit, A., Delhaye, S., Tempé, J. and Morel, G. (1970) Recherches sur les guanidines des tissus de crown-gall. Mise en évidence d'une relation biochimique spécifique entre les souches d'*Agrobacterium tumefaciens* et les tumeurs qu'elles induisent. *Physiol. Vég.* **8**: 205–213.

Pohlman, R.F., Genetti, H.D. and Winans, S.C. (1994) Common ancestry between IncN conjugal tranfer genes and macromolecular export systems of plant and animal pathogens. *Mol. Microbiol.* **14**: 655–668.

Rogowsky, P.M., Powell, B.S., Shirasu, K. *et al.* (1990) Molecular characterization of the *vir* regulon of *Agrobacterium tumefaciens*: complete nucleotide sequence and gene organization of the 28.63 kbp regulon cloned as a single unit. *Plasmid* **23**: 85–106.

Scherer, G. (1978) Nucleotide sequence of the O gene and of the origin of replication in bacteriophage λ DNA. *Nucleic Acids Res.* **5**: 3141–3156.

Schneider, I.R. and White, R.M. (1976) Tobacco ringspot virus codes for the coat protein of its satellite. *Virology* **70**: 244–246.

Shirasu, K. and Kado, C.I. (1993) Membrane location of the Ti plasmid VirB proteins involved in the biosynthesis of a pilin-like conjugative structure on *Agrobacterium tumefaciens*. *FEMS Microbiol Lett.* **111**: 287–294.

Sikorski, R.S., Michaud, W., Levin, H.L. *et al.* (1990) Trans-kingdom promiscuity. *Nature* **345**: 581–582.

Sprinzl, M. and Geider, K. (1988) Transfer of the Ti plasmid from *Agrobacterium tumefaciens* into *Escherichia coli* cells. *J. Gen. Microbiol.* **134**: 413–424.

Stachel, S.E. and Zambryski, P.C. (1986) *Agrobacterium tumefaciens* and the susceptible plant cell: a novel adaptation of extracellular recognition and DNA conjugation. *Cell* **47**: 155–157.

Stalker, D.M., Thomas, C.M. and Helinski, D.R. (1981) Nucleotide sequence of the region of the origin of replication of the broad host range plasmid RK2. *Mol. Gen. Genet.* **181**: 8–12.

Stalker, D.M., Kolter, R. and Helinski, D.R. (1982) Plasmid R6K DNA replication. I. Complete nucleotide sequence of an autonomously replicating segment. *J. Mol. Biol.* **161**: 33–43.

Tempé J., Petit, A., Holsters, M. *et al.* (1977) Thermosensitive step associated with transfer of the Ti plasmid during conjugation: possible relation to transformation in crown gall. *Proc. Natl Acad. Sci. USA* **74**: 2848–2849.

Tol, H.V., Buzayan, J.M. and Bruening, G. (1991) Evidence for spontaneous circle formation in the replication of the satellite RNA of tobacco ringspot virus. *Virology* **180**: 23–30.

Tolun, A., and Helinski, D.R. (1982) Separation of the minimal replication region of the F plasmid into a replication origin segment and a trans-acting segment. *Mol. Gen. Genet.* **186**: 372–377.

Waters, V.L. and Guiney, D.G. (1993) Processes at the nick region link conjugation, T-DNA transfer and rolling circle replication. *Mol. Microbiol.* **9**: 1123–1130.

Winans, S.C. (1992) Two-way chemical signaling in *Agrobacterium*–plant interactions. *Microbiol. Rev.* **56**: 12–31.

Arabidopsis ecotypes resistant to crown gall tumorigenesis

7

Jaesung Nam and Stanton B. Gelvin

SUMMARY

This chapter shows that there is considerable variation among ecotypes of *Arabidopsis thaliana* in their susceptibility to crown gall disease. Differences in susceptibility are heritable and, in one ecotype, segregate as a single predominant locus. The recalcitrance of one ecotype occurs at a late step in T-DNA transfer: transient expression of a T-DNA-encoded *gusA* gene is efficient, but the ecotype is deficient in crown gall tumorigenesis, in transformation to kanamycin resistance, and in stable GUS expression. DNA blot analysis shows that after infection by *Agrobacterium tumefaciens*, less T-DNA is integrated into the genome of the recalcitrant ecotype than is integrated into the genome of a highly susceptible ecotype.

7.1 INTRODUCTION

Agrobacterium tumefaciens is a soil-borne phytopathogen that induces the neoplastic disease crown gall on many dicotyledonous plants and on some monocots and gymnosperms (DeCleene and DeLey, 1976). The basic mechanism of tumorigenesis by *A. tumefaciens* involves the transfer of T-DNA (transfer DNA) molecules from the bacterial tumor-inducing (Ti) plasmid into the plant cell. Integration of T-DNA into the plant genome and the expression of T-DNA-encoded genes in the plant result in overproduction of the plant growth-regulating hormones auxins and cytokinins. Crown gall tumors subsequently develop at infection sites by unregulated plant cell division (for reviews, see Gelvin, 1990, 1992; Hooykaas and Beijersbergen, 1994; Zupan and Zambryski, 1995).

By using a large number of *A. tumefaciens* strains and mutants, we now understand reasonably well many of the early events of crown gall tumorigenesis, including the induction of the *vir* (virulence) genes (Winans, 1992), processing of the T-DNA from the Ti plasmid (Stachel *et al.*, 1986; Filichkin and Gelvin, 1993), and the formation of potential channels for exporting the T-DNA through the bacterial membranes (Ward *et al.*, 1988; Thompson *et al.*, 1988; Kuldau *et al.*, 1990), possibly as a DNA–protein complex (the T-complex; Howard and Citovsky, 1990). In contrast, little is known about plant factors involved in crown gall tumorigenesis. There are at least three obstacles that *A. tumefaciens* must overcome to transform a plant cell. First, *A. tumefaciens* must transfer T-DNA into the cytoplasm of plant cells after the DNA has crossed the plant cell wall and plasma membrane. We now know that T-DNA enters the plant as a single-stranded DNA molecule (Yusibov *et al.*, 1994; Tinland *et al.*, 1994). Before genes encoded by the T-DNA can be expressed, the T-DNA must reach the plant nucleus. These events may be aided by nuclear localization signals (NLS) found in VirD2 and VirE2 proteins that may accompany the T-DNA into the plant cell (Herrera-Estrella *et al.*, 1990; Citovsky *et al.*, 1992; Howard *et al.*, 1992; Shurvinton *et al.*, 1992; Tinland *et al.*, 1992; Koukolikova-Nicola *et al.*, 1993; Rossi *et al.*, 1993; Citovsky *et al.*, 1994). Finally, the T-DNA must become stabilized by integration into the plant genome followed by replication. T-DNA integration occurs either by illegitimate recombination (Gheysen *et al.*, 1991; Mayerhofer *et al.*, 1991; Ohba *et al.*, 1995; Matsumoto *et al.*, 1990). Should any of these steps in the infection pathway fail, the result would be an abortive infection.

Several studies have identified naturally occurring variations in susceptibility to crown gall disease in a number of plant species, including cucurbits (Smarrelli *et al.*, 1986), pea (Robbs *et al.*, 1991), soybean (Owens and Cress, 1984; Bailey *et al.*, 1994; Mauro *et al.*, 1995) and grapevine (Szegedi and Kozma, 1984). The basic mechanism for variation in these species is still not known. Recently, a number of laboratories have used different ecotypes of *Arabidopsis thaliana* as a model plant system to investigate differential host response to various strains or races of bacteria (Bent *et al.*, 1992; Yu *et al.*, 1993), fungi (Fuchs and Sacristan, 1996) and viruses (Leisner and Howell, 1992). We wished to determine whether or not differences also exist among *Arabidopsis* ecotypes with regard to tumorigenesis by *A. tumefaciens*, and if so, the nature of such differences. We therefore developed an *in vitro* root inoculation assay and screened *Arabidopsis thaliana* ecotypes for susceptibility or resistance to crown gall disease. We have identified several ecotypes that are hyper-susceptible, as well as other ecotypes that are recalcitrant to transformation.

7.2   RESULTS

7.2.1   RESPONSE OF *ARABIDOPSIS THALIANA* ECOTYPES TO
A. TUMEFACIENS

We characterized the response of 36 *Arabidopsis* ecotypes to *A. tumefaciens* using an *in vitro* sterile root segment infection tumorigenesis assay. We had previously determined that *A. tumefaciens* A208 (harboring the nopaline-type Ti plasmid pTiT37) incited the most rapidly growing crown gall tumors. As a control, we used *A. tumefaciens* A136 (lacking a Ti plasmid). We classified the response of these ecotypes into four categories: hypersusceptible; intermediate; recalcitrant; and no response. Table 7.1 shows that a wide variation in ecotype susceptibility to *A. tumefaciens* exists. Ecotypes Aa-0, Be-0, Ms-0, Wei-0 and WS showed a very strong response to *A. tumefaciens* A208; tumors from these ecotypes developed large green teratomas. Most of the ecotypes examined showed an intermediate response. We scored ecotypes An-1, Ang-0, Bl-1, Bla-2, Cal-0, Dijon-G, Est, Petergof, UE-1 and M7323S as recalcitrant to tumorigenesis because these ecotypes showed almost no tumorigenesis response to *A. tumefaciens* A208. Usually UE-1 did not respond to inoculation. Occasionally, a few small yellow tumorous calli lacking teratomas developed. In contrast, infected root segments of ecotype Aa-0 consistently developed large green tumors with teratomas (Fig. 7.1A). In addition, ecotype UE-1 exhibited a greatly reduced tumorigenesis response, relative to ecotype Aa-0, when we inoculated bacteria onto a wounded flower stalk (Fig. 7.1B). We therefore concentrated our initial analyses on these two ecotypes.

7.2.2   METABOLIC ACTIVITY OF *ARABIDOPSIS THALIANA* ECOTYPES
AA-0 AND UE-1

Because plant growth rate at the time of inoculation is an important factor in determining gall formation and because plant cell division is required for transformation, we compared the rates of DNA and protein synthesis between the ecotypes Aa-0 and UE-1. We incubated root segments from each ecotype in liquid MS basal medium containing [^{3}H]-thymidine or [^{14}C]-amino acids. Each hour we collected five root segments of each ecotype, ground them in a small centrifuge tube, and determined the amount of incorporated [^{3}H]-thymidine or [^{14}C]-amino acids by scintillation counting to establish rates of DNA and protein synthesis, respectively. We did not find a major difference in the incorporation of these isotopes between the ecotypes Aa-0 and UE-1 (Fig. 7.2).

In *A. tumefaciens*, T-DNA processing and transfer to the plant cell are initiated following the induction of bacterial *vir* genes by certain phenolic

Table 7.1 Tumorigenesis of *Arabidopsis* root segments inoculated with *Agrobacterium tumefaciens* A136 and A208

Ecotype	Tumorigenesis[a]	
	A136	A208
Aa-0	−	+++
AG-0	−	+
An-1	−	+/−
Ang-0		−
Bl-1	−	+/−
Bla-2		−
Bla-6	−	++
Bla-10		++
Be-0	−	+++
Ber	−	++
Cal-0	−	−
Co-1	−	++
Co-2	−	++
Col	−	++
Col-0	−	++
C24	−	++
Cvi-0		++
Dijon-G	−	−
Enkheim-D	−	++
Enkheim-T	−	++
Est	−	+/−
Hodja	−	++
La-er	−	++
Li-0		++
Lip	−	++
Ms-0	−	+++
No-0	−	++
Oy-0	−	++
Petergof	−	+/−
RLD	−	++
Shahdara	−	++
UE-1		+
Wei-0	−	+++
Ws	−	+++
M7884S	−	+++[b]
M7323S	−	+/−

[a] +++ Large green teratomas; ++ medium yellow calli without teratomas; + small yellow calli; − no response.
[b] Shooting response instead of teratomas.

Fig. 7.1 Tumorigenesis of *Agrobacterium tumefaciens* A208 on root explants and flower bolts of *Arabidopsis* ecotypes Aa-0 and UE-1. **(A)** Sterile root segments of *Arabidopsis* ecotypes UE-1, Aa-0 and the F1 progeny of a cross between these two ecotypes were co-cultivated with *A. tumefaciens* A208 for two days, then transferred to MS basal medium containing timentin. The plates were photographed after 4 weeks. **(B)** Flower bolts of *Arabidopsis* ecotypes UE-1 and Aa-0 were inoculated with *A. tumefaciens* A208 and the plants photographed after 4 weeks. Arrows indicate the site of inoculation.

compounds and sugars secreted by wounded plant cells (Gelvin, 1992). We therefore investigated the ability of *Arabidopsis* ecotypes Aa-0, UE-1 and several other ecotypes to induce *vir* genes. We introduced a *virH::lacZ* fusion into *A. tumefaciens* and monitored β-galactosidase activity following co-cultivation of *A. tumefaciens* with *Arabidopsis* root

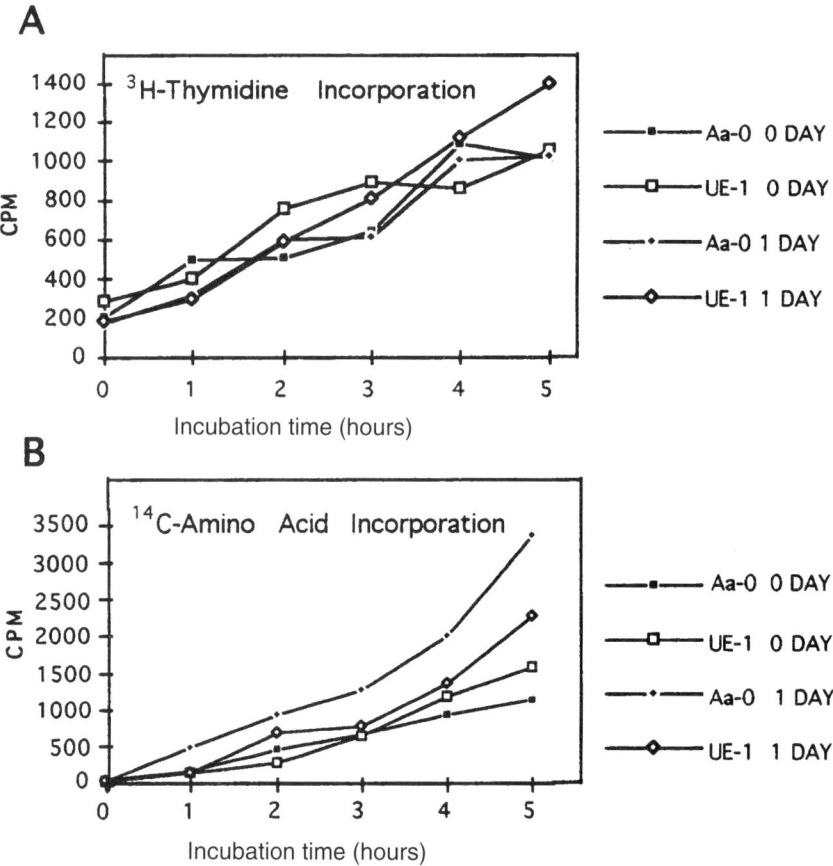

Fig. 7.2 Rates of macromolecular synthesis by *Arabidopsis* ecotypes Aa-0 and UE-1. Sterile root segments of *Arabidopsis* ecotypes Aa-0 or UE-1 were either first incubated for 1 day on CIM, or were incubated directly (0 day), in medium containing either **(A)** [³H]-thymidine or **(B)** a [¹⁴C]-amino acid mixture and the extent of isotope incorporation was determined by scintillation counting.

segments. There was no substantial difference in *vir* gene induction among any of these ecotypes (Fig. 7.3).

Because the growth of *Arabidopsis* roots is extremely sensitive to phytohormones and tumorigenesis occurs as a response of plant cells to these hormones, we compared the hormone sensitivity of Aa-0 and UE-1 by measuring the relative root growth of seedlings of these ecotypes grown vertically on plates containing different concentrations of phytohormones. The inhibition of primary root growth of the ecotypes Aa-0 and UE-1 was similar at several different concentrations

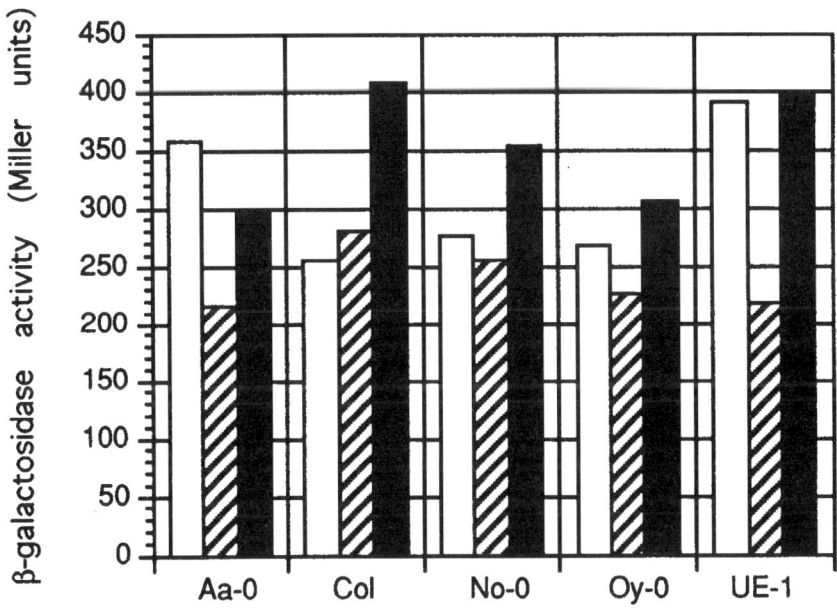

Fig. 7.3 Induction of *A. tumefaciens vir* genes by root exudates from various *Arabidopsis* ecotypes. Sterile root segments from *Arabidopsis* ecotypes Aa-0, Col, No-0, Oy-0, or UE-1 were co-cultivated either directly (□), or after 1 day (▨) or 3 days (■) incubation on CIM, with *A. tumefaciens* A348m:x219 (containing a *virH::lacZ* fusion). Bacterial cells were harvested after 24 hours and assayed for ß-galactosidase activity.

of kinetin, 2,4-D, indole-3-acetic acid and Callus Inducing Medium (CIM) (Fig. 7.4).

Thus, we could not attribute the difference in susceptibility to tumorigenesis by ecotypes Aa-0 and UE-1 to differences in the rates of macromolecular synthesis, the ability to synthesize and secrete compounds that induce *A. tumefaciens vir* genes, or sensitivity to phytohormones.

7.2.3   EFFICIENCY OF T-DNA TRANSFER TO CELLS OF THE *ARABIDOPSIS* ECOTYPES AA-0 AND UE-1

To investigate whether the difference in tumorigenesis between *Arabidopsis* ecotypes Aa-0 and UE-1 results from a deficiency in T-DNA transfer from *A. tumefaciens* to UE-1 cells, we investigated the transient T-DNA-mediated transfer to, and expression of, a *gusA* gene in the cells of these two ecotypes. We characterized β-glucuronidase (GUS) expression in *Arabidopsis* cells using both a qualitative histochemical staining assay and a quantitative fluorimetric assay. We introduced into *A. tumefaciens*

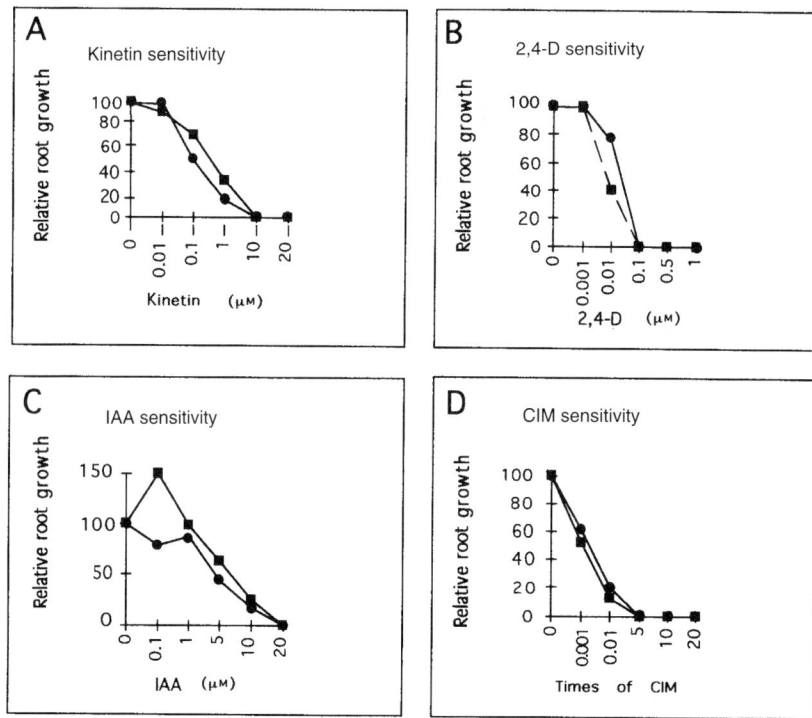

Fig. 7.4 Sensitivity of roots of *Arabidopsis* ecotypes to growth inhibition by various phytohormones. Sterile seedlings of *Arabidopsis* ecotypes Aa-0 (●) and UE-1 (■) were incubated on vertically positioned plates containing MS basal medium plus various concentrations of **(A)** kinetin, **(B)** 2,4-D, **(C)** IAA, or **(D)** on various dilutions of CIM. Root growth (units are percent relative to no hormone addition) was measured after 2 days.

A208 the binary T-DNA vector pBISN1 (Narasimhulu *et al.*, 1996). This vector contains a *gusA* gene, harboring an intron, under the control of a strong synthetic 'super-promoter' (Ni *et al.*, 1995). The intron in the *gusA* gene permits expression of GUS activity only in plant cells, but not in the bacteria. Root explants of the ecotypes Aa-0 and UE-1 showed approximately equal numbers and intensity of blue spots following infection by this *A. tumefaciens* strain for 4 days and histochemical staining with X-gluc (data not shown).

We additionally quantitated GUS activity in these infected roots using a fluorimetric MUG assay. After two days co-cultivation, we transferred the root segments onto CIM in the presence of antibiotics to kill the bacteria and measured GUS activity, using a MUG fluorescence assay, at various times. Root segments of both ecotypes Aa-0 and UE-1 first expressed detectable GUS activity 2 days after the start of co-cultivation (Fig. 7.5). For both ecotypes, GUS activity increased greatly 3 days after infection and then declined. In several repetitions of this experiment, roots of the ecotype UE-1 expressed approximately twice the GUS activity as did roots of the ecotype Aa-0. GUS activity in the roots of UE-1 continued to decrease during the course of this experiment and other repetitions of this experiment (data not shown). However, in the ecotype Aa-0, GUS activity increased after 25–30 days of the start of co-cultivation. GUS activity detected early after infection most probably represents transient expression of T-DNA that is not yet integrated into the plant genome (Janssen and Gardner, 1989; Liu *et al.*, 1992; Narasimhulu *et al.*, 1996). The increase in GUS activity seen in infected roots of ecotype Aa-0 after 25 days most likely represents expression of the *gusA* gene from T-DNA copies stabilized in *Arabidopsis* by integration into the plant chromosomes. These data indicate that the efficiency of T-DNA transfer to, and expression in, the plant nucleus are similar between the ecotypes Aa-0 and UE-1. We therefore conclude that the difference in tumorigenesis

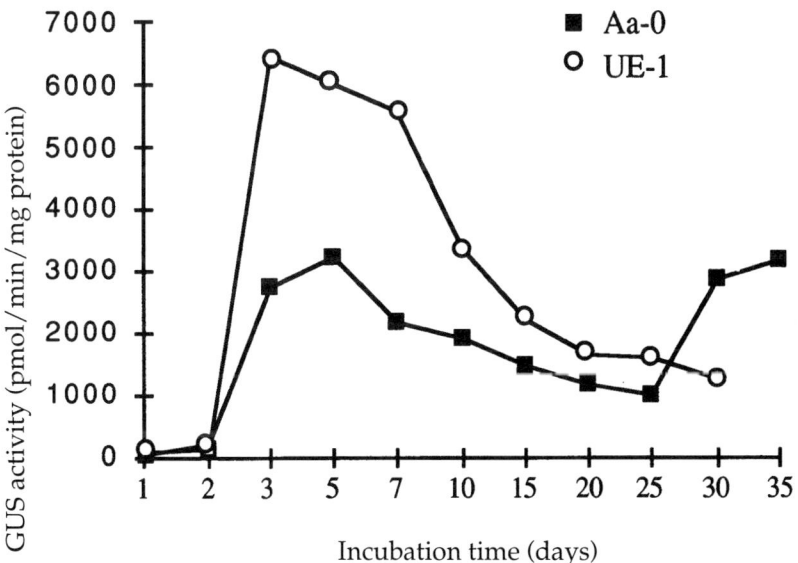

Fig. 7.5 Kinetics of GUS expression in *Arabidopsis*. Root segments of *Arabidopsis* ecotypes Aa-0 (■) and UE-1 (●) were infected with *A. tumefaciens* GV3101 (a disarmed nopaline-type strain) containing the plasmid pBISN1. After two days, the tissue was transferred to CIM containing timentin. After various periods of time, samples were assayed for GUS activity using a quantitative MUG fluorimetic assay. The times indicate days after initial infection.

between these ecotypes does not result from a deficiency of T-DNA transfer or nuclear targeting in the ecotype UE-1. These data further suggest that T-DNA can become stabilized in the genome of ecotype Aa-0 but not UE-1.

7.2.4   STABLE GUS EXPRESSION AND KANAMYCIN RESISTANCE IN INFECTED AA-0 AND UE-1 CELLS

The finding that ecotypes Aa-0 and UE-1 showed an approximately equal amount of transient GUS expression, but exhibited a different tumorigenesis response to *A. tumefaciens* infection, suggested to us that the ecotype UE-1 may be deficient in some step of tumorigenesis that involves the stabilization of T-DNA or its expression in infected plant cells. To test this hypothesis, we examined two additional transformation events that require stable integration and expression of T-DNA in the plant cells: stable GUS expression and kanamycin-resistant growth of calli derived from infected *Arabidopsis* root segments.

We co-cultivated root segments of the ecotypes Aa-0 and UE-1 for 2 days with A208(pBISN1), then transferred the root segments to CIM to induce calli without selection or to CIM containing kanamycin to select stable transformants. Only root segments from ecotype Aa-0 developed calli on CIM containing kanamycin (Fig. 7.6), although root segments of both ecotypes generated calli on CIM lacking kanamycin. We obtained the same results when we repeated this experiment using *A. tumefaciens* GV3101(pBISN1), a non-oncogenic strain. Additionally, we investigated stable GUS expression by X-gluc staining of calli derived from root segments grown on CIM without kanamycin selection. Calli from ecotype UE-1 showed only a few small blue spots, but most of the calli from ecotype Aa-0 displayed large patches of deep blue spots (data not shown).

7.2.5   EFFICIENCY OF T-DNA INTEGRATION INTO THE GENOMES OF ECOTYPES AA-0 AND UE-1

The results presented above suggest that the basis for the recalcitrance of UE-1 to tumorigenesis could be a deficiency of T-DNA integration into the plant genome. To investigate this possibility, we infected root segments of ecotypes Aa-0 and UE-1 with A208(pBISN1) and grew calli on CIM without selection. These calli were used to initiate cell suspensions grown in liquid CIM, which were composed of a mixture of transformed and non-transformed cells. We isolated DNA from these cells, separated high molecular weight plant DNA from any possible contaminating pBISN1 binary vector DNA by electrophoresis through a 0.6% agarose gel, and blotted the DNA onto a nylon membrane. When

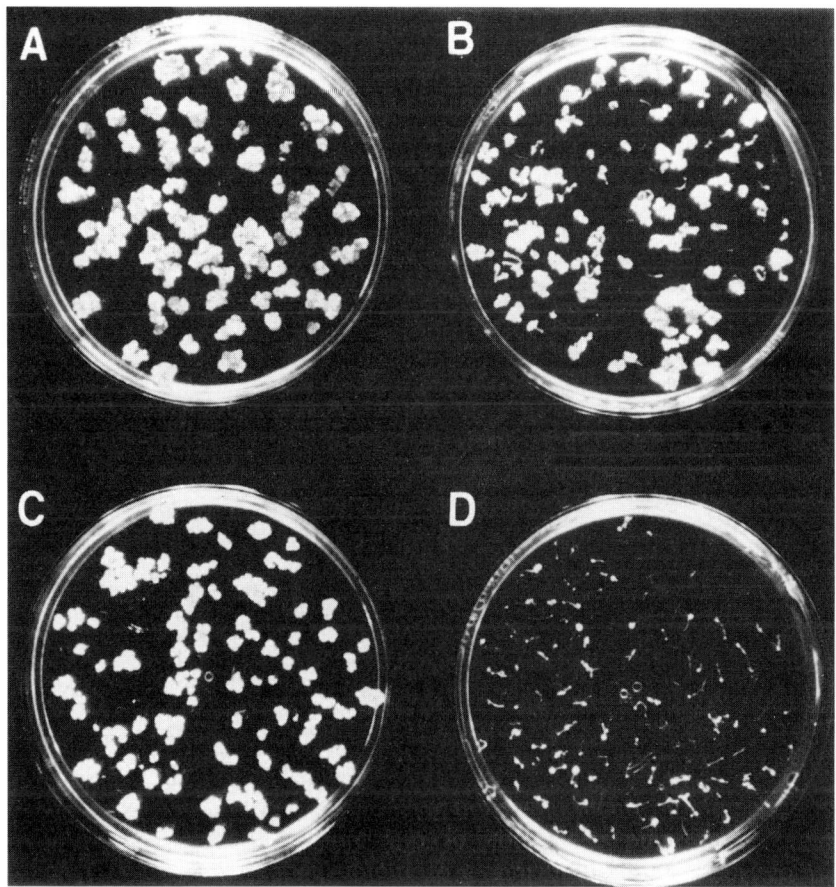

Fig. 7.6 Selection of kanamycin-resistant calli of *Arabidopsis*. Sterile root segments of *Arabidopsis* ecotypes Aa-0 (**A, B**) and UE-1 (**C, D**) were infected with *A. tumefaciens* A208 harboring the plasmid pBISN1. After 2 days, the roots were transferred to CIM either lacking (**A, C**) or containing (**B, D**) kanamycin, and the plates were incubated for 4 weeks.

hybridized with a *gusA* gene probe, high molecular weight DNA from ecotype Aa-0 showed a much stronger signal than did DNA from ecotype UE-1 (Fig. 7.7). Densitometric analysis of the autoradiogram indicated that this difference was approximately five-fold (data not shown). This result demonstrates that T-DNA integrated into the DNA of ecotypes Aa-0 and UE-1 to different extents. Control experiments indicated that the *Arabidopsis* DNA was not detectably contaminated with bacteria (Fig. 7.7B) and that each lane of the gel had an approximately equal loading of DNA (Fig. 7.7C).

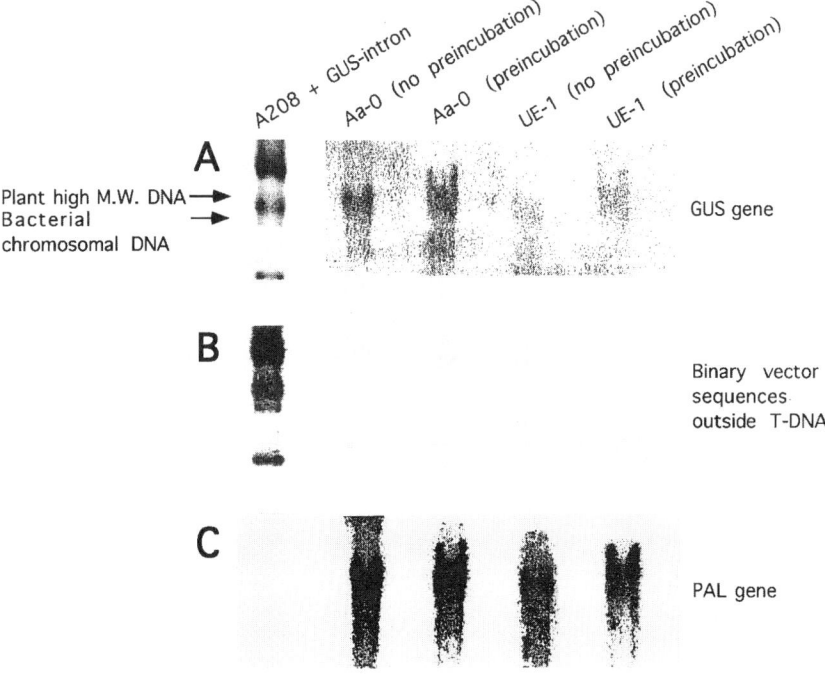

Plant high M.W. DNA →
Bacterial →
chromosomal DNA

A — GUS gene

B — Binary vector sequences outside T-DNA

C — PAL gene

Fig. 7.7 Integration of T-DNA into genomes of *Arabidopsis*. Root segments of *Arabidopsis* ecotypes Aa-0 and UE-1 were either incubated on CIM for 1 day, or infected directly with *A. tumefaciens* A208 containing the plasmid pBISN1. After 2 days, the roots were transferred to CIM containing timentin and calli were grown for 4 weeks. Calli were transferred to liquid CIM containing various antibiotics to kill the bacteria. After approximately 4 more weeks, high molecular weight plant DNA was isolated and subjected to DNA blot analysis. (**A**) *gusA* gene probe; (**B**) probe from the T-DNA binary vector replicon; (**C**) PAL gene probe.

7.2.6   GENETIC BASIS FOR DIFFERENCES IN SUSCEPTIBILITY TO CROWN
        GALL TUMORIGENESIS AND OTHER STABLE TRANSFORMATION
        PHENOTYPES

To determine whether susceptibility to tumorigenesis is a heritable trait in *Arabidopsis*, we crossed ecotypes Aa-0 and UE-1 and determined the pattern of susceptibility to tumorigenesis in subsequent generations. Susceptibility of inoculated root segments to tumorigenesis in the F1 generation is a dominant characteristic. These results were verified using a flower bolt inoculation assay (data not shown). We next determined the pattern of segregation of susceptibility to tumorigenesis in the F2 generation: we used a semi-quantitative flower bolt inoculation assay in

which we measured the weight of bolt segments surrounding the inoculation site 30 days after infection (Fig. 7.8A). Inoculation of flower bolts
of ecotype Aa-0 usually resulted in the development of tumors (Figs 7.1B
and 7.8A). Most of these tumors were large, weighing more than 31 mg
fresh weight. Inoculation of flower bolts of ecotype UE-1 generally
resulted in a wound response, with most tissue segments weighing less
than 15–20 mg fresh weight. Among the F2 segregants, approximately
25% of the tumors were small, approximately 25% were large and
approximately 50% were of intermediate size. This approximates a

A

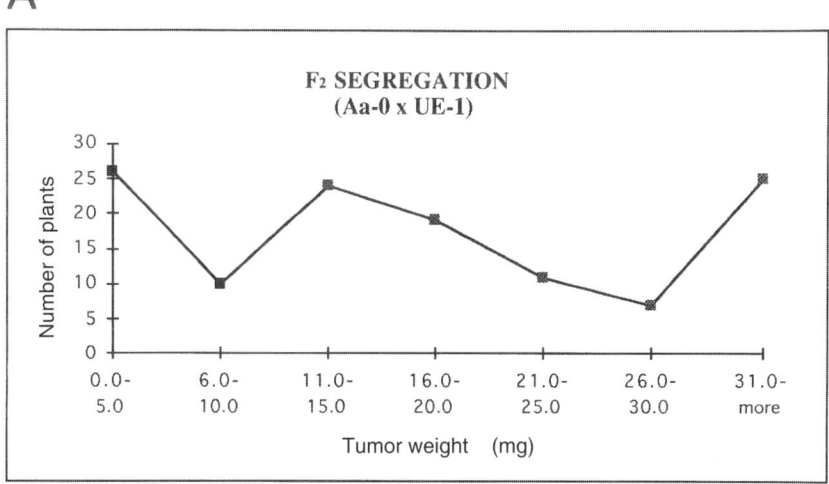

B

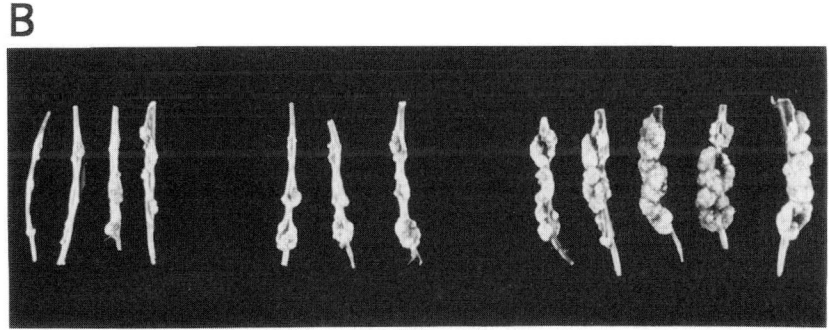

Fig. 7.8 Segregation of tumorigenesis susceptibility phenotype among the F2
progeny of a cross between *Arabidopsis* ecotypes Aa-0 and UE-1. The progeny
were inoculated on the flower bolt with *A. tumefaciens* A208. After 4 weeks, the
tumors were excised and their weights determined. **(A)** Distribution of tumor
weights; **(B)** small, medium sized and large tumors.

$1 : 2 : 1$ segregation ratio ($\chi^2 = 3.82$; $P > 0.1$) and suggests that susceptibility to tumorigenesis is a semi-dominant trait. This result apparently contradicts the root segment inoculation assay results (Fig. 7.1A) that indicate that susceptibility is completely dominant. We ascribe this incomplete dominance to physiological and/or environmental effects upon tumorigenesis using this stem inoculation assay. Alternatively, the results could indicate segregation of multiple loci that may modify the phenotype that we monitored.

In an attempt to distinguish between these possibilities, we used a different assay to determine the heritability of another stable phenotype: kanamycin resistance. We inoculated a large number of root bundle segments from individual plants with *A. tumefaciens* GV3101(pBISN1) and selected stable transformants on CIM containing kanamycin. We then calculated the percentage of root bundles from each plant that generated kanamycin-resistant calli. This was 21–40% for the recalcitrant ecotype UE-1 and greater than 81% for the susceptible ecotype Aa-0 (data not shown). More than 81% of the bundles of root segments of the F1 progeny were kanamycin resistant, indicating that the inheritance of this trait is completely dominant (data not shown). Among the 63 F2 progeny analyzed, 45 plants (71.4%) had root bundles that were highly kanamycin resistant (> 51% of the root bundles from each plant were kanamycin resistant), whereas 18 plants (28.6%) had root bundles that were mostly kanamycin susceptible (< 40% of the root bundles from each plant were kanamycin resistant). Thus, the ability to be transformed to kanamycin resistance segregates approximately $3 : 1$ ($\chi^2 = 0.34$; $P > 0.5$).

Taken together, these data indicate that susceptibility to crown gall tumorigenesis, and the ability to be transformed to kanamycin resistance, are heritable traits. The data further suggest that a single predominant locus determines the inheritance of these traits. Based on the quantitative nature of this trait, however, we cannot rule out the possible influence of other less predominant loci upon these phenotypes.

7.3   DISCUSSION

We have shown that there is considerable variation in susceptibility to crown gall tumorigenesis among a large number of *Arabidopsis thaliana* ecotypes. We further show that susceptibility is a heritable trait. From among the 36 *Arabidopsis* ecotypes screened, we selected the ecotype Aa-0 as the most susceptible and the ecotype UE-1 as recalcitrant to crown gall tumorigenesis and examined them in detail. Recalcitrance (or a total lack of response to infection) could result from a number of causes. These include defects in the ability of plant extracts to induce *A. tumefaciens vir* genes, defects in the ability of the bacteria to bind to the plant cell, deficiencies in the ability of the bacteria to transfer T-DNA to the plant cell, defects in T-DNA nuclear targeting or integration into the plant genome, a lack of ability to express T-DNA-encoded genes, or the plant's lack of response to the phytohormones whose synthesis is directed by T-DNA-encoded genes.

Hormone activated and dividing cells are more prone to transformation than are non-dividing cells (Bergmann and Stomp, 1992; Sangwan *et al.*, 1992) and metabolically active plant cells may produce molecules that induce the Ti-plasmid-encoded *vir* genes (Stachel *et al.*, 1985). We therefore first investigated whether any gross differences existed between ecotypes Aa-0 and UE-1 with regard to the biosynthesis of several classes of macromolecules, or their ability to produce chemicals that could induce *vir* genes. We saw little difference between these two ecotypes in the rates of DNA and protein biosynthesis as measured by [^{3}H]-thymidine and [^{14}C]-amino acid incorporation into macromolecules. In addition, there was little difference in the ability of exudates from roots of either of these ecotypes to induce a *virH::lacZ* fusion gene in *A. tumefaciens*. We therefore conclude that the difference in susceptibility to tumorigenesis does not result from major metabolic differences between these two ecotypes.

Phytohormone-insensitive mutants can be used to provide information on the mechanism of plant tumorigenesis. Lincoln *et al.* (1992) examined the response of the *Arabidopsis* auxin-resistant mutants *axr1* and *axr2* to infection by virulent strains of *A. tumefaciens* and *A. rhizogenes*. The mutants showed an attenuated response resulting in changes in the frequency of tumorigenesis as well as in tumor morphology. Because crown gall tumorigenesis results from a plant's response to auxins and cytokinins whose biosynthetic genes are encoded by the T-DNA (Gelvin, 1990), we examined the response of ecotypes Aa-0 and UE-1 to these hormones using a primary root growth inhibition assay. The similarity in sensitivity of these two ecotypes to externally applied hormones suggests that the differences in their susceptibility to crown gall tumorigenesis does not result from differences in their responses to these phytohormones.

To determine whether the recalcitrance of ecotype UE-1 to crown gall tumorigenesis resulted from the inability of the T-DNA to transfer to the plant nucleus (either because of a deficiency in bacterial binding to the plant cell, or because of a defect in T-DNA transfer to the plant or T-DNA nuclear transport), we conducted 'transient transformation' assays by monitoring the expression of a T-DNA-localized *gusA*-intron gene in inoculated *Arabidopsis* root segments. Using this assay, we could detect GUS activity as early as 2–3 days after infection, and infection of ecotype UE-1 resulted in approximately twice the GUS activity as infection of ecotype Aa-0 (Fig. 7.5). Thus, the ecotype UE-1 is at least as competent as is the ecotype Aa-0 in its ability to transfer T-DNA to the nucleus and convert the single-stranded T-strand to a double-stranded transcription-competent form. GUS expression in ecotype UE-1 was highly transient, however, whereas stable GUS expression was detected in ecotype Aa-0. The inability to transform ecotype UE-1 to kanamycin resistance efficiently (Fig. 7.6) further reflects the difficulty in stabilizing T-DNA-encoded traits in this ecotype.

The ability to transform ecotype UE-1 transiently, but not stably, suggested that in this ecotype T-DNA integration may be deficient. We tested this hypothesis directly by investigating the efficiency of T-DNA integration into the genomes of ecotypes Aa-0 and UE-1. We found that in unselected tissue derived from *Agrobacterium*-infected roots, approximately five times more T-DNA was integrated per microgram of high molecular weight plant DNA in ecotype Aa-0 than in ecotype UE-1 (Fig. 7.7A). Control experiments (Fig. 7.7B) indicated that the T-DNA signals that were detected on DNA blots did not derive from contaminating *Agrobacterium* cells. This is, to our knowledge, the first direct demonstration of natural variability in *Agrobacterium*-mediated plant transformation resulting from a deficiency in T-DNA integration. Experiments in our laboratory have suggested that the difficulty in transforming maize using *Agrobacterium* results from a deficiency in T-DNA integration in, but not T-DNA transfer to, this species (Narasimhulu *et al.*, 1996).

Genetic analysis of crosses between ecotypes Aa-0 and UE-1 indicated that in the F1 generation susceptibility to crown gall tumorigenesis is a dominant trait (Fig. 7.1A). Analysis of the F2 progeny suggested, however, that this susceptibility may be semi-dominant; i.e. the ability to be transformed to small, medium sized and large tumors segregated 1 : 2 : 1 (Fig. 7.8A). However, when we examined the ability of the F1 and F2 progeny to be transformed to kanamycin resistance, this trait segregated as though it were attributable to one major dominant locus. We suggest that the incomplete dominance regarding susceptibility to tumorigenesis results from physiological and/or environmental factors that affected the plants during the tumor growth period. We cannot, however, rule out the possibility that there are additional segregating loci that contribute to tumorigenesis.

ACKNOWLEDGMENTS

This work was supported in part by grants from the USDA (93-01215 and 95-37301-2040).

REFERENCES

Bailey, M.A., Boerman, H.R. and Parrott, W.A. (1994) Inheritance of *Agrobacterium tumefaciens*-induced tumorigenesis in soybean. *Crop Sci.* **34:** 514–519.

Bent, A.F., Innes, R.W., Ecker, J.R. and Staskawicz, B.J. (1992) Disease development in ethylene-insensitive *Arabidopsis thaliana* infected with virulent and avirulent *Pseudomonas* and *Xanthomonas* pathogens. *Mol. Plant–Microbe Interact.* **5**: 372 378.

Bergmann, B.A. and Stomp, A.M. (1992) Effect of host plant genotype and growth rate on *Agrobacterium tumefaciens*-mediated gall formation in *Pinus radiata*. *Phytopathol.* **82**: 1457–1462.

Citovsky, V., Warnick, D. and Zambryski, P. (1994) Nuclear import of *Agrobacterium* VirD2 and VirE2 proteins in maize and tobacco. *Proc. Natl Acad. Sci. USA* **91**: 3210–3214.

Citovsky, V., Zupan, J., Warnick, D. and Zambryski, P. (1992) Nuclear localization of *Agrobacterium* VirE2 protein in plant cells. *Science* **256**: 1802–1805.

DeCleene, M. and DeLey, J. (1976) Range of crown gall. *Bot. Rev.* **42**: 389–466.

Filichkin, S.A. and Gelvin, S.B. (1993) Formation of a putative relaxation intermediate during T-DNA processing directed by the *Agrobacterium tumefaciens* VirD1,D2 endonuclease. *Mol. Microbiol.* **8**: 915–926.

Fuchs, H., and Sacristan, M.D. (1996) Identification of a gene in *Arabidopsis thaliana* controlling resistance to clubroot (*Plasmodiaphora brassicae*) and characterization of the resistance response. *Mol. Plant Microbe Interact.* **9**: 91–97.

Gelvin, S.B. (1990) Crown gall disease and hairy root disease: a sledgehammer and a tackhammer. *Plant Physiol.* **92**: 281–285.

Gelvin, S.B. (1992) Chemical signaling between *Agrobacterium* and its plant host, in *Molecular Signals in Plant–Microbe Communications*, (ed. D.P.S. Verma), CRC Press, Boca Raton, pp. 137–167.

Gheysen, G., Villarroel, R. and Van Montagu, M. (1991) Illegitimate recombination in plants: a model for T-DNA integration. *Genes Develop.* **5**: 287–297.

Herrera-Estrella, A., Van Montagu, M. and Wang, K. (1990) A bacterial peptide acting as a plant nuclear targeting signal: the amino-terminal portion of *Agrobacterium* VirD2 protein directs a β-galactosidase fusion protein into tobacco nuclei. *Proc. Natl Acad. Sci. USA* **87**: 9534–9537.

Hooykaas, P.J.J. and Beijersbergen, A.G.M. (1994) The virulence system of *Agrobacterium tumefaciens*. *Annu. Rev. Phytopathol.* **32**: 157–179.

Howard, E. and Citovsky, V. (1990) The emerging structure of the *Agrobacterium* T-DNA transfer complex. *BioEssays* **12**: 103–108.

Howard, E., Zupan, J.R., Citovsky, V. and Zambryski, P.C. (1992) The VirD2 protein of *A. tumefaciens* contains a C-terminal bipartite nuclear localization signal: implications for nuclear uptake of DNA in plant cells. *Cell* **68**: 109–118.

Janssen, B.J. and Gardner, R.C. (1989) Localized transient expression of GUS in leaf discs following cocultivation with *Agrobacterium*. *Plant Mol. Biol.* **14**: 61–72.

Koukolikova-Nicola, Z., Raineri, D., Stephens, K. *et al.* (1993) Genetic analysis of the *virD* operon of *Agrobacterium tumefaciens*: a search for functions involved in transport of T-DNA into the plant cell nucleus and in T-DNA integration. *J. Bacteriol.* **175**: 723–731.

Kuldau, G.A., De Vos, G., Owen, J. *et al.* (1990) The *virB* operon of *Agrobacterium tumefaciens* pTiC58 encodes 11 open reading frames. *Mol. Gen. Genet.* **221**: 256–266.

Leisner, S.M. and Howell, S.H. (1993) Symptom variation in different *Arabidopsis thaliana* ecotypes produced by cauliflower mosaic virus. *Phytopathol.* **82**: 1042–1046.

Lincoln, C., Turner, J. and Estelle, M. (1992) Hormone-resistant mutants of *Arabidopsis* have an attenuated response to *Agrobacterium* strains. *Plant Physiol.* **98**: 979–983.

Liu, C.-N., Li, X.-Q. and Gelvin, S.B. (1992) Multiple copies of *virG* enhance the transient transformation of celery, carrot, and rice tissues by *Agrobacterium tumefaciens*. *Plant Mol. Biol.* **20**: 1071–1087.

Matsumoto, S., Ito, Y., Hosoi, T. *et al.* (1990) Integration of *Agrobacterium* T-DNA into a tobacco chromosome: possible involvement of DNA homology between T-DNA and plant DNA. *Mol. Gen. Genet.* **224**: 309–316.

Mauro, A.O., Pfeiffer, T.W. and Collins, G.B. (1995) Inheritance of soybean susceptibility to *Agrobacterium tumefaciens* and its relationship to transformation. *Crop Sci.* **35**: 1152–1156.

Mayerhofer, R., Koncz-Kalman, Z., Nawrath, C. *et al.* (1991) T-DNA integration: a mode of illegitimate recombination in plants. *EMBO J.* **10**: 697–704.

Narasimhulu, S.B., Deng, X.-B., Sarria, R. and Gelvin, S.B. (1996) Early transcription of *Agrobacterium* T-DNA genes in tobacco and maize. *Plant Cell* **8**: 873–886.

Ni, M., Cui, D., Einstein, J. *et al.* (1995) Strength and tissue specificity of chimeric promoters derived from the octopine and mannopine synthase genes. *Plant J.* **7**: 661–676.

Ohba, T., Yoshioka, Y., Machida, C. and Machida, Y. (1995) DNA rearrangement associated with the integration of T-DNA in tobacco: an example for multiple duplications of DNA around the integration target. *Plant J.* **7**: 157–164.

Owens, L.D. and Cress, D.E. (1984) Genotypic variability of soybean response to *Agrobacterium* strains harboring the Ti or Ri plasmids. *Plant Physiol.* **77**: 87–94.

Robbs, S.L., Hawes, M.C., Lin, H.-J. *et al.* (1991) Inheritance of resistance to crown gall in *Pisum sativum*. *Plant Physiol.* **95**: 52–57.

Rossi, L., Hohn, B. and Tinland, B. (1993) The VirD2 protein of *Agrobacterium tumefaciens* carries nuclear localization signals important for transfer of T-DNA to plants. *Mol. Gen. Genet.* **239**: 345–353.

Sangwan, R.S., Bourgeois, Y., Brown, S. *et al* (1992) Characterization of competent cells and early events of *Agrobacterium*-mediated genetic transformation in *Arabidopsis thaliana*. *Planta* **188**: 439–456.

Shurvinton, C.E., Hodges, L. and Ream, W. (1992) A nuclear localization signal and the C-terminal omega sequence in the *Agrobacterium tumefaciens* VirD2 endonuclease are important for tumor formation. *Proc. Natl Acad. Sci. USA* **89**: 1837–1841.

Smarrelli, J., Watters, M.T. and Diba, L.H. (1986) Response of various cucurbits to infection by plasmid-harboring strains of *Agrobacterium*. *Plant Physiol.* **82**: 622–624.

Stachel, S.E., Messens, E., Van Montagu, M. and Zambryski, P. (1985) Identification of the signal molecules produced by wounded plant cells that activate T-DNA transfer in *Agrobacterium tumefaciens*. *Nature (London)* **318**: 624–629.

Stachel, S.E., Timmerman, B. and Zambryski, P. (1986) Generation of single-stranded T-DNA molecules during the initial stages of T-DNA transfer from *Agrobacterium tumefaciens* to plant cells. *Nature (London)* **322**: 706–712.

Szegedi, E. and Kozma, P. (1984) Studies on the inheritance of resistance to crown gall disease of grapevine. *Vitis* **23**: 121–126.

Thompson, D.V., Melchers, L.S., Idler, K.B. *et al.* (1988) Analysis of the complete nucleotide sequence of the *Agrobacterium tumefaciens virB* operon. *Nucl. Acids Res.* **16**: 4621–4636.

Tinland, B., Koukolikova-Nicola, Z., Hall, M.N. and Hohn, B. (1992) The T-DNA-linked VirD2 protein contains two distinct functional nuclear localization signals. *Proc. Natl Acad. Sci. USA* **89**: 7442–7446.

Tinland, B., Hohn, B. and Puchta, H. (1994) *Agrobacterium tumefaciens* transfers single-stranded transferred DNA (T-DNA) into the plant cell nucleus. *Proc. Natl Acad. Sci. USA* **91**: 8000–8004.

Ward, J.E., Akiyoshi, D.E., Regier, D. *et al.* (1988) Characterization of the *virB* operon from an *Agrobacterium tumefaciens* Ti plasmid. *J. Biol. Chem.* **263**: 5804–5814.

Winans, S.C. (1992) Two-way chemical signaling in *Agrobacterium*- plant interactions. *Microbiol. Rev.* **56**: 12–31.

Yu, G.-L., Katagiri, F. and Ausubel, F.M. (1993) *Arabidopsis* mutations at the *RPS2* locus result in loss of resistance to *Pseudomonas syringae* strains expressing the avirulence gene *avrRpt2*. *Mol. Plant–Microbe Interact.* **6**: 434–443.

Yusibov, V.M., Steck, T.R., Gupta, V. and Gelvin, S.B. (1994) Association of single-stranded transferred DNA from *Agrobacterium tumefaciens* with tobacco cells. *Proc. Natl Acad. Sci. USA* **91**: 2994–2998.

Zupan, J.R. and Zambryski, P. (1995) Transfer of T-DNA from *Agrobacterium* to the plant cell. *Plant Physiol.* **107**: 1041–1047.

Evidence for the ancient transfer of Ri plasmid T-DNA genes between bacteria and plants

8

Corinne Fründt, Alain D. Meyer, Takanari Ichikawa and Frederick Meins, Jr

SUMMARY

One of the best-documented examples of gene transfer between prokaryotic and eukaryotic organisms in nature is the interaction of pathogenic bacteria in the genus *Agrobacterium* with plants. Like *A. tumefaciens*, *A. rhizogenes* (the causal agent of hairy root) transfers the T-DNA region of a root-inducing (Ri) plasmid from the bacterium to the plant cell nucleus. If plants are regenerated from the transformed cells, the T-DNA can be transmitted meiotically into the next seed generation. Sequences homologous to the core region of the T_L-DNA have been found in petunia, carrot, and six *Nicotiana* species. By Southern blot hybridization and PCR cloning, we show that homologs of the T_L-DNA genes *rolB*, *rolC* and *orf13*, called *trolB*, *trolC* and *torf13*, are present in tobacco (*Nicotiana tabacum* L. cv. 'Havana 425'). These *c-rol* genes are 70–80% identical in sequence to their bacterial counterparts. Modern cultivated tobacco is thought to be an amphidiploid hybrid of an ancestral species most closely related to *N. sylvestris* and *N. tomentosiformis*. *C-rol* sequences are present in *N. tomentosiformis* but not in *N. sylvestris*, suggesting that these sequences resulted from the ancient transfer of genes between bacteria and a progenitor of tobacco. The *c-rol* genes *trolB*, *trolC* and *torf13* are transcribed in tobacco and show developmental regulation at the steady-state mRNA level. These results are intriguing because they suggest that genes of bacterial origin introduced during evolution can have a function in a modern plant. We speculate that the transfer of genes between *Agrobacterium* and plants might have had a role in speciation.

8.1   INTRODUCTION

One of the best-documented examples of gene transfer between prokaryotic and eukaryotic organisms in nature is the interaction of the pathogenic rhizosphere bacteria in the genus *Agrobacterium* with plants (Baron and Zambryski, 1995). *A. tumefaciens* and *A. rhizogenes* cause neoplastic diseases on many dicotyledonous plants (reviewed in Sheng and Citovsky, 1996; Zupan and Zambryski, 1995). Pathogenicity depends upon a large plasmid, harbored by the bacteria, called the tumor-inducing (Ti) plasmid in *A. tumefaciens* and the root-inducing (Ri) plasmid in *A. rhizogenes*. In both cases, during the infection process part of the plasmid DNA is transferred from the bacterium to the plant-cell nucleus. The region of the plasmid transferred, called the T-DNA, is bordered by a conserved 25 bp direct repeat, which is necessary for excision from the bacterial plasmid and integration into the host genome. Foreign genes with plant expression signals placed between the borders of disarmed (i.e. non-virulent) T-DNA can be transmitted meiotically into the next seed generation of plants regenerated from the transformed cells. This important property of the T-DNA transfer system has been widely exploited to generate novel plant varieties with agronomically desirable transgenes (e.g. Flavell, 1995).

Certain plant species contain DNA sequences very similar to genes present in the T-DNA of the Ri plasmid (Spanò *et al.*, 1982; White *et al.*, 1982; Furner *et al.*, 1986; Aoki *et al.*, 1994; Meyer *et al.*, 1995). It has been proposed that these plant homologs, which we call cellular *rol* genes (*c-rol*), resulted from the ancient, horizontal transfer of genes between plants and an ancestor of *A. rhizogenes* (Furner *et al.*, 1986; Meyer *et al.*, 1995). Here, we briefly review the evidence for this hypothesis and provide results suggesting that *c-rol* genes function in modern plants.

8.2   EXPRESSION AND FUNCTION OF RI PLASMID T-DNA GENES IN PLANTS

Infection of plants with virulent strains of *A. rhizogenes* results in hairy root disease characterized by the formation of highly organized root teratomas at the infection site (reviewed in De Cleene and De Ley, 1981). Ri plasmid-transformed roots from several plant species form complete plants when cultured *in vitro* (Ackermann, 1977; Tepfer, 1984). These plants exhibit a characteristic 'hairy root syndrome' – wrinkled epinastic leaves, shortened internodes, reduced apical dominance and small flowers (Ackermann, 1977; Tepfer, 1984). Both the hairy root syndrome and the T-DNA are transmitted meiotically through successive seed generations.

During infection with the A4 agropine strain of *A. rhizogenes*, two regions of the Ri plasmid, T_L- and T_R-DNA, are integrated into the genome of the host cells (Chilton *et al.*, 1982; White *et al.*, 1982; Willmitzer *et al.*, 1982). Mutational analysis of the Ri plasmid (White *et al.*, 1985; Sinkar *et al.*, 1988) and the introduction of Ri genes alone or in combination into plants (Offringa *et al.*, 1986; Cardarelli *et al.*, 1987; Schmülling *et al.*, 1988; Spanò *et al.*, 1988; van Altvorst *et al.*, 1992) have shown that T_R-DNA carries two loci functionally and structurally homologous to the *tms1* and *tms2* loci of the Ti plasmid, which are responsible for the biosynthesis of the auxin indole-3-acetic acid by a metabolic pathway typical of microorganisms (Huffman *et al.*, 1984; White *et al.*, 1985; Offringa *et al.*, 1986). Genes important for the hairy root – *rolA*, *rolB*, *rolC*, *rolD*, *orf13* and *orf14* – are located in the T_L-DNA (White *et al.*, 1985; Cardarelli *et al.*, 1987; Spena *et al.*, 1987; Vilaine *et al.*, 1987; Schmülling *et al.*, 1988; Spanò *et al.*, 1988; Capone *et al.*, 1989). Of these genes, *orf13* and *orf14* show the highest nucleotide sequence conservation (Brevet and Tempé, 1988). Although the individual genes introduced into plants can give an aberrant phenotype, there is considerable interaction between *rol* genes and the effects observed depend on both the bacterial strain and the host. For example, the pRiA4 *rolB* alone is sufficient for induction of roots in tobacco (Cardarelli *et al.*, 1987); whereas in carrot root discs this *rolB* is only effective in the presence of either an exogenous supply of auxin or *orf13* and *orf14* (Capone *et al.*, 1989).

Combinations of the growth hormones auxin and cytokinin can regulate both the growth and organ formation by cultured tissues of many plant species (reviewed in Krikorian, 1995). Studies of *rol*-gene function have focused primarily on their possible role in the synthesis, metabolism and perception of these hormones. The expression of *rol* genes alone or in combination can mimic certain physiological and morphogenetic effects of auxins and cytokinins (Schmülling *et al.*, 1988; Walden *et al.*, 1993). For example, some *rol* genes have auxin-like effects in promoting root initiation, rate of root growth, and sensitivity of organs and cells to exogenous auxins (Shen *et al.*, 1988, 1990; Spanò *et al.*, 1988). Protoplasts from tobacco plants transformed with *rolA*, *rolB* and to a lesser degree with *rolC* show striking increases in the sensitivity of the membrane potential to auxin (Maurel *et al.*, 1991). Expression in *E. coli* has provided evidence that *rolB* encodes proteins with protein phosphotyrosine phosphatase (Fillippini *et al.*, 1996) and indoxyl-β-glucosidase activities (Estruch *et al.*, 1991b) and that *rolC* encodes a cytokinin-β-glucosidase activity (Estruch *et al.*, 1991a). Nevertheless, detailed studies have failed to show any effect of these *rol* genes on the accumulation or metabolism of auxins and cytokinins in transformants (Nilsson *et al.*, 1993; Faiss *et al.*, 1996). The available evidence suggests that if *rol* genes affect auxin or cytokinin physiology, their action is likely to be on the response of cells to these hormones.

8.3 EVIDENCE FOR PLANT HOMOLOGS OF *ROL* GENES

The earliest evidence for sequences homologous to Ri plasmid *rol* genes came from two reports in 1982 showing that hybridization signals were obtained with carrot (Spanò *et al.*, 1982) and *Nicotiana glauca* (White *et al.*, 1982) DNA under stringent conditions with Ri plasmid T-DNA probes. Based on these Southern blot hybridization studies, Ri plasmid T_L-DNA sequences were detected in six of 17 *Nicotiana* species tested as well as in petunia and carrot (Spanò *et al.*, 1982; Furner *et al.*, 1986).

The existence of the homologs *NgrolB*, *NgrolC*, *Ngorf13* and *Ngorf14* in *N. glauca* (Furner *et al.*, 1986; Aoki *et al.*, 1994) and the homologs *trolB*, *trolC* and *torf13* in tobacco (*Nicotiana tabacum* L. cv. 'Havana 425') (Meyer, 1995; Meyer *et al.*, 1995; Ichikawa *et al.*, unpublished; Fründt *et al.*, unpublished) have been demonstrated by cloning and sequencing. Expression and Southern blot hybridization studies indicate that the tobacco genome also contains homologs of *orf14* (Meyer, 1995). The region of homology in the *N. glauca* genome extends *ca.* 11 kb representing the core region of the Ri plasmid T_L-DNA with homologous genes present in the same arrangement. In addition, the *N. glauca* genome contains a partial inverted repeat of a *rolC* homolog interrupted by a stop codon generated by a frame shift mutation. Although the length of homology has not been established for tobacco, PCR cloning experiments indicate that the tobacco genome contains at least one copy each of regions homologous to *rolB-rolC* and *rolC-orf13* in the arrangement found for the Ri plasmid. A partial schematic map of the regions of homology between the T_L-DNA of Ri plasmid strain A4, *N. glauca* and tobacco is shown in Fig. 8.1.

Comparisons of the aligned coding region and deduced amino acid sequences of *rolB*, *rolC* and *orf13* homologs from the agropine Ri plasmid pRiA4 (Slighton *et al.*, 1986), the mannopine Ri plasmid pRi8196 (Hansen *et al.*, 1991), *N. glauca* and tobacco are shown in Table 8.1. Sequence identity ranged from 64% to 89% for DNA and 55% to 89% for protein, indicating that the homologs in the plasmid strains and plants are highly conserved. The relative conservation of the three genes was *orf13* > *rolB* > *rolC*. It is of interest that the divergence between Ri plasmid strains, between plants and between the plasmids and plants was comparable.

8.4 EXPRESSION OF *C-ROL* GENES

The first evidence for transcription of *c-rol* genes came from studies of interspecific hybrids of *Nicotiana*. Sexual hybrids made between *Nicotiana*

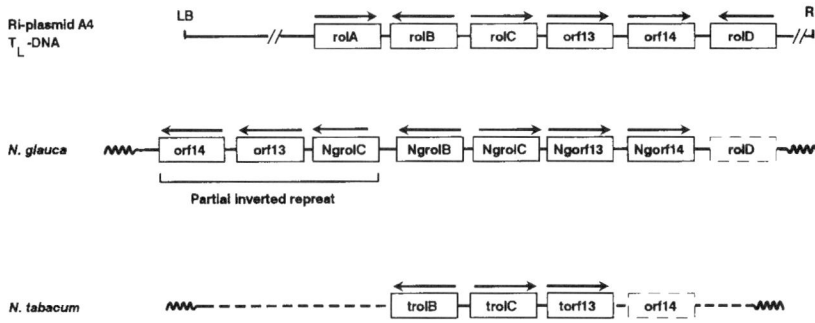

Fig. 8.1 A partial schematic map showing the arrangement of *c-rol* genes and their homologs in the T$_L$-DNA of Ri-plasmid pRiA4. Open reading frames (□) for *c-rol* genes are aligned relative to the nucleotide sequence of the homologous region of pRiA4 (Slightom *et al.*, 1986). Their direction of transcription is indicated by arrows. The structure of the *N. glauca* homologs *NgrolB*, *NgrolC*, *Ngorf13*, and *Ngorf14* (Furner *et al.*, 1986; Aoki *et al.*, 1994) and the tobacco homologs *trolB*, *trolC* and *torf13* (Meyer, 1995; Meyer *et al.*, 1995; Ichikawa *et al.*, unpublished; Fründt *et al.*, unpublished) were established by cloning and sequencing. The arrangement and length of homologous regions in the *N. glauca* genome were established by sequencing and Southern blot hybridization. In the case of tobacco, the arrangement of the cloned open reading frames is inferred from PCR-amplification of intergenic regions. The arrangement of the plant homologs not cloned (dashed boxes) is assumed based on alignment with pRiA4.

species in the groups designated 'plus' and 'minus' are tumor prone (Näf, 1958). For example, the hybrid *N. glauca* × *N.langsdorffii* (GGLL) forms teratomatous shooty tumors in response to wounding and spontaneously at the onset of flowering (reviewed in Bayer, 1983; Smith, 1988; Ichikawa and Syono, 1991). These teratomas are true tumors capable of sustained growth in culture on media without added auxin and cytokinin, which are required for the growth of normal tissues in culture. The *c-rol* genes in GGLL are derived from *N. glauca*; as judged from Southern blot hybridization these genes are not present in *N. langsdorffii* (Furner *et al.*, 1986). Although no transcription of *c-rol* genes has been reported for the *N. glauca* parent, transcripts of *NgrolB*, *NgrolC*, *Ngorf13* and *Ngrol14* have been detected by RNA blot hybridization in tissues of GGLL plants and in teratoma tissue (Ichikawa *et al.*, 1990; Aoki *et al.*, 1994). Moreover, reporter-gene studies with the *NgrolB* and *NgrolC* promoters indicate that these genes are transcriptionally regulated (Nagata *et al.*, 1995, 1996).

We have investigated expression of the tobacco *c-rol* genes *trolB*, *trolC* and *torf13* (Meyer, 1995; Meyer *et al.*, 1995; Ichikawa *et al.*, unpublished;

Table 8.1 Comparison of nucleotide and amino acid sequences of plant and Ri-plasmid *rol* gene coding regions

% Sequence identity[a]

Bacterium or plant	*rolB*				*rolC*				*orf13*			
	Ri plasmid		Plant		Ri plasmid		Plant		Ri plasmid		Plant	
	AGRO	MANN	Ng	TOB	AGRO	MANN	Ng	TOB	AGRO	MANN	Ng	TOB
Agropine (AGRO)[b]	100 (100)	78 (81)	85 (87)	80 (85)	100 (100)	74 (63)	82 (75)	74 (64)	100 (100)	87 (87)	89 (88)	79 (79)
Mannopine (MANN)[c]		100 (100)	75 (75)	72 (77)		100 (100)	82 (74)	64 (55)		100 (100)	87 (89)	82 (82)
N. glauca (Ng)[d]			100 (100)	76 (79)			100 (100)	69 (64)			100 (100)	79 (81)
Tobacco (TOB)[e]				100 (100)				100 (100)				100 (100)

[a] Nucleotide sequences of coding regions were aligned using the BESTFIT algorithm (Devereux et al., 1984) and were analyzed for nucleotide and deduced amino acid sequence identity. Data for the tobacco *rolB* homolog is for a PCR clone encompassing 233 nucleotides of the coding region (Meyer et al., 1995). Values in parentheses are for amino acid sequences.
[b] Slightom et al., 1986.
[c] Hansen et al., 1991.
[d] Furner et al., 1986; Aoki et al., 1994.
[e] Meyer et al., 1995 Ichikawa et al., unpublished; Fründt et al., unpublished.

Fründt *et al.*, unpublished). Northern blot analysis established that these genes are expressed in organs of mature, normal tobacco plants. In the case of *trolC*, the accumulation of mRNA showed a spatial distribution in the plant. High levels were found in the shoot tip, upper leaves and middle leaves of mature plants. No *trolC* mRNA was detected with comparable amounts of RNA from lower leaves, stem and roots. The accumulation of *trolC* mRNA was also regulated in cultured tissues by the growth hormones auxin and cytokinin: *trolC* was down-regulated by auxin and induced by cytokinin at the RNA level. Our results indicate that *c-rol* genes are expressed in tobacco plants and that expression at the RNA level is developmentally and hormonally regulated. Thus, *c-rol* gene expression is not an exclusive property of tumorous or tumor-prone cells.

8.5   EVIDENCE FOR THE ANCIENT TRANSFER OF *ROL* GENES BETWEEN PLANTS AND BACTERIA

Sequences homologous to *rol* genes have been reported for petunia, carrot and six *Nicotiana* species (Spanò *et al.*, 1982; White *et al.*, 1982; Furner *et al.*, 1986). The coding regions of the *rol* genes in the Ri plasmids, *N. tabacum* and *N. glauca*, are highly conserved and are present in the same arrangement. Therefore, it appears that horizontal transfer of genes occurred between plants and *A. rhizogenes* in the history of the genus. At present we cannot establish the direction of transfer. One attractive possibility is that the T-DNA genes are of plant origin, which would account for the finding that these genes have eukaryotic expression signals (Slightom *et al.*, 1986) and an intron in *rolA* (Magrelli *et al.*, 1984). On the other hand, *c-rol* genes could be the result of a second, more recent transfer from the bacterium to the plant. This seems plausible because the *N. glauca* homologs are present as an imperfect inverted repeat, which is frequently encountered in cells transformed with T-DNA (Jorgensen *et al.*, 1987); and under laboratory conditions complete plants containing *rol* genes can be readily regenerated from Ri plasmid-transformed cells (Tepfer, 1984). Similar regeneration events might have occurred, although rarely, from *A. rhizogenes*-infected tissues in nature. Finally, it seems quite unlikely that by chance a limited number of *Nicotiana* species were the progenitors of *rol* sequences distributed in Ri plasmids world wide and with similar oncogenic effects on a broad range of dicotyledonous plants.

The important question that arises is when the transfer of *rol* genes occurred. It is very unlikely that *c-rol* genes were introduced into plants in a contemporary transformation – for example, as a result of a laboratory accident. Homologs of *rol* genes have been reported for at least eight plant species, including several different isolates of *N. glauca* and several tobacco varieties from different sources. Moreover, cloned *c-rol* genes differ considerably in sequence from their T_L-DNA homologs in known strains of Ri plasmids.

The available evidence strongly suggests that the tobacco *c-rol* genes were introduced into an ancestor before the emergence of modern tobacco. Tobacco is an amphidiploid species derived by hybridization of ancestors most closely related to the present day species *N. sylvestris* and *N. tomentosiformis* (reviewed in Gerstel, 1976). Studies of sequence divergence suggest that hybridization of *N. tomentosiformis* and *N. sylvestris* progenitors occurred in South America less than 6 million years ago and probably in pre-Columbian times (Gerstel, 1976; Okamuro and Goldberg, 1985). Southern blot hybridization with probes for the highly conserved *trolC* and *torf13* coding regions and for the less conserved *trolB-trolC* intergenic region established that *c-rol* sequences are present in the *N. tomentosiformis* but not in the *N. sylvestris* genome. This provides strong evidence that tobacco *c-rol* DNA is of *tomentosiformis* origin and, hence, is the result of a transformation event which occurred before the emergence of modern tobacco.

Because the plasmid progenitors of *c-rol* genes are not known, sequence divergence cannot be used to establish the relationship between the *c-rol* sequences present in *N. glauca* and tobacco. Evolutionary considerations suggest these sequences resulted from a transformation event that occurred after the divergence of the *Cestroid* and *Petunioid* complexes of the genus *Nicotiana*. *N. glauca* is a member of the section *Paniculatae* whereas the *tomentosiformis* ancestor of tobacco is a member of the section *Tomentosae* (Goodspeed, 1954; Gerstel, 1976). Both sections are thought to be descended from the same *Cestroid* ancestral complex (Goodspeed, 1954; Gerstel, 1976). The *sylvestris* ancestor of tobacco in the section *Alatae* is descended from the *Petunioid* ancestral complex. *C-rol* sequences are only found in species derived from the *Cestroid* ancestral complex that gave rise to *N. glauca* and to the *tomentosiformis* parent of tobacco. Thus, the *c-rol* genes found in tobacco and *N. glauca* could have arisen either from a single transformation event in a common *Cestroid* ancestor or from independent transformation events after the *Tomentosae* and *Paniculatae* sections diverged 75–100 million years ago (Okamuro and Goldberg, 1985).

8.6   CONCLUDING REMARKS

The findings summarized here support the hypothesis that an ancient transfer of T_L-DNA sequences occurred between bacteria and progenitors of modern *N. glauca* and tobacco. Although we favor the hypothesis that modern *c-rol* genes are of bacterial origin, we do not exclude the possibility that T_L-DNA genes, which have eukaryotic expression signals, ultimately had their origin in some ancestral plant species. It is plausible that these genes were derived from plant 'pre-*rol*' genes with biological functions similar to those of plasmid *rol* genes. Presumably, pre-*rol* and their modern representatives in Ri plasmids have diverged considerably and, thus, might not be detected by cross-hybridization.

We speculate that the horizontal transfer of T_L-DNA genes might have had a role in the evolution of both *Agrobacterium* and the plant partner. This view is summarized in the model shown in Fig. 8.2. According to the model, pre-*rol* genes were transferred in pre-*Nicotiana* times from an early angiosperm to ancestral *Agrobacterium*, which became the progenitors of modern strains of *A. rhizogenes* carrying the Ri plasmid. Later, in pre-*tabacum* times T_L-DNA genes were introduced one or more times into *Nicotiana* hosts giving rise to species with *c-rol* genes. The fact that *c-rol* genes are expressed in a regulated fashion argues for the conclusion that these genes can have a physiological function in modern plants. The core of our hypothesis is that the rare, ancient transfer of genes from

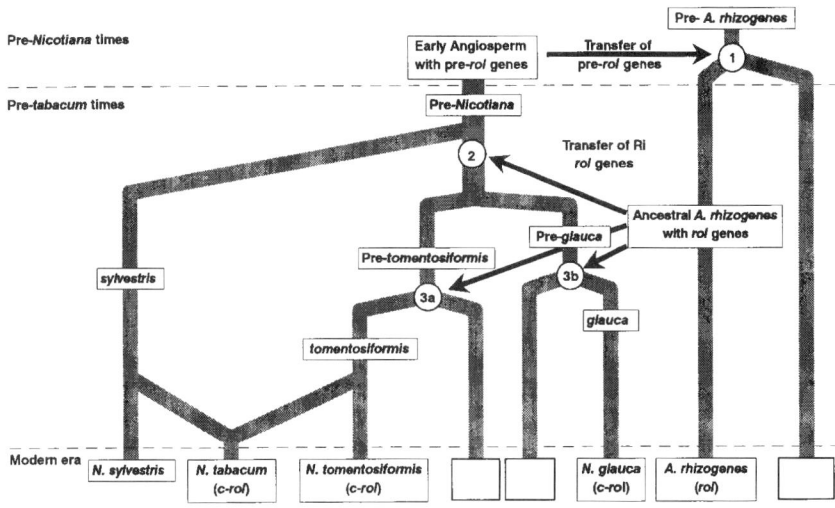

Fig. 8.2 A speculative model for the evolution of *N. glauca*, tobacco and *A. rhizogenes* based on the horizontal transfer of T_L-DNA genes. The model assumes that there was transfer of pre-*rol* genes from an early Angiosperm to a progenitor of *A. rhizogenes* (event 1), which gave rise to modern *A. rhizogenes* strains carrying *rol* sequences. Later, after the emergence of the genus *Nicotiana* but before the emergence of tobacco, one (event 2) or more (events 3a and 3b) transfers of genes from ancestral *A. rhizogenes* occurred to give the modern species *N. tabacum*, *N. tomentosiformis*, and *N. glauca* with *c-rol* genes. A major assumption of the model is that the ancient transfer of *rol* genes from bacteria to plants has adaptive significance and resulted in the emergence of new plant types with *c-rol* genes.

bacteria to plants might have generated novel variants that are progenitors of the modern species. It is plausible that features typical of *rol*-gene transformants such as reduced stature, altered flower and leaf development, decreased apical dominance, and increased rootiness (Ackermann, 1977; Tepfer, 1984; Frugis *et al.*, 1995) could have led to increased reproductive isolation, increased fitness and eventual speciation. If, as we propose, expression of *c-rol* genes has an adaptive significance, then altering their expression by sense and antisense transformation should have biological effects.

ACKNOWLEDGMENTS

We thank our colleagues Jurek Paszkowski and Andreas Hewelt for their critical comments. The work of A.D.M. was supported in part by a fellowship from the Emilia Guggenheim-Schnurr Stiftung der Basler Naturforschenden Gesellschaft.

REFERENCES

Ackermann, C. (1977) Pflanzen aus *Agrobacterium rhizogenes*-Tumoren an *Nicotiana tabacum*. *Pl. Sci. Lett.* **8**: 23–30.

Aoki, S., Kawaoka, A., Sekine, M. *et al.* (1994) Sequence of the cellular T-DNA in the untransformed genome of *Nicotiana glauca* that is homologous to ORFs 13 and 14 of the Ri plasmid and analysis of its expression in genetic tumors of *N. glauca* × *N. langsdorffii*. *Mol.Gen.Genet.* **243**: 706–710.

Baron, C. and Zambryski, P.C. (1995) Notes from the underground: highlights from plant–microbe interactions. *Trends Biotechnol.* **13**: 356–361.

Bayer, M.H. (1983) Genetic tumors: physiological aspects of tumor formation in interspecific hybrids, in *Molecular Biology of Plant Tumors*, (eds G. Kahl and J.S. Schell), Academic Press, New York, pp. 33–67.

Brevet, J. and Tempé, J. (1988) Homology mapping of T-DNA regions on three *Agrobacterium rhizogenes* Ri plasmids by electron microscope heteroduplex studies. *Plasmid* **19**: 75–83.

Capone, I., Spanò, L., Cardarelli, M. *et al.* (1989) Induction and growth properties of carrot roots with different complements of *Agrobacterium rhizogenes* T-DNA. *Plant Molec. Biol.* **13**: 43–52.

Cardarelli, M., Mariotti, D., Pomponi, M. *et al.* (1987) *Agrobacterium rhizogenes* T-DNA genes capable of inducing hairy root phenotype. *Mol. Gen. Genet.* **209475**: 475–480.

Chilton, M.-D., Tepfer, D.A., Petit, A. *et al.* (1982) *Agrobacterium rhizogenes* inserts T-DNA into the genomes of host plant root cells. *Nature* **295**: 432–434.

De Cleene, M. and De Ley, J. (1981) The host range of infectious hairy-root. *Bot. Rev.* **47**: 147–194.

Devereux, J., Haeberli, P. and Smithies, O. (1984) A comprehensive set of sequence analysis programs for the VAX. *Nucl. Acids Res.* **12**: 387–395.

Estruch, J.J., Chriqui, D., Grossmann, K. *et al.* (1991a) The plant oncogene *rolC* is responsible for the release of cytokinins from glucoside conjugates. *EMBO J.* **10**: 2880–2895.

Estruch, J.J., Schell, J. and Spena, A. (1991b) The protein encoded by the *rolB* plant oncogene hydrolyses indole glucosides. *EMBO J.* **10**: 3125–3128.

Faiss, M., Strnad, M., Redig, P. *et al.* (1996) Chemically induced expression of the *rolC*-encoded β-glucosidase in transgenic tobacco plants and analysis of cytokinin metabolism: *rolC* does not hydrolyze endogenous cytokinin glucosides *in planta*. *Plant J.* **10**: 33–46.

Fillippini, F., Rossi, V., Marin, O. *et al.* (1996) A plant oncogene as a phosphatase. *Nature* **379**: 499–500.

Flavell, R.B. (1995) Plant biotechnology R & D – the next ten years. *Trends Biotechnol.* **13**: 313–318.

Frugis, G., Caretto, S., Santini, L. and Mariotti, D. (1995) *Agrobacterium rhizogenes rol* genes induce productivity-related phenotypic modifications in 'creeping-rooted' alfalfa types. *Plant Cell Repts.* **14**: 488–492.

Furner, I.J., Huffman, G.A., Amasino, R.M. Amasino *et al.* (1986) An *Agrobacterium* transformation in the evolution of the genus *Nicotiana*. *Nature* **319**: 422–427.

Gerstel, D.U. (1976) Tobacco, in *Evolution of Crop Plants*, (ed. N.W. Simmonds), Longman, London, pp. 273–277.

Goodspeed, T.H. (1954) *The genus* Nicotiana, Chronica Botanica, Waltham, MA.

Hansen, G., Larribe, M., Vaubert, D. *et al.* (1991) *Agrobacterium rhizogenes* pRi8(196 T-DNA: mapping and DNA sequence of functions involved in mannopine synthesis and hairy root. *Proc. Natl Acad. Sci. USA* **88**: 7763–7767.

Huffman, G.A., White, F.F., Gordon, M.P. and Nester, E.W. (1984) Hairy-root-inducing plasmid: physical map and homology to tumor-inducing plasmids. *J. Bacteriol.* **157**: 269–276.

Ichikawa, T. and Syono, K. (1991) Tobacco genetic tumors. *Plant Cell Physiol.* **32**: 1123–1128.

Ichikawa, T., Ozeki, Y. and Syono, K. (1990) Evidence for the expression of the *rol* genes of *Nicotiana glauca* in genetic tumors of *N. glauca* × *N. langsdorffii*. *Mol. Gen. Genet.* **220**: 177–180.

Jorgensen, R., Snyder, C. and Jones, J.D.G. (1987) T-DNA is organized predominantly in inverted repeat structures in plants transformed with *A. tumefaciens* C58 derivatives. *Mol. Gen. Genet.* **207**: 471–477.

Krikorian, A.D. (1995) Hormones in tissue culture and micropropagation, in *Plant Hormones: Physiology, Biochemistry and Molecular Biology*, (ed. P.J. Davies), Kluwer Academic, Dordrecht, pp. 774–796.

Magrelli, A., Langenkemper, K., Dehio, C. *et al.* (1994) Splicing of the rolA transcript of *Agrobacterium rhizogenes* in *Arabidopsis*. *Science* **266**: 1986–1988.

Maurel, C., Barbier-Brygoo, H., Spena, A. *et al.* (1991) Single *rol* genes from the *Agrobacterium rhizogenes* TL-DNA alter some of the cellular responses to auxin in *Nicotiana tabacum*. *Plant Physiol.* **97**: 212–216.

Meyer, A.D. (1995) Plant genes with potential oncogenic functions. PhD Thesis, University of Basel, Basel.

Meyer, A.D., Ichikawa, T. and Meins, F. Jr (1995) Horizontal gene transfer: regulated expression of a tobacco homologue of the *Agrobacterium rhizogenes rolC* gene. *Mol. Gen. Genet.* **249**: 265–273.

Nagata, N., Kosono, S., Sekine, M. *et al.* (1995) The regulatory functions of the *rolB* and *rolC* genes of *Agrobacterium rhizogenes* are conserved in the homologous genes (Ngrol) of *Nicotiana glauca* in tobacco genetic tumors. *Plant Cell Physiol.* **36**: 1003–1012.

Nagata, N., Kosono, S., Sekine, M. *et al.* (1996) Different expression patterns of the promoters of the *NgrolB* and *NgrolC* genes during the development of tobacco genetic tumors. *Plant Cell Physiol.* **37**: 489–498.

Näf, U. (1958) Studies on tumor formation in *Nicotiana* hybrids. I. The classification of parents into two etiologically significant groups. *Growth* **22**: 167–180.

Nilsson, O., Crozier, A., Schmülling, T. *et al.* (1993) Indole-3-acetic acid homeostasis in transgenic tobacco plants expressing the *Agrobacterium rhizogenes rolB* gene. *Plant J.* **3**: 681–689.

Offringa, I.A., Melchers, L.S., Regensberg-Tunik, A.J.G. *et al.* (1986) Complementation of *Agrobacterium tumefaciens* tumor inducing *aux* mutants by genes from the T_R-region of the Ri plasmid of *Agrobacterium rhizogenes*. *Proc. Natl Acad. Sci. USA* **83**: 6935–6939.

Okamuro, J.K. and Goldberg, R.B. (1985) Tobacco single-copy DNA is highly homologous to sequences present in the genomes of its diploid progenitors. *Molec. Gen. Genetics* **198**: 290–298.

Schmülling, T., Schell, J. and Spena, A. (1988) Single genes from *Agrobacterium rhizogenes* influence plant development. *EMBO J.* **7**: 2621–2629.

Shen, W.H., Petit, A., Guern, J. and Tempé, J. (1988) Hairy roots are more sensitive to auxin than normal roots. *Proc. Natl Acad. Sci. USA* **85**: 3417–3421.

Shen, W.H., Davioud, E., David, C. *et al.* (1990) High sensitivity to auxin is a common feature of hairy root. *Plant Physiol.* **94**: 554–560.

Sheng, J. and Citovsky, V. (1996) *Agrobacterium*-plant cell DNA transport: have virulence proteins, will travel. *PC* **8**: 1699–1710.

Sinkar, V.P., Pythoud, F., White, F.F. *et al.* (1988) *rolA* locus of the Ri plasmid directs developmental abnormalities in transgenic tobacco plants. *Genes Devel.* **2**: 688–697.

Slightom, J.L., Durand-Tardif, M., Jouanin, L. and Tepfer, D. (1986) Nucleotide sequence analysis of TL-DNA of *Agrobacterium rhizogenes* agropine type plasmid. *J. Biol. Chem.* **261**: 108–121.

Smith, H.H. (1988) The inheritance of genetic tumors in *Nicotiana* hybrids. *J. Hered.* **79**: 277–283.

Spanò, L., Pomponi, M., Costantino, P. *et al.* (1982) Identification of T-DNA in the root-inducing plasmid of the agropine type *Agrobacterium rhizogenes* 1855. *Plant Molec. Biol.* **1**: 291–300.

Spanò, L., Mariotti, D., Cardarelli, M. *et al.* (1988) Morphogenesis and auxin sensitivity of transgenic tobacco with different complements of Ri T-DNA. *Plant Physiol.* **87**: 479–483.

Spena, A., Schmülling, T., Koncz, C. and Schell, J.S. (1987) Independent and synergistic activity of *rol A, B* and *C* loci in stimulating abnormal growth in plants. *EMBO J.* **6**: 3891–3899.

Tepfer, D. (1984) Transformation of several species of higher plants by *Agrobacterium rhizogenes*: sexual transmission of the transformed geneotype and phenotype. *Cell* **37**: 959–967.

van Altvorst, A.C., Bino, R.J., van Dijk, A.J. *et al.* (1992) Effects of the introduction of *Agrobacterium rhizogenes rol* genes on tomato plant and flower development. *Plant Sci.* **83**: 77–85.

Vilaine, F., Charbonnier, C. and Casse-Delbart, F. (1987) Further insight concerning the TL region of the Ri plasmid of *Agrobacterium rhizogenes* strain A4: transfer of a 1.9 kb fragment is sufficient to induce transformed roots on tobacco leaf fragments. *Mol. Gen. Genet.* **210**: 111–115.

Walden, R., Czaja, I., Schmülling, T. and Schell, J. (1993) *Rol* genes alter hormonal requirements for protoplast growth and modify expression of an auxin responsive promoter. *Plant Cell Repts.* **12**: 551–554.

White, F.F., Ghidossi, G., Gordon, M.P. and Nester, E.W. (1982) Tumor induction by *Agrobacterium rhizogenes* involves transfer of plasmid DNA to the plant genome. *Proc. Natl Acad. Sci. USA* **79**: 3193–3197.

White, F.F., Taylor, B.H., Huffman, G.A. *et al.* (1985) Molecular and genetic analysis of the transferred DNA regions of the root-inducing plasmid of *Agrobacterium rhizogenes*. *J. Bacteriol.* **164**: 33–44.

Willmitzer, L., Sanchez-Serrano, J., Buschfeld, E. and Schell, J. (1982) DNA from *Agrobacterium rhizogenes* is transferred to and expressed in axenic hairy root plant tissues. *Mol. Gen. Genet.* **186**: 16.

Zupan, J.R. and Zambryski, P. (1995) Transfer of T-DNA from *Agrobacterium* to the plant cell. *Plant Physiol.* **107**: 1041–1047.

Gene transfer from bacteria to mammalian cells

9

Patrice Courvalin, Sylvie Goussard and Catherine Grillot-Courvalin

SUMMARY

Transfer of genetic information between phylogenetically remote bacterial genera (Trieu-Cuot *et al.*, 1987), from bacteria to yeast (Heinemann and Sprague, 1989), and from bacteria to plants (Buchanan-Wollaston *et al.*, 1987) by plasmid conjugation has been described. However, direct DNA transfer from prokaryotes to mammalian cells has not yet been demonstrated. Certain bacterial species have evolved the ability to enter mammalian cells by inducing their own internalization (Falkow, 1991). We show that invasive strains of *Shigella flexneri* and *Escherichia coli*, that undergo lysis upon entry into mammalian cells because of impaired cell wall synthesis, can act as stable DNA delivery systems to their host. This direct gene transfer is efficient and of broad host cell range and the replicative or integrative vectors so delivered are stably inherited and expressed by the cell progeny. DNA delivery by abortive invasion of eukaryotic cells by bacteria is of potential interest for stimulation of mucosal immunity and for *in vivo* or *ex vivo* gene therapy of human diseases.

9.1 INTRODUCTION

The ability to invade epithelial cells is a key step in the life cycle of intracellular pathogenic bacteria (Falkow, 1991). The genes responsible for entry and intercellular dissemination of *Shigella flexneri* serotype 5 M90T are borne by the *c.* 200 kb virulence plasmid pWR100 (Sansonetti *et al.*,

1982). Entry into epithelial cells is governed by a 31 kb plasmid fragment composed of two divergently transcribed regions. The first one encodes the Ipa proteins that are the effectors of the entry process (Ménard *et al.*, 1993), and the second their dedicated secretion apparatus (Ménard *et al.*, 1994). Transfer of the plasmid to *Escherichia coli* K12 confers an invasive phenotype to the new host (Sansonetti *et al.*, 1983).

The peptidoglycan of bacterial cell walls is composed of a basic unit polymerized linearly and cross-linked laterally (Koch and Woeste, 1992). In Gram-negative bacteria, cross-linking involves *meso*-diaminopimelate residues, and diaminopimelate auxotrophs undergo lysis unless the amino acid is present in the culture medium. The first two steps in synthesis of diaminopimelate, which is not present in mammalian cells, are catalyzed by enzymes encoded by the *dapA* and *dapB* genes (Cohen and Saint Girons, 1987).

9.2   MATERIALS AND METHODS

9.2.1   STRAINS AND PLASMIDS

Allelic exchange of chromosomal genes of *E. coli* MM294 [*thi-1*, *endA1*, *hsdR17* (r_k^-, m_k^+) (Bachmann, 1987) to generate *E. coli* BM2710 [*thi-1*, *endA1*, *hsdR17* (r_k^-, m_k^+), *supE44*, Δ(*lac*)X74, Δ*dapA*Ω*cat*, *recA1*] was performed by successive transductions with bacteriophage P1 (Richaud *et al.*, 1993).

Non-conjugative plasmid pWR110 (pWR100::Tn5), which confers resistance to kanamycin (Sansonetti *et al.*, 1983) was transferred from *S. flexneri* M90T (Sansonetti *et al.*, 1982) to *E. coli* BM2710 by mobilization (Lambert *et al.* 1994) with plasmid pOX38Gm (Makris *et al.*, 1988). Plasmids pAT497 or pAT498 were then introduced by transformation into a transconjugant that had spontaneously lost pOX38Gm.

The components of pAT497 are:

- the 8.9 kb blunt-ended *Xba*I linearized p220.2 (Yates *et al.*, 1985);
- the filled-in 4.75 kb *Kpn*I-*Xba*I PSV40 *nls lacZ* SV40 pA fragment of pSVE1 Ω *nls lacZ*.

The latter plasmid consists of the 3.5 kb *Sal*I-*Bam*HI fragment of pMFG-NB (Ferry *et al.*, 1991) cloned (Sambrook *et al.*, 1989) into the 3.26 kb *Sal*I-*Bgl*II digested pSVE1. The components of pAT498 are:

- the 5352 bp blunt-ended *Hind*III-*Xba*I fragment of pRc/CMV (Invitrogen);
- the filled-in 3.5 kb *Bam*HI *nls lacZ* fragment of pMFG-NB;
- since P cmv is functional in *E. coli*, a 68 bp blunt-ended *Hinc*II-*Bam*HI fragment of pTB361 (Brockbank and Barth, 1993) containing the transcription terminator of gene 32 from bacteriophage T4 was inserted downstream.

In control experiments, the vectors were introduced by lipofection into the cell lines studied and stable transcipients producing β-galactosidase and resistant to hygromycin B or G418 were obtained.

9.2.2   DNA TRANSFER FROM BACTERIA TO MAMMALIAN CELLS

Recipient cells were cultured in six-well plates (Costar) in DMEM with 2 mM L-glutamine and 10% fetal calf serum (Myoclone +, Gibco), referred to as complete medium. They were incubated in 2 ml of complete medium per well at 5×10^4 cells/well for Cos, 1×10^5 cells/well for CHO and 2×10^5 cells/well for HeLa and A549 overnight at 37°C in 5% CO_2. *E. coli* strains were grown with shaking overnight at 30°C and then for 2 hours at 37°C, in 5 ml of Luria broth supplemented with 0.5 mM diaminopimelic acid and containing 100 µg ampicillin/ml, centrifuged and resuspended in 1 ml of complete medium. Serial dilutions of bacteria (from 1.5×10^6 to 2.5×10^7 CFU) were added in complete medium to five wells of each plate (5 ml final volume). The plates were centrifuged for 10 minutes at $500 \times g$ and incubated at 37°C for 1 hour. The adherent cells were washed three times with DMEM and 5 ml of complete medium with 20% fetal calf serum and 20 µg of gentamicin/ml were added. The plates were incubated for 40 to 44 hours at 37°C in 5% CO_2. Each set of experiments was performed in duplicate and one plate was screened on day 2 for β-galactosidase production (Sanes *et al.*, 1986). Cells from the other plate were selected for antibiotic resistance by adding hygromycin B (250, 600 and 700 µg/ml for HeLa, A549 and CHO, respectively) or G418 (800, 700 and 1000 µg/ml for HeLa, A549 and CHO, respectively) and incubated for 2 to 4 weeks.

RESULTS AND DISCUSSION

In preliminary experiments (data not shown), we have tested the capability of a *dapB* mutant of *S. flexneri* M90T (C. Parsot *et al.*, unpublished) to deliver *in vitro* replicative or integrative vectors into mammalian cells. When this bacterial host carrying either the pAT497 replicative or pAT498 integrative plasmids (Fig. 9.1) was used to infect cells in culture, direct synthesis of nuclear β-galactosidase was observed. In four independent experiments with each vector, β-galactosidase producing HeLa or Cos transcipients were obtained simply by incubating plated cells with bacteria for 1 hour. Transfer was not observed if the donor bacteria were devoid of the virulence plasmid pWR100.

To document further the conditions and main characteristics of the direct transfer of genetic information from prokaryotes to mammalian cells in a well-defined genetic background, we constructed *E. coli*

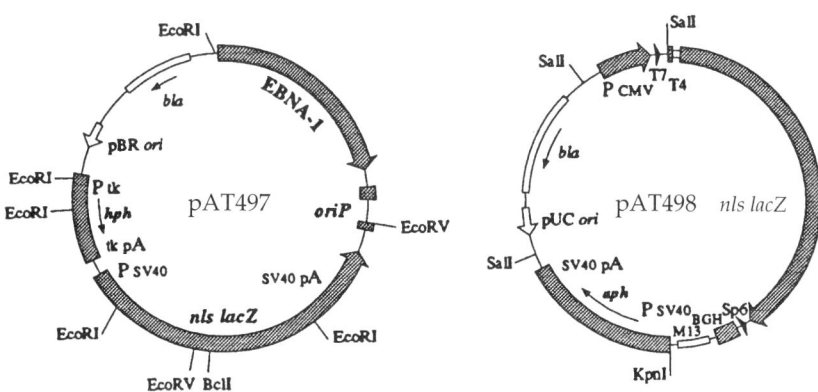

Fig. 9.1 Structure of the replicative plasmid pAT497 and of the integrative plas-mid pAT498. Plasmid pAT497 (13.65 kb) contains a single *Bcl*I site; plasmid pAT498 is 8.92 kb. Symbols for pAT497: pBR *ori*, replication origin of plasmid pBR322; *bla*, ß-lactamase gene; P tk and tk pA, promoter and polyadenylation site of Herpes simplex virus thymidine kinase; *hph*, hygromycin phosphotrans-ferase gene; EBNA-1, gene encoding the Epstein-Barr virus nuclear antigen; *oriP*, Epstein-Barr virus origin of replication; *nls*, sequence encoding the 21-amino acid nuclear localization sequence of the SV40 large tumor antigen; *lacZ*, ß-galactosidase gene; P SV40, SV40 virus early promoter; SV40 pA, SV40 virus polyadenylation site. Additional symbols for pAT498: pUC *ori*, pUC18 replication origin; P cmv, enhancer-promoter sequences of the immediate early gene of human cytomegalovirus; T7 and Sp6, promoters for transcription of sense and antisense RNA. T4, transcription terminator of gene 32 from bacteriophage T4 ; BGH, polyadenylation signal and transcription terminator of bovine growth hormone ; M13, bacteriophage M13 replication origin; *aph*, aminoglycoside phosphotransferase gene. Arrows indicate direction of transcription. Open and closed boxes indicate expression in bacteria and mammalian cells, respec-tively.

BM2710. This strain does not produce β-galactosidase and is a stable diaminopimelate auxotroph obtained by insertional mutation and dele-tion of the *dapA* gene. Plasmids pWR110, a Tn5-tagged derivative of pWR100 (Sansonetti *et al.*, 1983), pAT497 and pAT498 were then intro-duced into BM2710 and the bacteria were incubated for 1 hour with HeLa, CHO, A549 (a human pulmonary carcinoma cell line) or Cos cells; transcipients that produced β-galactosidase were detected after 2 days with efficiencies ranging from 2–5‰ to 1–3% for bacteria–mammalian cell ratios varying from 7.5 to 100 (Fig. 9.2; Table 9.1), depending on the cell line. In 2–4 weeks, selection in the presence of hygromycin B or G418 yielded clones of A549 and CHO cells that were stably resistant (Fig. 9.2). In control experiments, transfer of pAT497 and pAT498 from *E. coli* BM2710 that did not harbor pWR110 was not detected (Table 9.1).

Table 9.1 Transfer of genetic information from *E. coli* BM2710 to mammalian cells

Plasmid present in donor	β-Galactosidase production[a] in			
	HeLa	CHO	A549	Cos
pWR110 + pAT497	+/++[b]	++	+	+++
pAT497	0	0	0	0
pWR110 + pAT498	++	+++	+	++++
pAT498	0	0	0	0

[a] β-Galactosidase production was detected with 5-bromo-4-chloro-3-indolyl β-D-galactopyranoside (X-Gal; Sigma) (Sanes *et al.*, 1986).
[b] Values are from a minimum of three independent experiments: 0, no β-galactosidase producing cells per well; +, fewer than 20 β-galactosidase producing cells per well; ++, from 20 to 100 β-galactosidase producing cells per well; +++, from 100 to 500 β-galactosidase producing cells per well; ++++, more than 500 β-galactosidase producing cells per well.

DNA from six independently obtained G418 resistant A549 clones was purified and analyzed by Southern blot hybridization with a probe specific for the *aph* gene of integrative vector pAT498 (Fig. 9.3a). In every transcipient, a DNA fragment with electrophoretic mobility indistinguishable from that of pAT498 hybridized with the probe. Since DNA was prepared from cells grown for at least a month in the presence of G418, these results suggest that integration of pAT498 in the genome of the recipient cells occurred. Determination of the numbers of integrated copies of the vectors and of the insertion sites are currently under study.

Total DNA from hygromycin B-resistant transcipients obtained in four independent experiments was analyzed similarly using a probe made from the entire replicative vector (Fig. 9.3b). Three of the four transcipients had hybridization patterns indistinguishable from that of pAT497. The remaining clone was found to contain, in addition to pAT497, a plasmid dimer with a deletion in the *nls lacZ* portion; a minority of cells having acquired the vector did not produce β-galactosidase (data not shown). In contrast to that of the donor bacteria, the plasmid DNA extracted from the transcipients was not methylated, indicating that pAT497 had replicated in the recipient cells. DNA from the same four transcipients was introduced by electrotransformation into *E. coli* and plasmid DNA from 12 randomly selected transformants from each experiment was purified and analyzed by agarose gel electrophoresis after restriction endonuclease digestion (Fig. 9.3c). With the exception of three transformants corresponding to the transcipient that also contained a deleted pAT497 dimer, the profiles obtained were indistinguishable from that of pAT497, indicating that no major DNA rearrangements occurred during the trans-kingdom transfer process. This is in contrast to

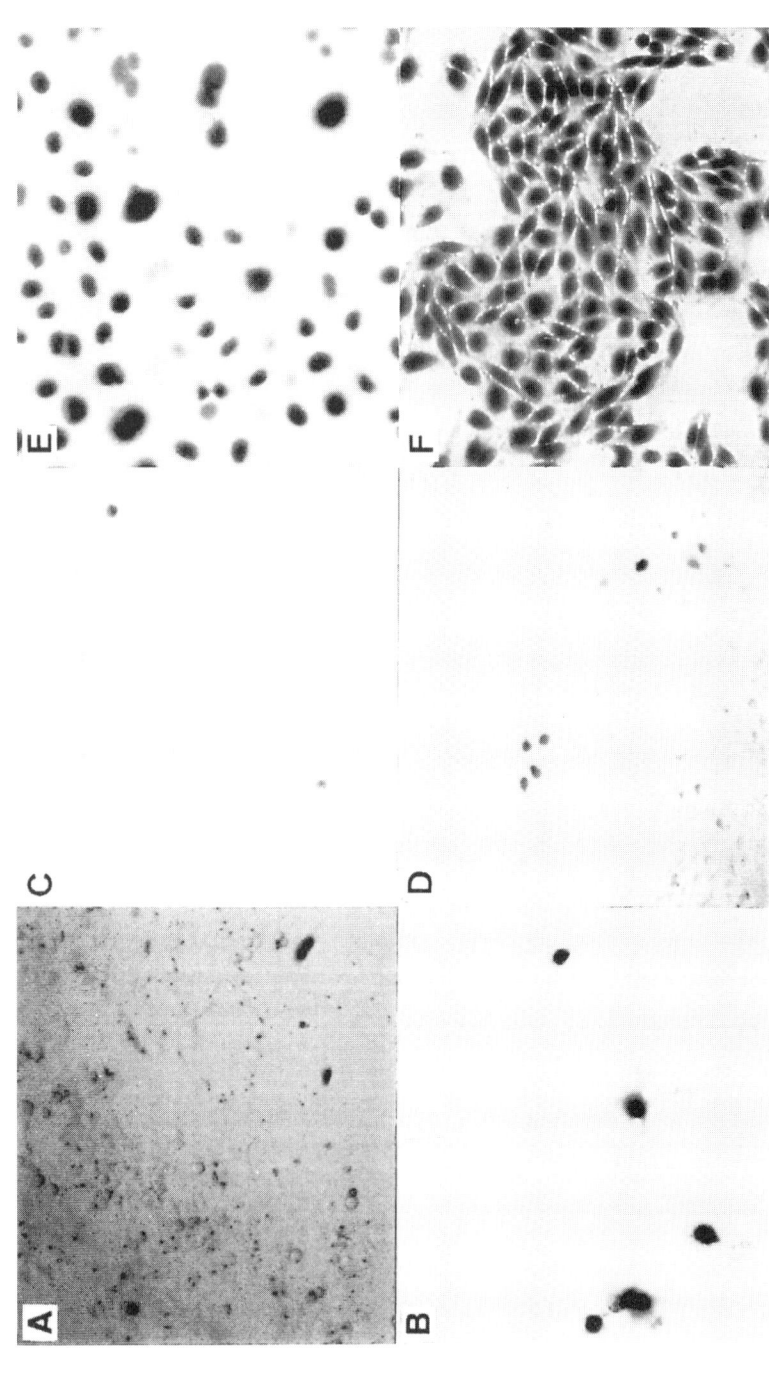

Fig. 9.2 β-Galactosidase expression in different cell types after plasmid transfer from *E. coli* BM2710. Detection of β-galactosidase [Sanes *et al.*, 1986] was performed at day 2 (A, B, C, D) or after 53 days (E) or 14 days (F) of culture in selective medium: (A) HeLa and pAT497; (B) Cos and pAT498; (C, E) A549 and pAT497; (D, F) CHO and pAT497.

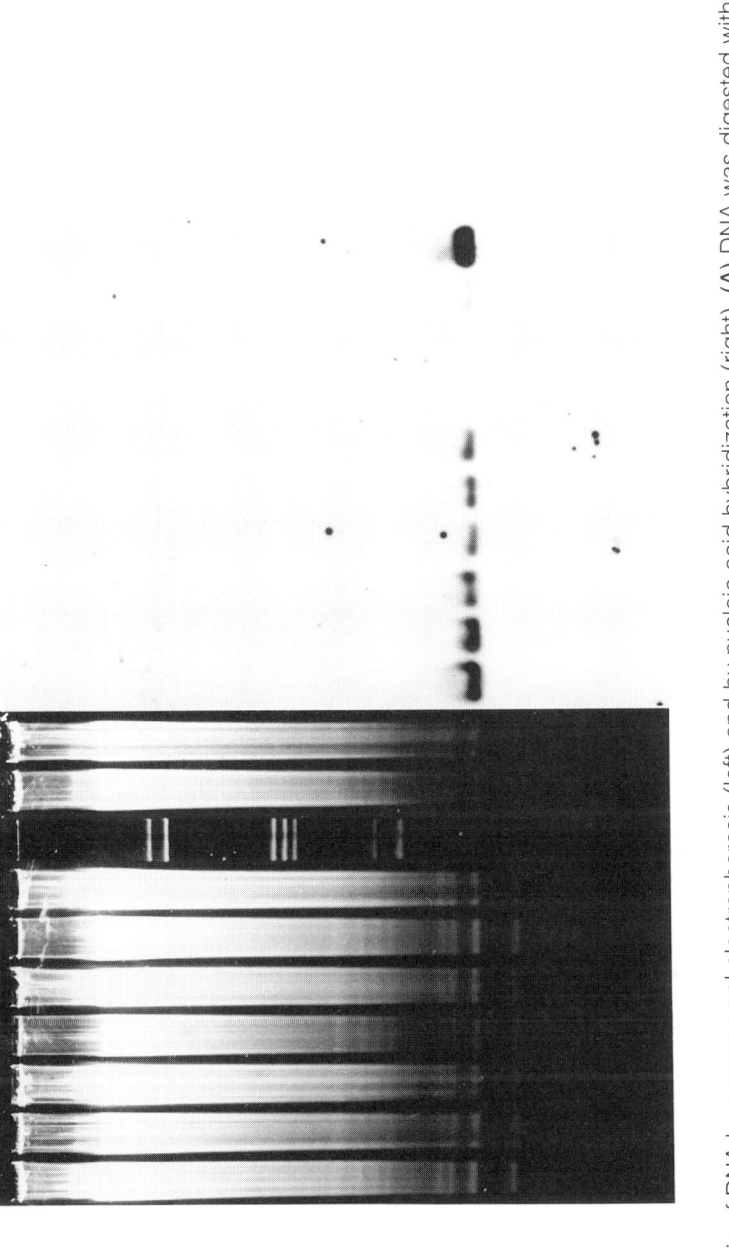

Fig. 9.3 Analysis of DNA by agarose gel electrophoresis (left) and by nucleic acid hybridization (right). (A) DNA was digested with *KpnI* + *SalI*. The resulting fragments were separated by agarose gel electrophoresis, transferred to a Nytran membrane and hybridized to the *in vitro* ^{32}P-labeled 1528 bp *KpnI-SalI* fragment of pAT498 as a probe (Sambrook *et al.*, 1989). Lanes 1 to 6, DNA of transcipients independently obtained from *E. coli* BM2710 (pWR110 + pAT498); Lane 7, DNA of A549 (negative control); Lanes 8 and 9, DNA of A549 mixed with 0.01 or 0.1 ng of pAT498 DNA, respectively (positive controls); bacteriophage λ DNA digested with *PstI* (λ) served as internal standard.

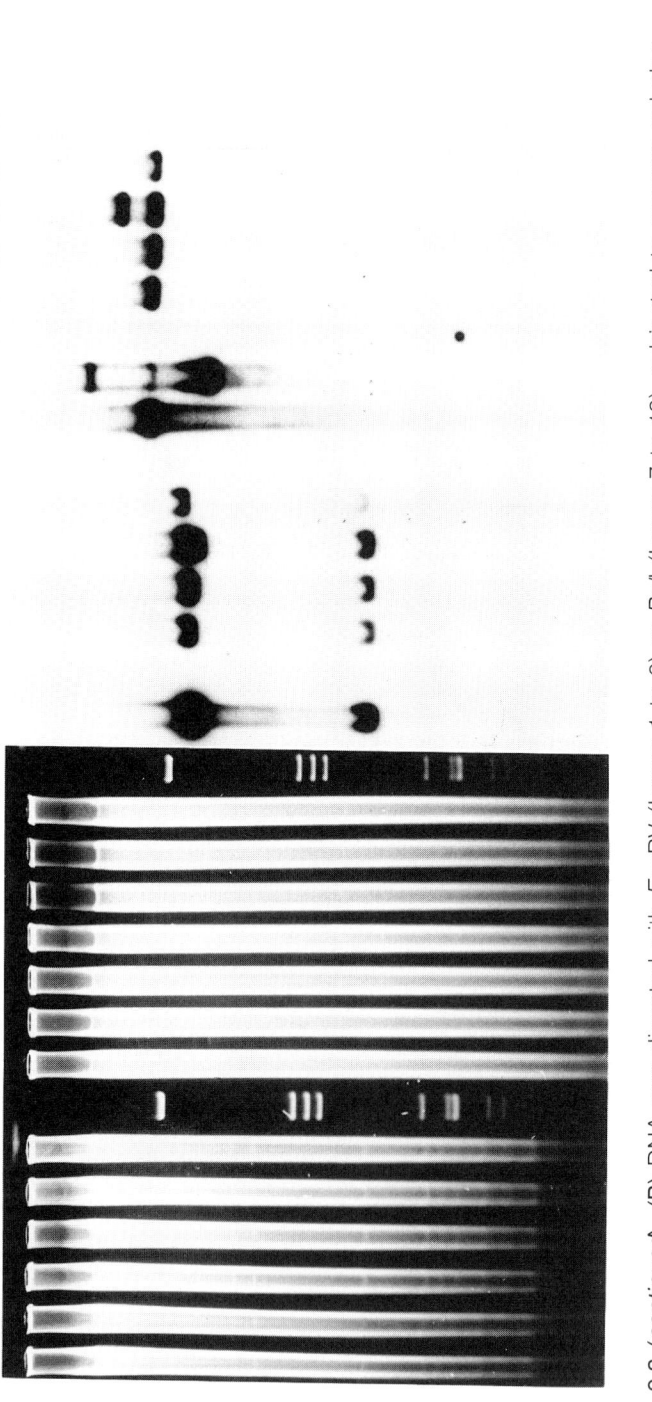

B

Fig. 9.3 (*continued*) (B) DNA was digested with *EcoRV* (Lanes 1 to 6) or *Bcl*I (Lanes 7 to 13), subjected to agarose gel electrophoresis, transferred to a Nytran membrane and hybridized to *in vitro* ^{32}P-labeled pAT497 DNA. Lane 1, DNA of A549 mixed with 0.5 ng of pAT497 DNA (positive control); Lanes 2 and 9, DNA of A549 (negative control); Lanes 3 and 10, 4 and 11, 5 and 12 and 6 and 13, DNA of four transcipients independently obtained from *E. coli* BM2710 (pWR110 + pAT497); λ, bacteriophage λ DNA digested with *Pst*I; Lane 7, DNA of A549 mixed with 0.5 ng of pAT497 purified from *E. coli* GM33 (*dam⁻*); Lane 8, DNA of A549 mixed with 0.5 ng of pAT497 purified from *E. coli* BM2710 (*dam⁺*). Restriction endonuclease *EcoRV* cuts pAT497 DNA in two fragments of 10.1 and 3.5 kb (Fig. 9.1). There is a single T↓GA*TCA recognition site for *Bcl*I in pAT497 (Fig. 9.1) which is cleaved (↓) only when one of the adenine residues (*) is not methylated. The transcipient in Lanes 5 and 12 also contains a deletion of a pAT497 dimer, as detailed below.

C

λ 1 2 3 4 5 6 7 8 9 10 11 12 13 14 15 16 17 18 λ

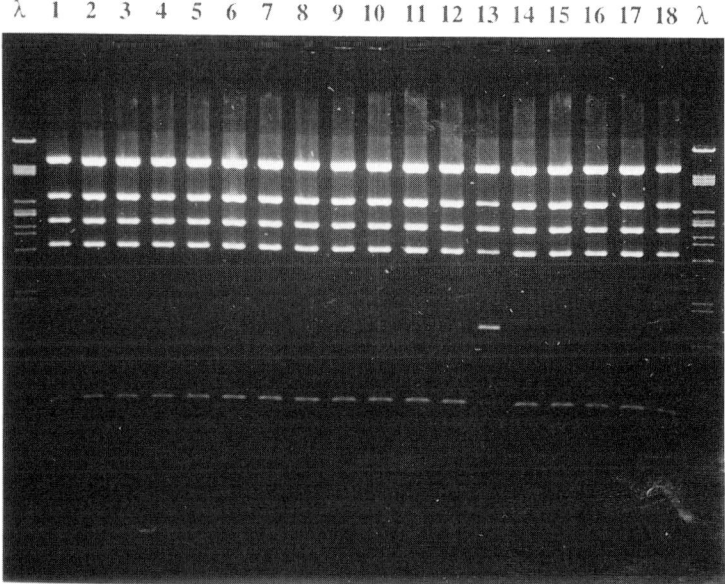

Fig. 9.3 (*continued*) (C) Analysis of plasmid DNA by agarose gel electrophoresis. DNA from the four A549 transcipients harboring pAT497 analyzed in (B) was introduced by electrotransformation into *E. coli* DH5α with selection on ampicillin (100 μg/ml). The plasmid DNA of 12 clones (four are shown) from each experiment was purified, digested with *Eco*RI, and analyzed by agarose gel electrophoresis. Lanes 1 and 18, pAT497 DNA purified from *E. coli* BM2710; Lanes 2 to 5, plasmid DNA from transformants obtained from transcipient in (B) Lane 3; Lanes 6 to 9, plasmid DNA from transformants obtained from transcipient in (B) Lane 4; Lanes 10 to 13, plasmid DNA from transformants obtained from transcipient in (B) Lane 5; nine transformants (as in Lanes 10, 11 and 12) harbored plasmids with *Eco*RI patterns indistinguishable from that of pAT497, whereas three transformants (as in Lane 13) contained a pAT497 dimer with a 3.9 kh deletion in the *nls lacZ* region, as shown by restriction endonuclease mapping. Lanes 14 to 17, plasmid DNA from transformants obtained from transcipient in (B) Lane 6; λ, bacteriophage λ DNA digested with *Pst*I.

the frequent molecular reorganization observed in pAT497, following transfer by lipofection (data not shown). Taken together, these observations suggest that transfer of genetic material to the host cells following suicidal invasion of bacteria is a process that delivers native intact DNA.

Transcipients from A549 were screened for the presence of viable *E. coli* BM2710 harboring pWR110 (Mounier *et al.*, 1992). Bacterial survivors (per well containing approximatively 2×10^5 mammalian cells) decreased from 500–1800 after 24 hours, to 80–110 after 48 hours and

0–30 after 72 hours, despite initial bacteria–mammalian cell ratios of 200 to 300 (data not shown).

The virulence plasmid of *S. flexneri* is very large and the extremely complex cluster required for cell invasion includes more than 30 genes. We have thus constructed a second-generation delivery system that combines the invasion of *Yersinia pseudotuberculosis* for cell entry and the haemolysin of *E. coli* or the listeriolysin of *Listeria monocytogenes* for lysis of the phagocytic vacuole. Preliminary experiments indicate that this binary, genetically defined system is easier to handle, of broad host range, less toxic for the recipients, and more efficient.

The results presented here demonstrate that genetic information can be transferred from invasive bacteria to mammalian cells. The gene transfer system described is undoubtedly more convenient than methods involving delivery of purified vector DNA. Using donors belonging to only two bacterial species, we were able to deliver DNA to various mammalian cell types, including CHO and Cos, that are extensively used for the *in vitro* study of gene expression, with efficiencies consistently equal to or greater than those obtained by lipofection (data not shown). This is likely to have many applications; for example, gene transfer from bacteria to cells of pulmonary origin should permit the delivery of the CFTR structural gene into cystic fibrosis tracheobronchial gland epithelial cells *in vitro* and in animal models.

Intracellular microorganisms invade a variety of distinct human cell types and this may allow targeting of the recipient cells by selection of the donor bacterium. Bacterial delivery systems are thus of interest in stimulation of mucosal immunity as well as for directed gene therapy procedures. This trans-kingdom process is also pertinent to the use of genetically engineered microorganisms and their release in the environment.

ACKNOWLEDGMENTS

We thank P. Sansonetti for many helpful suggestions and gift of strains; P. Marlière for help with P1 transduction; and P. Barth, O. Schwartz, M. Perricaudet and A. Israel for gift of plasmids. This work was supported by the Comité Consultatif des Applications de la Recherche of the Institut Pasteur and by the French Cystic Fibrosis Foundation.

REFERENCES

Bachmann, B.J. (1987) Derivations and genotypes of some mutant derivatives of *Escherichia coli* K-12, in *Escherichia coli and Salmonella typhimurium, Cellular and Molecular Biology* (eds K. Brooks Low, J.L. Ingraham, B. Magasanik *et al.*), American Society for Microbiology, Washington, DC, pp. 1190–1219.

Brockbank, S.M.V. and Barth, P.T. (1993) Cloning, sequencing, and expression of the DNA gyrase genes from *Staphylococcus aureus*. *J. Bacteriol.* **175**: 3269–3277.

Buchanan-Wollaston, V., Passiatore, J.E. and Cannon, F. (1987) The *mob* and *oriT* mobilization functions of a bacterial plasmid promote its transfer to plants. *Nature* **328**: 170–175.

Cohen, G.N. and Saint Girons, I. (1987) Biosynthesis of threonine, lysine, and methionine, in *Escherichia coli and Salmonella typhimurium, Cellular and Molecular Biology* (eds K. Brooks Low, J.L. Ingraham, B. Magasanik *et al.*), American Society for Microbiology, Washington, DC, pp. 429–444.

Falkow, S. (1991) Bacterial entry into eukaryotic cells. *Cell* **65**: 1099–1102.

Ferry, N., Duplessis, O., Houssin, D. *et al.* (1991) Retroviral-mediated gene transfer into hepatocytes *in vivo*. *Proc. Natl Acad. Sci. USA* **88**: 8377–8381.

Heinemann, J.A. and Sprague, G.F. Jr (1989) Bacterial conjugative plasmids mobilize DNA transfer between bacteria and yeast. *Nature* **340**: 205–209.

Koch, A.L. and Woeste, S. (1992) Elasticity of the sacculus of *Escherichia coli*. *J. Bacteriol.* **174**: 4811–4819.

Lambert, T., Gerbaud, G. and Courvalin, P. (1994) Characterization of transposon Tn1528 which confers amikacin resistance by synthesis of aminoglycoside 3'-O-phosphotransferase type VI. *Antimicrob. Agents Chemother.* **38**: 702–706.

Makris, J.C., Nordmann, P.L. and Reznikoff, W.S. (1988) Mutational analysis of insertion sequence *50* (IS*50*) and transposon *5* (Tn*5*) ends. *Proc. Natl Acad. Sci. USA* **85**: 2224–2228.

Ménard, R., Sansonetti, P.J., Parsot, C, and Vasselon, T. (1994) Extracellular association and cytoplasmic partitioning of the IpaB and IpaC invasins of *S. flexneri. Cell* **79**: 515–525.

Ménard, R., Sansonetti, P.J. and Parsot, C. (1993) Nonpolar mutagenesis of the *ipa* genes defines IpaB, IpaC, and IpaD as effector of *Shigella flexneri* entry into epithelial cells. *J. Bacteriol.* **175**: 5899–5906.

Mounier, J., Vasselon, T., Hellio, R. *et al.* (1992) *Shigella flexneri* enters human colonic Caco-2 epithelial cells through basolateral pole. *Infect. Immun.* **60**: 237–248.

Richaud, C., Mengin-Lecreulx, D., Pochet, S. *et al.* (1993) Directed evolution of biosynthetic pathways. *J. Biol. Chem.* **268**: 26827–26835.

Sambrook, J., Fritsch, E.F. and Maniatis, T. (1989) *Molecular Cloning: a Laboratory Manual*, 2nd edn, Cold Spring Harbor Laboratory Press, Cold Spring Harbor, NY.

Sanes, J.R., Rubenstein, J.L. and Nicolas, J.F. (1986) Use of a recombinant retrovirus to study post-implantation cell lineage in mouse embryos. *EMBO J.* **5**: 3133–3142.

Sansonetti, P.J., Hale, T.L., Dammin, G.J. *et al.* (1983) Alteration in the pathogenicity of *Escherichia coli* K-12 following the transfer of plasmid and chromosomal genes from *Shigella flexneri. Infect. Immun.* **39**: 1392–1402.

Sansonetti, P.J., Kopecko, D.J. and Formal, S.B. (1982) Involvement of a large plasmid in the invasive ability of *Shigella flexneri. Infect. Immun.* **35**: 852–860.

Trieu-Cuot, P., Carlier, C., Martin, P. and Courvalin, P. (1987) Plasmid transfer by conjugation from *Escherichia coli* to Gram-positive bacteria. *FEMS Microbiol. Letts* **48**: 289–294.

Yates, J.L., Warren, N. and Sugden, B. (1985) Stable replication of plasmids derived from Epstein–Barr virus in various mammalian cells. *Nature* **313**: 812–815.

Horizontal gene transfer in the host–parasite system *Absidia glauca–Parasitella parasitica* 10

J. Wöstemeyer, A. Wöstemeyer, A. Burmester and K. Czempinski

SUMMARY

The infection of the zygomycete *Absidia glauca* by the facultative myco-parasite *Parasitella parasitica* is accompanied by the formation of a limited cytoplasmic continuum between both partners. Nuclei of the parasite invade the host's mycelium. Simple genetic experiments show that many different auxotrophic mutants of *A. glauca* are complemented by acquiring *Parasitella*'s genetic material. We also showed at the molecular level that an artificial plasmid coding for neomycin resistance is efficiently transferred. In all cases genetic transfer depends on the formation of the typical infection structures.

Successful infection of *A. glauca* by *P. parasitica* requires that the partners belong to complementary mating types. This observation points towards a physiological relationship between parasitism and the sexual pathway. In the Mucoraceae, the family to which both fungi belong, sexual differentiation depends on the synthesis of the sex pheromone trisporic acid. In *Absidia glauca* trisporic acid is synthesized via the complementary action of both mating types. Also complementary combinations between *A. glauca* and *P. parasitica* produce trisporic acid, although this biochemical complementation does not induce the sexual reproductive pathway. Therefore we hypothesize that trisporic acid is also involved in host/parasite recognition and presumably in mediating the first steps during formation of infection structures. To check this assumption at the molecular level, we have cloned the gene for dihydromethyl-trisporic acid dehydrogenase, one of the key genes for the synthesis of trisporic acid. The gene and substantial parts of its chromosomal surrounding have been sequenced. Experiments towards construction of gene disruption or gene replacement mutants at this locus have been started. Such mutants should be impaired in both the formation of zygospores and parasitism.

10.1   INTRODUCTION

In 1924, the German botanist Hans Burgeff described the interaction between the facultative mycoparasite *Parasitella parasitica* (syn. *P. simplex*) and one of its many hosts, *Absidia glauca*. By careful microscopic observation he recognized that both partners of this interaction undergo fusion of their mycelia with each other. Nuclei of the parasite were seen invading the host's mycelium. He was not able to follow their fate or measure their activity. Today, *P. parasitica* is recognized as a fusion biotroph. Burgeff is probably the first to report an experimental system for studying gene transfer between organisms belonging to different species. Few people have read this paper, and those who did were interested in mycology, not in genetics or mechanisms of evolution. One of the few exceptions was Joshua Lederberg, who cited Burgeff in a review article in the 1950s and assumed that the transfer of genetic material might be associated with this kind of parasitism.

Biotrophic fusion parasites are not very widespread in the fungal world (reviewed in Jeffries and Young, 1994). At least two additional fusion biotrophs (Zygomycetes), *Absidia parricida* and *Chaetocladium brefeldi*, are strongly suspected of transferring genetic material to their hosts. For *Absidia parricida* preliminary experiments have been performed at the genetic level (A. Wöstemeyer, unpublished). *Chaetocladium brefeldi* has not, to our knowledge, been analyzed in this respect.

10.2   RESULTS AND DISCUSSION

10.2.1   THE INFECTION PATHWAY OF *PARASITELLA PARASITICA*

When complementary mating types of *A. glauca* and *P. parasitica* are grown on the same agar medium in a Petri dish, typical infection structures are formed in the aerial mycelium after several days of growth. The morphology of this differentiation pathway is complex (Fig. 10.1).

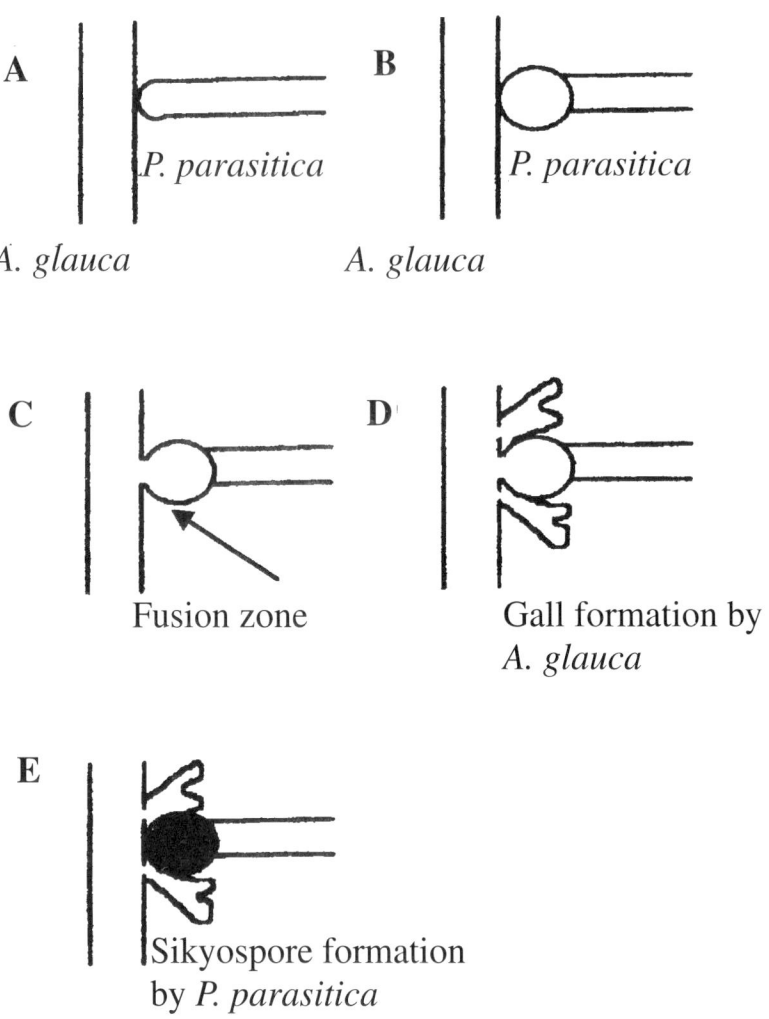

Fig. 10.1 Sikyotic parasitism of *Parasitella parasitica* on its host *Absidia glauca*. **(A)** Contact formation; **(B)** septum and bulb formation (primary sikyotic cell); **(C)** fusion between *Absidia glauca* and the primary sikyotic cell of *Parasitella parasitica*; **(D)** formation of the secondary sikyotic cell by *Parasitella parasitica* and a gall at the *A. glauca/P. parasitica* junction; **(E)** differentiation of a resting spore, the sikyospore, from the secondary sikyotic cell.

Infection starts with directed growth of a *Parasitella* hypha towards its host. Host hyphae can be any part of *A. glauca*'s aerial mycelium, including the sporangiophore. As soon as the contact is made, *P. parasitica*'s hyphal tip swells and cytoplasm (including nuclei) is transported into this bulb. Distal to this bulb a septum is formed. All fungi belonging to the Mucoraceae grow as a syncytium. Normally only the sporangio-phores for mitotic sporulation and the zygophores for the formation of sexual spores are delimited by septae. After septum formation *P. parasit-ica*'s infection bulb fuses with the host hypha. The enzymology of this process is not clear. We have looked for the induction of cell wall-lytic enzymes in compatible co-cultures of host and parasite: chitosanase, chitinase, glucanases and proteases. We have seen a rise in protease activity in such cultures, but this result needs confirmation (S. Hartwich, K.H. Riemay, unpublished).

Although the physiology of the fusion process is not understood, fusion is the prerequisite for delivering the parasite's genetic material into the host. At this stage roughly 20 *P. parasitica* nuclei invade *A. glauca*'s mycelium. By an unknown process a complex morphogenetic program is induced in the contact zone. *A. glauca* develops a branched, gall-like structure that surrounds *P. parasitica*'s secondary bulb. Burgeff has called it a secondary sikyotic cell; the Greek-derived word 'sikyotic' means 'like a cupping glass', which is an allusion to the assumed func-tion of this cell – the resorption of nutrients. After 3–4 days, the secondary sikyotic cell differentiates into a spore-like structure, the sikyospore, which may germinate under favorable conditions.

10.2.2   INFECTION LEADS TO EFFICIENT GENE TRANSFER

As an immediate consequence of the fusion event, nuclei of both partners come into close contact. The obvious question to address is: does this intimate intergeneric contact lead to a parasexual recombination event? We answered this question simply by infecting many different auxotrophic mutants of the potential recipient *A. glauca* with a compati-ble prototrophic donor strain of *P. parasitica*. The recipient survives the infection and forms normal sporangia containing hundreds of uninucle-ate spores, even close to the infection sites. After one sporulation cycle, the spores were washed from the agar medium and plated on minimal medium under conditions where only prototrophic *Absidia* offspring grow. Germination of the *Parasitella* spores was suppressed by low levels of neomycin. Between 0.1% and 1.5% prototrophic *A. glauca* spores were monitored directly on the infection plates. A mean value for nine inde-pendent experiments was 0.42%. This frequency is extraordinarily high for gene transfer systems: it exceeds the reversion rate of the recipient auxotrophs by more than 10^4 even in experiments with as little as 0.1% transfer. These results imply that intergeneric recombinants were formed as a consequence of infection. We coined the term para-recombinants for these hybrids, which alludes to both parasexuality and parasitism. The appearance of para-recombinants depends strictly on the formation of infection structures. Incompatible pairs of donor and recipient, e.g.

strains belonging to identical mating types, do not lead to prototrophic offspring (Kellner *et al.*, 1993).

At the molecular level gene transfer can be proven by transforming the donor, *P. parasitica*, with a plasmid which also replicates in the recipient (Fig. 10.2). We have constructed an autonomously replicating plasmid that confers neomycin resistance to both fungi (Burmester, 1992; Kellner *et al.*, 1993). After infection, neomycin-resistant *Absidia* recipients were selected. These para-recombinants contained the plasmid, as we have shown in Southern blot hybridization experiments and by retransformation into *Escherichia coli* (Kellner *et al.*, 1993).

10.2.3   TRANSFERRED MARKERS TEND TO BE UNSTABLE

In most instances para-recombinants have proven to be somewhat unstable. Figure 10.3 gives the loss frequencies of the *met*⁺ phenotype acquired

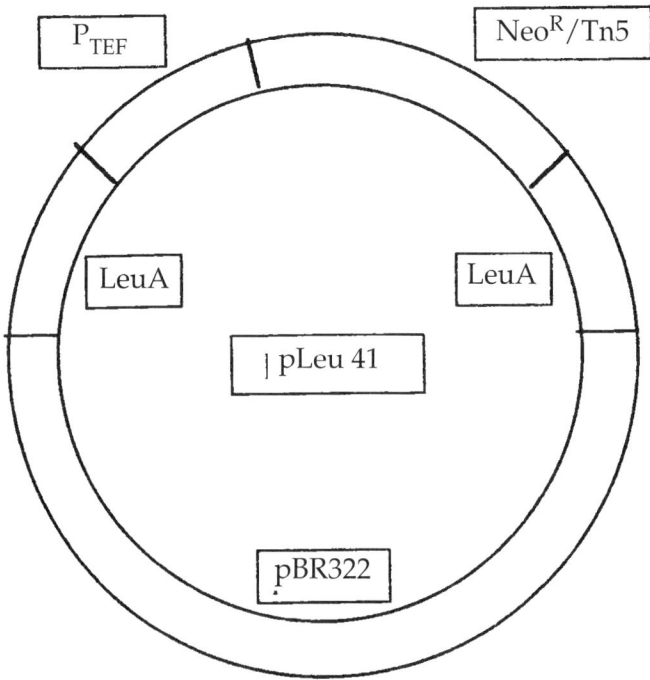

Fig. 10.2 pLeu41, an autonomously replicating plasmid conferring neomycin (Neo) resistance to *Parasitella parasitica* and *Absidia glauca*. *P. parasitica* was transformed by electroporation (Kellner *et al.*, 1993). A single spore Neo^R-isolate was used for infection of *A. glauca*. After one sporulation cycle Neo^R-colonies of *A. glauca* were selected and shown to contain the plasmid by Southern blot hybridization analysis and by transformation in *E. coli*.

after parasitic transfer. The initial culture had 0.2% of para-recombinants. For subsequent sporulation cycles, a single uninucleated spore was used for inoculation of agar media. After one week approximately 10^9 progeny spores were formed. Such a cycle corresponds to roughly 30 subsequent mitotic divisions ($2^{30} = 10^9$). Thus, the stability of para-recombinants is much higher than that of an ARS plasmid in yeast but lower than expected for an integrative event. We do not know the reason for the instability. This question is coupled with the mechanism of gene transfer. Two basic possibilities are obvious (Fig. 10.4).

1. After entering the host mycelium, *Parasitella* nuclei fuse with *Absidia* recipients. During subsequent mitotic divisions the resulting inter-generic hybrid nucleus stabilizes via several aneuploid intermediates. The karyotype of the final stabilized nucleus could depend on the selection procedure. To test this hypothesis we compared electrophoretic

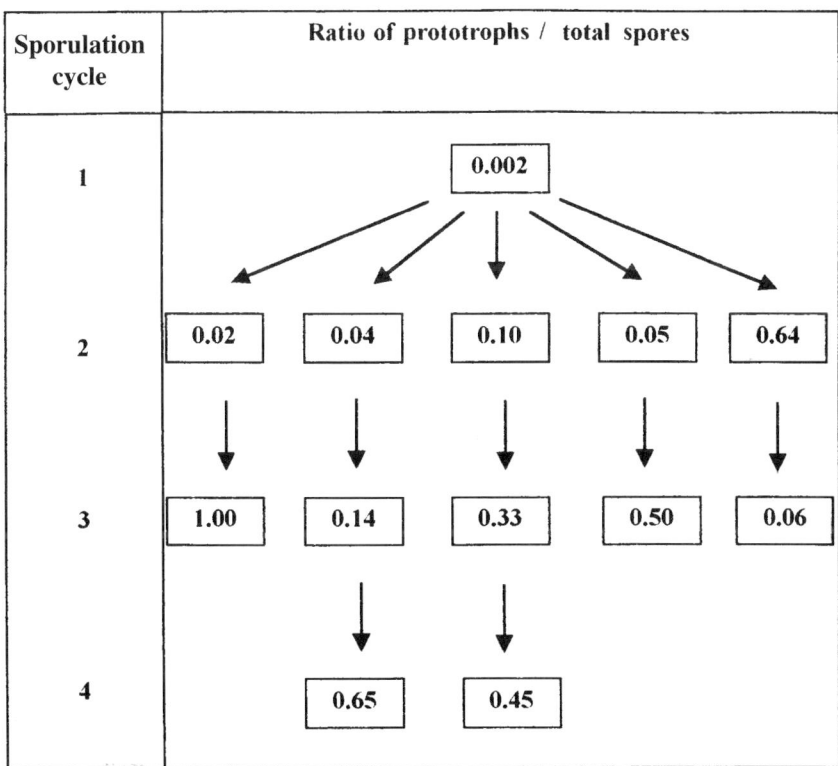

Fig. 10.3 Loss frequency of the *met*+-phenotype of a para-recombinant in subsequent mitotic sporulation cycles.

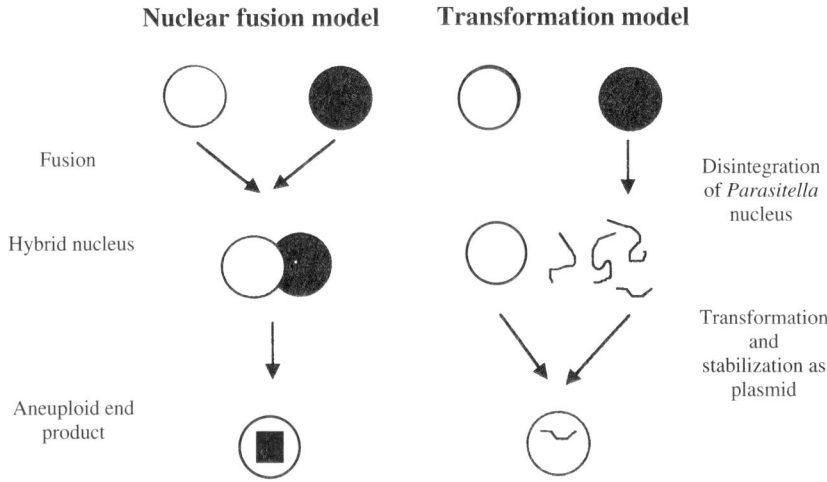

Fig. 10.4 Hypothetical mechanisms for the formation of para-recombinants.

karyotype patterns of donor and recipient with that of several para-recombinants. Although we frequently observed aberrations, we never obtained evidence for the introduction of a complete *Parasitella* chromosome into *Absidia*'s chromosomal complement.

2. After invading the host's cytoplasm, *Parasitella* nuclei disintegrate. The DNA may enter *Absidia* nuclei via a pathway analogous to transformation. In this case the amount of *Parasitella* DNA in recombinant nuclei should be lower than that after nuclear fusion. We know from hybridization studies with highly repetitive, interspersed *Parasitella* DNA that the amount of transferred DNA is below 2%. However, the data are insufficient to discriminate between the two alternatives for the mechanism of gene transfer (Fig. 10.4).

To account for the instability of transferred markers, there are additional considerations. Nuclear fusion could explain the instability, but the karyotype data do not support the idea of a hybrid intergeneric chromosome set. Transformation is appropriate if the DNA is established in the recipient by a mechanism other than integration. Presently, there is no clear-cut answer but we would like to propose a hypothesis that can easily be checked experimentally, as soon as a molecular probe is available for the transferred genetic character. It is possible that the *Parasitella* DNA circularizes within *Absidia* nuclei and is propagated as an extrachromosomal nuclear replicon. Several observations suggest this hypothesis. In contrast to ascomycetous fungi, *A. glauca* contains many small circular plasmids (Hänfler *et al.*, 1992). We have analyzed one of these plasmids down to the sequence level. We know that it codes for one protein, and that it has no chromosomal counterpart. This shows that the replication and the partitioning machinery of *A. glauca* copes well with small circular plasmids. For foreign DNA from *P. parasitica* the situation might be slightly different with respect to stability; this would explain the unstable phenotype.

10.2.4   SEXUAL AND PARASITIC INTERACTIONS MAY SHARE A
          COMMON RECOGNITION PATHWAY

The observation that successful infection depends on sexually complementary host/parasite combinations prompted us to check the possibility that trisporic acid, the sexual pheromone of mucoraceous fungi, mediates both differentiation programs. The most straightforward approach would be to analyze a defined mutant blocked in a step of trisporic acid biosynthesis, assessing both zygospore formation and development of sikyotic infection structures.

Trisporic acid is synthesized as a cooperative action of complementary mating types (Werkman, 1976; van den Ende, 1978) (Fig. 10.5). Starting from β-carotene, mating type specific precursors are produced, which are passed on to the partner, where they are processed to the active pheromone, trisporic acid. We have concentrated on the (–) type specific enzyme dihydromethyltrisporic acid dehydrogenase. For practical reasons we used the related fungus *Mucor mucedo* for the purification procedure. The enzyme was purified by gel filtration, affinity chromatography on Blue-Sepharose (Pharmacia) and gel electrophoresis. It was digested with endopeptidase Lys-C and approximately 30% of the protein was sequenced in cooperation with Dr Volker Kruft at the Max-Planck Institute in Berlin-Dahlem. Degenerate primers allowed us to identify a PCR-fragment that was colinear with part of the protein sequence (Czempinski *et al.*, 1996). Based on this probe we have isolated and sequenced a nearly complete cDNA clone and the chromosomal copy of the gene. We have now made constructs comprising the coding region and several kilobases of upstream and downstream sequences. After insertion of a cassette conferring neomycin resistance, we will use this plasmid for gene replacement experiments. The prospects for this experiment are good. The flanking regions may be long enough to allow homologous recombination at a reasonable rate (Arnau *et al.*, 1991; Burmester *et al*, 1990). The gene has only a single copy in both mating types, and also the flanking regions are conserved between the mating types according to Southern analysis data (Burmester and Borriss, unpublished).

With regard to more general and evolutionary relationships between parasitism and sex in fungi the *Parasitella* system offers many possibilities

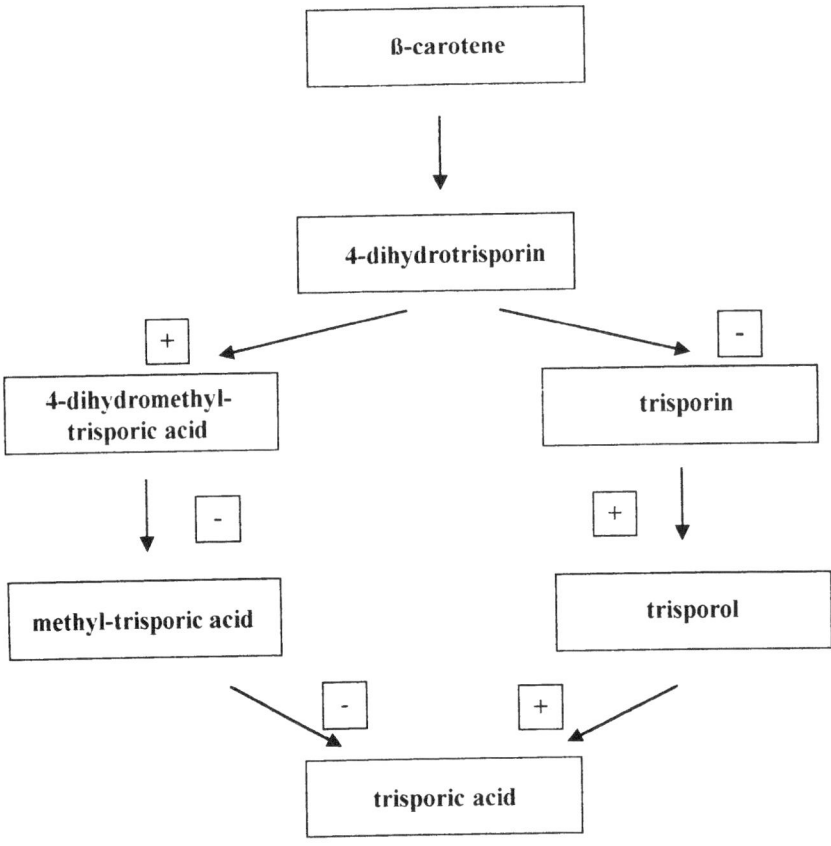

Fig. 10.5 Cooperative biosynthesis of the sex pheromone, trisporic acid, in *Mucor*-like fungi (van den Ende, 1978).

for rewarding studies. We have a highly efficient system where we can study the mechanisms of horizontal gene transfer in laboratory experiments. We do not know if this gene transfer has contributed to the evolutionary biology of mucoraceous fungi. In order to find out we will have to compare sequences of conserved genes within the host range of *P. parasitica* with those from without.

REFERENCES

Arnau, J., Jepsen, L.P. and Stroman, P. (1991) Integrative transformation by homologous recombination in the zygomycete *Mucor circinelloides*. *Mol. Gen. Genet.* **225**: 193–198.

Burgeff, H. (1924) Untersuchungen über Sexualität und Parasitismus bei Mucorineen I. *Botanische Abhandlungen* **4**: 1–135.

Burmester, A. (1992) Transformation of the mycoparasite Parasitella simplex to neomycin resistance. *Curr. Genet.* **21**: 121–124.

Burmester, A., Wöstemeyer, A. and Wöstemeyer, J. (1990) Integrative transformation of a zygomycete, *Absidia glauca*, with vectors containing repetitive DNA. *Curr. Genet.* **17**: 155–161.

Czempinski, K., Kruft, V., Wöstemeyer, J. and Burmester, A. (1996) Purification of 4-hydromethyl-trisporate dehydrogenase from *Mucor mucedo*, an enzyme of the sexual pheromone pathway, and cloning of the corresponding gene. *Microbiology* **141**: 2647–2654.

Hänfler, J., Teepe, H., Weigel, C. *et al.* (1992) Circular extrachromosomal DNA codes for a surface protein in the (+) mating type of the zygomycete *Absidia glauca. Curr. Genet.* **22**: 319–325.

Jeffries, P. and Young, T.W.K. (1994) *Interfungal Parasitic Relationships*, CAB International, Wallingford, pp. 143–146.

Kellner, M., Burmester, A., Wöstemeyer, A. and Wöstemeyer, J. (1993) Transfer of genetic information from the mycoparasite *Parasitella parasitica* to its host *Absidia glauca. Curr. Genet.* **23**: 334–337.

Van den Ende, H. (1978) Sexual morphogenesis in the phycomycetes, in *The Filamentous Fungi*, Vol. III, (eds J.E. Smith and D.R. Berry), Edward Arnold, London, pp. 257–274.

Werkman, B. (1976) Localization and partial characterization of a sex-specific enzyme in homothallic and heterothallic mucorales. *Arch. Microbiol.* **109**: 209–213.

PART TWO

Bacterial Gene Transfer
in Nature

Selective DNA uptake and DNA restriction as barriers to horizontal gene exchange by natural genetic transformation in *Pseudomonas stutzeri* JM300

11

Michael G. Lorenz, Birte Meyer, Marcus Wittstock, Stefan Graupner and Wilfried Wackernagel

SUMMARY

Mechanisms which could impede interspecific gene transfer by natural genetic transformation were investigated in *Pseudomonas stutzeri* JM300, a soil-inhabiting non-fluorescent pseudomonad. In particular, DNA uptake specificity and the effect of DNA restriction during transformation were studied. In DNA competition experiments, excess of heterologous DNA did not decrease transformation by homologous DNA whereas non-transforming homologous DNA did. This suggests discrimination of foreign DNA during uptake. In experiments on natural transformation by plasmids it was found that the insertion of a *P. stutzeri* chromosomal fragment into a *P. stutzeri–Escherichia coli* shuttle plasmid greatly increased the transformation frequency and made it a one-hit process. By using such plasmids isolated from *P. stutzeri* (modified) or from *E. coli* (unmodified) it was demonstrated that DNA restriction reduced the frequency of natural transformation by more than 10-fold. The findings suggest that selective DNA uptake together with DNA restriction make interspecific gene transfer in *P. stutzeri* JM300 a rare event. As a consequence, one would expect the species to evolve freely without the compensating effect of interspecies transformation on neutral sequence divergence.

11.1   INTRODUCTION

There is increasing evidence for horizontal transfer of chromosomal genes among bacteria in the environment. For instance, the human pathogens *Neisseria gonorrhoeae*, *N. meningitidis* and *Helicobacter pylori* and the soil organisms *Rhizobium meliloti* and *Bacillus subtilis* have been shown to have a non-clonal population structure (Istock *et al.*, 1992; Maynard Smith *et al.*, 1993; O'Rourke and Spratt, 1994; Go *et al.*, 1996). Frequent recombination promotes shuffling of alleles and thus generates genetic diversity within populations. Moreover, the recruitment of genes or parts of genes from other species can result in mosaic structures of the genes, which has been documented in several instances (Spratt *et al.*, 1989; Frosch and Meyer, 1992; Bowler *et al.*, 1994; Feil *et al.*, 1995; Rokbi *et al.*, 1995). Many of the organisms studied for their population structures and molecular organization of genes are naturally transformable, i.e. they can develop the physiological state of competence for active uptake of DNA and its integration into the recipient's chromosome (for review see Lorenz and Wackernagel, 1994). This, together with laboratory simulations of gene transfer in the environment (Chapter 13), suggests that transformation is an important mechanism of exchange of chromosomal genes within natural populations. Transformation between different species is generally lower than transformation within a species, for instance between four-fold and 10^5-fold in *Haemophilus* (Albritton *et al.*, 1984).

In principle, there may be three main mechanisms which would lead to sexual isolation in the sense of reduced interspecific (compared with intraspecific) transformation. First, recognition of a specific nucleotide sequence on the donor DNA, as has been observed in *Haemophilus influenzae* and *N. gonorrhoeae* (Sisco and Smith, 1979; Goodman and Scocca, 1988), results in the exclusion of heterologous DNA lacking the sequence from uptake into the cell. Second, differences in DNA restriction/modification specificities between donor and recipient may decrease transformation frequencies. Last, sequence divergence may hamper donor strand integration into the chromosome during homologous recombination. In the present study, *P. stutzeri* JM300, a transformable non-fluorescent soil pseudomonad, was characterized with respect to mechanisms of sexual isolation. The data suggest that specific DNA uptake and DNA restriction are barriers to interspecific transformation in this species.

11.2   RESULTS AND DISCUSSION

11.2.1   DISCRIMINATION OF HETEROLOGOUS DNA

DNA competition experiments were performed using transforming chromosomal DNA (*his*+) and competing chromosomal non-transforming homologous (*his*) or heterologous (calf thymus) DNA. The results (Fig. 11.1) showed that transformation of *P. stutzeri* JM300 was unaffected by an up to 20-fold mass excess of heterologous DNA, whereas transformation frequencies were reduced proportional to the mass excess by non-transforming homologous DNA. Apparently *P. stutzeri* JM300 is able to recognize its own DNA and to exclude heterologous DNA from uptake. This uptake of DNA from its own but not from other species corresponds to what has been observed in two human pathogens, *H. influenzae* and *N. gonorrhoeae* (Mathis and Scocca, 1982). In this respect, *P. stutzeri* JM300 differs from other free-living transformable bacteria such as *Acinetobacter calcoaceticus* or *Bacillus subtilis* (Lorenz and Wackernagel, 1994) which appear not to discriminate against heterologous DNA during uptake. Whether *P.stutzeri* recognizes a specific nucleotide sequence for selective DNA uptake, like *H.*

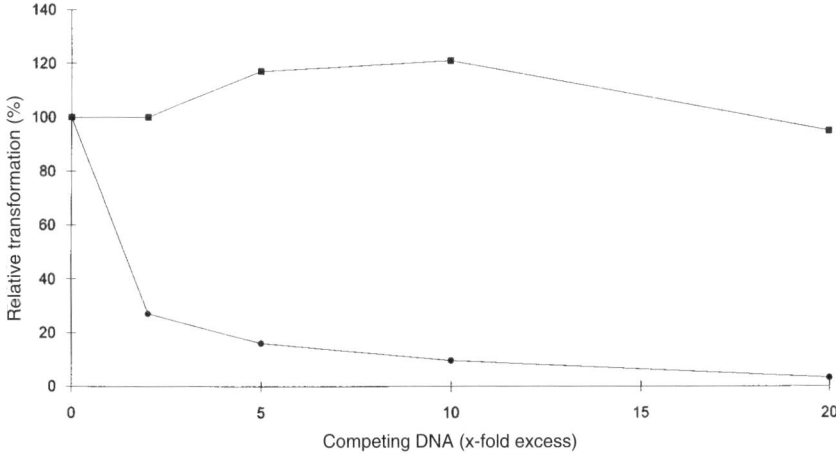

Fig. 11.1 Specific DNA uptake in *P. stutzeri* JM300 transformation. A competent culture of the *his*-1 mutant JM302 was transformed at a saturating concentration of wild-type DNA (1 µg/ml). Relative transformation is the transformation frequency obtained in the presence of competing DNA, either non-transforming homogamous (*his*-1) DNA (●) or heterologous (calf thymus) DNA (■), at the indicated mass excess divided by the transformation frequency obtained without competing DNA. Relative transformation of 1 corresponds to $1.2 \pm 0.2 \times 10^{-6}$ transformation frequency.

influenzae and *N. gonorrhoeae* (Sisco and Smith, 1979; Goodman and Scocca, 1988), is not yet known.

Despite the differences in eukaryotic and prokaryotic sex, DNA discrimination during uptake is reminiscent of prezygotic isolation mechanisms in eukaryotes. These mechanisms are efficient barriers to interspecific gene flow. Accordingly, DNA discrimination during DNA uptake may contribute to sexual isolation of *P. stutzeri* populations from other diverging populations.

11.2.2   DNA RESTRICTION

The experimental strategy for studying DNA restriction during natural transformation was to use a shuttle plasmid that can replicate in *P. stutzeri* JM300 and a restriction/modification deficient *Escherichia coli*, so that the plasmid DNA could be properly modified or non-modified without the need for a modification/restriction-deficient mutant of *P. stutzeri*. The construction of such a plasmid and the characterization of plasmid transformation in JM300 are described first before the effects of DNA restriction on natural transformation are discussed.

(a) Characterization of natural plasmid transformation of *P. stutzeri* JM300

A plasmid (pWI85, 16.4 kb) was constructed that consists of pKT210 as the vector (an RSF1010 derivative; Bagdasarian *et al.*, 1981) and an 8.5 kb insert of JM300 chromosomal DNA. This plasmid complements the *leu* auxotrophy of JM306. The characteristics of transformation by pWI85 were compared with the characteristics of transformation by the vector without insert and by chromosomal DNA (Fig. 11.2).

With pKT210, the tangent of the dose–response inclination angle was approximately 2, indicating a two-hit mechanism: two pKT210 molecules are required for transformation of a cell. Apparently, the mechanism of plasmid transformation in *P. stutzeri* is similar to the mechanism of other competent bacteria such as *B. subtilis*, *Streptococcus pneumoniae*, *H. influenzae* or *Acinetobacter calcoaceticus* (see Lorenz and Wackernagel, 1994). According to a proposed model (Canosi *et al.*, 1981), hybridization of two single strands of opposite polarity, originating from different plasmid molecules, and subsequent repair synthesis give a double-stranded molecule (see Fig. 11.4B).

With pWI85, the dependence of the transformation frequency on the amount of DNA followed a one-hit reaction (for both Leu$^+$ and Sm^r selection) and thus resembled transformations with chromosomal DNA (Fig. 11.2). Further, with saturating DNA amounts the transformation frequencies for pWI85 were 16 times (Sm^r selection) and 1635 times

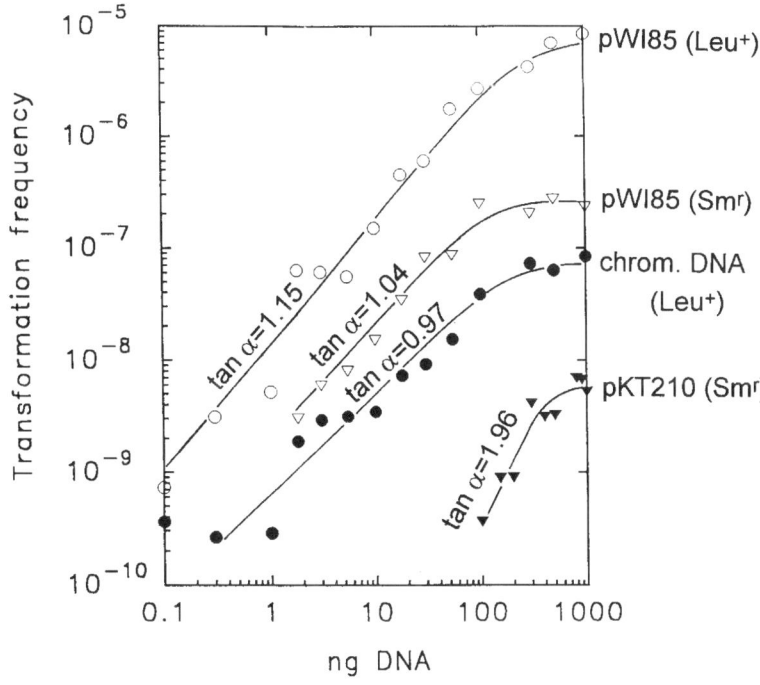

Fig. 11.2 Natural transformation of *P. stutzeri* JM300 by various amounts of chromosomal (Leu⁺ selection) and plasmid DNA. The plasmids used were pKT210 (Smʳ selection) and pWI85 carrying an insert of JM300 chromosomal DNA (Smʳ and Leu⁺ selection, respectively).

(Leu⁺ selection) higher than transformation frequencies with the vector, pKT210 (Fig. 11.2). It is concluded that the chromosomal insert increases the efficiency of plasmid transformation and makes it a one-hit process. Other results suggested that, after uptake, pWI85 interacts with the recipient's chromosome via the homologous region leading to enhanced transformation frequencies (data not shown). When co-transformation of a non-selected with a selected marker was examined and the transformants were tested for the presence of pWI85, it was found that at Leu⁺ selection the Smʳ marker was mostly lost and that Leu⁺ Smˢ cells did not contain pWI85 (Table 11.1). The results suggest that, following uptake, the Leu⁺ marker is mostly integrated into the chromosome by a replacement process. This is supported by the finding that all Leu⁺ transformants of the *recA* mutant APS121 had the non-selected Smʳ marker and contained a plasmid identical with pWI85 (Table 11.1). It is concluded that allele exchange at the chromosomal locus is the main route of transformation by plasmids with homology to the genome.

At Sm^r selection, nearly half of the JM306 transformants and most of the APS121 transformants were also Leu$^+$ and had pWI85. This together with the one-hit DNA concentration dependence of pWI85 transformation at Sm^r selection (Fig. 11.2) suggests a mode of homology-facilitated plasmid transformation similar to that of *B. subtilis*, *S. pneumoniae* and *H. influenzae* (Lopez *et al.*, 1982; Stuy and Walter, 1986; Canosi *et al.*, 1981). According to a model, which was detailed by Lopez *et al.* (1982), double-strand breaks are introduced into the plasmid upon its binding to the cell surface followed by uptake of a single strand. If the double-strand break occurs within the homology insert this would allow hybridization of both ends of the entering single-stranded molecule to the homologous region in the chromosome (see Fig. 11.4A). Gap closure would reconstitute a circular single-stranded molecule and subsequent synthesis of the complementary plasmid strand would produce a double-stranded circular replicon. This type of plasmid reconstitution appears to be the minor route of homology-facilitated plasmid transformation: the transformation frequency of the Sm^r marker, which stands for plasmid-containing cells (Table 11.1, column 6), was approximately 40 times lower than the transformation frequency of the Leu$^+$ marker (Fig. 11.2; saturating DNA amounts). The model of Lopez *et al.* (1982) considers occasional gene conversion events during plasmid reconstitution, which, by the finding of one plasmid-carrying Sm^r Leu$^-$ transformant, seems to hold true also for JM300 (Table 11.1, last column).

(b)   DNA restriction activity in *P. stutzeri* JM300

Plasmid pWI85 was isolated either from *P. stutzeri* (pWI85$_{Ps}$) or from the restriction-modification-deficient (r$^-$m$^-$) *E. coli* mutant AB1157 (pWI85$_{Ec}$) and used to electroporate *P. stutzeri* JM306 (*leu*). In this way, the plasmid is transferred in double-stranded form into the cells. Figure 11.3 shows that electrotransformation frequencies obtained with pWI85$_{Ec}$ were about 400 times lower than those with pWI85$_{Ps}$, irrespective of whether selection was for the Leu$^+$ or for the Sm^r marker. A similarly lower transformation with pWI85$_{Ec}$ DNA was observed in a *recA* mutant (Table 11.2, column 3). The results indicate a DNA restriction activity in *P. stutzeri* JM300.

(c)   DNA restriction during natural transformation

P. stutzeri JM300 was naturally transformed with pWI85$_{Ec}$ or pWI85$_{Ps}$. To find an effect, DNA restriction activity on the two routes of homology-facilitated plasmid transformation, i.e. allele exchange (main route) and plasmid reconstitution (minor route), transformations were performed using both the *recA$^+$* strain and the *recA* mutant (Table 11.2).

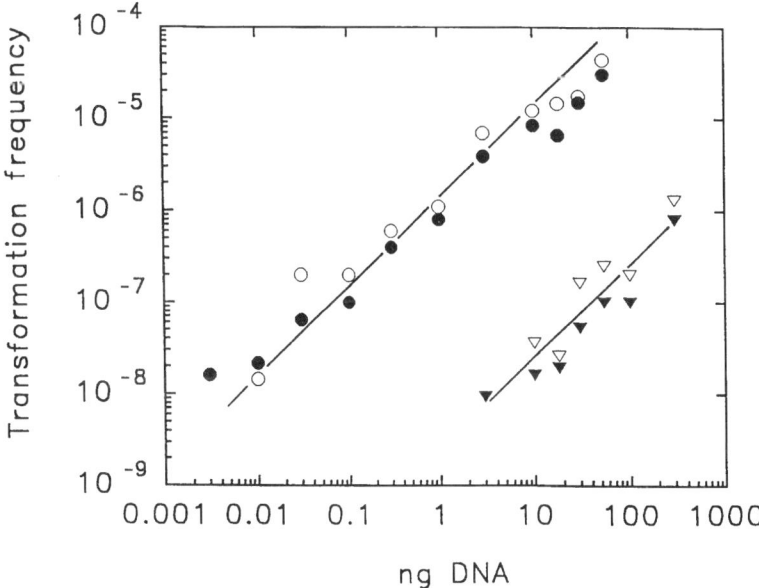

Fig. 11.3 DNA restriction in *P. stutzeri* JM300. Cells were electroporated with pWI85 isolated from either JM300 (○, ●) or *E. coli* AB1157 (r⁻m⁻; △, ▲). Open symbols indicate Leu⁺; filled symbols, Smʳ selection.

In the *recA⁺* strain, a 36 times lower transformation frequency by the chromosomal Leu⁺ marker with pWI85$_{Ec}$ than with pWI85$_{Ps}$ was noticed (Table 11.2, column 3). This indicates that DNA restriction acts on the main route of plasmid transformation. Possibly, during integration of the unmodified donor strand, repair processes led to regions in the chromosome with both strands unmodified. In these regions, DNA restriction would lead to a lethal double-strand break and thereby would reduce transformation frequency. The minor route of plasmid transformation appears also to be affected by DNA restriction. This can be concluded from the 3 (Leu⁺ selection) to 82 times (Smʳ selection) lower co-transformation frequencies obtained with pWI85$_{Ec}$ than with pWI85$_{Ps}$ (compare Table 11.2, columns 5 and 6, with Table 11.1, columns 2 and 3). Further, in the *recA* mutant, in which natural transformation occurs only by formation of plasmid replicons as indicated by the high co-transformation frequency of the Smʳ marker with Leu⁺, the transformation frequency was 130 times lower (Leu⁺ selection) with the unmodified than with the modified plasmid (Table 11.2, column 3). The model shown in Fig. 11.4 explains DNA restriction during natural transformation by plasmids on the basis that duplex unmodified regions are formed

Table 11.1 Characterization of *P. stutzeri* JM300 transformants obtained after transformation by plasmid pWI85

Competent strain	Co-transformation[a]		pWI85 present in transformants (total tested)[b]			
	% Sm[r]	%Leu[+]	Leu[+] Sm[r]	Leu[+] Sm[s]	Sm[r] Leu[+]	Sm[r] Leu[−]
JM306 (*leu, recA*[+])	12.7	49.3[d]	18 (18)	0 (18)	18 (18)	1 (18)
APS121 (*leu, recA*)[c]	100.0	78.0[d]	8 (8)	n.a.	5 (5)	0 (6)

[a] Selection was for Leu[+] and Sm[r], respectively; plate transformation was done according to Lorenz and Wackernagel (1991); transformants from experiments with various saturating and non-saturating DNA amounts (see Fig. 11.2) were tested for the non-selected marker by streaking colonies (50 to 245) on LB medium with streptomycin and on minimal succinate medium, respectively.

[b] First marker indicates selected marker.

[c] The strain is a *recA*::Tn5 insertion mutant of JM306 (Vosman and Hellingwerf, 1991).

[d] Most Sm[r] Leu[−] clones did not have the plasmid (see last column), presumably representing spontaneous Sm[r] mutants.

n.a. = not applicable.

Table 11.2 Effect of DNA restriction on transformation of *P. stutzeri*

Competent strain	Mode of transformation[a]	Relative transformation[b]		Co-transformation[c]		pW185 present in transformants (total tested)[d]			
		Leu⁺	Sm^r	%Sm^r	%Leu⁺	Leu⁺ Sm^r	Leu⁺ Sm^s	Sm^r Leu⁺	Sm^r Leu⁻
JM306 (*recA⁺*)	Natural	36	9	4.2	0.6	7 (7)	0 (37)	1 (1)	0 (22)
	Artificial	388	376	100.0	80.0	4 (4)	n.a.	4 (4)	0 (2)
APS121 (*recA*)	Natural	130	12	78.0	0[e]	8 (8)	0 (5)	n.a.	0 (5)
	Artificial	140	146	100.0	90.0	4 (4)	n.a.	8 (8)	0 (2)

[a] Natural transformation was done as in Table 11.1; artificial transformation was performed by electroporation according to Pemberton and Penfold, 1992.

[b] Transformation frequency with pWl85 isolated from *P. stutzeri* relative to the transformation frequency with pWl85 isolated from *E. coli*; arithmetic average of values in the range 3 to 54 ng (electroporation) or 1 to 100 ng plasmid DNA (natural transformation).

[c] Selection was for Leu⁺ and Sm^r, respectively; values are from transformations with pWl85 isolated from *E. coli*; colonies (50 to 245) from transformations with saturating and non-saturating DNA amounts were streaked to test for the non-selected plasmid marker.

[d] Clone analysis of transformations with pVl85 from *E. coli*; first marker indicates selected marker.

[e] The absence of Sm^r Leu⁺ transformants is not understood; perhaps the period for expression of the Sm^r marker was not sufficient to give rise to transformants.

n.a. = not applicable

at different steps and that the location of the restriction site is within the homology insert. The steps are:

- during the complementary DNA strand synthesis after homology-facilitated circle closure and DNA synthesis (Fig. 11.4A);
- during hybridization of complementary strands in the two-hit process of transformation by plasmids without homology to the chromosome (Fig. 11.4B).

A double-strand break introduced by restriction activity would abort natural plasmid transformation. Another model takes into consideration that DNA restriction enzymes cleave the entering single-stranded donor DNA, as observed in transfections with the *B. subtilis* phage SSP1 (discussed in Ganesan, 1982). This activity of restriction enzymes would affect both the main and the minor route of homology-facilitated plasmid transformation. DNA restriction would lead to fragmentation of entering linear single-stranded DNA, which would prevent homology-facilitated reconstitution of a replicon whereas integration of the chromosomal fragment of the plasmid into the genome would still result in transformants. The approximately 10 times lower sensitivity of natural transformation against DNA restriction than of artificial transformation in the *recA*+ strain (Table 11.2, column 3) may support this hypothesis. In any case, the results indicate that DNA restriction impedes natural transformation of *P. stutzeri* JM300 by DNA lacking JM300-specific modification.

11.3   CONCLUSIONS

The results shown here suggest selective DNA uptake and DNA restriction to be effective barriers to interspecific gene transfer by natural transformation in *P. stutzeri* JM300. In other experiments employing rifampicin resistance as a marker (Lorenz, unpublished) transformation of JM300 by DNA from donor strains that were less than 3% divergent from JM300 (determined by 16S rRNA gene nucleotide sequence comparisons) was more than 1000 times lower than transformation of *P. stutzeri* JM300 by its own DNA. For comparison, transformation between different *Bacillus* species, which were as much as 5% divergent, was only three times lower than intraspecies transformation (Roberts and Cohan, 1993). In *Bacillus*, which does not select against heterologous DNA during uptake, the effect of DNA restriction is rather mild and neutral sequence divergence appears to be the main cause for sexual isolation by transformation (Cohan *et al.*, 1991; Zawadzki *et al.*, 1995). Interspecific transformation in *P. stutzeri* JM300 is therefore far more bounded than in *Bacillus*. Following a theory of the consequences of neutral sequence divergence among species with respect to gene exchange (Cohan, 1995), *P. stutzeri* JM300 with its two pre-recombination barriers (DNA discrimi-

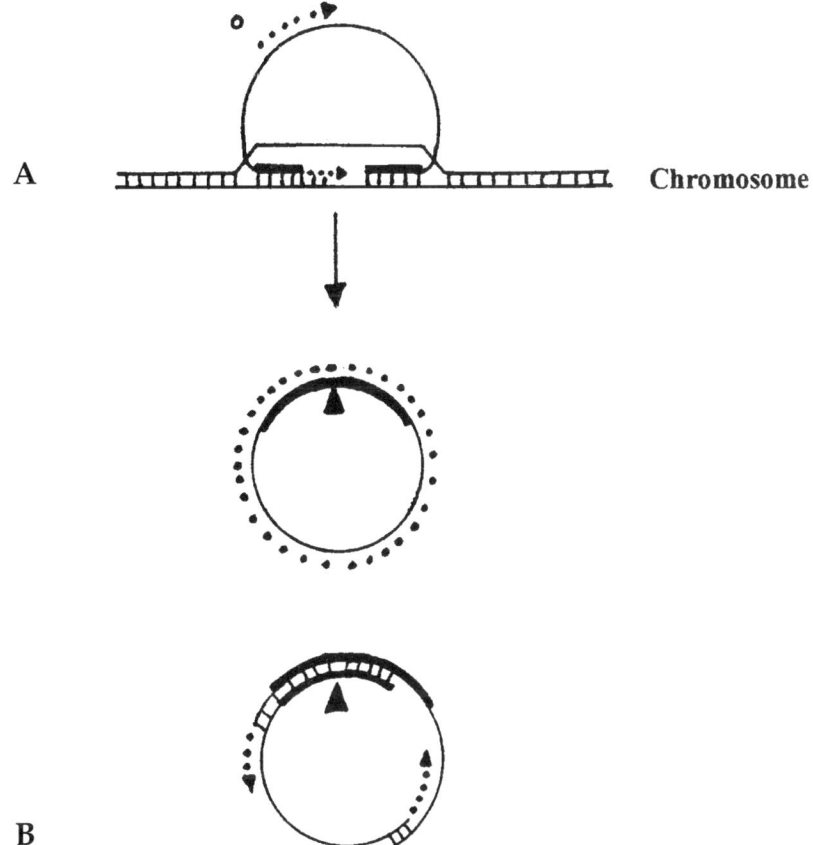

A   **Chromosome**

B

Fig. 11.4 Model of plasmid transformation and action of restriction in *P. stutzeri* JM300. (A) Reconstitution during homology-facilitated plasmid transformation. (B) Formation of a plasmid replicon from two linear single-stranded molecules with partially overlapping complementary stretches. For further details refer to the text. Dots and small arrows indicate DNA synthesis (o, plasmid origin of replication). Big arrows indicate DNA restriction at unmodified recognition sites (assumed to be located in the homology insert; bold-drawn areas).

nation, DNA restriction) would be a species which is subject to rare interspecific transformation and consequently, as far as transformation is the main gene transfer mechanism in this species, to free divergence.

ACKNOWLEDGMENTS

The work was supported by the Bundesminister für Bildung und Forschung and by the Fonds der Chemischen Industrie.

REFERENCES

Albritton, W.L., Setlow, J.K., Thomas, M. *et al.* (1984) Heterospecific transformation in the genus *Haemophilus*. *Mol. Gen. Genet.* **193**: 358–363.

Bagdasarian, M.R., Lurz, B., Rückert, F.C.H. *et al.* (1981) Specific-purpose plasmid cloning vectors. II. Broad host range, high copy number, RSF1010-derived vectors, and a host–vector system for gene cloning in *Pseudomonas. Gene* **16**: 237–247.

Bowler, L.D., Zhang, Q.-Y., Riou, J.-Y. and Spratt, B.G. (1994) Interspecies recombination between the *penA* genes of *Neisseria meningitidis* and commensal *Neisseria* species during the emergence of penicillin resistance in *N. meningitidis*: natural events and laboratory simulation. *J. Bacteriol.* **176**: 333–337.

Canosi, U., Iglesias, A. and Trautner, T.A. (1981) Plasmid transformation in *Bacillus subtilis*: effects of insertion of *Bacillus subtilis* DNA into plasmid pC194. *Mol. Gen. Genet.* **181**: 434–440.

Cohan, F.M. (1995) Does recombination constrain neutral divergence among bacterial taxa? *Evolution* **149**: 164–175.

Cohan, F.M., Roberts, M.S. and King, E.C. (1991) The potential for genetic exchange by transformation within a natural population of *Bacillus subtilis*. *Evolution* **45**: 1383–1421.

Feil, E., Carpenter, G. and Spratt, B.G. (1995) Electrophoretic variation in adenylate kinase of *Neisseria meningitidis* is due to inter- and intraspecies recombination. *Proc. Natl Acad. Sci. USA* **92**: 10535–10539.

Frosch, M. and Meyer, T.F. (1992) Transformation-mediated exchange of virulence determinants by co-cultivation of pathogenic neisseriae. *FEMS Microbiol. Lett.* **100**: 345-350.

Ganesan, A.T. (1982) Uptake, restriction, modification and recombination of DNA molecules during transformation in *B. subtilis*, in *Molecular Cloning and Gene Regulation in Bacilli*, (eds A.T.C. Ganesan and J.A. Hoch), Academic Press, New York, pp. 261–268.

Go, M.F., Kapur, V., Graham, D.Y. and Musser, J.M. (1996) Population genetic analysis of *Helicobacter pylori* by multilocus enzyme electrophoresis: extensive allelic diversity and recombinational population structure. *J. Bacteriol.* **178**: 3934–3938.

Goodman, S.D. and Scocca, J.J. (1988) Identification and arrangement of the DNA sequence recognized in specific transformation of *Neisseria gonorrhoeae. Proc. Natl Acad. Sci. USA* **85**: 6982–6986.

Istock, C.A., Duncan, K.E., Ferguson, N. and Zhou, X. (1992) Sexuality in a natural population of bacteria – *Bacillus subtilis* challenges the clonal paradigm. *Mol. Ecol.* **1**: 95–103.

Lopez, P., Espinosa, M., Stassi, D.L. and Lacks, S. (1982) Facilitation of plasmid transfer in *Streptococcus pneumoniae* by chromosomal homology. *J. Bacteriol.* **150**: 692–701.

Lorenz, M.G. and Wackernagel, W. (1991) High frequency of natural genetic transformation of *Pseudomonas stutzeri* in soil extract supplemented with a carbon/energy and phosphorus source. *Appl. Environ. Microbiol.* **57**: 1246–1251.

Lorenz, M.G. and Wackernagel, W. (1994) Bacterial gene transfer by natural genetic transformation in the environment. *Microbiol. Rev.* **58**: 563–602.

Mathis, L.S. and Scocca, J.J. (1982) *Haemophilus influenzae* and *Neisseria gonorrhoeae* recognize different specificity determinants in the DNA uptake step of genetic transformation. *J. Gen. Microbiol.* **128**: 1159–1161.

Maynard Smith, J., Smith, N.H., O'Rourke, M. and Spratt, B.G. (1993) How clonal are bacteria? *Proc. Natl Acad. Sci. USA* **90**: 4384–4388.

O'Rourke, M. and Spratt, B.G. (1994) Further evidence for the non-clonal population structure of *Neisseria gonorrhoeae*: extensive genetic diversity within isolates of the same electrophoretic type. *J. Gen. Microbiol.* **140**: 1285–1290.

Pemberton, J.M. and Penfold, R.J. (1992) High-frequency electroporation and maintenance of pUC- and pBR-based cloning vectors in *Pseudomonas stutzeri*. *Curr. Microbiol.* **25**: 25–29.

Roberts, M.S. and Cohan, F.M. (1993) The effect of DNA sequence divergence on sexual isolation in *Bacillus*. *Genetics* **134**: 401–408.

Rokbi, B., Maitre-Wilmotte, G., Mazarin, V. *et al.* (1995) Variable sequences in a mosaic-like domain of meningococcal *tbp2* encode immunoreactive epitopes. *FEMS Microbiol. Lett.* **132**: 277–283.

Sisco, K.L. and Smith, H.O. (1979) Sequence-specific DNA uptake in *Haemophilus* transformation. *Proc. Natl Acad. Sci. USA* **76**: 972–976.

Spratt, B.G., Zhang, Q.-Y., Jones, D.M. *et al.* (1989) Recruitment of a penicillin-binding protein gene from *Neisseria flavescens* during the emergence of penicillin resistance in *Neisseria meningitidis*. *Proc. Natl Acad. Sci. USA* **86**: 8988–8992.

Stuy, J.H. and Walter, R.B. (1986) Homology-facilitated plasmid transfer in *Haemophilus influenzae*. *Mol. Gen. Genet.* **203**: 288–295.

Vosman, B. and Hellingwerf, K.J. (1991) Molecular cloning and functional characterization of a *recA* analog from *Pseudomonas stutzeri* and construction of a *P. stutzeri recA* mutant. *Antonie van Leeuwenhoek J. Microbiol.* **59**: 115–123.

Zawadzki, P., Roberts, M.S. and Cohan, F.M. (1995) The log-linear relationship between sexual isolation and sequence divergence in *Bacillus* transformation is robust. *Genetics* **140**: 917–932.

Transformation in aquatic environments

12

Martin Day

SUMMARY

Transformation, one of three mechanisms that promote gene exchange in bacterial populations, is performed by many Gram-positive and Gram-negative species. It is an active, highly evolved and dedicated process, governed by chromosomal genes. 'Free' DNA, released from a living or a lysed 'dead' cell, is taken up by a competent recipient cell. The environment is awash with eukaryotic and prokaryotic DNA; thus mechanisms have evolved to identify the DNA as 'self' or 'non-self'. In some species this 'free' DNA has an internal sequence that allows it to be selectively identified and taken up. Most species examined have a restriction system that identifies 'non-self' species transforming DNA. The state and period of competence is governed, in most species, by the physiological status of the individual that may in turn be regulated by that of their population. This state is in turn regulated (induced or repressed) by a range of environmental parameters, such as nutrient status, ionic composition, temperature, pH, etc. The few studies performed in microcosms and *in situ* have resolved some of the practical difficulties and demonstrated gene exchange by transformation to be a measurable event. Thus, as for other transfer systems, the potential for contributing to evolution is clear; however, examples have yet to be identified.

12.1 INTRODUCTION

Phenotypic changes occur in bacteria due to the impact of the environment on the expression of genes within the genome. New phenotypes may arise through mutation or through some form of parasexual process providing homologous DNA for recombination. In the former case some alteration, via an insertion, duplication, deletion or transposition event, is made to the DNA sequence, resulting in a change to the cellular phenotype. In the latter case the DNA sequence is again changed but the change is confined to the sequence transferred and exchanged. For the sequence to become integrated it will normally have homology with the recipient genome and thus will insert by homologous recombination mediated by the recA protein. If the sequence has a replication origin (for example, it is a plasmid) and no homology with other gene sequences within the cell, then to become established it must be able to recombine internally. In this instance there are two scenarios for successful establishment. If the sequence has an internal direct repeat this provides the homology needed for recombination to allow for its construction as a replicon, but the recombinational process deletes some of the genetic information carried. Alternatively two plasmid molecules, transferred coincidentally and linearized at different sites, may then recombine to yield the original plasmid. Thus gene transfer processes are potentially mutagenic, as the genome may be altered from the wild-type as a result of the recombinational process.

12.2   GENE TRANSFER PROCESSES

To acquire an understanding of gene exchange in microbial systems it is important to view DNA transfer as potentially a combination of processes. Briefly there are three mechanisms involved. Conjugation requires the donor to have a plasmid or a transposon, which carries the genes to mobilize the donor DNA. It provides a mechanism that recognizes a recipient, establishes close physical contact and transfers DNA into it. To achieve success both cells must be physiologically active (Ippen-Ihler, 1989). Transduction requires the donor cell to release phage carrying host genomic DNA and this requires the donor to be active during the synthesis of the phage (Kokjohn, 1989). The recipient is closely related to the donor cell, but can be both temporally and spatially separated from the donor. Phage particles are inert and generally show some level of resistance to adverse environmental conditions (Stotzsky, 1989). Gene transfer by transformation may be achieved from free DNA or that released from a living or a 'lysed' dead cell. Persistence of DNA in the environment and its biological availability will determine to what degree it participates or retains a role in gene exchange (Leff et al., 1992). As a general rule a bacterium must be in a physiologically active state in order to be competent. This property depends on a set of chromosomally encoded genes (Hahn et al., 1987; Dubnau, 1991) which are coordinately expressed and physiologically regulated. This implies that the transfer process responds (like an enzyme system) to inducers and repressors, which enables cells to actively regulate their level of participation in transformation. Bacteria appear to be the only group of organisms capable of undergoing transformation naturally (Lorenz and Wakernagel, 1994). Although the process is distinct from conjugation and transduction, the genetic consequences are identical: the genome may be altered by the recombinational integration of a novel sequence.

12.3   TRANSFORMATION

Transformation is an active and dedicated process of gene exchange, governed by chromosomal genes, and allows the uptake of exogenous 'free' DNA selected by a competent cell. As far as this discussion is concerned, transformation is defined by those cells that carry out the process naturally and do not require laboratory pre-treatments to make them competent to participate in the exchange. Transformation, due to the presence of an extracellular DNA step, may be discriminated from conjugation and transduction because of its sensitivity to DNAase (Albritton *et al.*, 1982). The process was first described in 1928 (Griffith, 1928) and in 1944 its use became transiently important to the determination of the role of DNA in genetics (Avery *et al.*, 1944). It is not a mechanism of gene exchange present in all bacteria, but is apparently widespread amongst individual strains of many species. As a result transformation has not been generally envisaged as having a significant contribution to the genetics of microbial populations. A comparison of the species in Table 12.1 with their taxonomic positions shows an absence of transformation-proficient species in the *Crenarchaeota* (*Euryarchaeota*) and of the following groups in the division *Bacteria*: thermogales, fusobacterium, fibrobacter, spirochaetes and planctomyces/chlamydia. Within other groups and subgroupings the occurrence of transformable species is patchy. This reflects, in part, a lack of work in this area to establish the genetic capacities of many of these more exotic bacterial types. There are several stages to the transformation process, described in the following sections.

12.3.1   COMPETENCE

In most transformable bacteria the state of competence is transient (Fig. 12.1), only in *Neisseria gonorrhoeae* is it clearly constitutive (Sparling, 1966). In *Haemophilus influenzae* competence develops when the cells cease to divide (Smith *et al.*, 1981). In *Acinetobacter calcoaceticus* (Palmen *et al.*, 1992), *Azotobacter vinelandii* (Page and Sadoff, 1976) and *Anacystis nidulans* (Chauvat *et al.*, 1983) it occurs early to late log phase, and in the transition from log to stationary phase in *Bacillus subtilis* (Smith *et al.*, 1981), *Chlorobium limicola* (Ormerod, 1988) and *Pseudomonas stutzeri* (Carlson *et al.*, 1983). In *Agmenellum quadruplicatum*, *Deinococcus radiodurans* and

Table 12.1 Naturally transformable prokaryotic species (modified from Lorenz and Wackernagel, 1994)

Species isolated from terrestrial or aquatic habitats	Tf	Woess grouping
Protolithotrophic		
Agmenellum quadruplicatum	4.3×10^{-4}	
Anacystis nidulans	8.0×10^{-4}	
Chlorobium limicola	1.0×10^{-5}	CFB group
Nostoc muscorum	1.2×10^{-3}	cyanobacteria
Synechocystis sp. strain 6803	5.0×10^{-4}	cyanobacteria
Synechocystis sp. strain OL50	2.0×10^{-4}	cyanobacteria
Chemolithotrophic		
Thiobacillus thioparus	$10^{-3} - 10^{-2}$	
Thiobacillus sp. strain Y	1.7×10^{-3}	
Heterotrophic		
Achromobacter spp.	+	
Acinetobacter calcoaceticus	7.0×10^{-3}	γ; purple
Azotobacter vinelandii	9.5×10^{-2}	
Bacillus subtilis	3.5×10^{-2}	LGC
Bacillus licheniformis	1.2×10^{-2}	LGC
Deinococcus radiodurans	2.1×10^{-2}	
Lactobacillus lactis	2.3×10^{-5}	LGC
Mycobacterium smegmatis	$10^{-7} - 10^{-6}$	HGC
Pseudomonas stutzeri	7.0×10^{-5}	
Rhizobium meliloti	7.0×10^{-4}	
Streptomyces	+	HGC
Thermoactinomyces vulgaris	2.7×10^{-3}	
Thermus thermophilus	1.0×10^{-2}	
Thermus flavus	8.8×10^{-3}	
Thermus caldophilus	2.7×10^{-3}	
Thermus acquaticus	6.4×10^{-4}	
Vibrio sp. strain D19	2.0×10^{-7}	γ; purple
Vibrio sp. strain WJT-1C[d]	2.5×10^{-4}	γ; purple
Vibrio parahaemolyticus	1.9×10^{-9}	γ; purple
Methylotrophic		
Methylobacterium organophilum	5.3×10^{-3}	α; purple
Archaebacteria		
Methanobacterium thermoautotrophicum	+	Euryarchaeota
Methanococcus voltae	8.0×10^{-6}	Euryarchaeota
Clinical isolates of pathogenic species		
Campylobacter jejuni	2.0×10^{-4}	ε; purple
Campylobacter coli	1.2×10^{-3}	ε; purple
Haemophilus influenzae	7.0×10^{-3}	α; purple
Haemophilus parainfluenzae	8.6×10^{-3}	α; purple
Helicobacter pylori	5.0×10^{-4}	δ & ε; purple
Moraxella spp.	+	
Neisseria gonorrhoeae	1.0×10^{-4}	β; purple
Neisseria meningitidis	1.1×10^{-2}	β; purple
Straphylococcus aureus	5.5×10^{-6}	
Streptococcus pneumoniae	2.9×10^{-2}	HGC
Streptococcus sanguis	2.0×10^{-2}	HGC
Streptococcus mutans	7.0×10^{-4}	HGC

Tf = transformation frequency (chromosomal marker transformants/viable cell); + = transformation reported; CBF = cytophaga/flexibacteria/bacteroides group; HGC = high G+C Gram-positive; LGC = low G+C Gram-positive.

Mycobacterium smegmatis (Norgard and Imaeda, 1978; Page and Grant, 1987) competence occurs throughout the log phase and declines thereafter. The proportion of competent cells within a culture varies between species (Fig. 12.1). For example, it has been estimated that the competent fraction of *A. vinelandii* (Doran *et al.*, 1987) is 100%, while within *B. subtilis* (Smith *et al.*, 1981) and *A. calcoaceticus* (Palmen *et al.*, 1993) it is 10–25% and in *H. influenzeae* (Redfield, 1991) and *P. stutzeri* it is below 1% (Lorenz and Wackernagel, 1992). The different types of regulatory systems between the species may determine the level of competence achieved. For example, a critical cell concentration is required to induce competence in *S. pneumonia* and *B. subtilis* (Smith *et al.*, 1981).

12.3.2 MAINTENANCE OF THE COMPETENCE STATE

This is clearly a facet of the process which shows inconsistencies between species/strains and between environments. The *Vibrio* sp. (Frischer *et al.*, 1993) shows a capacity to retain competence for over 10 days at 29°C in artificial seawater and transfer frequencies fell only 200-fold over a 7-day period at 7°C (Lorenz *et al.*, 1992). It is possible that dying cells are providing a short pulse of nutrients that could 'kick' the competence period into action periodically. In the photosynthetic bacteria *A. nidulans* and *A. quadruplicatum* exclusion from light results in a loss of competence (Chauvat *et al.*, 1983; Essich *et al.*, 1990).

12.3.3 AVAILABILITY OF EXOGENOUS DNA

In a stable environment mortality and growth occur at much the same rate (Servais *et al.*, 1983). Thus free DNA is present due to normal processes of cell death and lysis. The spontaneous release of DNA from bacterial cells in the aquatic environment is also normal (Paul *et al.*, 1987; Paul and David, 1989) and is largely in the form required for transformation, i.e. it is linear and double stranded (Paul and Carlson, 1984; DeFlaun *et al.*, 1987; Karl and Bailiff, 1989; Paul and David, 1989). The second criterion is that the DNA has to be of a suitable size. For example, DNA between 10 and 60 kb gave equal frequencies with *P. stutzeri*, but decreased over 10-fold between 1 and 10 kb (Carlson *et al.*, 1983), while *A. calcoaceticus* gave a reduction in frequency proportional with size between 40 and 10 kb (Lonsdorf *et al.*, cited in Lorenz and Wackernagel, 1994).

12.3.4 DNA BINDING AND UPTAKE

Double-stranded DNA associates quickly with the external surface, especially in the presence of divalent ions, Mg^{2+} and Ca^{2+}, but monovalent

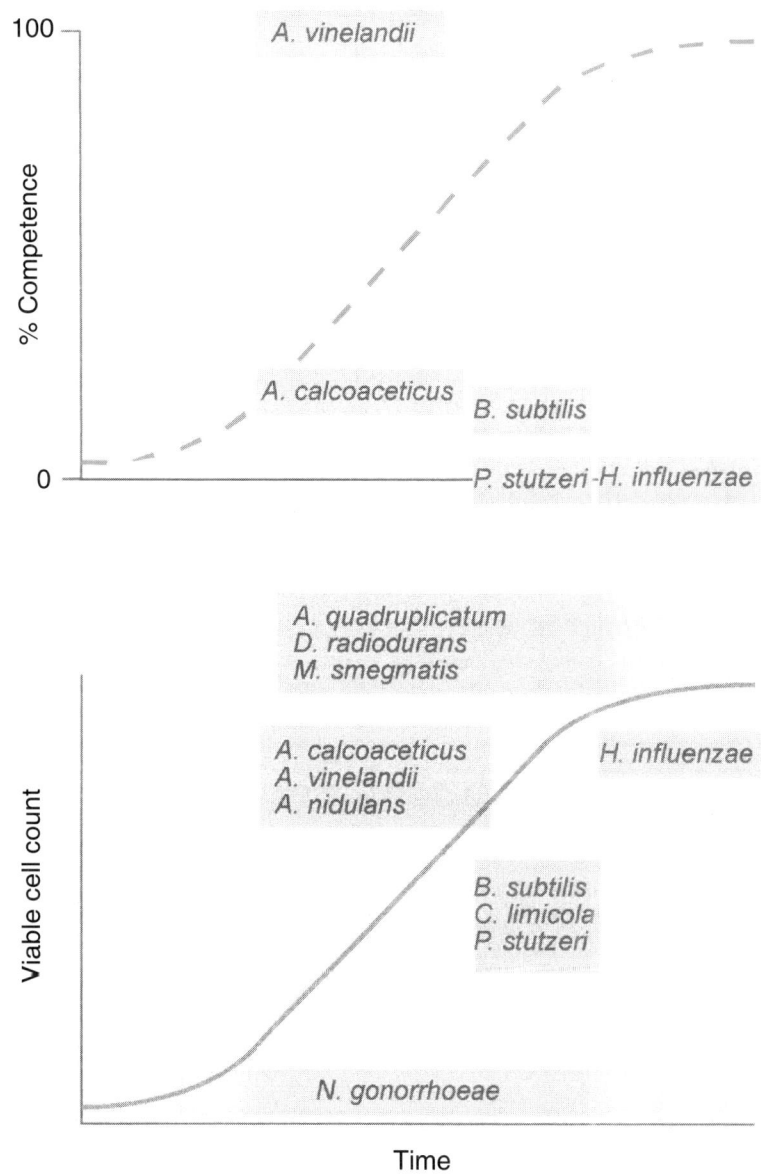

Fig. 12.1 A comparison between transformable species and the percentage of competent cells in a population with the stage in their growth cycle when they attain competence.

ions also have a role (Romanowski *et al.*, 1993). In both *B. subtilis* and *S. pneumoniae* it is held in a state that is resistant to gentle washing but still sensitive to DNAase. There are about 50 DNA uptake sites in *B. subtilis* and 30–80 in *S. pneumoniae* (Dubnau, 1991; Smith *et al.*, 1981). *H. influenzae* also takes up double-stranded DNA, but only from the same or closely related species, and specificity is determined by a DNA receptor that recognizes an 11 bp sequence distributed at 600 sites throughout the genome (Sisco and Smith, 1979; Dashman and Stotzky, 1986). There is evidence that AT-rich flanking regions increase transformation frequency 48-fold compared with GC flanking sequences (Danner *et al.*, 1982). DNA uptake by *N. gonorrhoeae* is similar and relies on a 10 bp sequence arranged as an inverted repeat and forms part of a transcriptional terminator (Goodman and Scocca, 1988).

DNA uptake is defined as the transition from DNAase-sensitive to resistant state and it is an energy-dependent process (Dubnau, 1991). In *B. subtilis* DNA can be recovered in the single-stranded form – the so-called eclipse state, because it has no transforming activity in a transforming experiment (Smith *et al.*, 1981). DNA enters the cell in the 3' to 5' direction and degradation of the opposite strand occurs concurrently at about 100 nucleotides/sec (Mejean and Claverys, 1993). Blebs or transformasomes are externally positioned membrane-derived vesicles, which in many species contain plasmid and chromosomal genes (Dorward and Garon, 1990). In *H. influenzae* these blebs are 80–100 nm in size; they appear during the onset of competence and are shed as the cell leaves this state. In *H. influenzae*, DNA with a specific 11 bp uptake sequence (Deich and Hoyer, 1982) binds with membrane vesicles and transformasomes (Goodgal, 1982), and competence is lost when the cell loses these organelles (Kahn *et al.*, 1983). In *H. influenzae* there appears to be a selection for chromosomal sequences because superhelical and nicked plasmids are not taken up as efficiently (Gromkova and Goodgal, 1979). Albritton *et al.* (1982) have shown that cell-to-cell transformation occurs in *H. influenzae* and it is relatively insensitive to DNAase. This species and many other Gram-positive and Gram-negative species produce extracellular vesicles that contain DNA.

12.3.5   STABILIZATION OF TRANSFORMED SEQUENCES

For a DNA sequence, transformed into a cell, to be stabilized, it needs to be integrated into a self-replicating element by the recombination system controlled by the *rec*A gene (Mahajan, 1988). Transformation is eliminated in *rec*A mutants of *A. calcoaceticus* and *P. stutzeri* (Vosman and Hellingwerf, 1991; Palmen *et al.*, 1992). Upon entry into the cytoplasm the single-stranded sequence displaces the corresponding homologous strand in the duplex and recombination occurs. In *E. coli* (King and

Richardson, 1986) 20–40 bp was the minimum length of sequence homology, compared with over 70 in *B. subtilis* (Khasanov *et al.*, 1992), which allowed a reasonable rate of *in vivo* recombination. The level of recA in competent cells of *B. subtilis* is 10-fold higher than normal (Lovett *et al.*, 1989) and on average 8–10 genes are integrated. The size of sequence displaced is 8.5 kb (Smith *et al.*, 1981; Dubnau, 1991), which is about 70% of that transferred (Dubnau, 1991).

Observations show that the frequency of transformation is lower for plasmid and phage molecules than for the single-hit response of chromosomal or plasmid dimers (Saunders and Guild, 1981). This indicates that the establishment of plasmid and phage genomes is achieved by the same mechanism, but requires a two-hit or second order dependency. This effectively means that at least two copies, linearized at different points of the plasmid or phage sequence, must enter the cell so that they can hybridize to form a duplex and circular molecule.

There are also marker-specific transformation frequencies, which vary over 20-fold, in *S. pneumoniae* (Gasc *et al.*, 1987). A comparison of transformation frequencies of a streptomycin-resistance mutation from *H. influenzae* to nine other *Haemophilus* spp. showed a 25 000-fold variation (Albritton *et al.*, 1984). *B. subtilis* was transformed (10^6-fold range) with DNA from a range of other *Bacillus* species (Harford and Mergeay, 1973). *A. vinelandii* was transformed by DNA from related species at 0.002–21% (Bishop *et al.*, 1977; Doran and Page, 1983; Page, 1985), but from other strains not at all (DeLey, 1992). Why should this be? A similar frequency bias was observed in transduction (Masters and Broda, 1971). In the host genome there are more origin sequences than terminal sequences in moderately to fast growing cells. This is due to further rounds of replication being initiated before the initial replication round has terminated. As phage only replicate in growing cells, this replication-produced bias in relative gene concentrations means that there are more transducing particles with sequences around the origin than at the terminus. This results in a differential in transduction frequencies between genes around the origin and those around the terminus. It is doubtful if this bias is completely responsible for the transformation frequency differences. Further questions can be asked. For example, can genera, species or strains, which cannot be transformed, provide DNA in a suitable state to be used by transformation-proficient ones?

The contribution that restriction and modification make to limiting the frequency of transfer remains to be determined. Restriction and modification (RM) is an enzymologically controlled process that permits the host cell to distinguish self from non-self DNA and degrade the latter (Meselson and Yuan, 1968). DNA that is enzymologically modified in a cell will survive the transfer to another similar cell. Potential hosts carrying different RM systems will thus discriminate incoming DNA and cleave those sequences not carrying the appropriate modification signals. The activity of an RM system can reduce the transfer frequency by 10^5-fold (Roberts, 1985). It is important in influencing the amount of DNA available for recombination when transfer is achieved by plasmid and phage. Therefore it will also be an important influence on transformation (Fry and Day, 1990a). Thus estimating the potential for gene exchange by transformation *in situ* remains confused, but probably no more so than for phage or plasmid-mediated transfers. Clearly factors other than taxonomic distance are relevant in determining frequencies. The brief description of transformation, in these four species, shows that there are similarities. Other species show that the process of transformation is superficially similar but it differs in detail. For example, *P. stutzeri* exhibits selectivity in DNA binding and uptake, but not to the level of *H. influenzae* and *N. gonorrhoeae* (Mathis and Scocca, 1982), and *A. calcoaceticus* shows no specificity in either uptake or binding (Lorenz *et al.*, 1992; Börsheim, 1993).

12.3.6   DNA AND THE AQUATIC ENVIRONMENT

DNA in the environment is both a nutrient and a source of genes for those strains able to take it up. Table 12.2 shows that it is degraded or hydrolyzed at a rapid rate (hours) in wastewater (Phillips *et al.*, 1989; Fibi *et al.*, 1991), seawater (Maeda and Taga, 1973, 1974; Bazelyan and Ayzatullin, 1979; Paul *et al.*, 1987; Turk *et al.*, 1992) and freshwater (Paul *et al.*, 1989). In marine sediments (Maeda and Taga, 1973, 1974; Novitzky, 1986) it is degraded at much lower rates. Although it may be protected from hydrolysis it can also become biologically inaccessible by binding to clays and other minerals (Lorenz and Wakernagel, 1994). DNA-degrading bacteria may form a substantial fraction (10^5/ml) in seawater populations (Maeda and Taga, 1974) and over 90% in other aquatic environments (Greaves and Wilson, 1970; Maeda and Taga, 1974). A longer

Table 12.2 High molecular weight DNA is present in aquatic systems (Lorenz and Wackernagel, 1994)

Habitat	Molecular size (k bp)	DNA concentration (µg/l)	half life (h)
Freshwater	ND	0.5–25.6	4.0–5.5
Estuarine	0.15–35.2	10–19	3.4–5.5
Offshore/ocean	0.24–14.3	0.2–1.9	4.5–83
Freshwater sediment	1.0–23.0	1.0	–
Marine sediment	–	–	140–235

ND = not detected; – = unknown.

half-life of 'free DNA', in aqueous solution, is supported by the observation that most DNA-degrading activity is associated with particles (Maeda and Taga, 1973; Lovett et al., 1989). However, as pBR322 can be degraded in 20 minutes (Phillips et al., 1989) in an aquatic microcosm, this indicates that the local environment and microbial population structure are important.

All bacteria examined so far excrete or release DNA during growth. It has been shown to occur with cells of Brucella spp., Flavobacterium spp., Alcaligenes spp., Micrococcus spp., Pseudomonas spp. (Catlin, 1956, 1960; Takahashi and Gibbons, 1957), Bacillus subtilis (Sinha and Iyer, 1971; Lorenz et al., 1991), P. cepacia and Bradyrhizobium japonicum (Paul and David, 1989) cells during normal growth, and these organisms are probably not unique. Thus the presence of free DNA in the environment may arise from several processes: cell lysis due to age, predation and phage replication (Börsheim, 1993) and normal physiological activity.

12.3.7   CELL PROXIMITY

Although DNA binds to particulate matter it is probable that more transformation occurs in habitats where active cells are close together. Total bacterial counts average 3×10^6/ml in water and 8×10^9/g in sediment (Van Es and Meyer-Riel, 1982). In the epilithon (the slimy microbial community on the surface of submerged surfaces) about 5×10^9 bacteria/ml may be present in film 200 µm thick (Fry and Day, 1990b).

12.3.8   LIFE IN COMPETENCE

In the natural environment bacteria live under nutrient limitation, with occasional bursts of excess. As a consequence individual cells experience 'short' periods of growth between periods of starvation (Roszak and Colwell, 1987; Kaprelyants et al, 1993). To enter the state of competence requires B. subtilis (Smith et al., 1981) and A. vinelandii (Page and Sadoff, 1976) to be in an active metabolic state, but this should not be equated with a nutrient-rich environment (Dubnau, 1991). Various studies have shown that competence depends on the imposition of a range of physiological stresses. For example, iron availability is one factor governing competence in A. vinelandii (Page, 1985) and A. nidulans (Lorenz and Wakernagel, 1992). When grown aerobically or anaerobically and starved of iron, they become competent (Page and von Tigerstrom, 1978). Some environments – for example, the rhizosphere – are nutrient rich and relatively iron free (O'Sullivan and O'Gara, 1992). Competence levels in A. vinelandii were higher when the organisms were starved of iron and molybdenum than when they were starved only of iron (Page, 1985). Divalent ions, like Ca^{2+}, also have a role in competence development in

A. vinelandii (Page and Doran, 1981). Logically regulation/induction of competence would not involve the build-up of a competence protein as the potential for dilution by diffusion, before a critical concentration was reached, would be too high.

12.4   MICROCOSM STUDIES

There have been very few microcosmic experiments. Paul and colleagues (1991) isolated a high frequency transformation mutant WJT-1C from the *Vibrio* strain D19 (Frischer *et al.*, 1990). This strain was transformed with pure DNA from a Tn5 mutant of the bhr plasmid RSF1010 (Paul *et al.*, 1991). In a sterile sediment and water microcosm, transformation only occurred at a low frequency (2×10^{-7}) in nonsterile water. Nutrient amendment raised the transformation frequency in water (3×10^{-5}), but had no effect on the sediment frequency ($> 1 \times 10^{-10}$; Paul *et al.*, 1991).

Transformation occurred, at 10-fold higher than the reversion rate, in *P. stutzeri* contained in a sterile marine water/sediment microcosm, amended with chromosomal DNA carrying a rifampicin-resistance gene (Stewart and Sinigalliano, 1990). Transformation was also observed in a variety of freshwater and marine sediments and prevented by DNAase. Interestingly under the same conditions the marine *Vibrio* sp. D19, a soil *P. stutzeri* and *V. parahaemolyticus* all performed poorly (Stewart and Sinigalliano, 1990).

Bacillus subtilis was hardly transformable in freshwater aquifer in the absence of Mg^{2+} (Romanowski *et al.*, 1993), but *Vibrio* sp. (Frischer *et al.*, 1993) did so in a marine environment and *A. calcoaceticus* (Fry *et al.*, 1992) did so in freshwater. This implies that, as species become genetically adapted to their habitats and ecological sites, their transformation activities have evolved in concert.

12.5   THE *IN SITU* ASSAY

DNA may be provided in various ways – pure, as a crude lysed cell lysate (Juni, 1972), as pasteurized cells or as whole viable cells (Williams, 1993). In Cardiff we have routinely adopted the following approaches. In laboratory transformation matings, early stationary phase cells were mixed and deposited on filters. These filters were placed on PCA and incubated, to allow growth through an appropriate phase for transformation to occur. The cells were suspended and enumerated for donors, recipients and transformants. Initially transformation occurring on selective media (post-experiment) proved to be a problem, particularly in whole-cell and *in situ* experiments. The addition of DNAase1 (0.5 µg/ml) inhibited transformation with lysates and between whole cells in liquid, confirming that the process was transformation. However, it did not always prevent transformation occurring on selective media in control matings with pasteurized donor cells or between whole cells (Williams, 1993).

Thus for *in situ* experiments an unambiguous protocol for monitoring *in situ* transfer events was required. Transformation was linearly inhibited by increasing EDTA concentrations (Fig. 12.2). EDTA ($300 \mu g/ml$) inhibits plate transformation, but also inhibits or delays cell growth (> 100 cells/ml; Fig. 12.3) on minimal media. At higher cell densities fewer effects were seen. When culture filtrates (0.5% w/v) were added to the media, or two plates were inoculated (one selective and the other nutrient based) and taped together so that they shared the same atmosphere, transformants grew at the rate of the controls. On this medium transformants were not detected in control experiments when donor and recipient were plated out together.

Laboratory experiments showed both temperature and pH to have importance. In *A. calcoaceticus* transformation was detected at temperatures as low as 2°C in the laboratory, at a frequency of 10^{-4} per recipient (Williams, 1993) and this increased linearly with temperature such that at 30°C it was $< 10^{-2}$ (Fig. 12.4). *In situ* experiments (Fig. 12.5) compare

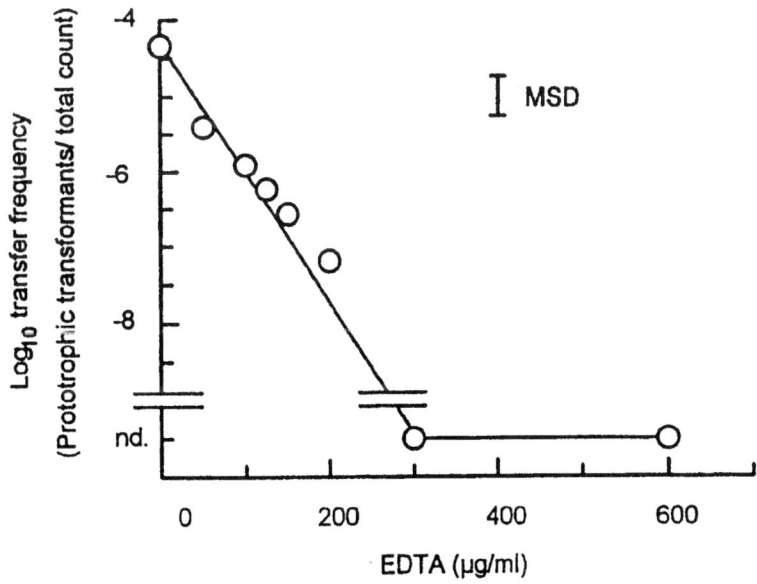

Fig. 12.2 The effect of EDTA on cell-to-cell transformation frequencies to prototrophy. HGW1501 and HGW1521 (pQM17) were mixed and plated on B22 agar plus EDTA ($300 \mu g/ml$) for 24 hours at 20°C. (MSD between $\log_{10}$ geometric means = 0.508 log units; nd = no transfer detected.)

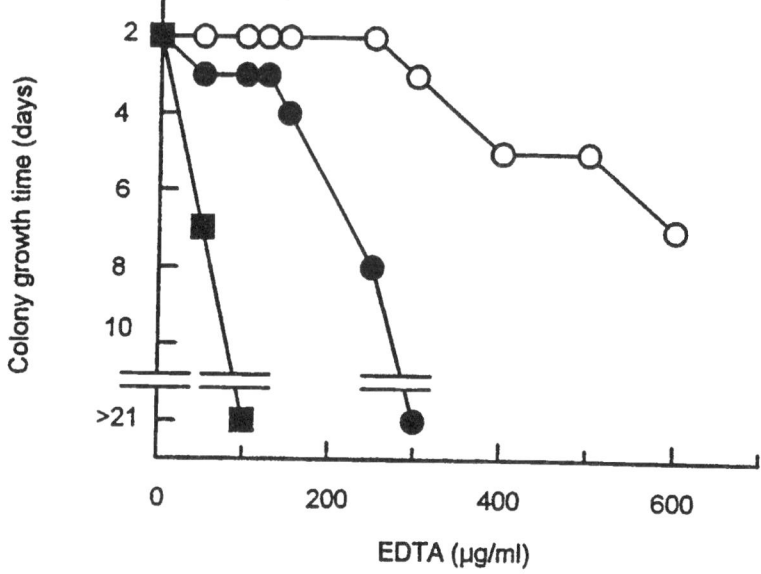

Fig. 12.3 The effect of EDTA and cell density on time taken for 100% colony formation (> 0.5 mm) of prototroph on minimal medium at 30°C. ■, 1 cell/cm²; ●, 10 cells/cm²; ○, 100 cells/cm².

frequencies of HGW1521 transformed to prototrophy with lysates and whole cells. The laboratory mutation rate to prototrophy for this gene is 5×10^{-9} on PCA (Table 12.3). Incubation *in situ* gives a 'reversion' frequency of 2×10^{-7} (Table 12.3; Fig. 12.5); thus either the *in situ* mutation rate is higher or there is DNA present in a suitable form available for the formation of prototrophic transformants. They could both contribute to the lower line (Fig. 12.5), which is the summation of 'reversion to prototrophy' and 'transformation by indigenous DNA' naturally present in the epilithon. The transformation frequency in the presence of exogenous DNA and 'donor' cells is significantly higher and clearly influenced by temperature. In the laboratory *A. vinelandii* transforms optimally between 26°C and 37°C (Page and Sadoff, 1976) and *P. stutzeri* increases exponentially between 12 and 20°C, with an optimum between 20°C and 37°C (Lorenz and Wackernagel, 1992). In a *Vibrio* sp. the transformation frequencies were high (between 15 and 33°C) but fell 100-fold as the temperature rose a further 4°C (Frischer *et al.*, 1993).

Both *A. vinelandii* and *P. stutzeri*, cultured in media with specific pH values, have a narrow peak of high transformation efficiency at around pH 7.0 (Lorenz and Wackernagel, 1992; Page, 1982; Page and Sadoff, 1976). The transfer frequency in *A. calcoaceticus* fell three orders of

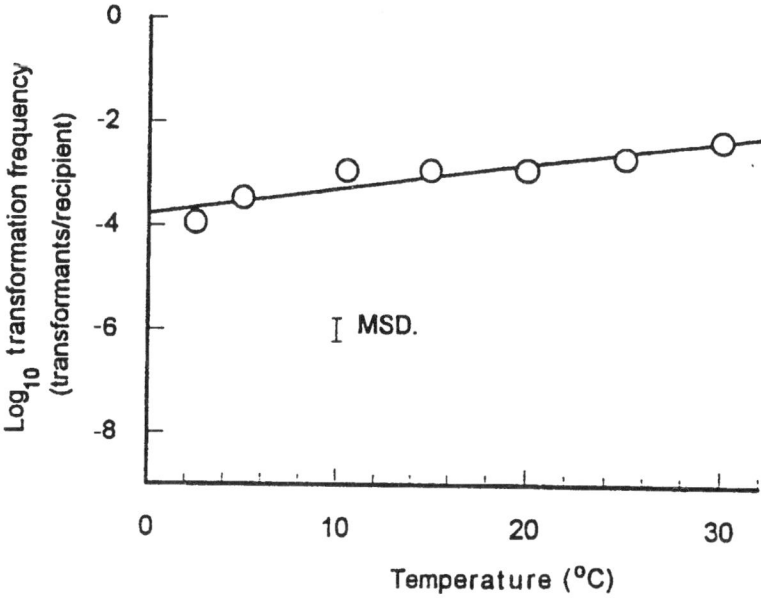

Fig. 12.4 The effect of temperature on the transformation frequency of HGW1521 (pQM17) to prototrophy by cell lysate. Each point is the mean of three replicates. MSD ($P = 0.05$) was 0.256 (log units).

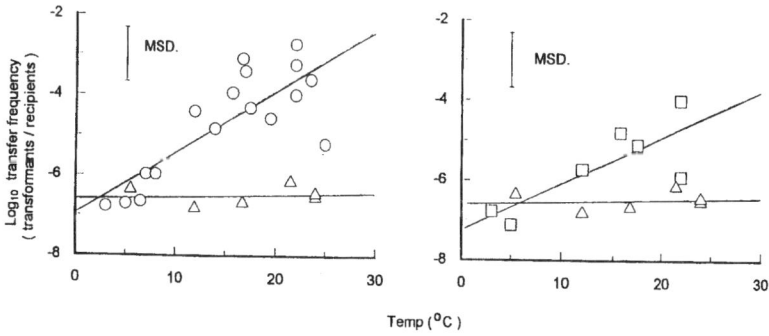

Fig. 12.5 Effect of temperature on transformation frequency *in situ* using cell lysates (O), whole cells ($\square$) and the indigenous microbial population ($\triangle$) as the source of DNA for transformation. The experiments below 20°C were performed at Cardiff and those above at Hillsborough river, USA. Each point represents the mean of three replicates. MSD = 1.36 log units.

Table 12.3 A comparison of transformation frequencies obtained from laboratory, microcosm and *in situ* matings

Phenotype	Source of DNA	Laboratory[a]	Microcosm sterile[a]	Microcosm non-sterile[a]	*in situ*[a]
his⁻	lysate	1×10^{-2} (M)	1×10^{-4} (M)	1×10^{-5} (M)	1×10^{-2} (A) 1×10^{-4} (E)
Hg^r	lysate	1×10^{-6} (M)			1×10^{-7} (A) ND (E)
his⁻	viable cells	1×10^{-4} (M)			$< 10^{-4}$ (A) ND (E)
Hg^r	viable cells	5×10^{-7} (M)			1×10^{-7} (A) ND (E)
his⁻	pasteurized cells	7×10^{-4} (M)			
Hg^r	pasteurized cells	1×10^{-6} (M)			
his⁻	no cells	5×10^{-9} (M)			2×10^{-7} (E)
Hg^r	no cells	2×10^{-9} (M)			1×10^{-7} (E)

[a] Representative frequencies for a histidine auxotrophic gene and mercury resistance, carried on the plasmid pQM17.
(A) = aquatic, experiments done in water; (E) = epilithon, experiments done in a biofilm; (M) = experiments done on a membrane.
ND = not detected, i.e. the frequency of transfer is at or below the expected reversion/mutation frequency.

magnitude (Palmen *et al.*, 1993) as the pH changed from 6.7 to 5.4. However, pH changes over this range with the same organism, in a different experimental system, had no effect (Rochelle *et al.*, 1988). As the pH of the culture medium of *S. pneumoniae* becomes more alkaline (pH 7.3–8.0) the competence period moves earlier in the growth phase and also at lower cell densities (Chen and Morrison, 1987), implying that the competence factor has more physiological activity in environments with higher pH values. Thus the relationship between the development of competence and environmental stimuli is not straightforward, and results with the same organism but using different protocols are likely to be different.

Finally, *B. subtilis* was hardly transformable in freshwater aquifer in the absence of Mg^{2+} (Romanowski *et al.*, 1993), but *Vibrio* spp. (Frischer *et al.*, 1990) did so in a marine environment and *A. calcoaceticus* (Williams, 1993) did so in freshwater. This is a simplistic analysis, but as the latter were both tested in a relevant environment, it suggests that as species have become genetically adapted to their habitats or ecological sites, their transformation activities have evolved in concert.

12.6   PLANNING *IN SITU* STUDIES

The way an *in situ* experiment is carried out is governed to a large extent by the environment chosen. The numbers of replicates needed may be more than can be physically handled, the design of selective media can confound and then there are the vagaries of weather, its effects on water flow rate, temperature, ionic concentrations, etc. and the 'small boy effect'. The latter occurs when somebody covertly watches you set up your *in situ* incubation; in Wales it is usually a small boy. When you leave the site, the 'small boy' proceeds to wade in and destructively examine the experiment. Thus there is a need to establish as simple and secure a procedure as is practically feasible and yet one that retains scientific quality.

In the experiments described, robust strains of *A. calcoaceticus* with good selectable phenotypes were used (Table 12.4). The source of transforming DNA (cells or lysate) may be deposited on a membrane, as may the recipient. The membranes are placed face to face and may be incubated either on agar or, if in a microcosm or *in situ*, on the surface of a stone (Bale *et al.*, 1988). When used *in situ* the stone might need to be marked and perhaps anchored.

The controls needed are simple. Firstly, set up parallel experiments: one without donors and the other without recipients. The strains used must have secondary unselected phenotypes – ones to distinguish them from indigenous bacteria that might grow on the selective media. Molecular procedures for further identification are excellent, if they are

Table 12.4 *Acinetobacter calcoaceticus* strains and their selectable phenotypes

Strain	Selectable phenotype
BD413	wild-type
HGW 1501	his⁻, rif^r, spc^r
HGW 1521 (pQM17)	his⁻, rif^r, spc^r, Hg^r
HGW 98 (pQM17)	met⁻, Hg^r

rif^r = rifampicin resistant; spc^r = spectinomycin resistant; Hg^r, mercury resistant; his⁻ histidine auxotroph; met⁻ methionine auxotroph.

economic in time, but generally they are best confined to confirmation of a proportion of putative transformants.

The choice of strains is critical. A combination of strains that provide the highest transfer rates is optimal. Use natural strains from the habitat if possible, as they have ecological relevance. An examination of the literature, especially in conjugation, shows a great deal of conservatism in this aspect. Choice has favored the use of well-known and established laboratory strains (e.g. *E. coli*) and genetic systems (e.g. RP4) in *in situ* experiments (e.g. Van Elsas *et al.*, 1988). In many cases this choice is apparently made without real consideration of the relevance of environment(s) whence the organisms and transfer systems were derived. The justification (largely unwritten) for this approach appears initially to place reliance on the information known about laboratory models. This means it is quicker to get into experiments as there is no need to identify and characterize a novel transfer system. The second (pessimistic) justification is that there is little faith in being able to identify a transfer system for the environment under investigation.

The use of as natural a system as possible is eminently sensible; it means that the work has direct and immediate ecological relevance. This provides a potential to manipulate the strains and conditions in various ways in later experiments. *In situ* transfer experiments do require a transfer system that provides a good and 'normal' unmodified transfer rate. Gene transfer frequencies obtained from experiments done *in situ*, in a natural microbial community, are widely reported to be depressed by several orders of magnitude compared with laboratory controls. This means that the higher the laboratory transfer frequency ($> 10^6$), the higher is the level of sensitivity available in *in situ*.

12.7   FREQUENCY

There are three sources of DNA for transformation available in different ways. Firstly there is the gradient in concentration of 'old and new' DNA, derived directly from the lysis of a range of organisms – both prokaryotes and eukaryotes. The 'DNA population' is historical; it has an age and a diversity which may not be temporarily consistent. In addition there is DNA, released directly from within the community of organisms. This DNA source is related to the immediate community structure, its members and their individual population sizes. Thus these sources of DNA will, it seems, contribute unequally, depending on the physiological status (health) of the community. In addition the evidence infers that the term 'transformable strains' may be a better descriptive term than 'transformable species', as not all members of a species are proficient in taking up DNA. Superimpose on this the daily or seasonal rhythms that contribute to periods of growth, starvation and death, together with the various regulatory constraints of the transformational process shown by different species. This may be confusing from our standpoint, and the situation might be the same at the molecular level.

Is it rational at present to attempt to estimate *in situ* transformation frequencies? It is difficult to see how a measurement designed for use in controlled conditions in the laboratory can be extrapolated directly and validly to a complex environment. Viable counts taken at '0' time and used to calculate transfer frequencies do not give the same frequencies when counts taken at 6, 8 or 24 hours are used. This is due to a combination of factors, such as mating time, cell population growth, death, ratio of cell types, repeated matings, changes in physiological status, temperature, selection (intended or not), etc. It is probably sensible to enumerate populations routinely before and after, and use the 'after' counts to calculate frequencies. At least the number that survived is known. We now need to determine what is happening to individual cells.

Therefore it is extremely uncertain what the significance of a transfer frequency calculated in an *in situ* experiment means to risk assessment analysis. To establish a frequency that has ecological significance requires that an ecological perturbation (some effect) be demonstrated. A simple analogy would be the determination of the critical dose relationship between the numbers of pathogenic viruses, or bacteria, required to produce disease in a human. Thus we are at the point where we wish to use or extrapolate *in situ* frequencies of gene transfer, but have little idea of their environmental relevance.

12.8   CONCLUSIONS

A general feature of gene transfer experiments in the laboratory is the maintenance of continuous environmental conditions. *In situ* experiments are characterized by an inherently uncontrollable discontinuous environment. The discontinuities in the environment at the microbial level are likely to lead to differing physiologies in adjacent cells, potentially influencing the efficiency and efficacy of genetic communication between them. So, can those data generated in the laboratory be expected to provide anything but an indication of transfer that will be observed *in situ*?

Logically the success of gene transfer by transformation will depend on:

- the donor and recipient and their immediate physiological states;
- the past, current and impending environmental conditions;
- the current density of the gene(s) in question;
- selection pressure;
- serendipity.

What are the chances of one transformant cell, in a biofilm population of millions, surviving to have selection pressure for the gene imposed upon it? The growth of the transformant within the population will depend on the benefits derived from the expression of the novel gene in response to the selective pressures imposed on the recombinant. It will be the amplification of the transformant that will provide evidence for successful *in situ* exchange. The challenge is now to extend relevant model laboratory systems into their environments and achieve a description of the ecological role of transformation.

ACKNOWLEDGMENTS

I would like to express my sincere thanks to my colleagues, Professor J. Fry and Drs K. Ashelford and H.G. Williams, for their 'scientific enthusiasm' throughout our continuing collaboration. Parts of the work reported here were supported by a Natural Environmental Research Council studentship, an EC contract (BIOT-CT91-0284) and a MAFF contract (CSA 3346).

REFERENCES

Albritton, W.L., Setlow, J.K. and Slaney, L. (1982) Transfer of *Haemophilus influenzae* chromosomal genes by cell-to-cell contact. *J. Bacteriol.* **152**: 1066–1070.

Albritton, W.L., Setlow, J.K., Thomas, M. *et al.* (1984) Heterospecific transformation in the genus *Haemophilus*. *Mol. Gen. Genet.* **193**: 358–363.

Avery, O.T., MacLeod, C.M. and McCarthy, M. (1944) Studies on the chemical nature of the substance inducing transformation in pneumococcal types. *J. Exp. Med.* **79**: 137–159.

Bale, M.J., Day, M.J. and Fry, J.C. (1988) Transfer and occurrence of large mercury resistance plasmids in epilithon. *Appl. Envir. Microbiol.* **54**: 972–978.

Bazelyan, V.L. and Ayzatullin, T.A. (1979) Kinetics of enzymatic hydrolysis of DNA in sea water. *Oceanology* **19**: 30–33.

Bishop, P.E., Dazzo, F.B., Appelbaum, E.R. *et al.* (1977) Intergeneric transfer of genes involved in the *Rizobium*–legume symbiosis. *Science* **198**: 938–940.

Börsheim, K.Y. (1993) Native marine bacteriophages. *FEMS Microbiol. Ecol.* **102**: 141–159.

Carlson, C.A., Pierson, L.S., Rosen, J.I. and Ingraham, J.L. (1983) *Pseudomonas stutzeri* and related species undergo natural transformation. *J. Bacteriol.* **153**: 93–99.

Catlin, B.W. (1956) Extracellular deoxyribonucleic acid of bacteria and a deoxyribonuclease inhibitor. *Science* **124**: 441–442.

Catlin, B.W. (1960) Transformation of *Neisseria meningitidis* by deoxyribonucleates from cells and from culture slime. *J. Bacteriol.* **79**: 579–590.

Chauvat, F., Astier, C., Vedel, F. and Joset-Espardellier, F. (1983) Transformation in the cyanobacterium *Synechococcus* R2: improvement of efficiency; role of pUH24 plasmid. *Mol. Gen. Genet.* **191**: 39–45.

Chen, J.D. and Morrison, D.A. (1987) Modulation of competence for genetic transformation in *Streptococcus pneumonia*. *J. Gen. Microbiol.* **133**: 1959–1967.

Danner, D.B., Deich, R.A., Sisco, K.L. and Smith, H.O. (1980) An eleven-base pari sequence determines the specificity of DNA uptake in *Haemophilus* transformation. *Gene* **11**: 311–318.

Danner, D.B., Smith, H.O. and Narang, S.A. (1982) Construction of DNA recognition sites active in *Haemophilus* transformation. *Proc. Natl Acad. Sci. USA* **79**: 2393–2397.

Dashman, T. and Stotzky, G. (1986) Microbial utilization of amino acids and a peptide bound on homoionic montmorillonite and kaolinite. *Soil Biol. Biochem.* **18**: 5–14.

DeFlaun, M.F., Paul, J.H. and Jeffrey, W.H. (1987) Distribution and molecular weight of dissolved DNA in subtropical estuarine and oceanic environments. *Mar. Ecol. Prog. Ser.* **38**: 65–73.

Deich, R.A. and Hoyer, L.C. (1982) Generation and release of DNA binding vesicles by *H. influenzae* during induction and loss of competence. *J. Bact.* **152**: 855–864.

DeLey, J. (1992) The proteobacteria: ribosomal RNA cistron similarities and bacterial taxonomy, in *The Prokaryotes* (eds A. Balows, H.G. Trüper, M. Dworkin *et al.*), Springer-Verlag, New York, pp. 2111–2140.

Doran, J.L., Bingle, W.H., Roy, K.L. *et al.* (1987) Plasmid transformation of *Azotobacter vinelandii* OP. *J. Gen. Microbiol.* **133**: 2059–2072.

Doran, J.L. and Page, W.J. (1983) Heat sensitivity of *Azotobacter vinelandii* genetic transformation. *J. Bacteriol.* **155**: 159–168.

Dorward, D.W. and Garon, C.F. (1990) DNA is packaged within membrane-derived vesicles of gram-negative but not gram-positive bacteria. *Appl. Environ. Microbiol.* **56**: 1960–1962.

Dubnau, D. (1991) Genetic competence in *Bacillus subtilis*. *Microbiol. Rev.* **55**: 395–424.

Essich, E., Sevens, E. Jr and Porter, R.D. (1990) Chromosomal transformation in the cyanobacterium *Agmenellum quadruphicarum*. *J. Bacteriol.* **172**: 1916–1922.

Fibi, M.R., Bröker, M., Schulz, R. *et al.* (1991) Inactivation of recombinant plasmid DNA from human erythropoietin-producing mouse cell line grown on a large scale. *Appl. Microbiol. Biotech.* **35**: 622–630.

Frischer, M.E., Thurmond, J.M. and Paul, J.H. (1990) Natural plasmid transformation in a high-frequency-of-transformation marine *Vibrio* strain. *Appl. Environ. Microbiol.* **56**: 3439–3444.

Frischer, M.E., Thurmond, J.M. and Paul, J.H. (1993) Factors affecting competence in a high frequency of transformation marine *Vibrio*. *J. Gen. Microbiol.* **139**: 753–761.

Fry, J.C. and Day, M.J. (1990a) Microbial ecology genetics and risk assessment, in *Release of Genetically Engineered and Other Microorganisms* (eds J.C. Fry and M.J. Day), Cambridge University Press, pp. 160–167.

Fry, J.C. and Day, M.J. (1990b) Plasmid transfer in the epilithon, in *Bacterial Genetics in Natural Environments* (eds J.C. Fry and M.J. Day), Chapman & Hall, London, pp. 55–80.Fry, J.C., Day, M.J. and Williams, H.G. (1992) Plasmid and chromosomal gene transfer by transformation in the aquatic environment, in *DNA Transfer and Gene Expression in Microorganisms* (eds E. Balla, G. Berencsi and A. Szentirmai), Intercept, pp. 111–121.

Gasc, A.M., Garcia, P., Baty, D. and Sicard, A.M. (1987) Mismatch repair during pneumococcal transformation of small deletions produced by site-directed mutagenesis. *Mol. Gen. Genet.* **210**: 369–372.

Goodgal, S.H. (1982) DNA uptake in *Haemophilus* transformation. *Annu. Rev. Genet.* **16**: 169–192.

Goodman, S.D. and Scocca, J.J. (1988) Identification and arrangement of the DNA sequence recognized in specific transformation of *Neisseria gonorrhoeae*. *Proc. Natl Acad. Sci. USA* **85**: 6982–6986.

Greaves, M.P. and Wilson, M.J. (1970) The degradation of nucleic acids and montmorillonite–nucleic-acid complexes by soil microorganisms. *Soil Biol. Biochem.* **2**: 257–268.

Griffith, F. (1928) The significance of pneumococcal types. *J. Hyg.* **27**: 113–159.

Gromkova, R. and Goodgal, S. (1979) Transformation by plasmid and chromosomal DNAs in *Haemophilus parainfluenzae*. *Biochem. Biophys. Res. Commun.* **88**: 1428–1434.

Hahn, J., Albans, M. and Dubnau, D. (1987) Isolation and characterization of Tn*917 lac*-generated competence mutants of *Bacillus subtilis*. *J. Bacteriol.* **169**: 3104–3109.

Harford, N. and Mergeay, M. (1973) Interspecific transformation of rifampicin resistance in the genus *Bacillus*. *Mol. Gen. Genet.* **120**: 151–155.

Ippen-Ihler, K. (1989) Bacterial conjugation, in *Gene Transfer in the Environment*, (eds S.B. Levy and R.V. Miller), McGraw-Hill, New York, pp. 33–72.

Juni, E. (1972) Interspecies transformation of *Acinetobacter*. Genetic evidence for a ubiquitous genus. *J. Bacteriol.* **112**: 917–931.

Kahn, M.E., Barany, F. and Smith, H.O. (1983) Transformasomes: specialized membraneous structures which protect DNA during *Haemophilus* transformation. *Proc. Natl Acad. Sci. USA* **80**: 6927–6931.

Kaprelyants, A.S., Gottschal, J.C. and Kell, D.B. (1993) Dormancy in nonsporulating bacteria. *FEMS Microbiol. Rev.* **104**: 271–286.

Karl, D.M. and Bailiff, M.D. (1989) The measurement and distribution of dissolved nucleic acids in aquatic environments. *Limnol. Oceanogr.* **34**: 543–558.

Khasanov, F.K., Zvingila, D.J., Zainullin, A.A. *et al.* (1992) Homologous recombination between plasmid and chromosomal DNA in *Bacillus subtilis* requires approximately 70 bp of homology. *Mol. Gen. Genet.* **234**: 494–497.

King, S.R. and Richardson, J.P. (1986) Role of homology and pathway specificity for recombination between plasmids and bacteriophage. *Mol. Gen. Genet.* **204**: 141–147.

Kokjohn, T.A. (1989) Transduction: mechanism and potential for gene transfer in the environment, in *Gene Transfer in the Environment*, (eds S.B. Levy and R.V. Miller), McGraw-Hill, New York, pp. 73–97.

Leff. L.G., McArthur, J.V. and Shimkets, L.J. Shimkets (1992) Information spiraling: movement of bacteria and their genes in streams. *Microb. Ecol.* **24**: 11 24.

Lorenz, M.G. and Wackernagel, W. (1992) DNA binding to various clay mineral and retarded enzymatic degradation of DNA in a sand/clay microcosm, in *Gene Transfer and Environment*, (ed. M.J. Gauthier), Springer-Verlag KG, Berlin, pp. 103–113.

Lorenz, M.G. and Wakernagel, W. (1994) Bacterial gene transfer by natural genetic transformation in the environment. *Microbiol. Rev.* **58**: 563–602.

Lorenz, M.G., Gerjets, D. and Wackernagel, W. (1991) Release of transforming plasmid and chromosomal DNA from two cultured soil bacteria. *Arch. Microbiol.* **156**: 319–326.

Lorenz, M.G., Reipschläger, K. and Wackernagel, W. (1992) Plasmid transformation of naturally competent *Acinetobacter calcoaceticus* in non-sterile soil extract and groundwater. *Arch. Microbiol.* **157**: 355–360.

Lovett, C.M., Love, P.E. and Yasbin, R.E. (1989) Competence-specific induction of the *Bacillus subtilis* RecA protein analog: evidence for dual regulation of a recombination protein. *J. Bacteriol.* **171**: 2318–2322.

Maeda, M. and Taga, N. (1973) Deoxyribonuclease activity in seawater and sediment. *Mar. Biol.* **20**: 58–63.

Maeda, M. and Taga, N. (1974) Occurrence and distribution of deoxyribonucleic acid hydrolyzing bacteria in seawater. *J. Exp. Mar. Biol. Ecol.* **14**: 157–169.

Mahajan, S.K. (1988) Pathway of homologous recombination in *Escherichia coli*, in *Genetic Recombination*, (eds G.R. Smith and R. Kucherlapati), American Society for Microbiology, Washington, DC, pp. 88–140.

Masters, M. and Broda, P. (1971) Evidence for the bidirectional replication of the *Escherichia coli* chromosome. *Nature* **232**: 137–140.

Mathis, L.S. and Scocca, J.J. (1982) *Haemophilus influenzae* and *Neisseria gonorrhoeae* recognize different specificity determinants in the DNA uptake step of genetic transformation. *J. Gen. Microbiol.* **128**: 1159–1161.

Mejean, V. and Claverys, J.-P. (1993) DNA processing during entry in transformation of *Streptococcus pneumoniae*. *J. Biol. Chem.* **268**: 5594–5599.

Meselson, M. and Yuan, R. (1968) DNA restriction enzyme from *E. coli*. *Nature* **217**: 1110–1114.

Norgard, M.V. and Imaeda, T. (1978) Physiological factors involved in the transformation of *Mycobacterium smegmatis*. *J. Bacteriol.* **133**: 1254–1262.

Novitsky, J.A. (1986) Degradation of dead microbial biomass in a marine sediment. *Appl. Environ. Microbiol.* **52**: 504–509.

Ormerod, J.G. (1988) Natural genetic transformation in *Chlorobium*, in *Green Photosynthetic Bacteria* (eds J.M. Olson, J.G. Ormerod, J. Amesz *et al.*), Plenum Press, New York, pp. 315–319.

O'Sullivan, D.J. and O'Gara, F. (1992) Traits of fluorescent *Pseudomonas* spp. involved in suppression of plant root pathogens. *Microbiol. Rev.* **56**: 662–676.

Page, W.J. (1982) Optimal conditions for competence development in nitrogen-fixing *Azotobacter vinelandii*. *Can. J. Microbiol.* **28**: 389–397.

Page, W.J. (1985) Genetic transformation of molybdenum-starved *Azotobacter vinelandii*: increased transformation frequency and recipient range. *Can. J. Microbiol.* **31**: 659–662.

Page, W.J. and Doran, J.L. (1981) Recovery of competence in calcium-limited *Azotobacter vinelandii*. *J. Bacteriol.* **146**: 33–40.

Page, W.J. and Grant, G.A. (1987) Effect of mineral iron on the development of transformation competence in *Azotobacter vinelandii*. *FEMS Microbiol. Lett.* **41**: 257–261.

Page, W.J. and Sadoff, H.L. (1976) Physiological factors affecting transformation of *Azotobacter vinelandii*. *J. Bacteriol.* **125**: 1080–1087.

Page, W.J. and von Tigerstrom, M. (1978) Induction of transformation competence in *Azotobacter vinelandii* iron-limited cultures. *Can. J. Microbiol.* **24**: 1590–1594.

Palmen, R., Vosman, B., Kok, R. *et al.* (1992) Characterization of transformation-deficient mutants of *Acinetobacter calcoaceticus*. *Mol. Microbiol.* **6**: 1747–1754.

Palmen, R., Vosman, B., Buijsman, B. *et al.* (1993) Physiological characterization of natural transformation in *Acinetobacter calcoaceticus*. *J. Gen. Microbiol.* **139**: 295–305.

Paul, J.H. and Carlson, D.J. (1984) Genetic material in the marine environment: implication for bacterial DNA. *Limnol. Oceanogr.* **29**: 1091–1097.

Paul, J.H. and David, A.W. (1989) Production of extracellular nucleic acids by genetically altered bacteria in aquatic-environment microcosms. *Appl. Environ. Microbiol.* **55**: 1865–1869.

Paul, J.H., Jeffrey, W.M. and DeFlaun, M.F. (1987) Dynamics of extracellular DNA in the marine environment. *Appl. Environ. Microbiol.* **53**: 70–179.

Paul, J.H., Jeffrey, W.H., David, A.W. *et al.* (1989) Turnover of extracellular DNA in eutrophic and oligotrophic freshwater environments of south west Florida. *Appl. Environ. Microbiol.* **55**: 1823–1828.

Paul, J.H., Frischer, M.E. and Thurmond, J.M. (1991) Gene transfer in marine water column and sediment microcosms by natural plasmid transformation. *Appl. Environ. Microbiol.* **57**: 1509–1515.

Paul, J.H., Jiang, S.C. and Rose, J.B. (1991) Concentration of viruses and dissolved DNA from aquatic environments by vortex flow filtration. *Appl. Environ. Microbiol.* **57**: 2197–2204.

Phillips, S.J., Dalgarn, D.S. and Young, S.K. (1989) Recombinant DNA in wastewater: pBR322 degradation kinetics. *J. Water Pollut. Control Fed.* **62**: 1588–1595.

Redfield, R.J. (1991) *Sxy*-1, a *Haemophilus influenzae* mutation causing greatly enhanced spontaneous competence. *J. Bacteriol.* **173**: 5612–5618.

Roberts, R.J. (1985) Restriction and modification enzymes and their recognition sequences. *Nucl. Acids Res.* **13**: 165–200.

Rochelle, P.A., Day, M.J. and Fry, J.C. (1988) Occurrence, transfer and mobilisation in epilithic strains of *Acinetobacter* of mercury-resistance plasmids capable of transformation. *J. Gen. Microbiol.* **134**: 2933–2941.

Romanowski, G., Lorenz, M.G. and Wackernagel, W. (1993) Plasmid DNA in a groundwater aquifer microcosm – adsorption, DNAse resistance and natural genetic transformation of *Bacillus subtilis*. *Mol. Ecol.* **2**: 171–181.

Roszak, D.B. and Colwell, R.R. (1987) Survival strategies of bacteria in the natural environment. *Microbiol. Rev.* **51**: 365–379.

Saunders, C.W. and Guild, W.R. (1981b) Pathway of plasmid transformation in pneumococcus: open circular and linear molecules are active. *J. Bacteriol.* **146**: 517–526.

Servais, P., Billen, G. and Vives-Rigo, J. (1983) Rate of bacterial mortality in aquatic environments. *Appl. Environ. Microbiol.* **49**: 1448–1454.

Sinha, R.P. and Iyer, V.N. (1971) Competence for genetic transformation and the release of DNA from *Bacillus subtilis*. *Biochim. Biophys. Acta* **232**: 61 71.

Sisco, K.L. and Smith, H.O. (1979) Sequence-specific DNA uptake in *Haemophilus* transformation. *Proc. Natl Acad. Sci. USA* **76**: 972–976.

Smith, H.O., Danner, D.B. and Deich, R.A. (1981) Genetic transformation. *Annu. Rev. Biochem.* **50**: 41–68.

Sparling, P.F. (1966) Genetic transformation of *Neisseria gonorrhoeae* to streptomycin resistance. *J. Bacteriol.* **92**: 1364–1371.

Stewart, G.J. and Sinigalliano, C.D. (1990) Detection of horizontal gene transfer by natural transformation in native and introduced species of bacteria in marine and synthetic sediments. *Appl. Environ. Microbiol.* **56**: 1818–1824.

Stotzky, G. (1989) Gene transfer among bacteria in soil, in *Gene Transfer in the Environment*, (eds S.B. Levy and R.V. Miller), McGraw-Hill, New York, pp. 165–222.

Takahashi, I. and Gibbons, N.E. (1957) Effect of salt concentration on the extracellular nucleic acids of *Micrococcus halodenitrificans*. *Can. J. Microbiol.* **3**: 687–694.

Turk, V., Rehnstam, A.S., Lundberg, E. and Hagström, A. (1992) Release of bacterial DNA by marine nanoflagellates, an intermediate step in phosphorus regeneration. *Appl. Environ. Microbiol.* **58**: 3744–3750.

Van Elsas, J.D., Trevors, J.T., Starodub, M.E. and Van Overbeek, L.S. (1988) Transfer of plasmid RP4 between pseudomonads after introduction into soil: Influence of spatial and temporal aspects of inoculation. *FEMS Microbial Ecol.* **73**: 1–12.

Van Es, R.B. and Meyer-Riel, L.A. (1982) Biomass and metabolic activity of heterotrophic marine bacteria. *Adv. Microbial Ecol.* **6**: 111–170.

Vosman, B. and Hellingwerf, K.J. (1991) Molecular cloning and functional characterization of a *recA* analog from *Pseudomonas stutzeri* and construction of a *P. stutzeri recA* mutant. *Antonie van leeuwenhoek J. Microbiol.* **59**: 115–123.

Williams, H.G. (1993) Plasmid and chromosomal gene transfer by transformation in river epilithon. PhD thesis, University of Wales, Cardiff.

Natural genetic transformation of bacteria in soil

13

*Wilfried Wackernagel, Johannes Sikorski,
Stephanie Blum, Michael G. Lorenz, and
Stefan Graupner*

SUMMARY

Results of recent studies are discussed which examine the potential of
horizontal gene transfer by natural genetic transformation among bacteria in soil. Previous experiments had indicated that, *in vitro*, DNA molecules adsorb rapidly to purified soil mineral materials and thereby gain
increased resistance to nucleolytic degradation. The persistence of DNA
is a prerequisite for transformation. New analytical microcosm experiments with non-sterile soil and radioactively labeled DNA show that
DNA rapidly distributes in soil by binding to soil particles and partially
remaining in soil solution. The latter fraction is degraded within a few
hours by soil-borne DNases, whereas the bound DNA can persist for
extended periods. The escape of DNA from nucleolysis by adsorption to
solid surfaces is proposed as a mechanism that would explain the previously observed long-term persistence of DNA in natural soil. Other soil
microcosm experiments show that competent cells of *Pseudomonas
stutzeri* JM300 find and take up DNA bound on particulate material
within the complex matrix of non-sterile brown earth, suggesting that
transformation can occur *in situ*. The examination of several transformation-deficient mutants of *P. stutzeri* JM300, including the cloning and
sequencing of a chromosomal DNA fragment which complements one of
the mutants, suggests that the DNA uptake potential depends both on
the formation of type 4 fimbriae and on other cellular functions in this

bacterium. It is discussed that development of competence of *P. stutzeri* JM300 is a regulated process responding to environmental signals, such as nutrient limitation, by an increasing DNA uptake capacity.

13.1   INTRODUCTION

Bacteria are the only organisms with the ability of natural genetic transformation. This horizontal gene transfer mechanism includes the active uptake of DNA from the environment, the integration of the DNA into the genome and the expression of gained genes. For this, cells have to develop the physiological state of competence which includes the expression of a set of specific genes located on the chromosome. Transformation is apparently an old genetic trait among bacteria, as concluded from the fact that transformability has been observed in many bacterial species from all major taxonomic groups including cyanobacteria, proteobacteria, Gram-positive bacteria and Archaea (Lorenz and Wackernagel, 1994). The DNA uptake capability has been conserved through evolution of bacteria probably because transformability provides an evolutionary potential for the species. Besides the possibility for gene acquisition, DNA uptake has been considered to play a role in gene regulation, nutrient supply, DNA repair and defense against bacteriophages. For all of these roles, arguments and experimental data have been presented (summarized in Lorenz and Wackernagel, 1994; Solomon and Grossman, 1996).

Among the transformable bacteria known at present, many are clinical isolates of human pathogens. However, more than half of the roughly 50 species are organisms living in aquatic habitats and soil (for references, see Lorenz and Wackernagel, 1994). In soil high numbers of bacterial cells (10^9 to 10^{10}/g of soil) and also high numbers of species are present. It was assumed previously that gene transfer by transformation does not occur in the environment. This assumption rested mainly on the ubiquitous presence of highly active extracellular microbial DNases. Enzymatic degradation makes DNA available to the cells as nutrients in the form of mono- and oligonucleotides. In support of this, Greaves and Wilson (1970) found that the majority of heterotrophic soil isolates produce extracellular DNases. On the other hand, high molecular size extracellular DNA has been detected in various environments like marine and freshwater, sediment (Crabb *et al.*, 1977; DeFlaun *et al.*, 1987; Paul *et al.*, 1989) and also soil (Steffan *et al.*, 1988; Spring *et al.*, 1992), suggesting that DNA may not always be degraded completely and instantaneously.

We can define conditions and prerequisites that would be essential for successful transformation events in a soil habitat. These are:

- that high molecular weight DNA is present in soil (which requires that extracellular DNA can persist for some time before its breakdown);

- that cells develop competence in soil (which is mostly a transient physiological state);
- that the conditions in soil are appropriate for cell–DNA interactions;
- that cells can take up DNA and inherit the genes in this habitat.

These conditions and prerequisites can be tested experimentally by using typical soil organisms and microcosms with soil material to simulate natural situations. The direct tracing of transformation events in the natural environment will require *in situ* experiments with genetically marked cells and DNA.

The recent interest in the possible ways in which bacteria are engaged in sexual processes in their habitat is particularly evident for natural genetic transformation through a variety of recent reviews on various aspects of this phenomenon. These include the genetics and regulation of the cellular components required for DNA uptake (Dubnau, 1991a, 1993; Grossman, 1995; Solomon and Grossman, 1996), the energetics of transformation (Palmen *et al.*, 1994), the translocation of DNA through bacterial cell walls (Dreiseikelmann, 1994), and transformation as a process occurring in the environment (Lorenz and Wackernagel, 1994; Paget and Simonet, 1994).

Here we summarize some observations on DNA stability in the environment, particularly in soil. We also discuss some of our recent findings on the availability of DNA in the soil habitat for competent cells, on the development of competence and on cellular functions involved in transformation of the soil bacterium *Pseudomonas stutzeri* JM300.

13.2   RESULTS AND DISCUSSION

13.2.1   PERSISTENCE OF DNA IN SOIL

Previous studies have documented the interesting finding that duplex DNA as a negatively charged polyelectrolyte can associate with the negatively charged surfaces of silicate minerals, including quartz and clay, when appropriate concentrations of free cations are present (Lorenz and Wackernagel, 1987, 1991). The DNA–mineral association was most effectively supported by divalent cations like Mg^{2+} or Ca^{2+} at millimolar concentrations, suggesting that ion bridging between negatively charged groups mediates the binding of DNA to surfaces (Lorenz and Wackernagel, 1987; Romanowski *et al.*, 1991). A further interesting observation was the increased resistance of the mineral-bound DNA to degradation by DNase I (Lorenz and Wackernagel, 1987, 1991; Romanowski *et al.*, 1991, 1993a; Khanna and Stotzky, 1992; Paget *et al.*, 1992). This protective effect was also documented in experiments where transforming DNA of *Acinetobacter calcoaceticus* bound on groundwater aquifer minerals was treated with an extracellular bacterial nuclease (produced by *Serratia marcescens*) (Ahrenholtz *et al.*, 1994). The DNA in solution lost its transforming activity much more rapidly than the DNA bound on mineral particles. These findings led to the hypothesis that DNA in bacterial habitats such as soil could be sufficiently stable to contribute to gene transfer by natural genetic transformation.

In fact, it was demonstrated in subsequent experiments that naked DNA introduced into non-sterile soil microcosms persisted for weeks and even months, while a slow continual breakdown was evident (Romanowski *et al.*, 1992). With a variety of sensitive monitoring methods, including dot blot and Southern transfer hybridization, PCR and electrotransformation (Romanowski *et al.*, 1992, 1993b), traces of DNA were quantitated after 60 days, constituting 0.01 to 0.5% of the input material. The observation of soil-specific biphasic kinetics of DNA degradation with typical slowly declining final slopes established the view that soils may contain a pool of extracellular DNA. The size of the pool would be governed by the release of DNA from the organisms in soil and the DNA degradative capacity of the soil.

Recently we have developed a mild method to recover DNA introduced into soil. This method allowed us to obtain, in separate fractions, the DNA that remained in the soil solution (non-adsorbed DNA) and the DNA that was bound on soil particulate material. The latter was recovered by a specific extraction buffer (Blum *et al.*, 1997). With this method, the fate of ^{3}H-thymidine-labelled DNA in non-sterile samples of different soils was followed. Rapid and extensive binding of DNA (at 23°C) to the soil particulate material was found. The DNA remaining non-adsorbed in the soil liquid was degraded within 24 hours to acid-soluble material by DNases indigenous to soil. The findings with one soil type (brown earth) are shown in Fig. 13.1. Similar observations were made with two other soils (loamy soil and podzol). In contrast, bound DNA recovered from soil consisted largely of acid-precipitable material even several days after introduction of the DNA into the soil (Fig. 13.1). The high molecular size of the recovered DNA was confirmed by Southern hybridization (data not shown). In other experiments it was found that DNA that desorbed spontaneously from soil particles into the soil solution was rapidly degraded to acid-soluble material (Blum *et al.*, 1997). These observations suggest that binding to surfaces of soil particles leads to prolonged persistence of extracellular DNA, whereas non-adsorbed DNA is rapidly degraded by soil-borne DNases. This protective mechanism can explain the previously documented long-term persistence of extracellular DNA in soil (Romanowski *et al.*, 1992, 1993b).

13.2.2   TRANSFORMATION IN SOIL

The question of whether the DNA associated with mineral particles is still available for uptake by competent bacteria has been addressed in

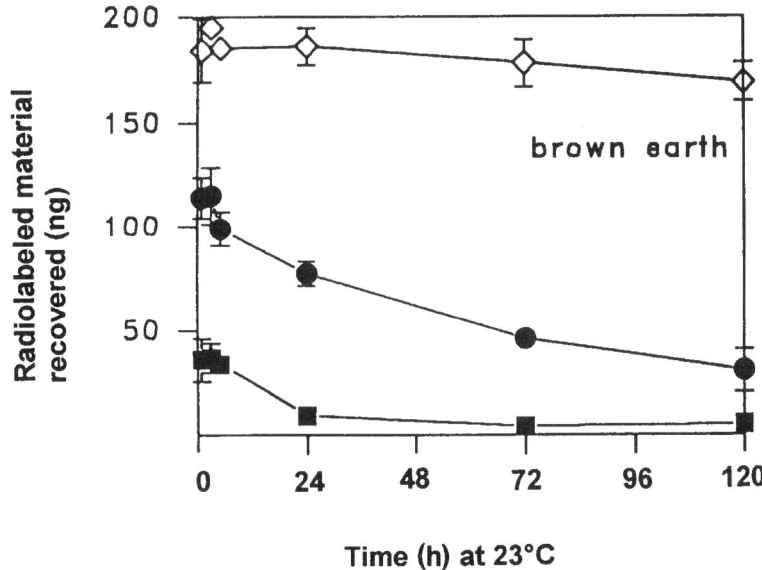

Fig. 13.1 Higher degradation velocity of DNA present in the soil solution (■) than of DNA bound on soil particles (●). Microcosms with 0.7 g of brown earth were loaded each with 300 ng of ³H-thymidine-labeled linear duplex DNA molecules of 41.6 kbp. After various periods (starting at 1 hour after loading), radiolabeled material present in the soil solution was washed out. Subsequently radiolabeled material bound on soil particles was recovered by a specific extraction buffer. In both fractions the amount of acid-precipitable material was quantitated. The total amount of radiolabeled material recovered from the microcosms is also shown (◊). Data of two experiments with deviations from the means are shown.

various microcosm studies, using DNA adsorbed on sea sand (quartz, feldspar, heavy minerals), on sand from groundwater aquifer or on different clay minerals. We have shown that sand-associated DNA is perfectly suitable to transform the soil bacteria *Bacillus subtilis* (Lorenz and Wackernagel, 1988; Lorenz *et al.*, 1988; Romanowski *et al.*, 1993a), *P. stutzeri* (Lorenz and Wackernagel, 1990) and *A. calcoaceticus* (Chamier *et al.*, 1993). Clay-associated DNA can also transform *B. subtilis* cells. In microcosms consisting of mixtures of sand plus clay it was observed that the sand-associated DNA was apparently more efficiently taken up by *B. subtilis* than DNA bound on clay (Lorenz and Wackernagel, 1993). Other experiments also suggest uptake of clay-bound DNA by *B. subtilis* (Khanna and Stotzky, 1992; Gallori *et al.*, 1994). Several experiments have focused on the effect of the chemical milieu in soil on transformation. It was found that *P. stutzeri* JM300 developed competence and was efficiently transformed in media prepared from extracts of various soils (Lorenz and Wackernagel, 1991, 1992). Also, competent cells of *A. calcoaceticus* were transformed in non-sterile soil extract and groundwater (Lorenz *et al.*, 1992), even when DNA was associated with chemically dirty mineral material recovered from a natural groundwater aquifer (Chamier *et al.*, 1993).

In a further step experiments were designed to examine whether competent cells can find and take up DNA bound on the particulate material of a non-sterile soil. Experimentally, transforming DNA was added to brown earth in a microcosm. After a DNA-adsorption period of 1 hour at 23°C the non-adsorbed DNA was removed by washing the soil with soil solution, followed by the introduction of competent *P. stutzeri* JM302 (*his*) cells. Subsequently, transformants were found among the *P. stutzeri* cells recovered from the soil and their frequency was as high as in transformation experiments in soil extract medium. The transformation frequency in soil was linearly dependent on the amount of DNA in the soil (between 10 and 600 ng of DNA added per microcosm; Fig. 13.2). Details on the kinetics and other characteristics of transformation in soil will be presented by Sikorski, Lorenz and Wackernagel (unpublished data). The data suggest that adsorbed DNA is available to competent cells in the complex structured soil environment and that the physico-chemical situation in soil does not impede finding of DNA and uptake by cells.

13.2.3   DEVELOPMENT OF COMPETENCE IN THE ENVIRONMENT

In *B. subtilis*, competence requires the expression of more than 30 genes which appear to be under complex regulation (Dubnau, 1993; Grossman, 1995). Environmental influences can be important stimuli for competence development. Examples are provided by *B. subtilis* and *Azotobacter vinelandii*, which hardly develop competence in complex (broth) media (Page and Sadoff, 1976; Dubnau, 1991b), while *Acinetobacter calcoaceticus* develops competence in broth and minimal medium (Palmen *et al.*, 1993). Nutrients are probably mostly scarce in the soil habitat. With *P. stutzeri* it was observed that single nutrient limitation (limitation of the N-, P- or organic carbon source) can drastically stimulate competence development in synthetic medium (Lorenz and Wackernagel, 1991) as well as in soil extract medium (Lorenz and Wackernagel, 1991, 1992). To put it another way, the chemical environment of soil itself, as represented by the soil extract medium, appears not to contain substances inhibitory to competence development and DNA uptake. Recent preliminary experiments showed that non-competent cells introduced into a microcosm with non-sterile brown earth became competent when a carbon and energy source was added to the soil (Sikorski, Lorenz and Wackernagel, unpublished data).

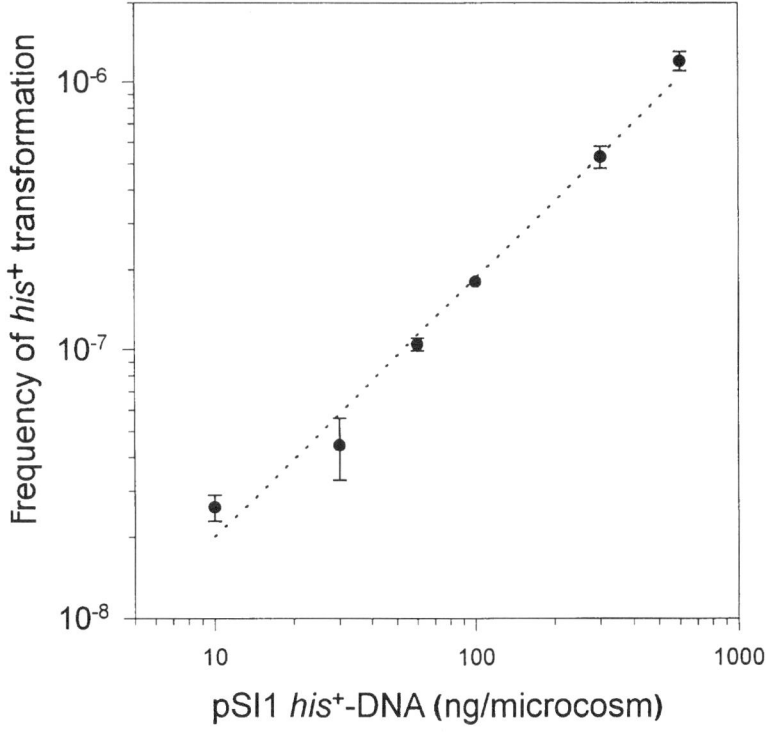

Fig. 13.2 Transformation of *P. stutzeri* JM302 (*his*, Km[R], Nal[R]) cells in non-sterile brown earth by *his*[+] DNA (pSI1) bound on particulate material of the soil matrix. Microcosms, each with 0.7 g of soil, were loaded with the indicated amount of DNA for 1 hour at 23°C (Lorenz and Wackernagel, 1987). The non-adsorbed DNA was washed out with soil solution and 2×10^9 competent cells were introduced. After 4 hours at 23°C the contents of the microcosms were treated with DNaseI (1 mg/ml) which degraded all extracellular DNA (data not shown). The frequency of transformants (*his*[+]) was determined among the *P. stutzeri* cells recovered on medium with antibiotics. The reversion frequency of the *his* marker was 5.6×10^{-9}. Data of two experiments with deviations from the means are shown.

13.2.4   A REGULATORY GENE FOR COMPETENCE IN *P. STUTZERI*

The stimulation of competence by nutrient limitation prompted us to examine competence genes of *P. stutzeri* JM300 in order to find clues to regulatory processes. We have isolated several dozens of transformation-deficient mutants from *P. stutzeri* JM302 (*his*) mutagenized by transposon insertions (Simon *et al.*, 1989). A DNA fragment from the wild-type was cloned which complemented one of the mutants. Nucleotide sequence analysis revealed an ORF highly similar to a gene, being a member of a two-component transcriptional regulatory system described in *P. aeruginosa* (Hobbs *et al.*, 1993). An amino acid sequence identity of 83% with the *pilR* gene product was found which is a transcription factor activated through phosphorylation by a sensor protein kinase (*pilS*). Recently part of the ORF coding for PilS protein was also identified in *P. stutzeri*. These observations suggest that transformability in *P. stutzeri* JM300 is under transcriptional activation control. It is hypothesized that competence in *P. stutzeri* JM300 is regulated by a signal recognition system for eliciting adaptive responses to changes of the environment. Limitation of certain nutrients could be the environmental signal sensed by the cells.

13.2.5   COMPETENCE AND TYPE 4 FIMBRIAE

The *pilR* gene of *P. aeruginosa* is a transcriptional activator of *pilA*, the gene for the protein subunit of type 4 fimbriae (Hobbs *et al.*, 1988, 1993). We asked whether the *P. stutzeri* strain used in these studies produces type 4 fimbriae. This is the case as concluded from the plaque formation of the type 4 fimbriae-specific phage PO4 (Bradley, 1973) on wild-type *P. stutzeri* JM300. However, the phage did not plate on the transformation-deficient mutant with the defective *pilRS* regulatory system. In accordance with the observation in *P. aeruginosa* (Hobbs *et al.*, 1993), this indicates that the transformation-deficient *P. stutzeri* mutant does not produce fimbriae. This suggests a role of type 4 fimbriae in the transformability of *P. stutzeri*.

When all the transformation-deficient mutants were screened for fimbriae production by plating of phage PO4, 44% turned out to have no fimbriae. They are probably defective in steps involved in production, processing or assembly of pilin. These data indicate that formation of type 4 fimbriae in *P. stutzeri* JM300 is an essential prerequisite for transformability, but that other cellular functions represented by the phage PO4-sensitive transformation-deficient mutants are also required. In two other naturally transformable bacteria, *Neisseria gonorrhoeae* and *Moraxella* spp., type 4 fimbriae are also associated with the competence for DNA uptake (Bøvre and Frøholm, 1972; Biswas *et al.*, 1977; Rudel *et al.*, 1995). In *B. subtilis* several proteins with some sequence similarity to pilins and pilin-assembling proteins are required for transformation (Dubnau, 1991a; Chung and Dubnau, 1995).

ACKNOWLEDGMENTS

The work of the laboratory has been supported by grants from the Deutsche Forschungsgemeinschaft and the Bundesministerium für Bildung und Forschung (BMBF) and by the Fonds der Chemischen Industrie.

REFERENCES

Ahrenholtz, I., Lorenz, M.G. and Wackernagel, W. (1994) The extracellular nuclease of *Serratia marcescens*: studies on the activity *in vitro* and effect on transforming DNA in a groundwater aquifer microcosm. *Arch. Microbiol.* **161**: 176–183.

Biswas, G.D., Sox, T., Blackman, E. and Sparling, P.F. (1977) Factors affecting genetic transformation of *Neisseria gonorrhoeae*. *J. Bacteriol.* **129**: 983–992.

Blum, S., Lorenz, M.G. and Wackernagel, W. (1997) Mechanism of retarded DNA degradation and prokaryotic origin of DNases in nonsterile soils. *Appl. Microbiol.* **20**: 513–521.

Bøvre, K. and Frøholm, L.O. (1972) Competence in genetic transformation related to colony type and fimbriation in three species of *Moraxella*. *Acta Path. Microbiol. Scand.* Sect. B. **80**: 649–659.

Bradley, D.E. (1973) Basic characterization of a *Pseudomonas aeruginosa* pilus-dependent bacteriophage with a contractile tail. *J. Virol.* **12**: 1139–1148.

Chamier, B., Lorenz, M.G. and Wackernagel, W. (1993) Natural transformation of *Acinetobacter calcoaceticus* by plasmid DNA adsorbed on sand and groundwater aquifer material. *Appl. Environ. Microbiol.* **59**: 1662–1667.

Chung, Y.S. and Dubnau, D. (1995) ComC is required for the processing and translocation of ComGC, a pilin-like competence protein of *Bacillus subtilis*. *Mol. Microbiol.* **15**: 543–551.

Crabb, W.D., Streips, U.N. and Doyle, R.J. (1977) Selective enrichment for genetic markers in DNA released by competent cultures of *Bacillus subtilis*. *Mol. Gen. Genet.* **155**: 179–183.

DeFlaun, M.F., Paul, J.H. and Jeffrey, W.H. (1987) Distribution and molecular weight of dissolved DNA in subtropical estuarine and oceanic environments. *Mar. Ecol. Progr. Ser.* **38**: 65–73.

Dreiseikelmann, B. (1994) Translocation of DNA across bacterial membranes. *Microbiol. Rev.* **58**: 293–316.

Dubnau, D. (1991a) Genetic competence in *Bacillus subtilis*. *Microbiol. Rev.* **55**: 395–424.

Dubnau, D. (1991b) The regulation of genetic competence in *Bacillus subtilis*. *Mol. Microbiol.* **5**: 11–18.

Dubnau, D. (1993) Genetic exchange and homologous recombination, in *Bacillus subtilis and Other Gram-positive Bacteria*, (eds A.L. Sonenshein, J.A. Hoch and R. Losick), American Society for Microbiology, Washington, DC, pp. 555–584.

Gallori, E., Bazzicalupo, M., Dal Canto, L. *et al.* (1994) Transformation of *Bacillus subtilis* by DNA bound on clay in non-sterile soil. *FEMS Microbiol Ecol.* **15**: 119–126.

Greaves, M.P. and Wilson, M.J. (1970) The degradation of nucleic acids and montmorillonite–nucleic-acid complexes by soil microorganisms. *Soil Biol. Biochem.* **2**: 257–268.

Grossman, A.D. (1995) Genetic networks controlling the initiation of sporulation and the development of genetic competence in *Bacillus subtilis*. *Annu. Rev. Genet.* **29**: 477–508.

Hobbs, M., Dalrymple, B., Delaney, S.F. and Mattick, J.S. (1988) Transcription of the fimbrial subunit gene and an associated transfer RNA gene of *Pseudomonas aeruginosa*. *Gene* **62**: 219–227.

Hobbs, M., Collie, E.S.R., Free, P.D. *et al.* (1993) PilS and PilR, a two-component transcriptional regulatory system controlling expression of type 4 fimbriae in *Pseudomonas aeruginosa*. *Mol. Microbiol.* **7**: 669–682.

Khanna, M. and Stotzky, G. (1992) Transformation of *Bacillus subtilis* by DNA bound on montmorillonite and effect of DNase on the transforming ability of bound DNA. *Appl. Environ. Microbiol.* **58**: 1930–1939.

Lorenz, M.G. and Wackernagel, W. (1987) Adsorption of DNA to sand and variable degradation rates of adsorbed DNA. *Appl. Environ. Microbiol.* **53**: 2948–2952.

Lorenz, M.G. and Wackernagel, W. (1988) Impact of mineral surfaces on gene transfer by transformation in natural bacterial environments, in *Risk Assessment for Deliberate Releases* (ed. W. Klingmüller), Springer-Verlag, Berlin Heidelberg, pp. 110–119.

Lorenz, M.G. and Wackernagel, W. (1990) Natural genetic transformation of *Pseudomonas stutzeri* by sand-adsorbed DNA. *Arch. Microbiol.* **154**: 380–385.

Lorenz, M.G. and Wackernagel, W. (1991) High frequency of natural genetic transformation of *Pseudomonas stutzeri* in soil extract supplemented with a carbon/energy and phosphorus source. *Appl. Environ. Microbiol.* **57**: 1246–1251.

Lorenz, M.G. and Wackernagel, W. (1992) Stimulation of natural genetic transformation of *Pseudomonas stutzeri* in extracts of various soils by nitrogen or phosphorus limitation and influence of temperature and pH. *Microb. Releases* **1**: 173–176.

Lorenz, M.G. and Wackernagel, W. (1993) Transformation as a mechanism for bacterial gene transfer in soil and sediment – studies with a sand/clay microcosm and the cyanobacterium *Synechocystis* OL50, in *Trends in Microbial Ecology* (ed. R.P.-A. Guerrero), Spanish Society for Microbiology, Barcelona, pp. 325–330.

Lorenz, M.G. and Wackernagel, W. (1994) Bacterial gene transfer by natural genetic transformation in the environment. *Microbiol. Rev.* **58**: 563–602.

Lorenz, M.G., Aardema, B.W. and Wackernagel, W. (1988) Highly efficient genetic transformation of *Bacillus subtilis* attached to sand grains. *J. Gen. Microbiol.* **134**: 107–112.

Lorenz, M.G., Reipschläger, K. and Wackernagel, W. (1992) Plasmid transformation of naturally competent *Acinetobacter calcoaceticus* in non-sterile soil extract and groundwater. *Arch. Microbiol.* **157**: 355–360.

Page, W.J. and Sadoff, H.L. (1976) Physiological factors affecting transformation of *Azotobacter vinelandii*. *J. Bacteriol.* **125**: 1080–1087.

Paget, E. and Simonet, P. (1994) On the track of natural transformation in soil. *FEMS Microbiol Ecol.* **15**: 109–118.

Paget, E., Jocteur-Monrozier, L. and Simonet, P. (1992) Adsorption of DNA on clay minerals: protection against DNaseI and influence on gene transfer. *FEMS Microbiol. Lett.* **97**: 31–40.

Palmen, R., Vosman, B., Buijsman, P. *et al.* (1993) Physiological characterization of natural transformation of *Acinetobacter calcoaceticus*. *J. Gen. Microbiol.* **139**: 295–305.

Palmen, R., Driessen, A.J.M. and Hellingwerf, K.J. (1994) Bioenergetic aspects of the translocation of macromolecules across bacterial membranes. *Biochim. Biophys. Acta Bio-Energetics* **1183**: 417–451.

Paul, J.H., Jeffrey, W.H., David, A.W. *et al.* (1989) Turnover of extracellular DNA in eutrophic and oligotrophic freshwater environments of southwest Florida. *Appl. Environ. Microbiol.* **55**: 1823–1828.

Romanowski, G., Lorenz, M.G. and Wackernagel, W. (1991) Adsorption of plasmid DNA to mineral surfaces and protection against DNase I. *Appl. Environ. Microbiol.* **57**: 1057–1061.

Romanowski, G., Lorenz, M.G., Sayler, G. and Wackernagel, W. (1992) Persistence of free plasmid DNA in soil monitored by various methods, including a transformation assay. *Appl. Environ. Microbiol.* **58**: 3012–3019.

Romanowski, G., Lorenz, M.G. Lorenz and Wackernagel, W. (1993a) Plasmid DNA in a groundwater aquifer microcosm – adsorption, DNase resistance and natural genetic transformation of *Bacillus subtilis*. *Mol. Ecol.* **2**: 171–181.

Romanowski, G., Lorenz, M.G. and Wackernagel, W. (1993b) Use of polymerase chain reaction and electroporation of *Escherichia coli* to monitor the persistence of extracellular plasmid DNA introduced into natural soils. *Appl. Environ. Microbiol.* **59**: 3438–3446.

Rudel, T., Facius, D., Barten, R. *et al.* (1995) Role of pili and the phase-variable PilC protein in natural competence for transformation of *Neisseria gonorrhoeae*. *Proc. Natl Acad. Sci. USA* **92**: 7986–7990.

Simon, R., Quandt, J. and Klipp, W. (1989) New derivatives of transposon Tn5 suitable for mobilization of replicons, generation of operon fusions and induction of genes in Gram-negative bacteria. *Gene* **80**: 161–169.

Solomon, J.M. and Grossman, A.D. (1996) Who's competent and when: regulation of natural genetic competence in bacteria. *Trends Genet.* **12**: 150–155.

Spring, S., Amann, R., Ludwig, W. *et al.* (1992) Phylogenetic diversity and identification of non-culturable magnetotactic bacteria. *Syst. Appl. Microbiol.* **15**: 116–122.

Steffan, R.J., Goksoyr, J., Bej, A.K. and Atlas, R.M. (1988) Recovery of DNA from soils and sediments. *Appl. Environ. Microbiol.* **54**: 2908–2915.

The importance of pseudolysogeny to *in situ* bacteriophage–host interactions

14

Robert V. Miller and Steven Ripp

SUMMARY

Bacteriophages occur in high numbers in aquatic environments and are significant mediators of microbial survival, evolution and activity. *In situ* studies demonstrate the importance of bacteriophages in horizontal gene transfer via transduction, and an important potential for transduction as a mediator of bacterial evolution in the natural environment has been demonstrated. However, physiological and molecular interactions between microbial populations and their phages *in situ* have been largely ignored. Our current knowledge of phage–host interactions rely on studies performed with well-fed laboratory-grown host organisms. Such studies do not predict the high titers of bacterial viruses that are maintained in the environmental setting. We have investigated phage–host interactions under nutrient-depleted conditions commonly encountered in natural environments. We chose a small freshwater lake as our test site and have employed both continuous-culture microcosms and *in situ* studies to explore phage–bacterial interactions in nutrient-depleted and starved hosts. As a result of these studies, we encountered the phenomenon of pseudolysogeny, which refers to a phage–host interaction where the viral nucleic acid is neither established as a true prophage nor does it elicit a lytic infection of the host. Rather, the phage nucleic acid simply resides within the cell in a non-active form. We hypothesize that, due to the cell's starved state, there is insufficient energy available for the phage to initiate either a temperate or lytic life cycle. Upon nutrient addition, the pseudolysogenic response is resolved into either a temperate or lytic response. Sequestering of phage genomes in the pseudolysogenic state effectively increases phage half-lives, leading to maintenance of viruses that otherwise would not exist in the environment. Pseudolysogeny may, in part, explain the large phage populations observed in aquatic environments.

14.1 INTRODUCTION

During the last two decades, virus-mediated horizontal transfer of genetic material (transduction) has been recognized as a potentially significant gene exchange mechanism among bacteria in aquatic habitats (Baross *et al.*, 1978; Morrison *et al.*, 1978; Saye *et al.*, 1987, 1990; Ripp and Miller, 1995). Both plasmid and chromosomal DNA can be transferred among bacteria in these environments. In addition, the occurrence of transduction among bacteria has been shown to act to maintain novel phenotypes in a bacterial population that would otherwise be eliminated from the gene pool if horizontal gene transfer did not occur (Replicon *et al.*, 1995).

However, transduction has only recently been considered as an important environmental phenomenon because the numbers of bacterial viruses in environmental ecosystems were regarded as insignificant until, in 1989, Oivind Bergh and co-workers at the University of Bergen in Norway reported viral populations in ocean waters to be three to seven times higher than previously assumed (Bergh *et al.*, 1989). Observations of samples using transmission electron microscopy revealed as many as 1×10^8 bacterial viruses/ml of water. However, our current knowledge of viral behaviors in environmental ecosystems does not support such high numbers. We have been investigating potential mechanisms that would increase the effective half-lives of viruses in nature. Such changes would increase viral reservoirs and lead to higher numbers of phage particles being observed. We have previously demonstrated that a process referred to as pseudolysogeny occurs in nature among viruses that only demonstrate lytic growth patterns in the laboratory. Pseudolysogeny may, in part, be responsible for maintaining high-density viral populations in the environment.

Pseudolysogeny describes a phage–host interaction in which the phage, upon infecting its host, does not initiate either a lysogenic or a lytic response but, rather, simply exists within the cell in an unstable, non-active form (Baess, 1971). This response occurs because the host cell is starved and cannot provide the phage with the necessary energy required for its expression. Cellular starvation in environmental ecosystems is commonplace. When a nutrient source is provided, however, the phage acquires its essential energy and is activated to initiate virion formation and subsequent cell lysis. The overall result of pseudolysogeny is an extension of the viral genome's half-life, which may, perhaps, explain the occurrence of large environmental phage populations.

In the study reported here, we wished to determine whether viruses that establish true lysogeny in their hosts under laboratory conditions would be able to do so in the environment. Alternatively, these viruses might also be forced into the pseudolysogenic state due to the starved condition of the host cells they infected.

14.2   RESULTS AND DISCUSSION

14.2.1   EFFECT OF STARVATION ON BACTERIOPHAGE–HOST INTERACTIONS

We added yeast extract to a starved culture containing *Pseudomonas aeruginosa* strains PAO515 (Rella and Hass, 1982) and RM132 (Saye *et al.*, 1990). Strain RM132 is lysogenic for bacteriophage F116 (Miller *et al.*, 1974) and PAO515 serves as a recipient cell sensitive to infection by phage F116. The culture medium consisted of *Pseudomonas* minimal medium lacking sodium citrate (PMM-c) but with yeast extract added at a very low concentration of $1 \times 10^{-5}\%$ (Replicon *et al.*, 1995). After 55 days of starvation, a nutrient spike (yeast extract at a final concentration of $1 \times 10^{-5}\%$) was added to the culture, simulating the feast–famine conditions typically encountered by microorganisms in the natural environment. Within 1–2 days of this spike, a small increase in the phage-to-bacterium ratio (PBR) and in the percentage of cells exhibiting lysogenic characteristics was seen (Fig. 14.1). Nutrient spikes continued on a weekly basis up to 110 days, with PBR and lysogen (measured by determining the number of clones releasing phage virions) increases occurring after each nutrient spike. No increases were observed in sister cultures that were not spiked with nutrient. These data illustrate that nutrient availability dramatically affects phage–host interactions and that, under starvation conditions, the establishment of true lysogeny and the production of virions are minimal at best.

14.2.2   INITIAL EVIDENCE OF A PSEUDOLYSOGENIC STATE

We next established a microcosm culture containing strain PAO515 in PMM-c. We removed three submicrocosm cultures from this main microcosm after a 40-day period of starvation (Fig. 14.2). To each submicrocosm were added phage F116 virions, which were allowed to infect PAO515 cells over a 6-hour period, after which all free virions were

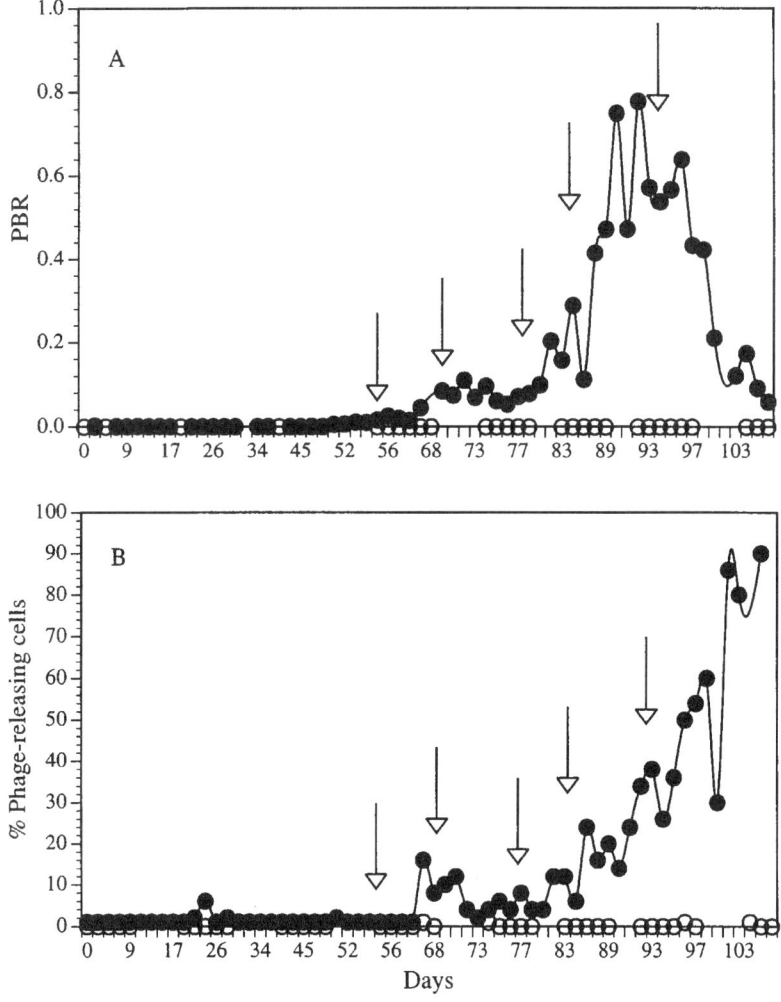

Fig. 14.1 (A) Phage-to-bacterium ratio (PBR) and (B) percentage of cells actively releasing phage in a microcosm containing PMM-c and yeast extract at 1 × 10⁻⁵%. Microcosms were inoculated with PAO515 and spiked with nutrient at times indicated by arrows. Bacteriophage counts were determined by filtering samples through 0.45 μm syringe filters to remove bacterial cells. The remaining phages were then diluted and 0.1 ml of the desired dilution combined with 0.1 ml of mid-exponential PAO1 (prototrophic; Miller and Ku, 1978) cells and 2.5 ml λ top agar (Miller and Ku, 1978). This mixture was poured on to LB (Miller and Ku, 1978) agar plates and the resulting PFU were counted following overnight incubation at 37°C. Viable counts were determined on LB agar plates and the PBR was determined. Cells actively releasing phages were identified by replica-plating 100 colonies per sample on to a top-agar overlay of PAO1 and incubating overnight at 37°C. A zone of lysis surrounding a colony was indicative of the spontaneous release of virions. ○: no nutrient added; ●: nutrient added.

removed by centrifugation (Saye *et al.*, 1987). In the first submicrocosm, yeast extract was added 24 hours after phage addition; in the second, yeast extract was added simultaneously with phage addition; in the third, no nutrient was added. Our hypothesis of pseudolysogeny is that a phage infects a starved cell but, due to the cell's starved state, there is not enough energy available for the phage to replicate itself. Therefore, enhanced virion production and increased frequency of phage-releasing clones should only be seen in the first submicrocosm, where the phages were allowed to infect and subsequently reside within the cell in a non-active state. Once this occurs, the addition of nutrient 24 hours later will supply the required energy for the pseudoprophages to activate themselves for either the establishment of true lysogeny or lytic growth. Figure 14.2 shows that this does indeed occur. Simultaneous phage and nutrient addition had little effect on phage production and lysogeny establishment except for an expected small initial burst of phages. After this occurs, nutrient supplies are again exhausted and no further phage production develops. However, when phage virions were allowed to interact with the host before nutrient was added, a larger number of phages were produced and the frequency of true lysogeny was greatly increased over the results obtained with other treatments.

To verify our theory further, we performed another experiment in which strains PAO515 and RM132 were combined in a batch culture of PMM-c and allowed to starve over a 27-day period. Subcultures were removed on days 6, 12, 20 and 27. They were centrifuged to remove free phage particles and norfloxacin was added to eliminate viable RM132 cells and thus prevent further phage production in the culture (Saye *et al.*, 1987). The remaining PAO515 cells were spiked with yeast extract at various intervals and assayed for their degree of phage production and frequency of lysogenized cells capable of releasing phage (Table 14.1). The submicrocosms removed earliest, and thus containing the least-starved cells, displayed immediate phage production. Although subcultures removed at later times exhibited increased lag times and required successively more nutrient spikes before phage and lysogen numbers increased, increases in these parameters occurred even in the 27-day culture (a total incubation time of 43 days). These data demonstrate that phage genomes are capable of a prolonged existence in a state of limbo within their starved hosts. We refer to such phage genomes as preprophages. As cellular starvation levels increase, so does the quantity of energy required to revitalize the cell and, in turn, activate the preprophage. From Table 14.1, we can see that as the days of incubation advance and the degree of starvation increases, successively larger amounts of energy are necessary for phage production and lysogen establishment.

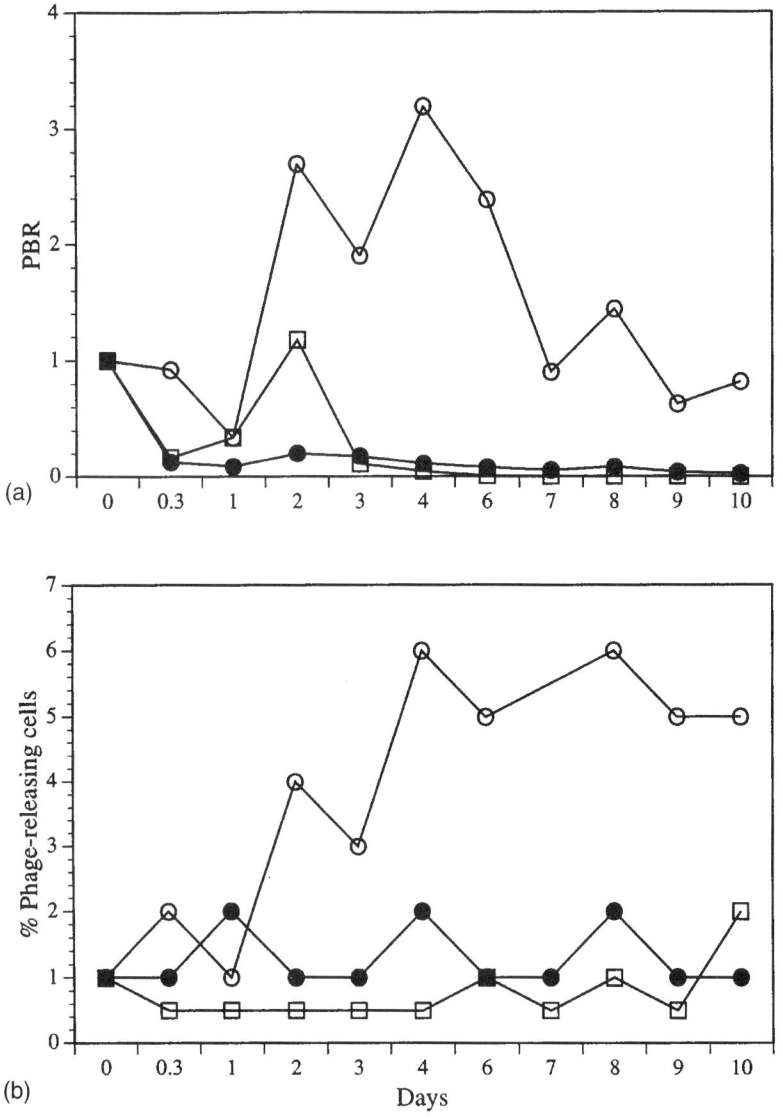

(a)

(b)

Fig. 14.2 (A) Phage-to-bacterium ratios (PBR) and (B) percentage of cells actively releasing phage in submicrocosms removed from a main batch culture inoculated with PAO515. ○, bacteriophage F116 was added 24 hours after submicrocosm was removed, allowed to infect for 6 hours and then removed by centrifugation. Yeast extract at a final concentration of $1 \times 10^{-5}\%$ was added 24 hours later. ●, bacteriophage F116 was added and removed as explained above. No yeast extract addition occurred. □, bacteriophage F116 was added and removed 6 hours later. Nutrient was added simultaneous to phage addition.

Table 14.1 Viable counts (CFU/ml), phage-to-bacterium ratios (PBR) and percentage of cells releasing phage virions in submicrocosms removed from a main microcosm after 6, 12, 20 or 27 days

Day	Nutrient added	Day 6			Day 12			Day 20			Day 27		
		CFU/ml	PBR ($\times 10^{-4}$)	% Cells releasing phage	CFU/ml	PBR ($\times 10^{-4}$)	% Cells releasing phage	CFU/ml	PBR ($\times 10^{-4}$)	% Cells releasing phage	CFU/ml	PBR ($\times 10^{-4}$)	% Cells releasing phage
0		5.7×10^6	0.09	<1	1.4×10^7	0.09	<1	3.1×10^7	0.01	<1	4.4×10^7	0.009	<1
1	X	1.5×10^6	2	<1	8.8×10^6	0.17	<1	1.7×10^7	0.01	<1	1.9×10^7	0.005	<1
2		5.0×10^4	54	<1	1.3×10^6	2	<1	2.3×10^6	15	<1	1.4×10^7	0.007	<1
4		2.9×10^4	107	<1	5.9×10^5	16	<1	2.6×10^6	3	<1	1.1×10^7	0.10	<1
6	X	9.7×10^4	16	<1	1.6×10^5	27	<1	2.8×10^6	1.0	<1	1.0×10^7	0.15	<1
8		8.3×10^4	87	<1	1.7×10^4	176	<1	3.4×10^6	1.6	<1	1.0×10^6	3	<1
10		2.1×10^5	62	11	1.0×10^3	1500	<1	2.0×10^4	650	<1	6.7×10^5	6	<1
12	X	1.0×10^6	61	72	1.5×10^4	180	<1	2.0×10^2	3150	<1	6.7×10^5	3	<1
14		1.0×10^6	8	65	1.0×10^5	3	<1	5.0×10^4	64	90	1.3×10^4	85	4
16		1.6×10^7	22	81	1.0×10^6	1	97	8.4×10^6	0.48	96	6.7×10^3	537	30
18		1.4×10^7	71	98	1.6×10^7	0.07	93	–	–	–	–	–	–
20		2.3×10^7	10	90	–	–	–	1.1×10^7	0.06	96	–	–	–

Submicrocosms were filtered to remove virions and norfloxacin was added to terminate production of new virions by induction of RM132 cells. Nutrient (1×10^{-5}% yeast extract, final concentration) was added at times indicated by X. Dashes (–) indicate that no sample was taken.

14.2.3   CONTINUOUS CULTURE

Continuous culture chemostats were used to maintain cultures of PAO515 and RM132 under idealized starvation conditions for extended periods of time (Replicon *et al.*, 1995). Again, PMM-c was used as the culture medium with yeast extract added at a final concentration of $1 \times 10^{-5}\%$ as a low-level carbon and energy source. Chemostats were run at a generation time (turnover time) of 14 hours. Samples were removed and assayed for the number of cells actively releasing phage (i.e. true lysogens) and, through colony hybridization, the total number of cells containing phage F116 DNA. The difference between these two values represents the number of cells that contain F116 DNA but do not actively release phage (pseudolysogens). As can be seen in Table 14.2, a large proportion (18–83%) of the population of cells that contained F116 DNA appeared to exist in a pseudolysogenic state. This is probably an under-estimation of the true proportion of pseudolysogenic cells in the starved chemostat because many of the cells releasing phage were most likely activated due to adsorption of nutrients during incubation on the plating medium.

Those colonies exhibiting pseudolysogeny were streaked onto Luria–Bertani agar plates to produce individual clonal colonies, which were then colony hybridized with F116 DNA. In so doing, we discovered that only a small number of cells in the original pseudolysogenic colonies actually contained phage F116 DNA. Thus, it appears that the

Table 14.2 Frequency of pseudolysogens and true lysogens occurring in an idealized starved chemostat culture

Day	Fraction of cells containing phage genome[a]	% Activated[b]	% Not activated[b]
0	0/266	0	0
7	23/2000	1	0
10	51/850	17	83
15	403/1200	79	21
23	189/450	81	19
28	88/130	82	18
34	112/215	81	19

[a] Cells containing phage DNA were determined by colony hybridization. Probes were constructed by isolating F116 DNA according to Silhavy *et al.* (1984) and labeling with a Genius Nonradioactive Labeling Kit (Boehringer Mannheim Biochemicals). Colony hybridizations were carried out as outlined in the Genius protocol (Anonymous, 1992) using Magnagraph 0.45 μm (mean pore size), 85 mm diameter nylon membranes (Micron Separations).
[b] Percentage of cells containing the phage genome that are either activated or not activated to release phage.

preprophage is not a stable entity within the pseudolysogenized cell. Presumably, due to the cell's starved state, there is not enough energy available for the infecting phage genome to become stably established in the host cell. Therefore, this preprophage is not replicated in synchrony with the host genome – which is similar to the process of abortive transduction, where the transduced allele does not integrate within its host cell's genome (Arber, 1994). Only one of the two daughter cells from each division cycle acquires the exogenote allele.

14.2.4   *IN SITU* INCUBATED MICROCOSMS

In situ incubations were performed at Lake Sanborn, a small, semi-oligotrophic freshwater lake near Stillwater, Oklahoma, with a mean depth of approximately 2 m (Heath and Francko, 1988). Microcosms consisted of one-liter Lifecell tissue culture chambers (Fenwal Laboratories, Deerfield, IL) which were filled with 500 ml filter-sterilized (0.2 μm) lake water and inoculated with PAO515 and RM132 (Saye *et al.*, 1990). Half of the microcosms were incubated on the lake surface. The other half were incubated on the bottom of the lake at a depth of 2 m. Some microcosms periodically received nutrient spikes of yeast extract at a final concentration of $1 \times 10^{-5}\%$.

Only the microcosm incubated on the lake's surface and spiked with yeast extract exhibited large concentrations of cells infected with phage (Table 14.3). Of these, a high percentage contained phage DNA but were not actively releasing virions. The other microcosm on the surface which was not nutrient spiked produced few phage-containing cells and very few pseudolysogenic cells. As was seen in the laboratory experiments, nutrient addition plays a vital role in phage–host interactions.

The microcosms incubated on the lake bottom produced virtually no phage-releasing cells. However, approximately 1–2% of the population routinely exhibited pseudolysogenic characteristics. In the nutrient-spiked microcosm, this value had risen to 20% by the end of the experiment, when cells were most starved. The lack of phage-producing cells in these microcosms was most likely due to the lack of exposure to ultraviolet (UV) radiation, which has previously been shown to induce F116 prophages to lytic growth (Kidambi *et al.*, 1996; Kokjohn and Miller, 1985). Regan *et al.* (1992) have shown that the biologically effective DNA-damaging dose of UVB (activating solar UV wavelengths) only penetrates to 2 m or less in estuarine waters. Our microcosms, incubated on the lake bottom, resided at a depth of 2 m in very cloudy water and were unlikely to be affected by solar UV light.

To test whether starvation eliminates the ability of UV to induce F116 prophage from true lysogens, we exposed phage F116-infected cells to $5\,J/m^2$ of UV-C radiation at various times over a 36-day starvation period (Fig. 14.3). Prophages remained inducible throughout this time.

Table 14.3 *In situ* microcosms incubated in Lake Sanborn[a] either on the lake surface or on the lake bottom at a depth of 2 m (yeast extract was periodically added to one of each of the surface and bottom microcosms)

	Surface						Bottom					
	No nutrient added			Nutrient added			No nutrient added			Nutrient added		
Day	Fraction of cells containing phage genome	Activated[b] (%)	Not activated (%)	Fraction of cells containing phage genome	Activated (%)	Not activated (%)	Fraction of cells containing phage genome	Activated (%)	Not activated (%)	Fraction of cells containing phage genome	Activated (%)	Not activated (%)
0	0/840	0	0	0/580	0	0	4/660	0	100	0/910	0	0
6	6/630	100	0	9/660	0	100	9/900	0	100	5/440	0	100
7	20/1600	100	0	3/520	0	100	1/93	0	100	7/500	0	100
9	9/1200	100	0	1/180	0	100	22/1100	0	100	1/140	0	100
12	3/530	100	0	7/350	0	100	6/870	0	100	11/2100	0	100
14	5/900	100	0	11/183	93	7	4/380	0	100	3/580	0	100
21	53/3000	50	50	323/449	86	14	8/795	0	100	14/1900	0	100
28	6/120	80	20	377/670	93	7	–	–	–	2/160	0	100
35	5/860	100	0	261/287	81	19	30/1200	0	100	8/240	0	100
41	1/110	100	0	365/406	78	22	2/210	0	100	81/800	20	80

a Lakewater sampling and *in situ* incubations were performed at Lake Sanborn, a small (5 ha), semi-oligotrophic freshwater lake near Stillwater, Oklahoma, with a mean depth extending to approximately 2 m and very low abiogenic turbidity (Heath and Francko, 1988). Phosphate has been shown to be the limiting nutrient in this environment (Heath and Francko, 1988). Temperature (25 ± 3°C), conductivity (90 ± 30 µS), and relevant weather conditions were obtained at the time of each sampling.

b Percentage of cells which contain the phage genome that are either activated or not activated to release phage.

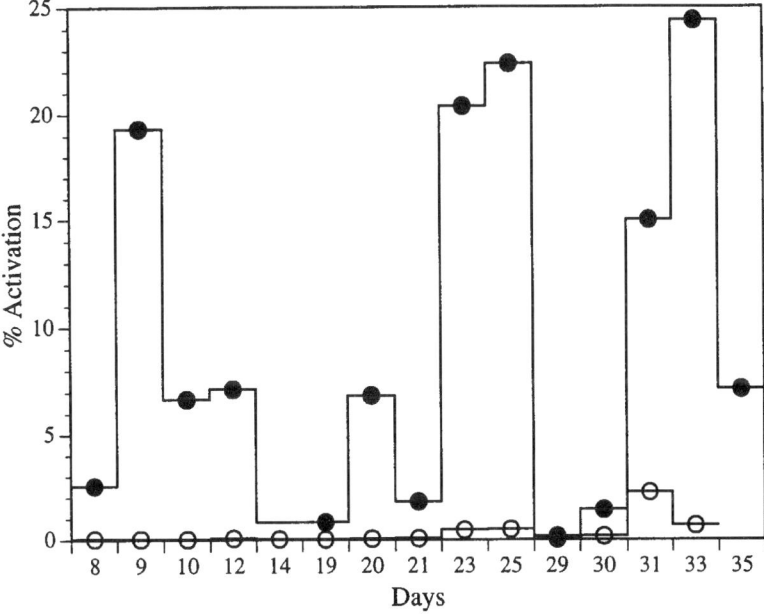

Fig. 14.3 Induction of starved F116-infected cells by ultraviolet (UV) light. ○, cells not exposed; ●, cells exposed to 5 J/m² UV-C.

14.3   CONCLUSIONS

We have shown in this study that pseudolysogenic relationships between temperate bacteriophages and their hosts are possible and actually appear quite prevalent in environmental ecosystems where the effects of nutrient limitation augment the response. These results are consistent with earlier observations we have made on pseudolysogeny in environmental interactions between lytic bacteriophages and their hosts (Ripp and Miller, 1997). As a direct result, bacteriophages are able to survive within their hosts for extended periods. This prolonged life span of phage genomes in the environment may provide an expanded natural reservoir of viruses leading to the large phage populations reported by other investigators (Bergh *et al.*, 1989; Proctor & Fuhrman, 1990; Paul *et al.*, 1991; Bratbak and Heldal, 1993; Hennes and Suttle, 1995). Our laboratory has previously shown that transduction (the transfer of genetic material by bacteriophages) is routinely encountered in natural ecosystems and therefore must be taken into consideration when assessing the risks involved with gene transfer between genetically engineered microorganisms and the indigenous microbial population (Miller and

Levy, 1989; Miller *et al.*, 1992). Pseudolysogeny may increase potential risk by increasing the probability of exposure of the natural population to transduction events, thereby increasing the potential for horizontal gene transfer between introduced and natural populations of bacteria.

ACKNOWLEDGMENTS

This work was supported by cooperative agreement CR820060 with the Environmental Research Laboratory, United States Environmental Protection Agency, Gulf Breeze, FL.

REFERENCES

Anonymous (1992) *The Genius System User's Guide for Filter Hybridization, Version 2.0*, Boehringer Mannheim Corporation, Indianapolis, Indiana.

Arber, W. (1994) Bacteriophage transduction, in *Encyclopedia of Virology*, Vol. 1, (eds R.G. Webster and A. Granoff), Academic Press, UK, pp. 107–113.

Baess, I. (1971) Report on a pseudolysogenic mycobacterium and a review of the literature concerning pseudolysogeny. *Acta Path. Microbiol. Scand.* **79**: 428–434.

Baross, J.A., Liston, J. and Morita, R.Y. (1978) Incidence of *Vibrio parahaemolyticus* bacteriophages and other *Vibrio* bacteriophages in marine samples. *Appl. Environ. Microbiol.* **36**: 492–499.

Bergh, O., Borsheim, K.Y., Bratback, G. and Heldal, M. (1989) High abundance of viruses found in aquatic environments. *Nature.* **340**: 467–468.

Bratbak, G. and Heldal, M. (1993) Total count of viruses in aquatic environments, in *Handbook of Methods in Aquatic Microbial Ecology* (eds P. Kemp, B.F. Sherr, E.B. Sherr and J.J. Cole), Lewis Publishers, Boca Raton, FL, pp. 135–138.

Heath, R.T. and Francko, D.A. (1988) Comparison of phosphorous dynamics in two Oklahoma reservoirs and a natural lake varying in abiogenic turbidity. *Can. J. Fish. Aquat. Sci.* **45**: 1480–1486.

Hennes, K.P. and Suttle, C.A. (1995) Direct counts of viruses in natural waters and laboratory cultures by epifluorescence microscopy. *Limnol. Oceanogr.* **40**(6): 1050–1055.

Kidambi, S.P., Booth, M.G., Kokjohn, T.A. and Miller, R.V. (1996) *recA*-dependence of the response of *Pseudomonas aeruginosa* to UVA and UVB irradiation. *Microbiol.* **142**: 1033–1040.

Kokjohn, T.A. and Miller, R.V. Miller (1985) Molecular cloning and characterization of the *recA* gene of *Pseudomonas aeruginosa*. *J. Bacteriol.* **163**: 568–572.

Miller, R.V. and Ku, C.M.C. (1978) Characterization of *Pseudomonas aeruginosa* mutants deficient in the extablishment of lysogeny. *J. Bacteriol.* **134**: 875–883.

Miller, R.V. and Levy, S.B. (1989) Horizontal gene transfer in relation to environmental release of genetically engineered microorganisms, in *Gene Transfer in the Environment*, (eds S.B. Levy and R.V. Miller), McGraw-Hill, New York, pp. 405–420.

Miller, R.V., Pemberton, J.M. and Richards, K.E. (1974) F116, D3, and G101: temperate bacteriophages of *Pseudomonas aeruginosa*. *Virology* **59**: 566–569.

Miller, R.V., Ripp, S., Replicon, J. *et al.* (1992) Virus-mediated gene transfer in freshwater environments, in *Gene Transfer and Environment*, (ed. M.J. Gauthier), Springer-Verlag, Berlin, pp. 51–62.

Morrison, W.D., Miller, R.V. and Sayler, G.S. (1978) Frequency of F116 mediated transduction of *Pseudomonas aeruginosa* in a freshwater environment. *Appl. Environ. Microbiol.* **36**: 724–730.

Paul, J.H., Jiang, S.C. and Rose, J.B. (1991) Concentration of viruses and dissolved DNA from aquatic environments by vortex flow filtration. *Appl. Environ. Microbiol.* **57**: 2197–2204.

Proctor, L.M. and Fuhrman, J.A. (1990) Viral mortality of marine bacteria and cyanobacteria. *Nature* **343**: 60–62.

Regan, J.D., Carrier, W.L., Gucinski, H. *et al.* (1992) DNA as a solar dosimeter in the ocean. *Photochem. Photobiol.* **56**: 35–42.

Rella, M., and Hass, D. (1982) Resistance of *Pseudomonas aeruginosa* PAO to nalidixic acid and low levels of B-lactam antibiotics: mapping of chromosomal genes. *Antimicrob. Agents Chemother.* **22**: 242–249.

Replicon, J., Frankfater, A. and Miller, R.V. (1995) A continuous culture model to examine factors that affect transduction among *Pseudomonas aeruginosa* strains in freshwater environments. *Appl. Environ. Microbiol.* **61**: 3359–3366.

Ripp, S. and Miller, R.V. (1995) Effects of suspended particulates on the frequency of transduction among *Pseudomonas aeruginosa* in a freshwater environment. *Appl. Environ. Microbiol.* **61**(4): 1214–1219.

Ripp, S. and Miller, R.V. (1997) The role of pseudolysogeny in bacteria–host interactions in a natural freshwater environment. *Microbiology* **143**: 2065–2070.

Saye, D.J., Ogunseitan, O., Sayler, G.S. and Miller, R.V. (1987) Potential for transduction of plasmids in a natural freshwater environment: effect of donor concentration and a natural microbial community on transduction in *Pseudomonas aeruginosa*. *Appl. Environ. Microbiol.* **53**: 987–995.

Saye, D.J., Ogunseitan, O.A., Sayler, G.S. and Miller, R.V. (1990) Transduction of linked chromosomal genes between *Pseudomonas aeruginosa* during incubation *in situ* in a freshwater habitat. *Appl. Environ. Microbiol.* **56**: 140–145.

Silhavy, T.J., Berman, M.L. and Enquist, L.W. (1984) *Experiments with Gene Fusions*, Cold Spring Harbor Laboratories, Cold Spring Harbor, NY.

Archeological footprints of horizontal gene transfer: mosaic cell surface proteins in *S. pyogenes* and *S. pneumoniae*

15

Susan K. Hollingshead, Debra E. Bessen and David E. Briles

SUMMARY

The flux of the host/parasite interaction depends in part on the natural selection that drives changes in cell surface antigens. Serological change allows escape from host adaptive immune response. In the case of protein antigens, the primary record of serological change is in the nucleotide sequences encoding the antigens. Thus, nucleotide sequence variation among gene alleles encoding serotype-specific cell surface proteins reveals the archeological footprints shaped by evolutionary forces that accumulated serologic change in these molecules over time. The footprints show that a major force in operation was horizontal gene transfer.

   This concept is illustrated with examples arising from two gene families composed of mosaic genes: the PspA protein family in *Streptococcus pneumoniae* and the M protein family in *S. pyogenes*. Both PspA proteins and M proteins are microbial cell surface antigens that elicit protective immunity in humans and both families consist of serologically varied proteins encoded by mosaic genes. But the differing population structure of these two distinct bacterial species influences the evolution of the gene families and the pattern of mosaicism seen in each case. For *S. pneumoniae*, individual mosaic alleles of a single PspA protein show evidence for multiple recent recombinatorial events which may result from genetic exchange events occurring rather frequently. This is shown by a sliding window analysis of intragenic segments of the *pspA* genes. Frequent horizontal genetic exchange is consistent with recent data supporting a panmictic or epidemic population structure in this species. For *S. pyogenes*, individual mosaic alleles encoding a single M protein and even *emm* gene clusters carrying up to three specific *emm* alleles are often associated with a single clonal strain which may be widely distributed through time and geographical location. In this case, although the mosaic genes themselves and the varied gene clusters were initially formed in part by recombination following horizontal gene transfer, they are not currently in random association with each other or with the chromosomal background of the strains in which they are expressed.

15.1 INTRODUCTION

Over the last several years, our general concept of the population structure(s) of bacteria has been changing, leading to an increased awareness of the impact of population structures on the mode of bacterial gene evolution. Bacteria are primarily haploid organisms that divide by fission to reproduce clones of themselves, a process which dictates a vertical transmission of genes and produces what we call a clonal population structure. However, bacteria are also capable of the lateral transfer of genetic information by genetic transfer mechanisms such as transformation, transduction or conjugation, which produces the horizontal transmission of genes or gene segments. Bacterial species can differ in their relative frequencies of vertical versus horizontal transmission of information; the difference is a function of their natural population structures. For example, when the frequency of horizontal transmission of genetic information is high relative to the frequency of vertical transmission, there is a population structure that has been termed panmictic (Lenski, 1993; Maynard Smith *et al.*, 1993). In panmictic populations, changes due to genetic exchange are more common than changes resulting from mutation. When vertical transmission frequency dominates, there is a clonal population structure and changes due to mutation are more common than changes resulting from recombination and exchange. Each bacterial species/population occupies a specific wavelength along the spectrum that ranges from totally clonal structures to totally panmictic structures. The particular wavelength occupied has profound implications for the spread of newly acquired traits throughout the members of that bacterial population.

Population structure in bacteria has generally been assessed through the use of data sets that measure allelic variation in multiple different gene loci. Data sets used are often based upon multilocus enzyme electrophoresis, in which alleles are distinguished by electrophoretic variation, but any type of data set dealing with the distribution of gene alleles can be analyzed in similar ways (Lenski, 1993) and useful tests for this purpose have been proposed (Brown *et al.*, 1980; Souza *et al.*, 1992; Maynard Smith *et al.*, 1993; Tibayrenc *et al.*, 1993). In testing the distribution of multiple allelic pairs, some combinations are found to be frequently associated while others are only rarely associated. If recombination occurred constantly the prediction would be that all pairs are randomly associated. The measurement of the frequency of the rare associations observed relative to those expected under free recombination is called linkage disequilibrium. Organisms with a clonal population structure are characterized by a great deal of linkage disequilibrium, while organisms with a panmictic population structure are in linkage equilibrium.

Whether a population is in linkage equilibrium (panmictic structure) or disequilibrium (clonal structure) has a major impact on the congruence of phylogenies at different gene loci. If panmictic, then the phylogenies at different gene loci will be non-congruent. If the population shows significant linkage disequilibrium and is clonal, then the phylogenies of different gene loci will often be congruent. The phylogeny of a specific gene locus is also likely to be congruent with the phylogeny of the overall genome as measured by a technique such as multilocus enzyme electrophoretic analysis. In a clonal population, strains that share a particular virulence trait are more likely to share other virulence traits in common because these traits are shared by all strains of that clonal type. In a panmictic population, the presence of one trait says very little about the likelihood of any other trait because of the linkage equilibrium.

The degree of linkage equilibrium or the population structure may also impact the evolution occurring at an individual gene loci. In this case, the impact is again realized through the relative contributions of recombination or of mutation to the phylogeny. In the panmictic structure, individual gene trees may show a process in which the contributions of recombination have outweighed those of mutations. This can result in the inability to observe the true clonal frame in some cases and thus to identify synonymous versus non-synonymous substitutions. Mosaicism is the result, in which different gene segments have different evolutionary histories.

Mosaic genes are not as common in organisms which have a clonal population structure. Most individual genes will be within a single clonal frame and the distances between alleles will reflect mostly substitutions. Under these conditions, the number of synonymous changes accumulates to become greater than non-synonymous changes in most genes. The rarer gene loci that may exhibit mosaicism in a clonal population are frequently found to be those where there has been intense selective pressure and where recombination events have obscured the clonal frame in these limited regions. For example, in *Escherichia coli* and *Salmonella*, which have a clonal population structure, mosaicism is associated only with certain highly selected gene loci such as *rfb*, encoding the determinants of O-polysaccharide structures. Gene trees which are discordant with other gene trees on the chromosome have also been observed in the loci near to *rfb* such as the *gnd* gene, where its presence has been attributed to bystander effect of recombination events in the nearby selected gene loci.

In the panmictic population structure, mosaic genes are more readily observed as genes in which regions of high similarity are interspersed with regions of dissimilarity. In *Neisseria gonorrhoeae*, the study of mosaic genes involved in the development of resistance to penicillin illuminated the process of how mosaic genes form. The penicillin-binding proteins (PBPs) of penicillin-sensitive organisms are not noticeable mosaics, but penicillin-resistant organisms have appeared in the 50 years since there has been selective pressure for resistance through the use of penicillin. The PBPs in penicillin-resistant organisms have a mosaic structure in which some of the segments of sequence dissimilarity are attributed to horizontal gene transfer from the PBPs of related organisms that were of higher natural resistance to penicillin.

The process was facilitated because both *N. gonorrhoeae* and *S. pneumoniae* organisms have a panmictic population structure. Another attribute of panmictic population structure is the rapidity with which it allows a new marker to spread horizontally throughout the species. Panmictic species would be expected to have much more rapid horizontal spread than clonal species. In the case of penicillin resistance, the mixia has also contributed to the rapid dissemination of the resistant PBPs throughout the world and the rapid rise among all strains circulating in a single community. Incidence of penicillin-resistant pneumococci has changed in the United States from being negligible in 1992 to up to 40% of all isolates in 1996 (Hofmann *et al.*, 1995; Doren *et al.*, 1996) and the highly resistant penicillin alleles are now associated with multiple capsular serotypes and different genetic backgrounds (Butler *et al.*, 1995, 1996).

Our own studies have focused on virulence loci other than drug resistance genes providing a view of the evolution of such genes. The paper reviews studies of evolution in two cell surface protein families in the genus *Streptococcus*. Both gene families exhibit mosaicism but the extent of this mosaicism and its realization in the serological changes of the encoded proteins differ greatly as a consequence of the different population structures of their host species. *Streptococcus pneumoniae* has a primarily panmictic population structure and *S. pyogenes* appears to be closer to a clonal population structure.

15.2 RESULTS AND DISCUSSION

The streptococci are Gram-positive spherical bacteria that grow in chains or pairs and are widely distributed in nature. Although most streptococcal species are non-pathogenic, there are several species that are important pathogens of humans. Two of these, *Streptococcus pyogenes* and *S. pneumoniae*, are examined here. *S. pyogenes*, or the group A streptococci, colonize the human nasopharynx or the epidermis, causing primary infections of pharyngitis and impetigo. Serious autoimmune conditions such as acute rheumatic fever and post-streptococcal glomerulonephritis can follow primary group A streptococcal infection. *S. pneumoniae*, or pneumococci, also initially colonize the human nasopharynx and are a major cause of otitis media and sinusitis. Frequent escape to the lower respiratory tract or the bloodstream can lead to acute bacterial pneumonia, meningitis or bacteremia.

Strains of *S. pneumoniae* are commonly identified and distinguished from each other by the specificity of their polysaccharide capsule. Capsules of differing serotypes each have a unique polysaccharide composition that is determined by allelic differences in the type-specific genes located in the *cps* locus (Guidolin *et al.*, 1994; Dillard *et al.*, 1995a, b). There are at least 90 capsular types upon which most epidemiologic investigations in pneumococci have been based. Because of the focus on immune responses to cell surface capsular polysaccharide, surface proteins in *S. pneumoniae* have been studied in detail only recently.

A surface protein called pneumococcal surface protein A (PspA) is a protection-eliciting antigen (McDaniel *et al.*, 1991; Tart *et al.*, 1996) that is present on the surface of all pneumococci (Crain *et al.*, 1990). It is attached to the surface by a complex choline-binding site near the C-terminus which is composed of ten 20 amino acid repeats (Yother *et al.*, 1992). Although its exact function is not known, PspA appears to slow the clearance of pneumococci from the blood in a mouse model system (Briles *et al.*, 1996). Antibodies to PspA are able to protect against either nasal colonization or blood sepsis in mouse models (Wu *et al.*, 1997).

The PspA proteins on different strains differ in size and in antigenic composition as judged by reactivity with a panel of monoclonal antibodies recognizing different epitopes (Crain *et al.*, 1990). A panel of seven antibodies distinguishes over 48 different PspA antigenic types. Despite this variability, a rabbit polyclonal antisera made against one PspA molecule is reactive with all PspA molecules; thus there is a considerable degree of shared immunodeterminants among PspA molecules. The protective immunity elicited by PspA is not specific for one PspA antigenic type as the immunization with one PspA molecule has been shown to be capable of eliciting protection against pneumococcal strains of different capsular types and different PspA types (identified by monoclonal patterns of reactivity) (McDaniel *et al.*, 1991).

15.2.1   MOSAIC STRUCTURE OF *PSPA* GENES AND DIVERSITY OF *PSPA* SEQUENCES

The variability in PspA combined with the knowledge that antibodies recognizing this protein are protective suggests that this gene locus is under intense selective pressure for change. Recently, 23 *pspA* genes (from strains of 14 capsular types and 13 serologic PspA types) were partially sequenced to address the genetic diversity present at this locus (Fig. 15.1) (Hollingshead *et al.*, submitted). Oligonucleotide primers LSM13 and SKH2 were used to amplify a DNA fragment ranging in size from 1200 to 1800 base pairs and containing the α-helical charged region of the *pspA* gene. Direct sequencing of each PCR-generated product was performed by automated DNA sequencing.

The strains were chosen to represent the broadest diversity of PspA types identified by their reactivity patterns with the panel of seven MAb; this included five PspAs that failed to react with any of the seven. The sequenced segments of each gene included those regions encoding: (A) the signal peptide and first 100 amino acids of the mature protein, (B) the major protection-eliciting domain (PE) or C-terminal third of the α-helical/charged region, and (C) the central proline-rich region. The relationships among the amino acid sequences were examined individually for each of these three windows, A, B and C. These relationships are indicated in the dendrograms in Fig. 15.1.

15.2.2   CLADES OF THE MAJOR PROTEIN-ELICITING REGION OF *PSPA*

The B region was the focus for strain comparison because it contains a major protection-eliciting domain of PspA. Among the 25 sequenced PspAs there were six major groups (clades) based on the sequence similarities within the B region. Within a clade, amino acid sequences are < 10% different, whereas between clades they may be as much as 80% different (mean 48%). Sequences were aligned with the aid of the program Pileup in the University of Wisconsin Genetics Computer Group package, and similar alignments were obtained with Clustal W. The dendrogram in Fig. 15.1 shows each of these five groups as a cluster or main branch of the tree.

When the relationships among sequences from windows A and C were compared with those in window B, it was clear that the groupings based on sequence similarity in these windows would be quite different (Fig. 15.1; Table 15.1). For example, an individual molecule in clade 1

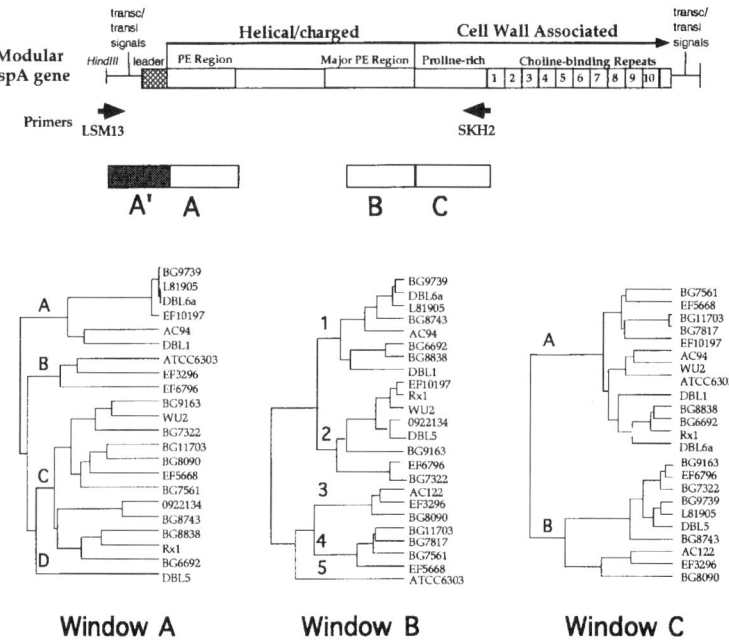

Fig. 15.1 Sliding window sequence analysis of 23 *pspA* genes. The top of the diagram shows a modular *pspA* gene with some of the major domains present in all *pspA* genes: leader region, α-helical/charged region, proline-rich region, choline-binding repeats, major protection-eliciting (PE) region, and PE region. The boxed regions labeled A', A, B and C depict segments that were sequenced and analyzed. The dendrograms below the picture depict the amino acid sequence similarities for an individual sequence region — either A, B or C. Each window indicates approximately 100 amino acids. In Window B, numerals 1–5 indicate groups (clades) based on similarities, encoding a major protection-eliciting domain. In Windows A and C, the letters A–D indicate groups in the dendrograms, based on sequence identities in those windows.

(based on window B) may be more closely related to a strain in another clade when the window of sequence data encoding the amino-terminal 100 amino acids (A) or the proline-rich domain (C) is considered. This indicates that the clonal frame differs for these different windows of sequence within the same gene locus and strongly suggests that much of the diversity of PspA has arisen from genetic recombination between different *pspA* alleles. The recombination is extensive because windows of sequence vary in 30–50% of the pairwise cases (Table 15.1). This evidence of extensive recombination is consistent with the type of combinatorial diversity seen in the expression of different epitopes on PspAs (Crain *et al.*, 1990).

Table 15.1 Comparison of three sequence windows in pspA gene

Strain name	B region	A region	C region
BG9739	1	A	. B
DBL6A	1	A	A
L81905	1	A	B
BG8743	1	C	B
AC94	1	A	A
BG6692	1	C	A
BG8838	1	C	A
DBL1	1	A	A
Rx1	2	C	A
E134	2	C	n.d.
EF10197	2	A	A
EF6796	2	B	B
BG9163	2	C	B
DBL5	2	D	B
WU2	2	C	A
EF3296	3	B	B
BG8090	3	B	C
AC122	3	n.d.	B
EF5668	4	B	A
BG7561	4	B	A
BG7817	4	n.d.	A
BG11703	4	B	A
ATCC6303	5	B	A

Studies of the distribution of PspA variants assessed by monoclonal Ab (Crain *et al.*, 1990) and IgA protease variants assessed by type-specific polyclonal antisera (Lomholt, 1995) have both provided evidence of natural horizontal gene transfer, indicating that pneumococci are panmictic. A primary contributor to the mixia may be the ability to undergo transformation in natural settings. *S. pneumoniae* is well known for its uptake of DNA from the environment by the process of transformation. Pneumococci in their natural habitat of the human nasal passages are frequently found in mixed colonizations and thus have the opportunity for horizontal gene exchange. The interspecies genetic exchange of a gene fragment for a penicillin-binding protein from *S. mitis* and another unknown species has been well documented in studies of the acquisition of penicillin resistance (Dowson *et al.*, 1993). The rapid spread of penicillin resistance associated with particular PBP alleles from strains of one clonal type to another further suggests a high frequency of horizontal transfer among the pneumococci in nature (Kell *et al.*, 1993; Barnes *et al.*, 1995).

A high frequency of transformation in natural settings would explain our sequence results examining diversity in the *pspA* gene family. Although our sample size of 25 genes is relatively small, we could easily detect evidence for over 20 intragenic recombinatorial changes that would change the clonal frame of the sequence being examined. In fact it is quite difficult to detect the windows of sequence which are sharing a similar clonal frame because these are apparently small – less than 100 amino acids in some cases. In a sample of similar size containing diverse *emm* genes discussed below, it would be hard to detect a single recombination event of this type. A high frequency of genetic exchange also explains the cross-reactivity of PspA molecules. Although two individual PspAs may share only 50% amino acid identity, this identity seems to occur in short blocks which are often large enough to encode individual antigenic determinants. Thus two quite different PspAs are often recognized by the same monoclonal antibody.

The impact of the high frequency of genetic exchange on the *pspA* gene family has then been to create a group of surface proteins that are both serotypically diverse and antigenically cross-reactive at the same time. This may allow the development of some natural resistance to pneumococcal infections in the form of antibodies to PspA. Although little is yet known on this subject, humans do make natural antibodies to PspA, and adults may have higher levels than children.

15.2.3   CELL SURFACE IN *S. PYOGENES*

Although members of the same genera, the group A streptococci and the pneumococci are quite different in the antigenic characteristics of their cell surfaces. *Streptococcus pyogenes* bear the Lancefield group A antigen in the cell wall – hence the designation, group A streptococci. The capsule in group A streptococci is identical to human hyaluronic acid and thus it is poorly antigenic in humans. Instead, immunologic typing in this species has focused on an antigenically heterogeneous antiphagocytic protein called M protein. M typing has been the tool for intraspecies serotypic analysis of group A streptococcal strains and protective immunity to group A streptococcal infection is M serotype-specific. The M protein family is a large family of tissue- and plasma-factor binding proteins and hundreds of genes and partial genes encoding the members of this family have been sequenced.

Transformation has never been demonstrated to occur in *S. pyogenes* and this species has many of the hallmarks of a clonal population structure. Identical electrophoretic types (ETs) have been recovered from different geographical sites worldwide and in isolates recovered over a large span of time (> 50 years) (Musser *et al.*, 1991; Kapur *et al.*, 1995). Although linkage disequilibrium values have not been calculated, there is an apparent association of certain M serotypes with only one or a small number of multilocus enzyme genotypes – another hallmark of a clonal population structure (Haase *et al.*, 1994).

15.2.4   MOSAIC STRUCTURE OF BOTH *EMM* GENES AND *EMM* GENE CLUSTERS

Because the *emm* gene locus is expected to be a locus that would experience intense selective pressure for change, it is not surprising that the *emm* genes show a mosaic structure with the mosaic gene regions having responsibility for the serotypically distinct regions of protein in family members. However, although there is a mosaic structure, it is interesting that the genes of any one mosaic type (equivalent to one serotype in this system) may frequently be sampled across the world geographically and as well over a span of 50 years. Very little sequence variation exists among the genes of a single serotype or single mosaic type (Hollingshead *et al.*, 1987; Musser *et al.*, 1995). Moreover, particular serotypes are most frequently associated with the same ET type (Hollingshead, unpublished data).

The evolution of the M protein family has involved a combination of: size variation and spontaneous antigenic changes occurring by intragenic recombination in repeat regions (Hollingshead *et al.*, 1987, 1991); gene duplication (Hollingshead *et al.*, 1994); and horizontal genetic exchanges among a multi-gene family (Bessen and Hollingshead, 1994). The last of these events is the focus for the following discussion.

The *emm* genes in group A streptococci are present in a locus called the *mga* regulon. The small cluster of *emm* genes in this locus lie just downstream from a positive regulator (Mga). In many strains, there is a single *emm* gene in this locus; while in other strains, there are up to three members of the *emm* gene family present and the number and type of *emm* genes can be determined by PCR using carefully chosen primers (Hollingshead *et al.*, 1993, 1994) (Fig. 15.2). Sequence analysis of the genes and their position in the clusters indicate a process of gene evolution by duplication in the formation of the gene clusters. Once formed, the clusters could be the site for gene replacement events that could occur by horizontal gene exchange, as will be discussed below.

15.2.5   FREQUENCY OF HORIZONTAL EXCHANGE – INSIGHTS FROM MAPPING OF *EMM* GENE CLUSTERS

While our previous communications and those of other laboratories (Bessen and Hollingshead, 1994; Podbielski *et al.*, 1994; Whatmore and

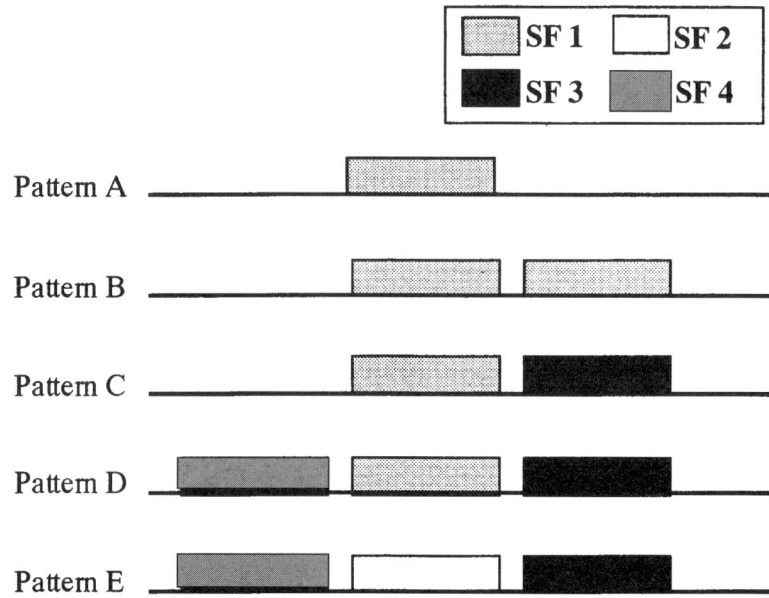

Fig. 15.2 Chromosomal patterns of *emm* genes. Four *emm* gene subfamilies (SF) are defined based on nucleotide sequence differences within the last 429 base pairs of *emm* genes (Hollingshead *et al.*, 1994). This segment encodes a cell-wall spanning C-terminus for M proteins. Nearly all isolates display one of five major patterns (**A–E**) at the chromosomal locus for *emm* genes which are recognized by PCR mapping using subfamily-specific primers to identify genes present in the cluster at the chromosomal site (Hollingshead *et al.*, 1993).

Kehoe, 1994a,b) have demonstrated that horizontal genetic exchanges that involve members of the *emm* gene superfamily have occurred, it is important to consider the frequency with which these events occur. Several lines of evidence support the idea that horizontal transfers of genetic information occur rather infrequently in the M protein family.

First, when isolates of the same M serotype are examined which were collected over a 50-year time span and over many geographical locations, the *emm* gene within these different strains is found to have very high overall sequence similarity (> 96% nucleotide identity). Sequence differences seen are frequently a difference in the number of repeats when a repeat motif is present, and occasional nucleotide substitutions. Moreover, these isolates of the same serotype almost invariably have a similar architecture to their chromosomal locus (*emm* gene cluster pattern). If the *emm6* gene is found in pattern A type clusters in one strain, it will almost always be found in a pattern A type

Table 15.2 Summary of *emm* family gene clusters examined in 287 strains[a]

Pattern	Number of isolates	Number of serotypes[b]	Serotypes
A–C	120	28	1, 3, **4**, 5, 6, 8, 12, 14, 17, 18, 19, 23, 24, 26, 29, 30, 31, 33, 38, 39, 43, 48, 51, 54, 55, 57, 59, 67
D	46	11	23, 28, 32, 33, 36, 42, **49**, 52, 53, 54, **56**
E	121	28	2, **4**, 8, 9, 11, 13, 15, 22, 25, 28, 34, 35, 37, 44, 46, 48, **49**, 50, **56**, 58, 60, 61, 62, 63, 65, 66, 67, 76

[a] Strain collection characteristics: (1) 57 different M serotypes represented. (2) Streptococcal class (I or II) has been tested for 149 of the 287 strains with approximately half of the strains studied being class I and opacity-factor-negative (OF⁻) and the other half being class II and OF-positive. (3) Within the collection are isolates from blood (3), from nasopharynx (88), from skin (65), from patients with toxic shock like syndrome (9), rheumatic fever (3), with acute glomerulonephritis (16), or with severe invasive disease (19). (4) Strains were isolated from a variety different geographical regions.
[b] In which at least one isolate had this *emm* gene cluster pattern. Bold type indicates serotypes with more than one pattern.

cluster in other strains as well. This is shown in some of the data included in Table 15.2.

Out of 57 total serotypes in the study depicted in Table 15.2, there was an opportunity to examine two or more independently collected strains for 40 of those serotypes. In only six out of 120 cases (5%) was there any difference in the *emm* gene cluster pattern for two strains of the same serotype. In 122 of these strains, the ET type of strain was available as well. In this case, for 86% of the time, strains of the same M protein serotype were also of the same ET type.

Genetic exchanges are observed in this data as the movement of a DNA segment encoding the determinant for serotype from one recognizable chromosomal background to another. Backgrounds are recognized in one of two ways:

• by PCR analyses to identify the pattern or architecture of the chromosomal locus in which the *emm* gene is situated;
• by analysis of chromosomal genotype based on multilocus enzyme electrophoresis at 11 polymorphic alleles (ET type).

Changes with regard to either assay of chromosomal 'background' would detect essentially a gene replacement event; in the latter case the replacement event might involve the exchange of an entire gene cluster.

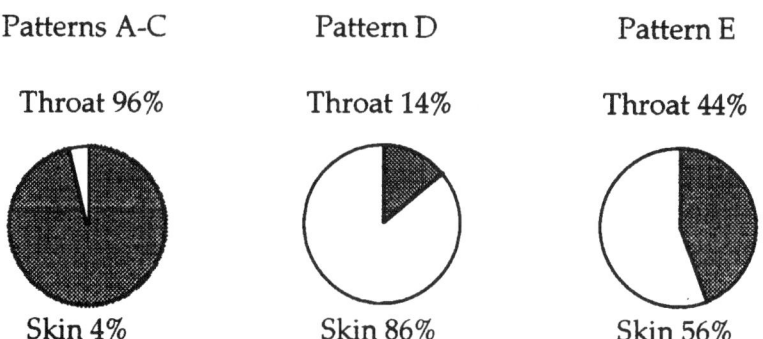

Patterns A-C   Pattern D   Pattern E

Throat 96%   Throat 14%   Throat 44%

Skin 4%   Skin 86%   Skin 56%

Fig. 15.3 Association between chromosomal patterns of *emm* genes and tissue site for infection: 105 *S. pyogenes* isolates from uncomplicated streptococcal disease and isolated from either the nasopharyngeal mucosa or the impetigo lesion in skin were examined for *emm* chromosomal gene cluster pattern. There were 53 patterns A–C strains, 21 pattern D strains and 31 pattern E strains. The tissue site distribution was significantly biased for A–C pattern strains (96% throat, $P < 0.001$, χ^2) and D pattern strains (96% skin, $P < 0.001$, χ^2), but was relatively balanced for E pattern strains.

The evidence that genetic exchange is infrequent in the M protein family is similar to evidence that recombination is relatively rare in *E. coli*. When one considers the assortment of individual binding domains that would be possible if frequent mixing by recombination occurred, then it is striking that relatively few of these combinations have formed. This can be seen in Table 15.2, where any particular serotype designation is associated with only one or, in a few cases, two *emm* gene cluster patterns.

An additional indication that there is limited horizontal gene exchange among group A streptococci is the long-held recognition that there are strains of streptococci highly adapted for human epidermis and other strains highly adapted for the human nasopharyngeal mucosa (Fig. 15.3). Acute rheumatic fever only follows throat infections and there is an association of certain M serotypes with rheumatic fever. Because colonization of a particular site is unlikely to be the result of M protein type, it seems likely that the epidermal strains and the nasopharyngeal strains share multigenic differences that determine habitat for colonization.

In our studies of the *emm* gene cluster patterns in diverse group A streptococcal strains we have found that there is a very strong association of *emm* gene cluster pattern and tissue site of isolation (Fig. 15.3). Strains that have cluster patterns A–C are almost exclusively from the nasopharynx, while those with cluster pattern D are almost exclusively from epidermal sites. A third group of strains (pattern E) can be found in either site (Bessen *et al.*, 1996). The strong asymmetry in this distribution pattern is reminiscent of the concept of linkage disequilibrium. The exchange of information content between group A streptococcal strains has not been frequent enough to break up this association between cluster patterns and tissue site.

REFERENCES

Barnes, D.M., Whittier, S., Gilligan, P.H. *et al.* (1995) Transmission of multidrug-resistant serotype 23F Streptococcus pneumoniae in group day care: evidence suggesting capsular transformation of the resistant strain *in vivo*. *J. Infect. Dis.* **171**(4): 890–896.

Bessen, D.E. and Hollingshead, S.K. (1994) Allelic polymorphism of *emm* loci provides evidence for horizontal gene spread in group A streptococci. *Proc. Natl Acad. Sci. USA* **91**: 3280–3284.

Bessen, D.E., Sotir, C.M., Readdy, T.L. and Hollingshead, S.K. (1996) Genetic correlates of throat and skin isolates of group A streptococci. *J. Inf. Dis.* **173**: 896–900.

Briles, D.E., Tart, R.C., Wu, H.-Y. *et al.* (1996) Systemic and mucosal protective immunity to pneumococcal surface protein A. *NY Acad. Sci.* **797**: 118–126.

Brown, A.H.D., Feldman, M.W. and Nevo, E. (1980) Multilocus structure of natural populations of *Hordeum spontaneum*. *Genetics* **96**: 523–536.

Butler, J.C., Breiman, R.F., Lipman, H.B. *et al.* (1995) Serotype distribution of *Streptococcus pneumoniae* infections among preschool children in the United States, 1978–1994: implications for development of a conjugate vaccine. *J. Infect. Dis.* **171**: 885–889.

Butler, J.C., Hofmann, J., Cetron, M.S. *et al.* (1996) The continued emergence of drug-resistant *Streptococcus pneumoniae* in the United States: an update from the Centers for Disease Control and Prevention's Pneumococcal Sentinel Surveillance System. *J. Infect. Dis.* **171**: 885–889.

Crain, M.J., Waltman, W.D.II, Turner, J.S. *et al.* (1990) Pneumococcal surface protein A (PspA) is serologically highly variable and is expressed by all clinically important capsular serotypes of *Streptococcus pneumoniae*. *Infect. Immun.* **58**: 3293–3299.

Dillard, J.P., Caimano, M., Kelly, T. and Yother, J. (1995a) Capsules and cassettes: genetic organization of the capsule locus of *Streptococcus pneumoniae*, in *Genetics of Streptococci, Enterococci and Lactococci*, Vol. 85, (eds J. Feretti, M.S. Gilmore, T.R. Klaenhammer and F. Brown), Karger, Dev Biol Stand, Basel, pp. 261–265.

Dillard, J.P., Vandersea, M.W. and Yother, J. (1995b) Characterization of the cassette containing genes for type 3 capsular polysaccharide biosynthesis in *Streptococcus pneumoniae*. *J. Exp. Med.* **181**: 973–983.

Doren, G.V., Brueggemann, A., Holley, H.P.J. and Rauch, A.M. (1996) Antimicrobial resistance of *Streptococcus pneumoniae* recovery from outpatients in the United States during the winter months of 1994 to 1995: results of a 30 center national surveillance study. *Antimicrob. Agents Chemother.* **40**: 1208–1213.

Dowson, C., Coffey, T., Kell, C. and Whiley, R. (1993) Evolution of penicillin resistance in *Streptococcus pneumoniae*; the role of *Streptococcus mitis* in the formation of a low affinity PBP2B in *S. pneumoniae*. *Mol. Micro* **9**: 635–643.

Guidolin, A., Morona, J.K., Morona, R. *et al.* (1994) Nucleotide sequence analysis of genes essential for capsular polysaccharide biosynthesis in *Streptococcus pneumoniae* Type 19F. *Infect. Immun.* **62**: 5384–5396.

Haase, A.M., Melder, A., Mathews, J.D. *et al.* (1994) Clonal diversity of *Streptococcus pyogenes* within some M-types revealed by multilocus enzyme electrophoresis. *Epidemiol. Infect.* **113**: 455–461.

Hofmann, J., Cetron, M.S., Farley, M.M. *et al.* (1995) Prevalence of drug-resistant *Streptococcus pneumoniae* in Atlanta. *New Engl. J. Med.* **333**: 481–486.

Hollingshead, S.K., Fischetti, V.A. and Scott, J.R. (1987) Size variation in group A streptococcal M protein is generated by homologous recombination between intragenic repeats. *Mol. Gen. Genet.* **207**: 196–203.

Hollingshead, S.K., Jones, K.F. and Fischetti, V.A. (1991) Isolation of spontaneous antigenic variants of group A streptococci that exhibit antigenic drift in the M6.1 protein, in *Genetics and Molecular Biology of Streptococci, Lactococci and Enterococci*, (eds G.M. Dunny, P.P. Cleary and L.L. McKay), American Society for Microbiology, Washington, DC, pp. 174–178.

Hollingshead, S.K., Readdy, T.L., Yung, D.L. and Bessen, D.E. (1993) Structural heterogeneity of the *emm* gene cluster in group A streptococci. *Mol. Microbiol.* **8**(4): 707–717.

Hollingshead, S.K., Arnold, J., Readdy, T.L. and Bessen, D.E. (1994) Molecular evolution of a multi-gene family in group A streptococci. *Mol. Biol. Evol.* **11**(2): 208–219.

Hollingshead, S.K., Becker, R. and Briles, D.E. (submitted) Diversity of the protection-eliciting region of PspA: the presence of clades and evidence for past recombination.

Kapur, V., W. M. Sischo, R. S. Greer, T. S. Whittam, and J. M. Musser. (1995. Molecular population genetic analysis of the streptokinase gene of *Streptococcus pyogenes*: mosaic alleles generated by recombination. *Mol. Microbiol.* **16**(3): 509–519.

Kell, C.M., Jordens, J.Z., Daniels, M. *et al.* (1993) Molecular epidemiology of penicillin-resistant pneumococci isolated in Nairobi, Kenya. *Infect Immun.* **61**(10): 4382–4391.

Lenski, R.E. (1993) Assessing the genetic structure of microbial populations. *Proc. Natl Acad. Sci. USA* **90**: 4334–4336.

Lomholt, H. (1995) Evidence for recombination and an antigenically diverse immunoglobulin A1 protease among strains of *Streptococcus pneumoniae*. *Infect. Immun.* **63**: 4238–4243.

Maynard Smith, J., Smith, N.H., O'Rourke, M. and Spratt, B.G. (1993) How clonal are bacteria? *Proc. Natl Acad. Sci. USA* **90**: 4384–4388.

McDaniel, L.S., Sheffield, J.S., Delucchi, P. and Briles, D.E. (1991) PspA, a surface protein of *Streptococcus pneumoniae*, is capable of eliciting protection against pneumococci of more than one capsular type. *Infect. Immun.* **59**: 222–228.

Musser, J.M., Hauser, A.R., Kim, M.H. *et al.* (1991) *Streptococcus pyogenes* causing toxic-shock-like syndrome and other invasive diseases: clonal diversity and pyrogenic exotoxin expression. *Proc. Natl Acad. Sci. USA* **88**:2668–2672.

Musser, J.M., Kapur, V., Szeto, J. *et al.* (1995) Genetic diversity and relationships among *Streptococcus pyogenes* strains expressing serotype M1 protein: recent intercontinental spread of a subclone causing episodes of invasive disease. *Infect. Immun.* **63**(3): 994–1003.

Podbielski, A., Krebs, B. and Kaufhold, A. (1994) Genetic variability of the *emm*-related genes of the large *vir* regulon of group A streptococci: potential intra- and intergenomic recombination events. *Mol. Gen. Genet.* **243**(6): 691–699.

Souza, V., Nguyen, T.T., Hudson, R.R. *et al.* (1992) Hierarchical analysis of linkage disequilibrium in Rhizobium populations: evidence for sex? *Proc. Natl Acad. Sci. USA* **89**: 8389–8393.

Tart, R.C., McDaniel, L.S., Ralph, B.A. and Briles, D.E. (1996) Truncated *Streptocccus pneumoniae* PspA molecules elicit cross-protective immunity against pneumococcal challenge in mice. *J. Infect. Dis.* **173**: 380–386.

Tibayrenc, M., Neubauer, K., Barnabe, C. *et al.* (1993) Genetic characterization of six parasitic protozoa: parity between random-primer DNA typing and multilocus enzyme electrophoresis. *Proc. Natl Acad. Sci. USA* **90**(4): 1335–1339.

Whatmore, A.M. and Kehoe, M.A. (1994a) Horizontal gene transfer in the evolution of group A streptococcal *emm*-like genes: gene mosaics and variation in Vir regulons. *Mol. Microbiol.* **11**(2): 363–374.

Whatmore, A.M. and Kehoe, M.A. (1994b) Non-congruent relationships between variation in *emm* gene sequences and the population genetic structure of group A streptococci. *Mol. Microbiol.* **14**(4): 619–631.

Wu, H.-Y., Nahm, M., Guo, Y. *et al.* (1997) Intranasal immunization of mice with PspA (pneumococcal surface protein A) can prevent intranasal carriage, pulmonary infection, and sepsis with *Streptococcus pneumoniae. J. Infect. Dis.* **175**: 839–846.

Yother, J., Handsome, G.L. and Briles, D.E. (1992) Truncated forms of PspA that are secreted from *Streptococcus pneumoniae* and their use in functional studies and cloning of the *pspA* gene. *J. Bact.* **174**: 610–618.

Roles of horizontal transfer in bacterial evolution

16

Jeffrey Lawrence and John Roth

SUMMARY

Gene loss and reacquisition may be key aspects of bacterial evolution. This was suggested by the history of B_{12} metabolism in enteric bacteria, which includes loss of multiple functions and reacquisition of genes from a foreign source. Many bacterial genes are located in cotranscribed clusters or operons; together, the genes in an operon generally provide a single function or selectable phenotype. Conditionally dispensable functions are usually encoded by operons; essential genes are less likely to be clustered. Operon formation may be driven by gene loss (by mutation during periods of dispensability) and reacquisition (by horizontal acquisition of small chromosome fragments followed by selection). Clustered genes can be cotransferred horizontally and therefore can spread faster than identical unclustered alleles; thus clustered alleles have higher fitness. Gene clustering may provide no immediate fitness benefit to the host organism and can be considered a selfish property of genes. Recently acquired genes may show atypical patterns of base composition and codon usage bias. With time, such sequences ameliorate toward the patterns of the new host. From the degree of sequence amelioration, one can estimate the time at which a sequence was introduced. We estimate that 31 kb of foreign DNA are introduced and substantially fixed in the *Escherichia coli* genome every million years; a corresponding amount of DNA is presumably lost. We predict that the genomes of *Salmonella enterica* and *E. coli* each include sequences – up to 30% of each genome – that are absent from the other genome. Horizontal transfer may drive bacterial speciation since it allows an organism to acquire a well-developed capability suddenly.

16.1   INTRODUCTION

The closely related enteric bacteria *Escherichia coli* (typified by strain K12) and *Salmonella enterica* (typified by its most intensively studied serovar, Typhimurium) diverged from a common ancestor about 100 million years ago (Ochman and Wilson, 1987, 1988). The base composition of each genome is approximately 50% GC; typical homologous genes are 85% identical and encode proteins that are 93% identical (Sharp, 1991). The genome of each organism – about 4800 kb in size – encodes nearly 5000 genes, and the locations of shared genes on their respective chromosomes show that their genetic maps have been strongly conserved (Sanderson *et al.*, 1995; Berlyn *et al.*, 1996). Extensive traditional genetic and biochemical analyses – in laboratory settings – suggest that the two species are metabolically and mechanistically similar (Neidhardt *et al.*, 1996). Despite the apparent similarities in genome composition, genetic map and metabolism, recent genetic and physical characterizations of these taxa suggest striking differences between these species.

Investigations of population structure demonstrate that *E. coli* and *S. enterica* are classical clonal prokaryotes which propagate primarily by binary fission (Selander and Levin, 1980; Ochman and Selander, 1984; Whittam and Ake, 1992; Selander *et al.*, 1996; Whittam, 1996). This view has expanded to incorporate evidence of occasional intraspecific recombination; that is, transfers of small segments of the chromosome among conspecific strains (Milkman and Bridges, 1990, 1993; Guttman and Dykhuizen, 1994a, b). Evidence also exists for interspecific, horizontal transfer of larger chromosomal regions, frequently including entire operons (Syvanen, 1994; Lawrence and Roth, 1996b). Some of these foreign genes can be recognized by an atypical base composition and unusual codon usage bias (Whittam and Ake, 1992; Lawrence and Ochman, 1996).

S. enterica and *E. coli* differ substantially in their abilities to synthesize and use the cofactor B_{12}. Only *Salmonella* can synthesize B_{12} *de novo* and perform the B_{12}-dependent degradation of propanediol (Lawrence and Roth, 1996a); the anaerobic degradation of propanediol by *S. enterica* may depend on the presence of particular electron acceptors: tetrathionate, thiosulfate and/or sulfite (Roth *et al.*, 1996). The gene clusters that encode these sulfur reduction pathways (*tth*, *phs* and *asr*, respectively) are found only in *S. enterica*. We believe that the evolution of B_{12} metabolism in these organisms may have been critical to divergence of *S. enterica* and *E. coli* and may reflect general principles of bacterial evolution. The exemplified processes of gene loss and gene acquisition by horizontal transfer provide the potential for rapid evolutionary change.

16.2   RESULTS AND DISCUSSION

16.2.1   HORIZONTAL TRANSFER AND B_{12} METABOLISM IN *SALMONELLA*

Cobalamin (B_{12}) is a huge cofactor (MW=1580) which is synthesized *de novo* by *S. enterica* but not by *E. coli*. Most of the B_{12} biosynthetic genes of *S. enterica* are located in a single operon (*cob*) containing 20 genes (Jeter and Roth, 1987; Roth *et al.*, 1993). A B_{12} transport system is encoded outside of the *cob* operon, and additional unlinked genes encode enzymes which contribute to B_{12} synthesis but also play additional metabolic roles (Roth *et al.*, 1996). We estimate that nearly 1% of the *S. enterica* genome is dedicated to acquisition of B_{12}. The ability to make B_{12} is characteristic of virtually all serovars of *S. enterica*, suggesting that it is essential to their shared lifestyle and raising the question of what selective pressures maintain this large genetic investment (Lawrence and Roth, 1996a).

A paradox is presented by the role of B_{12} in *Salmonella*. Coenzyme B_{12} supports degradation of ethanolamine and propanediol, employing genes of the *eut* (Roof and Roth, 1988, 1989) and *pdu* operons (Jeter, 1990), respectively. Since the B_{12} synthetic operon (*cob*) is coinduced with the *pdu* operon by propanediol, using a single regulatory protein, PocR (Bobik *et al.*, 1992), we conclude that propanediol degradation is the principal physiological role for B_{12} in *S. enterica*. However, propanediol and ethanolamine are only catabolized when oxygen is present as an electron acceptor; the standard alternative electron acceptors (nitrate or fumarate) do not support respiration of either carbon source. The oxygen requirement for use of B_{12} is paradoxical because *Salmonella* synthesizes B_{12} only in the absence of oxygen; therefore, aerobic utilization of propanediol and ethanolamine depends on exogenous B_{12}. We assume that natural anaerobic conditions exist under which *S. enterica* can simultaneously synthesize and employ B_{12}.

Although *E. coli* lacks the capability to synthesize B_{12} *de novo* (Lawrence and Roth, 1995), most other enteric bacteria do synthesize this cofactor (Lawrence and Roth, 1996a). However, non-*Salmonella* enteric bacteria (e.g. species of *Citrobacter* and *Klebsiella*) appear to synthesize B_{12} by a pathway that is different from that found in *Salmonella* (Fig. 16.1). *Klebsiella* synthesizes B_{12} under both aerobic and anaerobic growth conditions, and that synthesis is not stimulated by propanediol (Lawrence and Roth, 1996a). In addition, *Klebsiella* possesses a B_{12}-dependent glycerol dehydratase not found in *Salmonella* (Fig. 16.1). These broad phenotypic differences are reflected at the level of DNA sequence – the genomes of *Klebsiella* and other enteric bacteria do not contain sequences that are closely related (> 70% identity) to the *Salmonella cob*

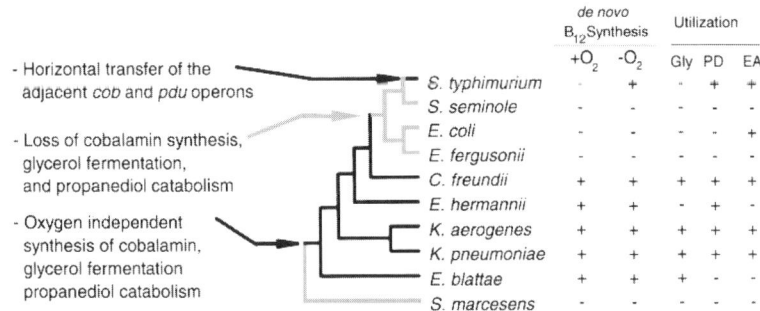

Fig. 16.1 B$_{12}$ metabolic phenotypes among enteric bacteria. Points of loss and gain are indicated on the phylogeny. Relevant B$_{12}$ metabolic phenotypes are provided. Gly, glycerol fermentation; PD, propanediol degradation; EA, ethanolamine degradation. Data are after Lawrence and Roth (1996a).

genes; typical chromosomal genes among these taxa are more than 80% identical. DNA sequence analysis of the adjacent *Salmonella cob* and *pdu* operons revealed unusual base composition and codon usage bias, suggesting that this region was introgressed from a foreign donor (Lawrence and Ochman, 1996; Ochman and Lawrence, 1996). The exogenous origin of the *Salmonella cob* and *pdu* operons is apparently responsible for the lack of hybridization between isofunctional regions of the genomes of these closely related bacteria.

We infer that the history of B$_{12}$ metabolism among enteric bacteria involves multiple acts of gene loss and at least one reacquisition by horizontal transfer (Fig. 16.2). The common ancestor of *E. coli* and *S. enterica* lost the ability to synthesize B$_{12}$, the ability to degrade propanediol and the ability to degrade glycerol (Lawrence and Roth, 1995); these losses are reflected by the continued absence of these metabolic pathways from *E. coli*. Two of these abilities were restored to *Salmonella*, which acquired (by horizontal transfer) a single chromosome fragment containing the *cob* operon (encoding a distinct pathway for B$_{12}$ synthesis), the *pdu* operon (providing for propanediol catabolism) and the *pocR* gene (providing for coregulation of the two operons).

16.2.2 SELFISH OPERONS – HORIZONTAL TRANSFER AND GENE CLUSTERING

This history of B$_{12}$ metabolism suggests that gene loss and reacquisition by horizontal transfer may be important aspects of bacterial population biology. Like the *cob* and *pdu* operons, many bacterial genes are needed only under particular conditions. If selection for a metabolic function is

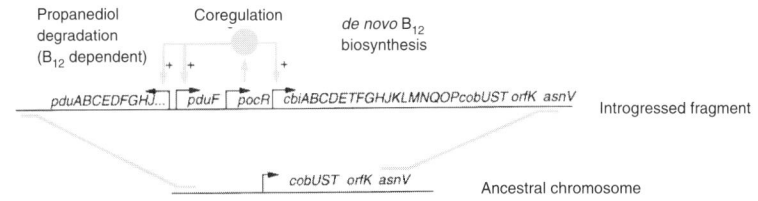

Fig. 16.2 The introgression of a single fragment bearing the *cob* and *pdu* operons into the genome ancestral to the *S. enterica* lineage. Gray lines represent the proposed sites of introgression. The composition of the insertion is inferred from the current structure of the corresponding region in the *E. coli* chromosome (Lawrence and Roth 1995).

relieved for long periods, the genes providing that function can be lost by mutation. Sufficiently weak (or infrequent) selection would result in the loss of these genes from bacterial populations by mutation and genetic drift (Lawrence and Roth, 1996b). Once multiple genes contributing to a single function have been lost, reimposition of selection is fatal unless all of the missing genes can be reacquired by a horizontal transfer event. Restoration of a single one of the missing functions does not provide a selectable phenotype and the introduced allele is unlikely to be fixed in the population. Since bacterial mechanisms of gene transfer generally transmit small chromosome segments, organisms in which related genes are closely linked are more likely to serve as successful donors.

A striking feature of bacterial genomes is the operon, a cluster of cotranscribed genes providing for a single selectable function. Several models for the origins of gene clusters have been suggested previously:

- The **Natal model**, in which gene clusters originate *in situ* by gene duplication and divergence. In this model, gene proximity is a historical property and provides no direct benefit to the individual.
- The **Fisher model**, whereby gene clusters are formed due to selection for maintenance of coadapted gene complexes, providing a benefit to the individual in the context of a genetically variable, freely recombining population.
- The **Coregulation model**, whereby gene clusters facilitate coordinate expression and regulation, providing a selective benefit to the individual.

We have described a new model, the **Selfish Operon model**, in which gene clusters allow dissemination of functionally related genes via horizontal transfer (Lawrence and Roth, 1996b). In this model, physical proximity provides no selective benefit to the individual organism, but does

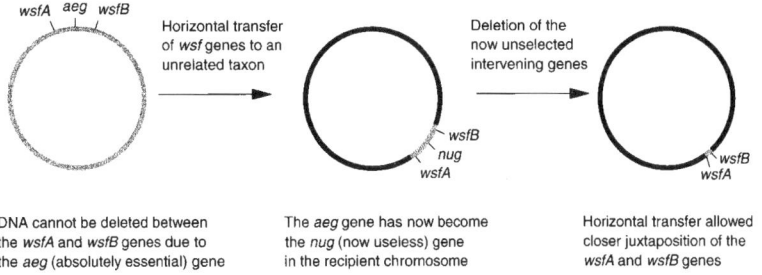

Fig. 16.3 Model for the transfer of gene clusters. Circles represent bacterial chromosomes; *wsfA* and *wsfB*, genes for a weakly selected function; *aeg*, absolutely essential gene; *nug*, now useless gene. After Lawrence and Roth (1996b).

enhance the fitness of the gene cluster itself, since clusters can be efficiently inherited horizontally as well as vertically. We suggest that the Selfish Operon model is more likely to explain the evolution of gene clusters in bacteria than other models.

We proposed that horizontal transfer allows the stepwise formation of gene clusters and their subsequent integration into operons. Known gene transfer mechanisms preferentially mobilize small segments of chromosome; thus, even weak linkage of genes (occurring by chance) enhances the probability of their simultaneous horizontal transfer. Following transfer of a chromosomal segment into a new host, any of the foreign genes not under selection are subject to subsequent loss; this process allows the genes providing the selectable phenotype to become progressively more tightly linked, and more subject to subsequent horizontal transfer (Fig. 16.3).

We suggest that cotranscription of gene clusters facilitates their horizontal spread. A cotranscribed gene cluster requires only a single promoter, which could be provided by the recipient chromosome at the site of integration. A cluster of independently transcribed genes would function only if the new host could recognize all of the needed promoters. Therefore, cotranscribed gene clusters (operons) may have been selected for their promiscuity – the ability to function in a variety of bacterial hosts. Since the gene cluster provides no benefit to the host genome, the clustered state can be considered a selfish property of the constituent genes. Operons have a selective advantage over identical independent genes in a clustered state and an even greater advantage over the same alleles in an unclustered state; genes in operons are most likely to propagate horizontally.

The Selfish Operon model contrasts sharply with the most widely held view that operons were formed via selection for coregulation. The Selfish Operon and Coregulation models are distinguished in several ways:

1. The Selfish Operon model predicts that dispensable functions are more likely to be encoded by clustered genes since these functions are more subject to periodic loss and reacquisition. This is found to be true when one inspects the genomes of *E. coli* and other bacteria.
2. Some gene clusters include multiple transcription units. These cases are not consistent with the Coregulation model but are easily explained as a selfish cluster. For example, the *cob/pdu* regulon comprises four tightly linked but independently transcribed genetic units which contribute to a single selectable phenotype.
3. The Selfish Operon model provides for slow progressive clustering and subsequent integration of clustered genes into operons. With the Coregulation model, clustering cannot occur by a progressive series of steps since no selective benefit is provided until after transcription units are successfully fused.

We believe that the organization of the *S. enterica cob* and *pdu* operons exemplifies the predictions the Selfish Operon model. The arrangement of genes in the *cob/pdu* operon cluster includes hierarchical levels of selfish gene clusters. The 20 cotranscribed *cob* genes can provide cobalamin synthesis when mobilized to recipient genomes. Similarly, the adjacent *pdu* operon can be transferred into foreign genomes to confer the ability to degrade propanediol. Together, the adjacent *cob* and *pdu* operons form a selfish regulon, providing the functions of propanediol degradation and synthesis of the cofactor required for that process. Additional transcription units in this cluster provide for propanediol transport (the *pduF* gene) and for the regulation of all four transcription units (the *pocR* gene). The organization of this DNA fragment allows horizontal transfer to mediate in a single step the introduction of complex metabolic pathways into novel genetic backgrounds.

16.2.3   ASSESSING THE RATE OF HORIZONTAL TRANSFER

We outlined above how horizontal transfer (loss and reacquisition) might drive the evolution of gene clusters and operons. Horizontal transfer of such functional units may also have a profound effect on the evolution and divergence of bacterial genomes. To assess the role of such transfers one must be able to identify foreign sequences in a bacterial genome, and determine their time of introgression.

Although bacterial species display wide variation in their overall base composition, the genes within a particular species are relatively similar in character (Sueoka 1961, 1962; Ochman and Lawrence, 1996).

The equilibrium base composition reflects a balance between natural selection and directional mutation pressures (Sueoka, 1988, 1992, 1993); equilibrium base compositions vary between 25% GC and 75% GC among bacterial taxa. Genes within any particular genome tend to share a similar overall base composition and a characteristic GC content for each of the three codon positions within coding sequence (Muto and Osawa, 1987). Increases of genomic GC content are reflected in large increases in GC content at third codon positions, and in more modest increases at first and second codon positions (Fig. 16.3). These Muto and Osawa relationships are robust and are characteristic of native and of fully ameliorated genes in a genome.

Genes that do not reflect the overall characteristics of their resident genome are likely to be of foreign origin – that is, they reflect the equilibrium base composition of their donor genomes rather than that of the current host. Thus, a recently horizontally transferred group of genes can often be identified by an atypical base composition and by an unusual codon usage bias. Using these and related criteria, scrutiny of the E. coli genome has provided estimates of the amount of horizontally transferred DNA in the genome. Based on the GC contents of 500 genes, Whittam and Ake (1992) estimated that 6% of the E. coli chromosome was acquired by horizontal transfer; but, in an analysis of codon usage patterns, Médigue et al. (1991) concluded that as many as 16% of E. coli genes were of novel origin. Combining these approaches, Lawrence and Ochman (1996) estimated that at least 13% of E. coli genes were acquired by horizontal transfer. All of these measures support the hypothesis that horizontal transfer has introduced a substantial number of genes into the E. coli genome. Yet all of these methods probably underestimate foreign introgressions since they can not identify genes introduced from genomes with a base composition similar to that of E. coli.

Over time, directional mutation pressure will cause horizontally transferred genes to ameliorate – that is, to adjust their nucleotide compositions to resemble the equilibrium base composition of their host genome. Since amelioration is a balance between selection and directional mutation pressures, nucleotide sites experiencing different selective pressures will ameliorate at different rates. For example, each codon position experiences different selective pressures due to the degeneracy of the genetic code; therefore, first, second and third codon positions will ameliorate at different rates (Fig. 16.4).

These differences in the rates of amelioration at each codon position furnish a property unique to horizontally transferred genes and provide a means of estimating the time a foreign gene has resided in a new genome. Recently transferred genes show the patterns of nucleotide composition typical of the donor genome; the base compositions of the codon positions reflect this equilibrium. Fully ameliorated genes show the nucleotide

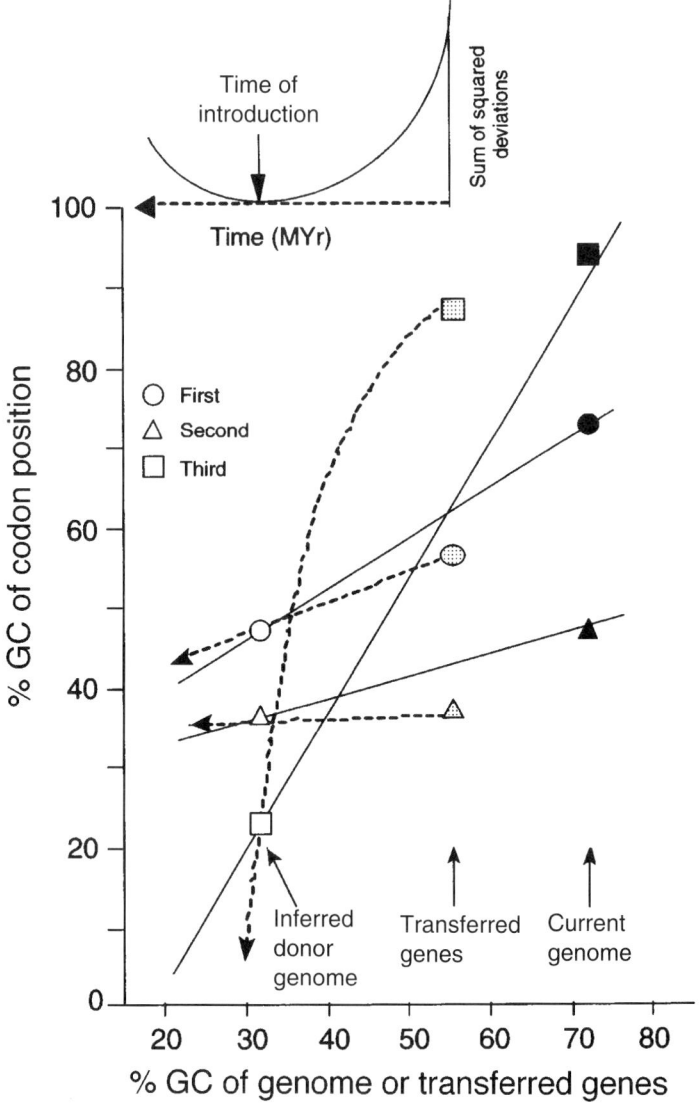

Fig. 16.4 Application of reverse amelioration to estimate the duration of intro-gressed genes in a novel genetic background. GC contents of the hypothetical recipient genome, and of genes suspected of being acquired by horizontal trans-fer, are superimposed on curves predicting the relationships derived by Muto and Osawa (1987). Dashed lines depict the projected GC contents of each codon position predicted by reverse amelioration algorithms. The graph in the upper left displays the least-squares analysis of the deviation of the values predicted by amelioration from the Muto and Osawa relationships. Time of intro-gression into the recipient genome is taken as the point of smallest least-squares deviation from the these relationships. From Lawrence and Ochman (1996).

compositions of the recipient genome. However, partially ameliorated genes show a nucleotide composition that does not resemble that of either the donor or the recipient genome. Most importantly, these sequences do not show the positional GC contents predicted by the Muto and Osawa relationships; genes undergoing amelioration are not in equilibrium.

Using both the equilibrium base composition of the new host and the rate of nucleotide substitution, one can estimate how long a gene has been ameliorating (Lawrence and Ochman, 1996) – that is, when the genes were last in equilibrium. These data provide the likely time of introduction of a horizontally transferred sequence. This method has been applied to the identifiably foreign portions of the *E. coli* genome (13%) which are assumed to have been horizontally transferred. The average entry time for these sequences was 25 million years ago (Fig. 16.5). The age distribution of these fragments suggests that about 31 kb of novel DNA is stably introduced into the *E. coli* genome every million years (Lawrence and Ochman, 1996). Since bacterial genomes are not growing constantly in size, this gain of DNA is probably offset by a corresponding deletion of chromosomal genes; the lost sequences could include both ancestral sequences and previously transferred foreign sequences. Therefore, the *E. coli* lineage has gained and lost more than 3 Mbp of DNA – or 60% of its current genome size – since its divergence from the *Salmonella* lineage 100 million years ago.

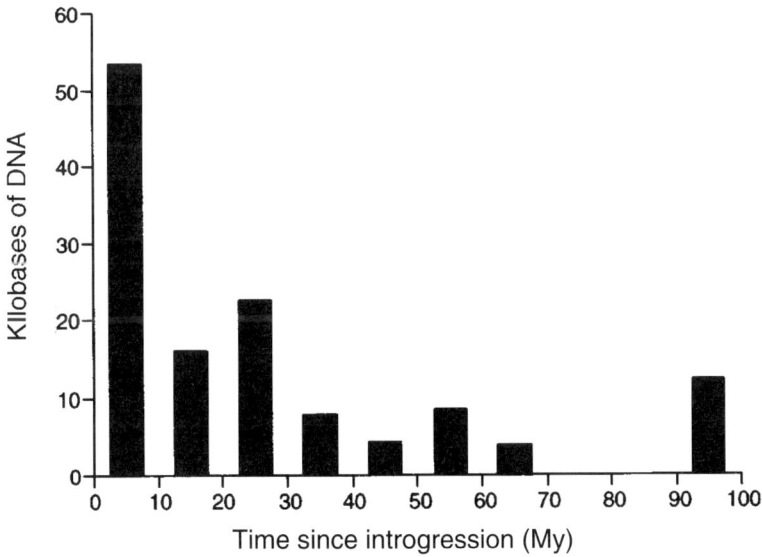

Fig. 16.5 Times of introgression of DNA fragments into the *E. coli* chromosome. The amount of DNA reflects cumulative pooled sequences. From Lawrence and Ochman (1996).

16.2.4   HORIZONTAL TRANSFER AND GENOME ORGANIZATION

The method for timing sequence introgression was applied to the *cob* and *pdu* operons of *S. enterica*, and led to the conclusion that these sequences have been ameliorating for 71 million years (Lawrence and Ochman, 1996). This estimate suggests that introduction of the B_{12} synthesis followed the divergence of *Salmonella* and *E. coli* from their common ancestor (100 million years), yet preceded the radiation of the known *Salmonella* serovars (50 million years). This timing agrees with the observation that *cob* sequences are absent from all isolates of *E. coli*, but are present in all serovars of *S. enterica* (Lawrence and Roth, 1996a).

Moreover, this timing suggests that the acquisition of the *cob/pdu* regulon may have played a role in establishing the identity of the nascent *Salmonella* lineage. As detailed above, all *S. enterica* isolates synthesize B_{12} and employ the cofactor for propanediol degradation; both of these functions are absent from *E. coli*. The importance of these pathways is demonstrated by the fact that 1% of the *Salmonella* genome is dedicated to B_{12} acquisition and an additional 1% is involved in functions requiring B_{12}. In addition, the pathways for the reduction of tetrathionate (*tth*), the reduction of thiosulfate (*phs*) and anaerobic sulfite reduction (*asr*) are all thought to be involved in *Salmonella*'s B_{12}-dependent anaerobic use of ethanolamine and propanediol, and all are absent from *E. coli*. Ability to synthesize B_{12} and degrade propanediol is the basis of a widely used test for identification of Salmonellae (Rambach, 1990).

Many functions distinguishing *E. coli* and *S. enterica* lineages were acquired by horizontal transfer. Aside from the operons mentioned above, other pathways specific to *E. coli* or *S. enterica* are of recognized foreign origin. For example, the *E. coli lac* operon – long a diagnostic criterion for distinguishing these taxa – is of exogenous origin (Buvinger *et al.*, 1984). Virulence genes of *S. enterica* are clearly of foreign origin (Groisman and Ochman, 1993; Ochman and Groisman, 1995; Barinaga, 1996). There are analogous functions in these two taxa – such as phosphonate utilization (Wanner and Metcalf, 1992; Metcalf and Wanner, 1993; Jiang *et al.*, 1995) – which employ non-homologous operons that were acquired independently by these two species from different foreign sources. Large-scale differences between the chromosomes are evident in the 'chromosome loops' by which the genetic maps differ (Riley and Krawiec, 1987; Riley and Sanderson, 1990); many 'loops' contain horizontally transferred genes (Fig. 16.6).

Considering the rate of horizontal transfer, and the large number of loci with acknowledged foreign ancestry, it may be surprising that the genetic maps of *E. coli* and *S. enterica* appear so similar. This may be explained by the fact that most genes appearing on these genetic maps were identified by mutations which affect growth under the common set

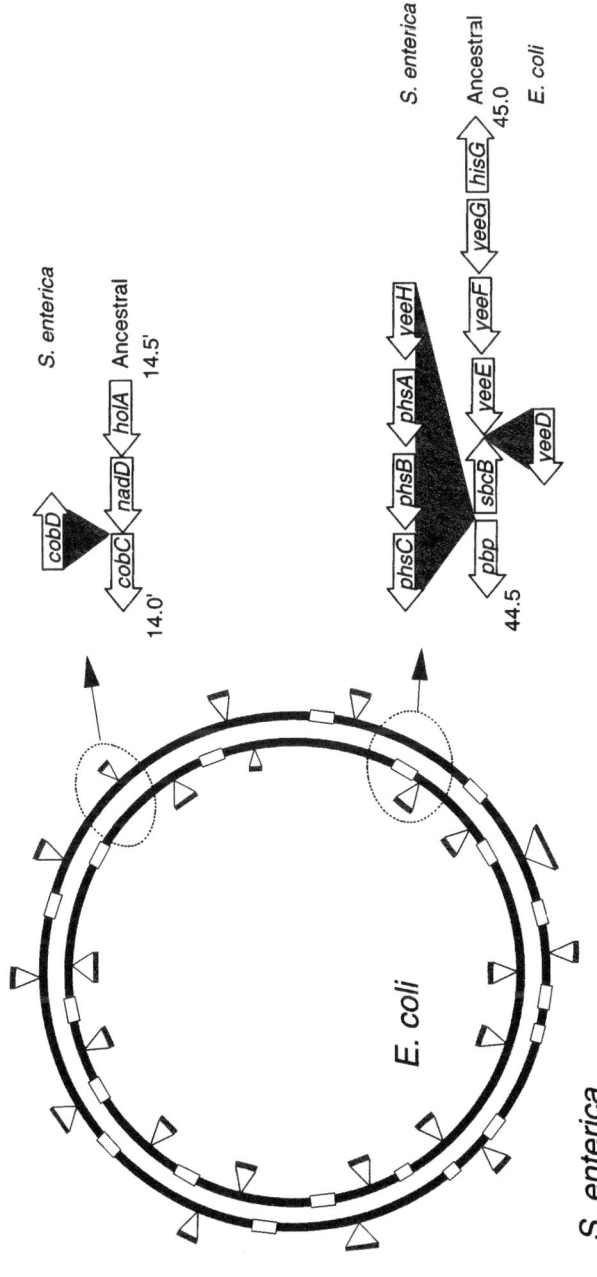

Fig. 16.6 Representation of the *E. col*: (inside circle) and *S. enterica* (outside circle) chromosomes. Boxes in the circles represent deleted ancestral sequences; triangles represent introgressed DNA. The two representative regions include the *sbcB* gere region (minute 44) and the *cobC* gene region (minute 15) described in the text.

of laboratory conditions used for culture of both organisms (Sanderson *et al.*, 1995; Berlyn *et al.*, 1996). We believe the perceived similarity between the genetic maps has been overestimated because of this history. A common set of housekeeping genes is required for growth under traditional laboratory conditions, so it is not surprising that the genetically detected genes comprise similar sets. Hence, the genetic maps have overestimated the similarity of the two genomes by ignoring the species-specific genes that are scattered among the known shared genes.

We predict that DNA sequences of the *S. enterica* and *E. coli* genomes will reveal mosaic chromosomes comprised of three classes of genes (Fig. 16.6):

- shared sequences present in both species and derived from their common ancestor;
- ancestral sequences that were deleted in one of the two species;
- horizontally acquired genes unique to one species.

Aside from the *cob, pdu* and *lac* operons described above, comparison of these chromosomes reveals many other examples of such fine-scale variation. For example, the region adjacent to the *his* operon shows evidence for two insertion/deletion events (Fig. 16.6); the region around the *cobC* gene also shows evidence for gene loss or gain, in this case a single gene. In many cases, the observed differences involve a small number of genes – many with unknown phenotypes – that can only be detected by direct comparison of genomic DNA sequences.

Based on our analysis of rate of gene gain and loss by *E. coli* – and the assumption that *S. enterica* has experienced a similar process – we predict that 15% to 30% of each genome will comprise sequences absent from the other taxon (Table 16.1). While this prediction can only be fully tested by comparing complete genome sequences, DNA hybridization data suggest that this estimate may be reasonable. Whole-genome reassociation studies determined that the *E. coli* and *S. enterica* genomes are 45% 'related' (Brenner and Cowie, 1968; Brenner *et al.*, 1969, 1972; Brenner and Falkow, 1971). This figure reflects two classes of genes in each genome: more than half of their genomes comprise shared sequences with 20% unpaired bases (80% identical); and the remaining portion of each genome – from 25% to 45% – is unique (Brenner and Falkow, 1971; Brenner *et al.*, 1972). The percentage identity of shared sequence corresponds well with the known levels of nucleotide divergence (Ochman and Wilson, 1988; Sharp, 1991), and the prediction of > 25% unique sequences corresponds closely with the values we have predicted from the amelioration analyses (Table 16.1).

Table 16.1 Contributions to the mosaic genomes of *E. coli* and *S. enterica*

Contributor	Lineage	
	E. coli	*S. enterica*
Starting genome size[a]	5000 kb	5000 kb
Insertion		
Rate of insertion	31 kb/Myr	31 kb/Myr
Separation	100 Myr	100 Myr
Total inserted DNA	3100 kb	3100 kb
Deletion		
Ancestral DNA	600 kb	600 kb
Inserted DNA	2500 kb	2500 kb
Total deleted DNA	3100 kb	3100 kb
Final Genome size	5000 kb	5000 kb
Shared DNA[b]	3900 kb	3900 kb
Unique ancestral DNA[c]	500 kb	500 kb
Unique inserted DNA	600 kb	600 kb

[a] Proposed genome size of the most recent common ancestor of the *E. coli* and *S. enterica* lineages.
[b] Estimated amount of DNA shared between *E. coli* and *S. enterica*.
[c] DNA present in the common ancestor of *E. coli* and *S. enterica*, but lost in one taxon.

16.2.5   HORIZONTAL TRANSFER AND BACTERIAL SPECIATION

Most definitions of biological species are based upon assessments of intraspecific gene flow – that is, sexual recombination among conspecific organisms (Vrba, 1985; King, 1993). Conspecific populations exchange genes with each other more often than with organisms of other species. Thus, speciation events create two reproductively isolated groups from one set of conspecific individuals. These concepts reflect the eukaryotic focus of most treatments of speciation. Bacterial species and bacterial speciation are more difficult to define in these terms because most bacteria are asexual organisms (Maynard Smith *et al.*, 1993; Maynard Smith, 1996). However, it is clear that isolates of these organisms can be unambiguously assigned to non-overlapping groups. The characteristics that define each group reflect sets of selective pressures inherent in their individual life styles. In this way, bacteria conform to an ecological species concept (Van Valen, 1976; Wiley, 1978) in that a bacterial species occupies a definable ecological niche that provides selection for essential, niche-specific functions.

We suggest that sympatric bacterial speciation – that is, the origin of new bacterial species that occupy definable ecological niches – is strongly influenced by horizontal transfer. Horizontal transfer is estimated to introduce sequences at a rate of 31 kb per million years into the *E. coli* genome (Lawrence and Ochman, 1996). Based on this rate, the *E. coli* chromosome has gained and lost over 3 Mbp of DNA (roughly 60% of the size of its current genome) since diverging from *Salmonella*. These sequence introductions include selfish operons, which can provide a novel selectable function upon introduction. In comparison, application of the synonymous substitution rate – which approximates the genomic mutation rate – yields approximately 22 000 point mutations, or 22 kb of variant DNA, occurring in the *E. coli* genome per million years. Although quantitatively similar amounts of variation are introduced through mutation and through horizontal transfer, the types of genetic information furnished by these processes are qualitatively very different. Nucleotide substitutions arising by mutation – or by intragenic recombination among strains, which is thought to generate about the same level of genic diversity as point mutations (Milkman and Bridges, 1990, 1993; Guttman and Dykhuizen, 1994a,b) – would provide modest changes in the encoded phenotypes and would only rarely confer novel characteristics on the organism.

Horizontally transferred sequences can provide a novel function immediately upon introgression. Selfish operons conferring novel metabolic capabilities (e.g., the *lac*, *cob* and *pdu* operons) allow the recipient taxa to explore novel ecological niches that were previously unavailable (or unexploitable). Selfish operons allow effective, competitive exploration of new niches since the encoded functions have been previously selected to perform these functions in the donor organism; the functions are only novel to the recipient organism. In contrast, novel functions arising by mutational processes allow only slow, initially inefficient exploitation of a new resource.

While it is clear that homologous genes with distinct functions evolve through duplication and divergence, the evolution of novel functions by mutation would seem to be an inefficient way to begin exploitation of a novel resource; success in a novel environment would depend upon the absence of any experienced competitor. We propose that novel functions evolve in a context where selection for function is not critical – that is, the novel function is not required for life. In this manner, functions can evolve under weak selection (in a donor species) and allow niche exploitation under strong selection following horizontal transfer. In effect, we suggest that many apparently paralogous genes in bacteria will prove to be orthologues acquired by horizontal transfer. Hence, we suggest that horizontal transfer serves to decouple the process of metabolic differentiation from the process of speciation.

REFERENCES

Barinaga, M. (1996) A shared strategy for virulence. *Science* **272**: 1261–1263.

Berlyn, M.K.B., Low, K.B. and Rudd, K.F. (1996) Linkage map of *Escherichia coli* K-12, edition 9, in *Escherichia coli and Salmonella: Cellular and Molecular Biology*, 2nd edn, (eds F.C. Neidhardt, R. Curtiss III, J.L. Ingraham *et al.*), ASM Press, Washington, DC, pp. 1715–1902.

Bobik, T.A., Ailion, M. and Roth, J.R. (1992) A single regulatory gene integrates control of vitamin B_{12} synthesis and propanediol degradation. *J. Bacteriol.* **174**: 2253–2266.

Brenner, D.J. and Cowie, D.B. (1968) Thermal stability of *Escherichia coli–Salmonella typhimurium* deoxyribonucleic acid duplexes. *J. Bacteriol.* **95**: 2258–2262.

Brenner, D.J. and Falkow, S. (1971) Molecular relationships among members of the Enterobacteriaceae. *Adv. Genet.* **16**: 81–118.

Brenner, D.J., Fanning, G.R., Johnson, K.E. *et al.* (1969) Polynucleotide sequence relationships among members of Enterobacteriaceae. *J. Bacteriol.* **98**: 637–650.

Brenner, D.J., Fanning, G.R., Skerman, F.J. Skerman and Falkow, S. (1972) Polynucleotide sequence divergence among strains of *Escherichia coli* and closely related organisms. *J. Bacteriol.* **109**: 953–965.

Buvinger, W.E., Lampel, K.A., Bojanowski, R.J. and Riley, M. (1984) Location and analysis of nucleotide sequences at one end of a putative *lac* transposon in the *Escherichia coli* chromosome. *J. Bacteriol.* **159**: 618–623.

Groisman, E.A. and Ochman, H. (1993) Cognate genes govern invasion of host epithelial cells by *Salmonella typhimurium* and *Shigella flexneri*. *EMBO J.* **12**: 3779–3787.

Guttman, D.S. and Dykhuizen, D.E. (1994a) Clonal divergence in *Escherichia coli* as a result of recombination, not mutation. *Science* **266**: 1380–1383.

Guttman, D.S. and Dykhuizen, D.E. (1994b) Detecting selective sweeps in naturally occurring *Escherichia coli*. *Genetics* **138**: 993–1003.

Jeter, R. (1990) Cobalamin-dependent 1,2-propanediol utilization by *Salmonella typhimurium*. *J. Gen. Microbiol.* **136**: 887–896.

Jeter, R., Olivera, B.M. and Roth, J.R. (1984) *Salmonella typhimurium* synthesizes cobalamin (vitamin B_{12}) *de novo* under anaerobic growth conditions. *J. Bacteriol.* **159**: 206–216.

Jeter, R.M. and Roth, J.R. (1987) Cobalamin (vitamin B_{12}) biosynthetic genes of *Salmonella typhimurium*. *J. Bacteriol.* **169**: 3189–3198.

Jiang, W., Metcalf, W.W., Lee, K.S. and Wanner, B.L. (1995) Molecular cloning, mapping, and regulation of *Pho* regulon genes for phosphonate breakdown by the phosphonatase pathway of *Salmonella typhimurium* LT2. *J. Bacteriol.* **177**: 6411–6421.

King, M. (1993) *Species Evolution*, Cambridge University Press, Cambridge.

Lawrence, J.G. and Ochman, H. (1997) Amelioration of bacterial genomes: rates of change and exchange. *J. Mol. Evol.* **44**: 383–397.

Lawrence, J.G. and Roth, J.R. (1995) The cobalamin (coenzyme B_{12}) biosynthetic genes of *Escherichia coli*. *J. Bacteriol.* **177**: 6371–6380.

Lawrence, J.G. and Roth, J.R. (1996a) Evolution of coenzyme B_{12} synthesis among enteric bacteria: evidence for loss and reacquisition of a multigene complex. *Genetics* **142**: 11–24.

Lawrence, J.G. and Roth, J.R. (1996b) Selfish operons: horizontal transfer may drive the evolution of gene clusters. *Genetics* **143**: 1843–1860.

Maynard Smith, J. (1996) Population genetics: an introduction, in *Escherichia coli and Salmonella typhimurium: Cellular and Molecular Biology*, 2nd edn, (eds F.C. Neidhardt, R. Curtiss III, J.L. Ingraham *et al.*), American Society for Microbiology, Washington, DC, pp. 2685–2690.

Maynard Smith, J., Smith, N.H., O'Rourke, M. and Spratt, B.G. (1993) How clonal are bacteria? *Proc. Natl Acad. Sci. USA* **90**: 4384–4388.

Médigue, C., Rouxel, T., Vigier, P. *et al.* (1991) Evidence of horizontal gene transfer in *Escherichia coli* speciation. *J. Mol. Biol.* **222**: 851–856.

Metcalf, W.W. and Wanner, B.L. (1993) Evidence for a fourteen-gene, *phnC* to *phnP* locus for phosphonate metabolism in *Escherichia coli*. *Gene* **129**: 27–32.

Milkman, R. and Bridges, M.M. (1990) Molecular evolution of the *E. coli* chromosome. III. Clonal frames. *Genetics* **126**: 505–517.

Milkman, R. and Bridges, M.M. (1993) Molecular evolution of the *E. coli* chromosome. IV. Sequence comparisons. *Genetics* **133**: 455–468.

Muto, A. and Osawa, S. (1987) The guanine and cytosine content of genomic DNA and bacterial evolution. *Proc. Natl Acad. Sci. USA* **84**: 166–169.

Neidhardt, F.C., Curtiss III, R., Ingraham, J.L. *et al.* (1996) *Escherichia coli and Salmonella: Cellular and Molecular Biology*, 2nd edn, ASM Press, Washington, DC.

Ochman, H. and Groisman, E.A. (1995) The evolution of invasion in enteric bacteria. *Can. J. Microbiol.* **41**: 555–561.

Ochman, H. and Lawrence, J.G. (1996) Phylogenetics and the amelioration of bacterial genomes, in *Escherichia coli and Salmonella typhimurium: Cellular and Molecular Biology*, 2nd edn, (eds F.C. Neidhardt, R. Curtiss III, J.L. Ingraham *et al.*) American Society for Microbiology, Washington, DC, pp. 2627–2637.

Ochman, H. and Selander, R.K. (1984) Evidence for clonal population structure in *Escherichia coli*. *Proc. Natl Acad. Sci. USA* **81**: 198–201.

Ochman, H. and Wilson, A.C. (1987) Evolutionary history of enteric bacteria, in *Escherichia coli and Salmonella typhimurium: Cellular and Molecular Biology*, (eds F.C. Neidhardt, J.L. Ingraham, K.B. Low *et al.*), American Society for Microbiology, Washington, DC, pp. 1649–1654.

Ochman, H. and Wilson, A.C. (1988) Evolution in bacteria: evidence for a universal substitution rate in cellular genomes. *J. Mol. Evol.* **26**: 74–86.

Rambach, A. (1990) New plate medium for facilitated differentiation of *Salmonella* spp. from *Proteus* spp. and other enteric bacteria. *Appl. Env. Microbiol.* **56**: 301–303.

Riley, M. and Krawiec, S. (1987) Genome organization, in *Escherichia coli and Salmonella typhimurium: Cellular and Molecular Biology*, (edis F.C. Neidhardt, J.L. Ingraham, K.B. Low *et al.*), American Society for Microbiology, Washington, DC, pp. 967–981.

Riley, M. and Sanderson, K.E. (1990) Comparative genetics of *Escherichia coli* and *Salmonella typhimurium*, in *The Bacterial Chromosome*, (eds M. Riley and K. Drlica), American Society for Microbiology, Washington, DC, pp. 95–96.

Roof, D.M. and Roth, J.R. (1988) Ethanolamine utilization in *Salmonella typhimurium*. *J. Bacteriol.* **170**: 3855–3863.

Roof, D.M. and Roth, J.R. (1989) Functions required for vitamin-B_{12} dependent ethanolamine utilization in *Salmonella typhimurium*. *J. Bacteriol.* **171**: 3316–3323.

Roth, J.R., Lawrence, J.G. and Bobik, T.A. (1996) Cobalamin (coenzyme B_{12}): synthesis and physiological significance. *Annu. Rev. Microbiol.* **50**: 137–181.

Roth, J.R., Lawrence, J.G., Rubenfield, M. *et al.* (1993) Characterization of the cobalamin (vitamin B_{12}) biosynthetic genes of *Salmonella typhimurium*. *J. Bacteriol.* **175**: 3303–3316.

Sanderson, K.E., Hessel, A. and Rudd, K.E. (1995) Genetic map of *Salmonella typhimurium*, Edition VIII. *Microbiol. Rev.* **59**: 241–303.

Selander, R.K. and Levin, B.R. (1980) Genetic diversity and structure in *Escherichia coli* populations. *Science* **210**: 545–547.

Selander, R.K., Li, J. and Nelson, K. (1996) Evolutionary genetics of *Salmonella enterica*, in *Escherichia coli and Salmonella typhimurium: Cellular and Molecular Biology*, 2nd edn, (eds F.C. Neidhardt, R. Curtiss III, J.L. Ingraham *et al.*), American Society for Microbiology, Washington, DC, pp. 2691–2702.

Sharp, P.M. (1991) Determinants of DNA sequence divergence between *Escherichia coli* and *Salmonella typhimurium*. *J. Mol. Evol.* **33**: 23–33.

Sueoka, N. (1961) Variation and heterogeneity of base composition of deoxyribonucleic acids: a compilation of old and new data. *J. Mol. Biol.* **3**: 31–40.

Sueoka, N. (1962) On the genetic basis of variation and heterogeneity in base composition. *Proc. Natl Acad. Sci. USA* **48**: 582–592.

Sueoka, N. (1988) Directional mutation pressure and neutral molecular evolution. *Proc. Natl Acad. Sci. USA* **85**: 2653–2657.

Sueoka, N. (1992) Directional mutation pressure, selective constraints, and genetic equilibria. *J. Mol. Evol.* **34**: 95–114.

Sueoka, N. (1993) Directional mutation pressure, mutator mutations, and dynamics of molecular evolution. *J. Mol. Evol.* **37**: 137–153.

Syvanen, M. (1994) Horizontal gene flow: evidence and possible consequences. *Ann. Rev. Genet.* **28**: 237–261.

Van Valen, L. (1976) Ecological species, multispecies, and oaks. *Taxon* **25**: 223–239.

Vrba, E.S. (1985) *Species and Speciation*, Transvaal Museum, Pretoria, South Africa.

Wanner, B.L. and Metcalf, W.W. (1992) Molecular genetic studies of a 10.9-kb operon in *Escherichia coli* for phosphonate uptake and biodegradation. *FEMS Microbiol. Lett.* **79**: 133–139.

Whittam, T.S. (1996) Genetic variation and evolutionary processes in natural populations of *Escherichia coli*, in *Escherichia coli and Salmonella typhimurium: Cellular and Molecular Biology*, 2nd edn, (eds F.C. Neidhardt, R. Curtiss III, J.L. Ingraham *et al.*), American Society for Microbiology, Washington, DC, pp. 2708–2720.

Whittam, T.S. and Ake, S. (1992) Genetic polymorphisms and recombination in natural populations of *Escherichia coli*, in *Mechanisms of Molecular Evolution*, (eds N. Takahata, N. and A.G. Clark), Japan Scientific Society Press, Tokyo, pp. 223–246.

Wiley, E.O. (1978) The evolutionary species concept reconsidered. *Systematic Zoology* **27**: 17–26.

Evolutionary evidence for recombination among bacteria in nature: *Escherichia coli*

17

Roger Milkman, Diane Cryderman, Melissa McKane, Kerri McWeeny and Elisabeth A. Raleigh

SUMMARY

Mosaic patterns of variation in bacterial genomes can be explained by the cutting of incoming DNA molecules by restriction endonucleases, which are polymorphic in *Escherichia coli*, followed by shortening by exonucleases. Results of conjugational crosses and backcrosses resemble those of corresponding transductions. Not unexpectedly, the results of reciprocal conjugations differ.

17.1   INTRODUCTION

Extensive observations of mosaic patterns in bacterial genomes indicate that recombination in nature is widespread, and the variety of mechanisms discovered and elaborated over the past half-century have now opened a rich field of explicit exploration.

The evolutionary evidence – genomic mosaicism – is easily understood in a general context of classical and molecular population genetics. To begin with, a purely clonal model of *E. coli* evolution is contradicted by the observation that different chromosomal regions have different phylogenies (Dykhuizen and Green, 1991; Guttman and Dykhuizen, 1994b). Thus the periodic selection model (Atwood *et al.*, 1951), which introduced the idea of clonal sweeps (Guttman and Dykhuizen, 1994a), now accommodates recombinational replacements in the initially clonal chromosome (persisting as a **clonal frame**) (Milkman and McKane Bridges, 1990, 1993; Milkman and McKane, 1995; Milkman, 1996a, b). The clone, whose vast spread was motivated by a very rare, broadly favorable mutant allele, is now seen to become a **meroclone**, whose members share clonal frames, derived from the clonal ancestor, but punctuated in diverse places by other clonal segments, which have different ancestry (Milkman and McKane Bridges, 1993; Milkman and McKane, 1995; Milkman, 1996a, b). Analysis of the ECOR (*E. coli* **R**eference) strains (Ochman and Selander, 1984; see also Herzer *et al.*, 1990) indicates three major meroclones in *E. coli* and probably a few more.

The details of the recombination processes that produce this dynamic state are now being sought in the experimental study of transduction and conjugation, as well as in the refinement of inferences about other means of gene transfer. These inferences are drawn from the comparison of genomes and encouraged by description of various mechanisms in this symposium.

Studies of transduction and conjugation in the ECOR strains of *E. coli*, which are of recent natural origin, are motivated by modern questions that are in turn framed in a vocabulary that owes a great deal to the techniques of modern molecular biology – sequencing, PCR and restriction analysis – as well as the relatively venerable multi-locus enzyme electrophoresis (MLEE). An interesting case in point is that, having wished to move some arabinose genes from *E. coli* B into K-12, Boyer (1964) published the results of his efforts (see also Arber and Morse, 1965; Boyer, 1966). He had had trouble with the conjugation; he figured that maybe bacteria restricted other bacteria as well as phages, noted that the problems disappeared when he backcrossed some of the rare successful exconjugants to K-12, concluded that strains B and K-12 (the two major laboratory strains then and now) have different restriction-modification genes, and correctly localized them near *thr*. The degree to which we have become able to resolve natural genetic variation in *E. coli* – in terms of sequences, restriction-modification systems and surface antigens – is of course due in no small part to Boyer's subsequent efforts.

17.2   RESULTS AND DISCUSSION

The following recent experiments show that both transduction and conjugation in *E. coli* in nature can mediate the transfer of DNA in small enough pieces to confer a specific advantage without bringing in counterproductive excess baggage. These results will be seen to contrast with those of crosses made between marker variants of a single strain, usually K-12 (Smith, 1991).

1. 18 P1 transductants from ECOR 47 into K-12 were analyzed, exploiting the two strains' sequence differences (which range from 1 to 4%) via the analysis of a sizeable series of contiguous or closely spaced 1500 bp PCR fragments, using restriction enzymes that distinguished the respective donor and recipient DNAs. The donor DNA was found to be **abridged** (shortened relative to the size of the entrant molecule, which would have been between 80 and 100 kb, and often incorporated as discrete segments) (McKane and Milkman, 1995). This matched expectations based on the existence of a high degree of natural polymorphism in restriction-modification systems (DuBose *et al.*, 1988; Sharp *et al.*, 1992; Milkman and McKane Bridges, 1993; King and Murray, 1994; Barcus *et al.*, 1995). These results and the following three sets are summarized in Table 17.1.

2. The two transductants with the largest continuous donor DNA stretches (25 and 20 kb, respectively) but otherwise identical to K-12 were each backtransduced to the K-12 recipient. Essentially no abridgement was seen, confirming a major role for restriction-modification.

3. Transduction of ECOR 47 into ER2476, a modified version of the recipient strain (deletion of all **known** restriction-modification systems), resulted in greatly reduced abridgement.

4. Backtransduction into ER2476 of an ECOR 47→ER2476 transductant containing a 61 kb segment of donor DNA produced 30 analyzed products with essentially no abridgement. This extended the observable range, and it also reduced the likelihood that the absence of abridgement in the earlier backtransduction was due simply to an insufficient extent of mismatched DNA (Modrich, 1991).

5. Conjugation of ECOR 47 into the same K-12 recipient, with selection at the same site, produced transconjugants that were strikingly similar to the transductants in the number of gaps introduced in the incorporated donor DNA. The average length of the incorporated donor segments was much greater (2×), but not by as much as that of the respective entrant molecules (presumably often > 10×) (Table 17.2).

6. Corresponding backconjugations showed essentially no abridgement (Table 17.3).

7. A reciprocal conjugation, with K-12 as donor and ECOR 47 as recipient, showed far less abridgement (Table 17.4) than the ECOR → K-12 cross detailed in Table 17.2. Since the mismatch level was essentially identical, mismatch level is excluded from a major role in abridgement. [The term 'essentially identical' is used rather than 'identical' because the K-12 donor had been modified to contain ECOR 47 DNA in the PCR regions from 1276 kb to 1335 kb (Blattner *et al.*, 1997)

Table 17.1 Original transductants: ECOR 47 → K12trpA33

PCR Fragment Location, kb (from Blattner et al., 1997)	27.5	27.9	28.1					28.3					28.6		28.8	29.3
	1276	1297	1305	1306	1308	1310	1312	1314	*	1316	1319	1327	1328	1330	1335	1362
(1) 47K–4	–	–	–	–	–	–	–	D	D	D	D	D	D	D	D	–
9	–	D	D	D	D	D	D	D	D	D	–	D	D	–	D	–
15	–	–	D	D	D	D	D	D	D	D	D	–	–	–	–	–
17	–	–	D	D	D	D	D	D	D	–	–	–	–	–	–	–
7	–	–	–	–	–	–	–	–	D	D	D	DΔ	–	–	–	–
13	–	–	–	–	–	–	/d	D	D	D	D	–	–	–	–	–
3	–	–	–	–	–	–	–	D	D	d/	D	–	–	–	–	–
11	–	–	–	–	–	–	–	D	D	d/	D	–	–	–	–	–
16	–	–	–	–	–	–	–	/d	D	D	–	–	–	–	–	–
18	–	–	–	–	–	–	–	D	D	–	–	–	–	–	–	–
(2) 6	–	D	D	D	d/	D	D	D	D	D	–	–	–	–	–	–
12	–	–	D	D	D	D	D	/d	D	/d	–	–	–	–	–	–
1	–	–	–	–	–	–	–	D	D	D	D	D	–	–	–	–
14	–	–	d/	–	–	/d	D	D	D	–	D	–	–	–	d/	–
5	–	–	–	–	/d	–	d	–	D	D	D	–	–	–	–	–
(3) 8	–	–	–	–	–	–	–	–	D	D	D	d/	D	D	D	–
2	–	/d	–	–	D	D	D	D	D	D	D	–	–	–	–	–
(4) 10	–	–	–	–	–	–	–	–	D	/d	/d	–	D	–	–	–

Table 17.1 *continued*

Map position (min.)	27.5	27.9	28.1						28.3						28.6	28.8	29.3
PCR Fragment Location, kb (from Blattner et al., 1997)	1276	1297	1305	1306	1308	1310	1312	1314	*	1316	1319	1327	1328	1334	1305	1335	1362
Donor: 47K-4	—	—	—	D	D	D	D	D	D	D	D	D	D	D	D	D	—
Backtr. 13	—	—	—	D	D	D	D	/d	D	D	D	D	D	D	D	D	—
Backtr. #2	—	—	—	D	D	D	D	/d	D	D	D	D	D	—	D	—	—
Backtr. #3	—	—	—	D	D	D	D	—	D	D	D	D	—	—	—	—	—
Donor: 47K-9	—	—	D	D	D	D	D	D	D	D	D	D	D	D	—	—	—
Backtr. 15	—	—	D	D	D	D	D	D	D	D	D	D	D	D	—	—	—

Backtransductants → K12

Transduction to 'restrictionless' recipient: 47 → ER2476

Location #	27.5	27.9	28.1						28.3						28.6	28.8	29.3
47R-10	—	—	D	D	D	D	D	D	D	D	D	D	D	D	D	D	D
11	—	—	D	D	D	D	D	D	D	D	D	D	D	D	D	D	D
12	—	—	D	D	D	D	D	D	D	D	D	D	D	D	D	D	D
14	—	—	D	D	—	D	D	D	D	D	D	D	D	D	—	D	D
1	—	—	—	—	—	—	—	D	D	—	D	D	D	D	D	—	D
16	—	—	—	—	—	—	—	D	D	—	D	D	D	D	D	D	D

Backtransductants

(2)

3	–	–	D	D	D	D	–	D	D	D	D	D	D	D	D	D	D	D	D	D	–	–	D	D	D	–	–
13	–	–	–	–	D	–	D	D	D	D	D	D	D	D	D	D	D	D	D	D	–	–	–	D	–	–	–
9	–	–	D	D	D	–	–	–	D	D	D	D	D	D	D	D	D	D	D	D	D	–	–	–	D	–	–
4	–	–	–	–	–	–	–	–	D	D	D	D	D	D	D	D	D	D	D	D	D	–	–	–	–	–	–
7	–	–	–	–	D	–	–	–	–	–	–	D	D	D	D	D	D	D	D	D	D	–	–	–	–	–	–
8	–	–	–	–	–	–	–	–	–	–	–	d/	D	D	D	D	D	D	D	D	D	D	–	–	–	–	–
18	–	–	–	–	–	–	–	–	–	–	–	–	–	–	–	–	D	D	D	D	D	–	–	–	–	–	–
2	–	–	–	–	–	–	–	–	–	–	–	–	–	–	–	–	D	D	–	–	–	–	–	–	–	–	–
19	–	–	–	d/	–	–	–	–	–	–	–	–	–	–	–	–	D	–	–	–	–	–	–	–	–	–	D

(2)

5	–	–	–	–	–	D	D	D	D	D	D	D	D	D	D	D	D	/d	D	D	D	–	–	–	–	–	D
17	–	–	–	–	D	D	D	–	D	D	D	D	D	D	D	/d	D	D	d/	D	–	–	–	–	D	–	–
20	–	–	–	D	D	–	–	–	D	D	d/	–	D	D	D	D	D	D	D	D	–	D	–	–	–	–	–
15	–	–	–	–	–	–	–	–	–	D	D	–	–	–	–	–	–	–	–	–	–	–	–	–	–	–	–
6	–	d/	–	–	–	–	–	–	D	–	–	–	–	–	–	–	–	–	–	–	–	–	–	–	–	–	–

Cross
47 → ER2476
↓

10 | – | – | D | D | D | D | D | D | D | D | D | D | D | D | D | D | D | D | D | D | – | D | D | D | D | – | D

Backcross
47R10→ER2476

9	–	–	D	D	D	D	D	D	D	D	D	D	D	D	D	D	D	D	D	D	–	D	D	D	D	–	D
4	–	–	D	D	D	D	D	D	D	D	D	D	D	D	D	D	D	D	D	D	–	D	D	D	D	–	–
4	–	–	D	D	D	D	D	D	D	D	D	D	D	D	D	D	D	D	D	D	–	D	–	D	D	–	–
6	–	–	D	D	D	D	D	D	D	D	D	D	D	D	D	D	D	D	D	D	–	D	–	–	–	–	–
1	–	–	D	D	D	D	D	D	D	D	D	D	D	D	D	d/	D	D	–	–	–	–	–	–	–	–	–
2	–	–	D	D	D	D	D	D	D	D	D	D	D	D	D	D	D	D	–	–	–	–	–	–	–	–	–
1	–	–	–	D	D	D	D	D	D	D	D	D	D	D	D	D	D	D	–	–	–	–	–	–	–	–	–
2	–	–	D	D	D	D	D	D	D	D	D	D	D	D	D	D	D	D	–	–	–	–	–	–	–	–	–

1 | – | – | D | D | D | D | D | D | D | – | D | D | D | D | D | – | – | – | – | – | – | – | – | – | – | – | –

(2)

Bold **numbers** represent the **number** of **individuals** in an **identical set**; –, recipient DNA; D, donor DNA; **D**, selected marker; *, whose physical map location is about 1315 kb; Δ, a small deletion; d/, donor DNA at left-hand sites only; /d, donor DNA at right only.
The transductants are arranged in order of the number of donor DNA segments (indicated in parentheses), and within each category by total extent of donor DNA.
One minute ≅ 47 kb.

Table 17.2 ECOR 47 → K12 W3110 *trpA33* transconjugants selected on minimal medium

Position, min	22	24	25	27.5	28.3														29.3		29.8
PCR	1	1	1	1	1	1	1	1	1	1	1	1	*	1	1	1	1	1	1	1	1
Fragment	0	1	1	2	2	3	3	3	3	3	3	3		3	3	3	3	3	3	3	3
Location,	3	1	8	7	9	0	0	0	1	1	1	1		1	1	2	2	3	6	3	8
kb	3	5	4	6	7	5	6	8	0	1	2	4		6	9	7	8	0	2	5	5

No. of donor fragments # (1)

Fragment	22	24	25	27.5	1297	1305	1306	1308	1310	1311	1312	1314	*	1316	1319	1327	1328	1330	1362	1335	1385
HAZ-12	–	–	–	D	D	D	D	–	D	D	D	–	D	D	D	D	D	D	D	D	–
4	–	–	–	–	D	D	D	–	D	D	D	–	D	D	D	D	D	D	D	D	–
8	–	–	–	–	D	D	D	–	D	D	D	–	D	D	D	D	D	D	D	D	–
11	–	–	–	–	–	–	–	–	–	–	–	D	D	D	D	D	–	–	D	–	D
21	–	–	–	D	D	D	D	D	D	D	D	D	D	D	D	D	–	–	D	–	–
3	–	–	–	–	–	D	D	D	D	D	D	D	D	D	D	D	D	D	D	D	–
7	–	–	–	–	D	D	D	D	D	D	D	D	D	D	D	–	–	–	D	–	–
16	–	–	–	–	–	–	–	–	–	–	–	–	D	–	–	–	–	–	D	–	–
13	–	–	–	–	–	–	–	–	–	–	–	–	D	–	–	–	–	–	D	–	–
25	–	–	–	–	–	–	–	–	–	–	–	–	D	–	–	–	–	–	D	–	–
29	–	–	–	–	D	D	D	D	D	D	D	D	D	D	D	D	–	D	D	–	–
22	–	–	–	–	D	D	D	D	D	D	D	D	D	D	D	D	D	D	D	–	–
6	–	–	–	–	–	–	–	–	–	–	–	–	D	–	–	–	–	–	–	–	–
14	–	–	–	–	–	–	–	–	–	–	–	–	D	D	D	–	–	–	–	–	–
24	–	–	–	D	–	–	–	–	–	D	D	D	D	D	D	D	D	–	–	–	–
15	–	–	–	–	–	–	–	–	–	–	–	D	D	D	D	D	D	D	–	–	–
26	–	–	–	–	–	–	–	–	–	–	–	–	D	D	D	–	–	–	–	–	–

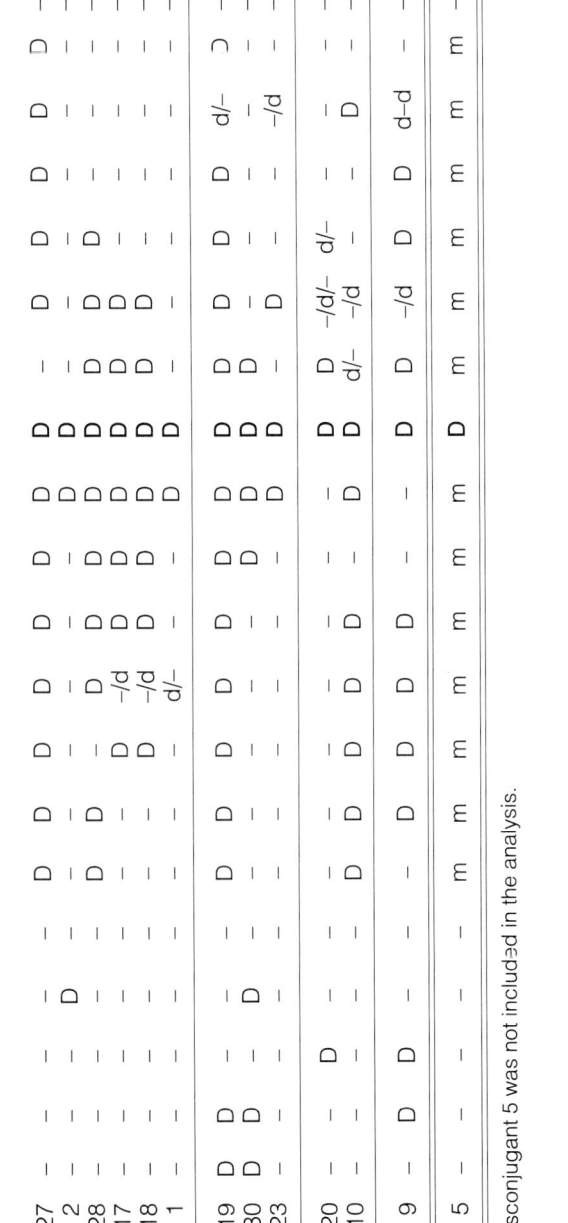

m = mix. Transconjugant 5 was not included in the analysis.

Table 17.3 JAZ: (ECOR 47 → K12W3110trpA33 Transconjugant HAZ−12) → K12 W3110 trpA33 Back-transconjugants

PCR Fragment Location, kb	1115	1184	1276	1297	1305	1306	1308	1311	1312	1314	*	1316	1319	1327	1328	1330	1335	1362	1385
											#								
[DONOR: HAZ-12]	–	–	D	D	D	D	D	D	D	D	D	D	D	D	D	D	D	D	–]
JAZ-1–3, 5–28, 30	–	D	D	D	D	–	–	D	D	D	D	D	D	D	D	D	D	D	D
4	–	–	–	–	–	D	–	D	D	D	D	D	D	D	D	D	D	D	D
29	–	–	–	–	–	–	–	–	–	–	D	D	D	D	D	D	D	D	D

Table 17.4 Conjugation: K12 → ECOЯ 47 [(HAZ-12) DCHF8 → DC47A3R]

min 90	22	24	25	27.5	28									29					30		
kb	1033	1115	1184	1276	1297	1305	1306	1308	1310	1311	1312	1314	*	1316	1319	1327	1328	1330	1133	1362	1385
Transc.																					
12	⌂	⌂	⌂	⌂	⌂	⌂	⌂	⌂	⌂	⌂	⌂	⌂	⌂	⌂	⌂	⌂	⌂	⌂	⌂	⌂	⌂
13	⌂	⌂	⌂	⌂	⌂	⌂	⌂	⌂	⌂	⌂	⌂	⌂	⌂	⌂	⌂	⌂	⌂	⌂	⌂	⌂	⌂
9	⌂	⌂	⌂	⌂	⌂	⌂	⌂	⌂	⌂	⌂	⌂	⌂	⌂	⌂	⌂	⌂	⌂	⌂	—	⌂	—
4	—	—	⌂	⌂	⌂	⌂	⌂	⌂	⌂	⌂	⌂	⌂	⌂	⌂	⌂	⌂	⌂	⌂	—	—	⌂
3	—	—	⌂	⌂	⌂	⌂	⌂	⌂	⌂	⌂	⌂	⌂	⌂	⌂	⌂	⌂	⌂	⌂	—	—	—
23	—	—	⌂	⌂	⌂	⌂	⌂	⌂	⌂	⌂	⌂	⌂	⌂	⌂	⌂	⌂	⌂	⌂	—	—	—
25	—	—	⌂	⌂	⌂	⌂	⌂	⌂	⌂	⌂	⌂	⌂	⌂	⌂	⌂	⌂	⌂	⌂	—	—	—
27	—	—	⌂	⌂	⌂	⌂	⌂	⌂	⌂	⌂	⌂	⌂	⌂	⌂	⌂	⌂	⌂	⌂	—	—	—
15	—	—	⌂	⌂	⌂	⌂	⌂	⌂	⌂	d/—	⌂	⌂	⌂	⌂	⌂	—	—	⌂	—	—	—
6	—	—	—	⌂	⌂	⌂	⌂	⌂	⌂	⌂	⌂	⌂	⌂	⌂	⌂	⌂	⌂	⌂	—	⌂	—
8	—	—	—	⌂	⌂	⌂	⌂	⌂	⌂	⌂	⌂	⌂	⌂	⌂	⌂	⌂	⌂	⌂	—	⌂	—
14	—	—	—	⌂	⌂	⌂	⌂	⌂	⌂	⌂	⌂	⌂	⌂	⌂	⌂	⌂	⌂	⌂	—	⌂	⌂
26	—	—	—	⌂	⌂	⌂	⌂	⌂	⌂	⌂	⌂	⌂	⌂	⌂	⌂	⌂	⌂	⌂	—	⌂	⌂
28	—	—	—	⌂	⌂	⌂	⌂	⌂	⌂	⌂	⌂	⌂	⌂	⌂	⌂	⌂	⌂	⌂	—	⌂	⌂
2	—	—	—	—	⌂	⌂	⌂	⌂	⌂	⌂	⌂	⌂	⌂	⌂	⌂	⌂	⌂	⌂	—	⌂	⌂
18	—	—	—	⌂	⌂	⌂	⌂	⌂	⌂	⌂	⌂	⌂	⌂	⌂	⌂	⌂	⌂	⌂	—	—	—
1	—	—	—	⌂	⌂	⌂	⌂	⌂	⌂	⌂	⌂	⌂	⌂	⌂	⌂	⌂	⌂	⌂	—	—	—

Table 17.4 Conjugation: K12 → ECOR 47 [(HAZ-12) DCHF8 → DC47A3R]

min	90	22	24	25	27.5	28											29					30		
kb	1033	1133	1115	1184	1276	1297	1305	1306	1308	1310	1311	1312	1314	*	1316	1319	1327	1328	1330	1335	1331	1333	1368	1385
Transc.																								
7		–	–	–	D	D	D	D	D	D	D	D	D	D	D	D	D	D	D	D	–	–	–	–
10		–	–	–	D	D	D	D	D	D	D	D	D	D	D	D	D	D	D	D	–	–	–	–
16		–	–	–	D	D	D	D	D	D	D	D	D	D	D	D	D	D	D	D	–	–	–	–
17		–	–	–	D	D	D	D	D	D	D	D	D	D	D	D	D	D	D	D	–	–	–	–
19		–	–	–	D	D	D	D	D	D	D	D	D	D	D	D	D	D	D	D	–	–	–	–
20		–	–	–	D	D	D	D	D	D	D	D	D	D	D	D	D	D	D	D	–	–	–	–
21		–	–	–	D	D	D	D	D	D	D	D	D	D	D	D	D	D	D	D	–	–	–	–
22		–	–	–	D	D	D	D	D	D	D	D	D	D	D	D	D	D	D	D	–	–	–	–
24		–	–	–	D	D	D	D	D	D	D	D	D	D	D	D	D	D	D	D	–	–	–	–
29		–	–	–	D	D	D	D	D	D	D	D	D	D	D	D	D	D	D	D	D	D	–	–
30		–	–	–	D	D	D	D	D	D	D	D	D	D	D	D	D	D	D	D	D	D	–	–
11		–	–	–	D	D	D	D	D	D	D	D	D	D	D	D	D	D	–	–	–	–	–	–
5		–	–	–	–	D	D	D	D	D	D	D	D	D	D	D	D	D	D	D	D	D	D	D

Transc.: transconjugant strain. D, donor DNA;–, recipient DNA; d/–, PCR fragment has donor DNA at left and recipient DNA at right. *,D denote selected marker (trpA+);| denotes counterselected marker, rpoB (rif).

Table 17.5 Summary of crosses

Cross type	Average range (kb)	Average maximum fragment length (kb)	Average number of interruptions per recombinant	Proportion of interrupted recombinants	Comments
Transduction					
ECOR 47 × K-12 n = 18	9.5	8	0.67	0.44	Evidently recipient strain contains restriction enzymes against which donor DNA is not protected
Backcross to K-12					
47K4 × K-12 n = 15	23.5	23.5	0	0	47-type DNA in donor was ~25 kb long
47K9 × K-12 n = 15	20	20	0	0	47-type DNA in donor was ~20 kb long
ECOR 47 × K-12ER2476 'Restrictionless' n = 20	35	31	0.25	0.25	Recipient is not fully restrictionless?
Backcross to ER2476					
47D10 × K-12ER2476 n = 30	32	32	0.03	0.03	47-type DNA in donor was ~61 kb long

Table 17.5 continued

Cross type	Average range (kb)	Average maximum fragment length (kb)	Average number of interruptions per recombinant	Proportion of interrupted recombinants	Comments
Conjugation					
ECOR 47 → K-12 $n = 29$	47 (28)[a]	21 (16.5)[a]	0.76	0.41	Consistent with transduction; see text
Backcross (47 → K-12) → K-12 $n = 30$	57	57	0	0	Consistent with backtransduction; 47-type DNA in donor was ~61 kb long
Reciprocal cross K-12 → 47 $n = 30$	115 (57)	115 (57)	0	0	Less abridgement than in ECOR 47→K12[b]

[a] Numbers in parentheses refer only to the 59 kb region between 1276 kb and 1335 kb, for comparison to the conjugation backcross, which involved a mismatch only in that region.
[b] K-12 → 47 crosses analyzed over a longer range do show interruptions of donor DNA, but far less frequently than 47 → K-12.

while the ECOR 47 recipient now had K-12 DNA in these same regions. The exact limits of the switch are not known, however. This switch enabled donor selection via *trp+* vs. the recipient's *trpA33*, just as in all the other crosses compared.]

A comparison of the summaries of the results of these respective crosses is given in Table 17.5. The working interpretation of these results is that the recipient strain's effects on the donor DNA will depend on the recipient strain's restriction-modification systems (Raleigh, 1992; Kelleher and Raleigh, 1995) that are lacking in the donor. This predicts that reciprocal crosses should frequently produce unequal results, since there appear to be several different polymorphic restriction-modification systems in *E. coli*. It was no surprise, then, that the reciprocal conjugations produced different levels of abridgement.

The availability of the closely grouped PCR fragments in the *trp* region had been essential to the analysis of the transductants of the 47 →K-12 crosses; elsewhere, the markers, which were spaced at intervals on the order of a minute, were not useful for comparisons between transductants and transconjugants in these specific crosses.

In addition, a small number of crosses were made to compare the effects of ECOR 56 and ECOR 72 as recipients to those of ECOR 47 (all with K-12 as donor), and the recipients appeared to differ. ECOR 56 cut the most, and ECOR 72 the least, but these differences were modest relative to the great contrast between K-12 → 47 and 47 → K-12.

These experiments have established some basic patterns of DNA incorporation in crosses between isolates of recent natural origin, as well as the role of restriction-modification systems in recombination.

ACKNOWLEDGMENTS

We thank Erich Jaeger and Ryan McBride for technical contributions and Dr E.A. Adelberg for noting Boyer's work. This work has been supported by NSF Grants DEB90-20173 and MCB95-20613.

REFERENCES

Arber, W. and Morse, M.L. (1965) Host specificity of DNA produced by *Escherichia coli*. VI. Effects on bacterial conjugation. *Genetics* 51: 137–148.

Atwood, K.C., Schneider, L.K. and Ryan, F.J. (1951) Selective mechanisms in bacteria. *Cold Spring Harbor Symp. Quant. Biol.* 16: 345–355.

Barcus, V.A., Titheradge, A.J.B. and Murray, N.E. (1995) The diversity of alleles at the *hsd* locus in natural populations of *Escherichia coli*. *Genetics* 140: 1187–1197.

Blattner, F.R., Plunkett III, G., Bloch, C. *et al.* (1997) The complete genome sequence of *Escherichia coli*. *Science* 277: 1453–1474.

Boyer, H. (1964) Genetic control of restriction and modification in *Escherichia coli*. *J. Bacteriol.* **88**: 1652–1660.

Boyer, H. (1966) Conjugation in *Escherichia coli*. *J. Bacteriol.* **91**: 1767–1772.

DuBose, R.F., Dykhuizen, D.E. and Hartl, D.L. (1988) Genetic exchange among natural isolates of bacteria: recombination within the *phoA* gene of *Escherichia coli*. *Proc. Natl Acad. Sci. USA* **85**: 7036–7040.

Dykhuizen, D. and Green, L. (1991) Recombination in *Escherichia coli* and the definition of biological species. *J. Bacteriol.* **173**: 7257–7268.

Guttman, D.S. and Dykhuizen, D.E. (1994a) Detecting selective sweeps in naturally occurring *Escherichia coli*. *Genetics* **138**: 993–1003.

Guttman, D.S. and Dykhuizen, D.E. (1994b) Clonal divergence in *Escherichia coli* as a result of recombination, not mutation. *Science* **266**: 1380–1383.

Herzer, P.J., Inouye, S., Inouye, M. and Whittam, T.S. (1990) Phylogenetic distribution of branched RNA-linked multicopy single-stranded DNA among natural isolates of *Escherichia coli*. *J. Bacteriol.* **172**: 6175–6181.

Kelleher, J.E. and Raleigh, E.A. (1995) On the regulation and diversity of restriction in *Escherichia coli*. *Gene* **157**: 229–230.

King, G. and Murray, N.E. (1994) Restriction enzymes in cells, not eppendorfs. *Trends in Microbiol.* **2**: 465–469.

McKane, M. and Milkman, R. (1995) Transduction, restriction and recombination patterns in *Escherichia coli*. *Genetics* **139**: 35–43.

Milkman, R. (1996a) Recombinational exchange among clonal populations, in *Escherichia coli and Salmonella: Cellular and Molecular Biology,* (eds F.C. Neidhardt, R. Curtiss III, J.L. Ingraham *et al.*), American Society for Microbiology, Washington, DC.

Milkman, R. (1996b) Recombination and DNA sequence variation in *E. coli*, in *Ecology of Pathogenic Bacteria: Molecular and Evolutionary Aspects*, (eds B.A.M. van der Zeijst, W.P.M. Hoekstra, J.D.A. van Embden and A.J.W. van Alphen), North-Holland, Amsterdam.

Milkman, R. and McKane, M. (1995) DNA sequence variation and recombination in *E. coli*, in *Population Genetics of Bacteria*, (eds S. Baumberg, J.P.W. Young, E.M.H. Wellington and J.R. Saunders), Cambridge University Press, Cambridge.

Milkman, R. and McKane Bridges, M. (1990) Molecular evolution of the *Escherichia coli* chromosome. III. Clonal frames. *Genetics* **126**: 505–517.

Milkman, R. and McKane Bridges, M. (1993) Molecular evolution of the *Escherichia coli* chromosome. IV. Sequence comparisons. *Genetics* **133**: 455–468.

Modrich, P. (1991) Mechanism and biological effects of mismatch repair. *Annu. Rev. Genet.* **25**: 225–253.

Ochman, H. and Selander, R.K. (1984) Standard reference strains of *E. coli* from natural populations. *J. Bacteriol.* **174**: 6886–6995.

Raleigh, E.A. (1992) Organization and function of the *mcrBC* genes of *Escherichia coli* K-12. *Molec. Microbiol.* **6**: 1079–1086.

Sharp, P.M., Kelleher, J.E., Daniel, A.S. *et al.* (1992) Roles of selection and recombination in the evolution of type I restriction-modification systems in enterobacteria. *Proc. Natl Acad. Sci. USA* **89**: 9836–9840.

Smith, G. (1991) Conjugational recombination in *E. coli*: myths and mechanisms. *Cell* **64**: 19–27.

The lysogenic conversion genes of coliphage P2 have unusually high AT content

18

*Richard Calendar, Sidney Yu, Heejoon Myung,
Virginia Barreiro, Richard Odegrip, Karin Carlson,
Laura Davenport, Gisela Mosig, Gail E. Christie
and Elisabeth Haggård-Ljungquist*

SUMMARY

The essential genes of temperate coliphage P2 are about 45% AT and
have the codon usage typical of the host. Three lysogenic conversion
genes have 67% AT and codon usage that is not typical for *E. coli*. These
genes are non-essential but confer selective advantage, because they
make lysogenic cells refractory to infection by other phages. In the
middle of the P2 genome is the *fun* gene, which makes lysogenic cells
sensitive to FUdR and unable to propagate phage T5. At the right end of
the P2 genome are two cotranscribed genes: the promoter proximal gene,
old, blocks the multiplication of lambdoid phages and encodes a nuclease, while the promoter distal gene, *tin*, prevents growth of T-even
phages. Upon infection with T4 wild-type, there is little or no *de novo*
phage DNA synthesis in bacteria containing P2 *tin* clones. T4 and T2
mutants that evade P2 *tin* restriction have normal DNA replication. They
are changed in gene *32*, which encodes an essential single-stranded DNA
binding protein. Asp codon 163 is changed to Asn or Gly. Asp163 lies on
the surface of the gp32 monomers, far away from the DNA-binding cleft,

suggesting that it is involved in protein–protein interactions. We suggest that the *old*, *tin* and *fun* genes have been acquired by horizontal transfer from a foreign genome.

18.1   INTRODUCTION

Temperate phage P2 encodes genes for DNA replication, capsid and tail synthesis, cell lysis and lysogenization. The base composition of these essential genes is about 45% A + T, like that of the *E. coli* host. In addition, P2 encodes some genes that are not essential either for lysogenization or for production of progeny (Bertani and Six, 1988). Among these, the *old* gene, which interferes with the growth of lambdoid phages, has an AT content of 67% (Haggård-Ljungquist *et al.*, 1989) (Fig. 18.1). The *old* gene is not universal in P2-like phages. P2's relatives, Wφ and HK239, interfere with lambda growth (Bertani and Bertani, 1971; Dhillon and Dhillon, 1973). Phage 186, which is part of the P2 family, does not block the growth of lambda when it is in the prophage state (Brumby *et al.*, 1996). No functional gene with this phenotype is found in any well characterized *E. coli* strain. Thus it seems to come from some other species.

Another non-essential P2 gene, *orf-30*, which lies between essential genes *S* and *V* (Fig. 18.1), has an AT content of 67%. This gene is transcribed in the opposite direction from its flanking genes (Linderoth *et al.*, 1994). Its function, if any, is not known.

With the complete sequence of P2 now determined, we have found two additional genes that may have come from a foreign genome. In scanning the sequence of the P2 genome (Fig. 18.1), the high AT content of the *old*, *tin*, *fun* and *orf-30* genes stands out. Partial denaturation mapping by Inman and Bertani (1969) shows AT-rich regions in the same locations. In all four of these genes, the AGPu Arg codons, which are rare in *E. coli*, constitute the majority of Arg codons (*old* = 8/14, *tin* = 8/11, *fun* = 15/23, *orf-30* = 5/7).

18.2   RESULTS AND DISCUSSION

18.2.1   THE *TIN* (*T-EVEN INTERFERENCE*) GENE

Between genes *A* and *old* is a region of previously unreported sequence. We cloned this region by cutting P2 DNA with *Sca*I (position 31590) and *Tth*111I (position 30133), and inserting the fragment into pUC19 cut with *Sma*I and *Pst*I to yield pYMD2. Staggered ends generated by some of these restriction enzymes were made blunt by filling or editing with Klenow fragment. Sequence analysis showed that this region of P2 contains an open reading frame that we call *tin* for reasons discussed

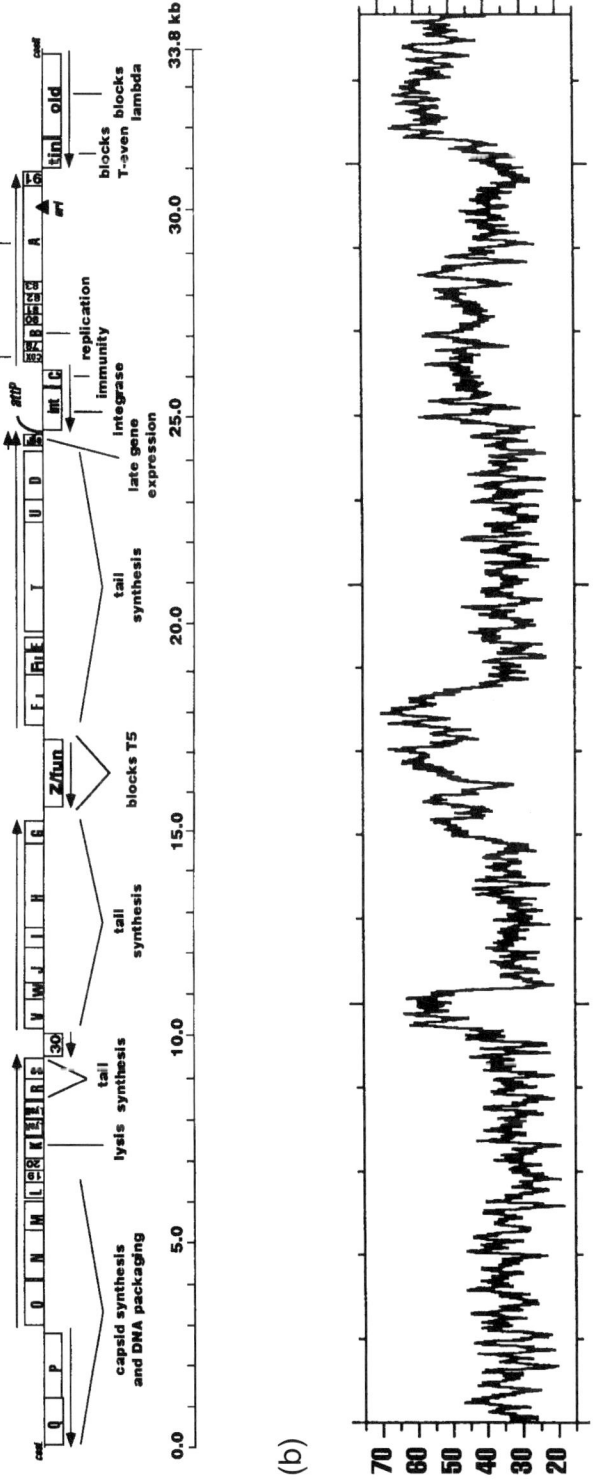

Fig. 18.1 Genetic and physical maps of bacteriophage P2. (a) Genetic map. The locations of genes and open reading frames (named by their locations in percentage from the left end) are shown by boxes above or below the line, indicating the coding strand used. The functions of the genes are shown below the boxes. Transcriptional units are shown by the arrows above or below the line. For further details see Linderoth *et al.* (1993). (b) AT content of the complete P2 genome. The plot was obtained using the Window and StatPlot programs of the Wisconsin Sequence Analysis Package from Genetics Computer Group, Madison, WI

below. The corresponding Tin protein is predicted to have 253 amino acid residues (Fig. 18.2). In P2 phage, *tin* is co-transcribed with the *old* gene from a promoter upstream of *old* (position 33425), as determined by S1 nuclease mapping. Thus the *tin* and *old* genes may have been inherited together by horizontal transfer. In the plasmid pYMD2 construct the entire *old* gene and its promoter are absent. We surmise that in this plasmid, *tin* is transcribed from a plasmid promoter, perhaps the *ori* promoter. No major 5' RNA end close to the *tin* gene was found by primer extension in pYMD2-bearing cells. We suspected that the *tin* gene might account for the interference with T-even phages that was reported for P2 prophage in *Shigella dysenteriae* (Bertani, 1953; Lederberg, 1957). Smith *et al.* (1969) reported that P2 prophage in *Shigella* prevented T2 DNA synthesis and limited RNA and protein synthesis to the first few minutes after infection. For reasons that are not known, P2 prophage inhibits growth of T-even phages less severely in *E. coli*; i.e. it reduces T4 plaque size and reduces T2 efficiency of plating by 100-fold. In contrast, the multicopy plasmid pYMD2, carrying only the P2 *tin* and *orf-91* genes (Liu *et al.*, 1993), prevents plaque formation by T-even phages (T2, T4 and T6) in *E. coli*. This effect is not due to the *orf-91* gene, which can be deleted by digestion with *Eco*RV (30683) and *Stu*I (30196), with no effect on T4 interference. There is no *de novo* synthesis of T4 DNA in cells that carry pYMD2, and little or no late protein synthesis (Mosig *et al.*, 1997). Taken together with the results of Smith *et al.* (1969), these results suggest that the P2 gene *tin* is responsible for P2's interference with T-even growth and that different degrees of interference in different situations might be due to different levels of *tin* expression. These inferences are supported by the following results. T2 and T4 suppressor mutants that can grow in *E. coli* B bearing pYMD2 are found with a frequency of less than 10^{-7}. When such mutants were isolated and crossed with amber mutants *A453* and *E315* in gene *32*, fewer than 1% wild-type recombinants resulted. Crosses between phages with distant mutations result in about 25% recombination; thus, the suppressor mutations are linked to gene *32*. Sequence analysis shows that in these mutants the codon for Asp163 of gp32, the T4 single-stranded DNA binding protein, is changed to an Asn or Gly codon (Mosig *et al.*, 1997). Thus, the target of interference in both *Shigella* P2 lysogens and in *E. coli* bearing a multicopy *tin*-containing plasmid is a single-stranded DNA binding protein that is essential for T4 DNA replication and recombination. Therefore the new P2 gene is called *tin*, for T-even interference, and the T4 mutants that overcome the effect of the *tin* gene are called Asp, for *a*borts *s*ensitivity to *P*2.

Asp163 of gp32 lies on the outside of this protein, away from the DNA-binding cleft (Shamoo *et al.*, 1995a, b), suggesting that it is involved in protein-protein contacts.

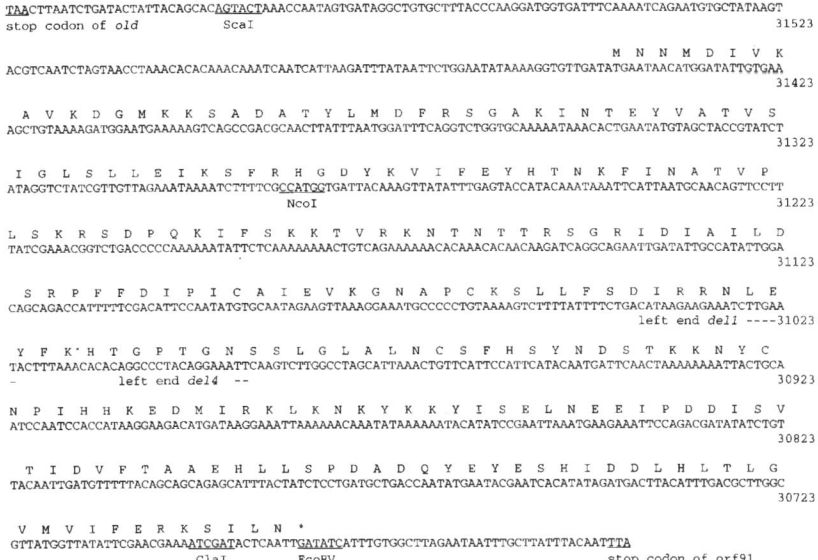

Fig. 18.2 DNA sequence of the P2 *tin* gene region and the deduced amino acid sequence of Tin. Note that the sequence shown goes from right to left on the genetic map, and the numbers indicate the distance from the left end. The stop codons for the flanking genes, *old* and *orf91*, are indicated. The possible ribosome binding site (RBS) for *tin* is underlined. The left ends of the *del1* and *del4* deletions are indicated by dotted lines; the right ends are located at nucleotide 33454 (between two 6 base pairs repeated sequences) and 33334 (between two 2 base pairs repeated sequences), respectively. Relevant restriction enzyme sites are shown below the DNA sequence. Universal and reverse primers were used to sequence from the *Cla*I, *Nco*I and *Sca*I sites shown, after subcloning of fragments. In addition, two primers were used:

DraCla=5'-CATACAATGATTCAACTAAA (30955-30936)
DraNco=5'-CTTGAATTTCCTGTAGGGCC (30988-31007).

The GenBank accession number for this sequence is X99628.

18.2.2   THE *FUN*(Z) GENE

P2-lysogenic *E. coli* are sensitive to 5-fluorodeoxyuridine (FUdR) and 5-fluorouracil (FU; Bertani, 1964). The *fun* gene, which causes this phenotype, maps between 15 kb and 18 kb on the P2 physical and genetic map (Linderoth *et al.*, 1993) (Fig. 18.1). The *fun-1* mutation abolishes sensitivity to FUdR and FU (Bertani and Levy, 1964). We now find that the *fun* gene causes the inability of T5 phage to form plaques on *Shigella dysenteriae* lysogenic for P2 (Bertani, 1953; Lederberg, 1957). *S. dysenteriae* lysogenic for P2 *fun-1* or P2 *del-2*, which deletes the *fun* gene, allows normal

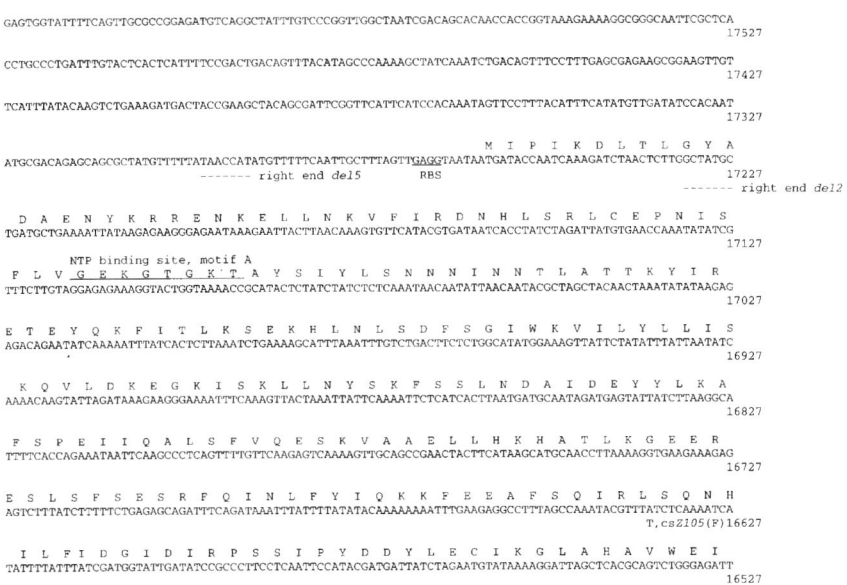

Fig. 18.3 DNA sequence of the P2 *fun*(*Z*) region and the deduced amino acid sequence of Fun(Z). Note that the sequence shown goes from right to left on the genetic map, and the numbers indicate the distance from the left end. The possible ribosome binding site (RBS) and the inverted repeat that might constitute a transcriptional teminator (*ter*) are underlined. The examined mutational nucleotide changes with their corresponding amino acid changes, are indicated below the sequence. The left and right ends of the *del2* and *del5* deletions are indicated by dotted lines. The location of the possible NTP binding site is indicated above the amino acid sequence.

The *fun* gene was amplified by PCR using primers 46.3R and 51.5L (see below) and cloned into the *Sma*I site of pUC18. Several clones were analyzed to check for PCR-induced mutations. To get the whole sequence we also made subclones. The DNA sequence was determined by the chain termination method of Sanger *et al.* (1977) using α-^{35}S-dATP (Amersham) and T7 sequenace (Pharmacia). The mutants were automatically sequenced using ABI PRISMA TM Dye Terminator Cycle Sequencing Ready Reaction Kit (Perkin Elmer) in a MiniCyclerTM apparatus (MJ Research), and the sequence reactions were analyzed on an ABI PRISMATM 377 DNA Sequencer (Perkin Elmer). The M13 'universal' and 'reverse' primers used were:

46.3R=5'-CACAATTCGGTCTGCGTCCC (15583-15602)
47.0R=5'-CAGAGAGGAATTCAAAGAAC (15980-15999)
48.3R=5'-CGTTGGGTGAATCCCAAGGG (16253-16272)
50.0L=5'-CTCATCACTTAATGATGCAA (16869-16850)
50.4L=5'-AACAATACGCTAGCTACAAC (17060-17041)
50.6R=5'-GTTGTAGCTAGCGTATTGTT (17041-17060)
51.5L=5'-GCGACAGAGCAGCGCTATGT (17324-17305)

The GenBank accession number for this sequence is X99627.

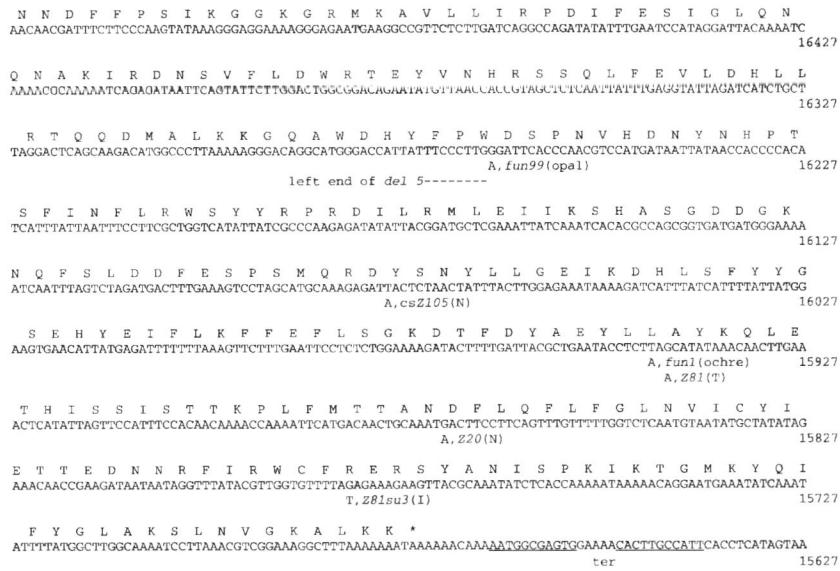

Fig. 18.3 *continued*

plaque formation by phage T5. Moreover, *E. coli* strain BL21(DE3), expressing the wild-type *fun* gene from plasmid pEE678, has a T5 plaquing efficiency of less than 10^{-5}. [The *fun*(Z) gene (nt 15583 to 17323) was amplified by the polymerase chain reaction and integrated, in the orientation that gives expression from the T7 promoter, into the filled-in *Bam*HI site of pET3 (Studier *et al.*, 1990), generating plasmid pEE678.] Thus, the *fun* gene has two phenotypes: one that allows for selection of the wild-type allele (T5 resistance) and another that allows for selection of the null mutant type (these mutants are resistant to fluorouracil). P2 prophage only interferes weakly with T5 in *E. coli* C, and not at all in *E. coli* B. Thus a higher level of expression of *fun* may be necessary to observe T5 resistance in certain strains.

Another apparent P2 gene maps in the same region as the *fun* gene. P2 Z gene mutations, originally defined by P2 Z20, cause cell killing and a concomitant clear plaque phenotype (Bertani, 1976). Sequence analysis has now allowed us to determine that *fun* and Z are the same gene. Figure 18.3 shows that there are 391 untranslated nucleotides between the start of gene *FI* and the translation start site for an open reading frame (ORF) encoding 528 amino acids, which is read in the opposite direction from the flanking P2 essential genes. There are 400 nucleotides between the end of this orf and the G gene. The *del5* deletion extends from 16267 to 17299, and there is no direct repeat to account for its formation. Since the *del5* deletion removes the start of this orf, and since *del5* is *fun*-defective (Nicoletti and Bertani, 1983), this ORF is likely to be the *fun* gene. The *fun-1* mutation creates a UAA stop codon in this orf, and *fun-99* creates a UGA stop codon. The *fun*-defective deletion *del2* affects the same orf; it extends from 17221 to 15294 and occurs between two 7 bp repeated sequences. The *Z20* and *Z81* (Bertani, 1976) carry mutations in this same ORF. A suppressor *Z81* (*suZ-3*; Bertani, 1976) also causes a missense mutation in the *fun* gene. The mutant *Z105*, which is cold sensitive (Bertani, 1978), has two missense mutations in the same orf that carries *fun* mutations (Fig. 18.3). Thus, *fun* and *Z* mutations lie in the same gene, which we now call *fun(Z)*. IPTG induction of T7 polymerase in strain BL21(DE3) carrying the *fun*-expressing plasmid pEE678 leads to overexpression of a 58 kDa protein, supporting the conclusion that this region only codes for one protein. The *fun-1*, *fun-99* and *del2* mutations cause chain termination and must be null or near-null mutations. The killing phenotype of the *Z* mutations suggests that they might cause an increase in activity of the Fun (Z) protein. The activity of the *fun* gene in *Z* mutants has been detected by FU sensitivity of P2-lysogenic cells superinfected by P2 Z mutants (L.E. Bertani, unpublished data). P2 *Z81 suZ-3* as prophage is not sensitive to FU; thus the suppressor causes loss of activity of the *fun(Z)* protein (L.E. Bertani, unpublished data). There is a possible ribosomal binding site, GAGG, preceding the start codon, and the gene is followed by an inverted repeat that looks like a good transcriptional terminator for transcription coming from the opposite direction, but it may act in both orientations. Near the amino terminus of the Fun (Z) protein is a Walker type A NTP binding motif (Walker *et al.*, 1982), suggesting that this protein might be part of a membrane transport system of the ABC type (Hyde *et al.*, 1990). The AT content of the *fun(Z)* region is 67%, starting about 300 bp upstream of the *fun(Z)* gene and extending to the *G* gene. Thus the *fun(Z)* gene may have been inherited from a foreign genome.

The phage interference properties of the *fun(Z)*, *old* and *tin* genes give P2 prophage and P2-lysogenic strains a selective advantage.

18.2.3   DISTRIBUTION OF NON-ESSENTIAL GENES IN THE P2 FAMILY

Sequence analysis shows that phage 186 has no *old*, *tin*, *fun* or *orf-30* gene (Brumby *et al.*, 1996; J. Barry Egan, personal communication). No other member of the P2 family has yet been shown to possess any of these genes by sequence analysis. P2 Hy *dis*, which is derived by recombination between P2 and a cryptic prophage in *E. coli* B (Cohen, 1959), is not homologous to P2 in the *old/tin* gene region (Chattoraj and Inman, 1973). This phage, along with Wø (Kerzman *et al.*, 1967), HK239 (Dhillon and Dhillon, 1973) and PK (Jesaitis and Hutton, 1963), have been reported to interfere with lambda phage (Table 18.1), suggesting the presence of *old* gene equivalents; most P2-related phages do not have such genes (T.S. Dhillon, E.W. Six and G. Bertani, personal communication). PK's interference with lambda is like that of P2: lambda *bio-1* can grow on both P2- and PK-lysogenic strains. Wø is different from P2 in this regard: although *bio-1* can grow on Wø-lysogenic strains, lambda mutants that overcome Wø's interference cannot grow on P2-lysogenic strains (Kerzman *et al.*, 1967). This asymmetric difference in lambda interference is not understood. P2 Hy1 *dis* prophage interferes with lambda growth, though not quite as strongly as does P2 prophage. Furthermore, P2 Hy1 *dis* also interferes with lambda *bio-1*; hence a non-*old*-based interference appears to be indicated (with an additional *old*-like interference not ruled out; E.W. Six and G. Bertani, personal communication). Wø also interferes with the growth of T-even phages (Pizer *et al.*, 1968). PK prophage does not interfere with plaque formation by T5 (E.W. Six, personal communication), and PK interferes only slightly with T4 phage plaque formation (G. Bertani, personal communication). P3 (Bertani, 1951) and 299 (Geisselsoder and Mandel, 1970) prophages do not interfere with plaque formation by lambda, T4 or T5 (E.W. Six and G. Bertani, personal communication).

18.2.4   THE ORIGINS OF HORIZONTALLY TRANSFERRED GENES

The source of P2's AT-rich genes is uncertain. It appears that they do not come from *E. coli*, since standard laboratory *E. coli* strains, such as K-12

Table 18.1 Growth of phages on lysogenic strains

Prophage	Infecting phage		
	Lambda	T-even	T5
P2	–(C, K)	–(T4, *Sh*)	–(*Sh*)
186	+(K)	+(T2, K)	+(K)
Wø	–(C, W)	–(T2, C, W)	nt
PK	–(C)	+(T4, *Sh*)	+(*Sh*)
P3	+(C)	+(T4, Sh)	+(*Sh*)
299	+(C)	nt	nt
P2 Hy *dis*	+/–(C)	+(T4, *Sh*)	–(*Sh*)
HK239	–(K)	nt	nt

+, denotes ability to make plaques, with efficiency of plaque formation (e.o.p.) near 1 plaque per virus particle.
–, denotes inability to make plaques, with e.o.p. less than 0.003, compared with a non-lysogenic strain.
C, K and W are strains of *Escherichia coli*.
Sh denotes *Shigella disenteriae*.
nt, not tested.

and C, are sensitive to lambdoid phages, T-even phages and T5. However, in a large collection of clinical coliform isolates, it has been found that the great majority did not give plaques with the phages in question (G. Bertani, personal communication). It is not known whether any of these phage resistances may be due to *old*, *tin*, *fun* or *orf30*. The lysogenic conversion genes of P2 might have come from another enteric bacterium. Since *Bacteroides* is a major component of the bacterial flora in feces, we have probed *Bacteroides* strains with labeled *fun(Z)*, *tin* and *old* DNA, but we found no hybridization signal. It is noteworthy that all of P2's lysogenic conversion genes are high in AT and none are high in GC. The AT-richness may be due either to selective pressure for AT-rich codons or to transfer from AT-rich organisms where the high AT content was selected in response to certain tRNA repertoires, restriction-modification systems or perhaps recombination requirements (Syvanen, 1994). At present, we do not know the origins of these genes. They could come from a plasmid, phage, transposon or integron, or from the chloroplasts of algae. A search of the non-redundant database at National Center for Biotechnology Information (NCBI) using the BLASTN program did not yield significant homologies. FastA and BLAST searches for the amino acid sequence of the Fun, Tin and Old proteins found no significant homologies. However, Koonin and Gorbalenya (1992) have noted limited homology between *old* and the recombination nuclease gene, *sbcC*. Although the *sbcC* gene from *E. coli* does not have an unusually high AT content (Naom *et al.*, 1989), the *Clostridium sbcC* gene is 65% AT, so the P2 *old* gene could be a closer relative of this gene.

ACKNOWLEDGMENTS

This work was supported by American Cancer Society research grant NP-869 to G.C., by reasearch grant AI-08722 from the National Institute of Allergy and Infectious Diseases to R.C., by NIH GM13221 to G.M. and by grants 72 and B96-13X-11577-01A from the Swedish Medical Research Council to E.H.L. and K.C., respectively.

REFERENCES

Bertani, G. (1951) Studies on lysogenesis. I. The mode of phage liberation by lysogenic *Eschericha coli*. *J. Bacteriol.* **62**: 293–300.

Bertani, G. (1953) Lysogenic versus lytic cycle of phage multiplication. *Cold Spring Harbor Symp. Quant. Biol.* **18**: 65–70.

Bertani, L.E. (1964) Lysogenic conversion by bacteriophage P2 resulting in an increased sensitivity of *Escherichia coli* to 5-fluorodeoxyuridine. *Biochim. Biophys. Acta* **87**: 631–640.

Bertani, L.E. (1976) Characterization of clear mutants belonging to the Z gene of bacteriophage P2. *Virology* **71**: 85–96.

Bertani, L.E. (1978) Cold-sensitive mutations in the Z gene of prophage P2 that result in increased sensitivity of the lysogens to a low molecular weight product of the host bacteria. *Molec. Gen. Genet.* **166**: 85–90.

Bertani, L.E. and Bertani, G. (1971) Genetics of P2 and related phages. *Advances in Genetics* **16**: 199–237.

Bertani, L.E. and Levy, J.A. (1964) Conversion of lysogenic *Escherichia coli* by non-multiplying, superinfecting bacteriophage P2. *Virology* **22**: 634–640.

Bertani, L.E. and Six, E.W. (1988) The P2-like phages and their parasite, P4, in *The Bacteriophages*, Vol. 2, (ed. R. Calendar), Plenum Press, New York, pp. 73-143.

Brumby, A.M., Lamont, I., Dodd, I.B. and Egan, J.B. (1996) Defining the SOS operon of coliphage 186. *Virology* **219**: 105–114.

Chattoraj, D.K. and Inman, R.B. (1973) Electron microscope heteroduplex mapping of P2 Hy *dis* bacteriophage DNA. *Virology* **55**: 174–182.

Cohen, D. (1959) A variant of phage P2 originating in *Escherichia coli*, strain B. *Virology* **7**: 112–126.

Dhillon, E.K.S. and Dhillon, T.S. (1973) HK239: a P2-related temperate phage which excludes *rII* mutants of T4. *Virology* **55**: 136–142.

Geisselsoder, J. and Mandel, M. (1970) Physical properties of phage 299. *Molec. Gen. Genet.* **108**: 158–166.

Haggård-Ljungquist, E., Barreiro, V., Calendar, R. *et al.* (1989) The P2 phage *old* gene: sequence, transcription and translational control. *Gene* **85**: 25–33.

Hyde, S.C., Emsley, P., Hartshorn, M.J. *et al.* (1990) Structural model of ATP-binding proteins associated with cystic fibrosis, multidrug resistance and bacterial transport. *Nature* **346**: 362–365.

Inman, R.B. and Bertani, G. (1969) Heat denaturation of P2 bacteriophage DNA: compositional heterogeneity. *J. Mol. Biol.* **44**: 533–549.

Jesaitus, M.A. and Hutton, J.J. (1963) Properties of a bacteriophage derived from *Escherichia coli* K-235. *J. Exp. Med.* **117**: 285–302.

Kerzman, G., Glover, S.W. and Aronovitch, J. (1967) The restriction of bacterio-phage lambda in *E. coli* strain W. *J. Gen. Virol.* **1**: 333–347.

Koonin, E. and Gorbalenya, A. (1992) The superfamily of UvrA-related ATPases includes three more subunits of putative ATP-dependent nucleases. *Protein Seq. Data Anal.* **5**: 43–45.

Lederberg, S. (1957) Suppression of multiplication of heterologous bacterio-phages in lysogenic bacteria. *Virology* **3**: 496–513.

Linderoth, N.A., Christie, G.E. and Haggård-Ljungquist, E. (1993) Bacteriophage P2: physical, genetic and restriction map, in *Genetic Maps – Locus Maps of Complex Genomes*, 6th edn, (ed. S.J. O'Brien), Cold Spring Harbor Laboaotry Press, Cold Spring Harbor, NY, pp. 1.62–1.69.

Linderoth, N.A., Julien, B., Flick, K.E. *et al.* (1994) Molecular cloning and charac-terization of bacteriophage P2 genes *R* and *S* involved in tail completion. *Virology* **200**: 347–359.

Liu, Y., Saha, S. and Haggård-Ljungquist, E. (1993) Studies of bacteriophage P2 DNA replication. The DNA sequence of the *cis*-acting gene *A* and *ori* region and construction of a P2 mini-chromosome. *J. Mol. Biol.* **231**: 361–374.

Mosig, G., Yu, S., Myung, H. *et al.* (1997) A novel mechanism of virus–virus inter-actions: bacteriophage P2 Tin protein inhibits phage T4 DNA synthesis by poisoning the T4 single-stranded DNA binding protein, gp32. *Virology* **230**: 72–81.

Naom, I.S., Morton, S.J., Leach, D.R.F. and Lloyd, R.G. (1989) Molecular organisation of *sbcC*, a gene that affects genetic recombination and the viability of DNA palindromes in *Escherichia coli* K-12. *Nucl. Acids. Res.* **17**: 8033–8045.

Nicoletti, M. and Bertani, G. (1983) DNA fusion product of phage P2 with plasmid pBR322: a new phasmid. *Molec. Gen. Genet.* **189**: 343–347.

Pizer, L., Smith, H.S., Miovic, M. and Pylkas, L. (1968) Effect of prophage W on the propagation of bacteriophages T2 and T4. *J. Virol.* **2**: 1339–1345.

Sanger, F., Nicklen, S. and Coulson, A.R. (1977) DNA sequencing with chain terminating inhibitors. *Proc. Natl Acad. Sci. USA* **74**: 5463–5857.

Shamoo, Y., Friedman, A.M., Parsons, M.R. *et al.* (1995a) Crystal structure of a replication fork single-stranded DNA binding protein (T4 gp32) complexed to DNA. *Nature* **376**: 362–366.

Shamoo, Y., Friedman, A.M., Parsons, M.R. *et al.* (1995b) Crystal structure of a replication fork single-stranded DNA binding protein (T4 gp32) complexed to DNA. *Nature* **376**: 616 (erratum).

Smith, H.S., Pizer, L.I., Pylkas, L. and Lederberg, S. (1969) Abortive infection of *Shigella dysenteriae* P2 by T2 bacteriophage. *J. Virol.* **4**: 162–168.

Studier, F.W., Rosenberg, A.H., Dunn, J.J. and Dubendorff, J.W. (1990) Use of T7 RNA polymerase to direct expression of cloned genes. *Meth. Enzym.* **185**: 60–89.

Syvanen, M. (1994) Horizontal gene transfer: evidence and possible consequences. *Annu. Rev. Genet.* **28**: 237.

Walker, J.E., Saraste, M., Runswick, M.J. and Gay, N.J. (1982) Distantly related sequences in the alpha and beta subunits of ATP synthase, myosin, kinases and other ATP-requiring enzymes and a common nucleotide binding fold. *EMBO J.* **1**: 945–951.

PART THREE

Eukaryotic Mobile Elements

Evidence for horizontal transfer of the *P* transposable element between *Drosophila* species

19

Jonathan B. Clark and Margaret G. Kidwell

SUMMARY

There is overwhelming support for the involvement of horizontal transfer in the origin of the *P* transposable elements in natural populations of *Drosophila melanogaster*. Presented here is a summary of the results of an extensive phylogenetic analysis of *P* element nucleotide sequences from the subgenus *Sophophora* of the genus *Drosophila*. The *P* element phylogeny is examined in the context of a phylogeny of the species in which these elements are found. In addition to identifying two unequivocal cases of horizontal transfer, the analysis suggests that a number of additional horizontal transfers may have occurred earlier in the evolution of this subgenus.

19.1 INTRODUCTION

Although difficult to document rigorously, there is some evidence that eukaryotic transposable genetic elements may be relatively more susceptible to horizontal gene transfer than genes that are strictly transmitted by Mendelian inheritance (Kidwell, 1993). It has been argued that, like other types of 'selfish DNA', it may be in the interest of transposable elements (TEs) to transfer to new hosts in order to survive over long periods of evolutionary time (Hurst *et al.*, 1992; Kidwell, 1992). TEs also have the obvious advantage of possessing the molecular machinery that facilitates such transfer.

The *P* family of transposable genetic elements is among the best-studied eukaryotic transposable element systems. First discovered in *Drosophila melanogaster*, they have been subsequently identified in other species of the genus *Drosophila* and related genera. *P* elements are members of the Class II transposable elements (Finnegan, 1989) that replicate by means of a DNA intermediate. Active, autonomous *P* elements are 2.9 kb in length and are flanked by 31 bp inverted repeats. They contain four exons designated as open reading frames (ORFs) 0, 1, 2 and 3, and encode two polypeptides involved in *P* element transposition: an 87 kDa transposase (encoded by all four exons) and a 66 kDa repressor protein (encoded by the first three exons) (Rio *et al.*, 1986; Misra and Rio, 1990). The majority of *P* elements in most species examined are, however, non-autonomous because of internal deletions or other mutations. They are unable to transpose unless the genome in which they reside carries at least one autonomous element. Some deleted non-autonomous elements may also act as repressors of transposition.

Earlier work in this laboratory on the population biology and evolution of *P* elements in *Drosophila melanogaster* suggested that the presence or absence of *P* elements in a strain was closely related to the length of time that had elapsed since collection of the strain from a natural population (Kidwell, 1979, 1983; Anxolabéhère *et al.*, 1988). Flies collected from worldwide natural populations in the late 1970s and early 1980s were invariably found to carry *P* elements. In contrast, *P* elements were absent from strains collected more than 20 years earlier from around the world. This is consistent with a hypothesis of recent invasion of *D. melanogaster* (Kidwell, 1979, 1983).

A recent *P* element invasion of the cosmopolitan species *D. melanogaster* is supported by a number of additional arguments. It was observed that complete *P* elements from geographically dispersed populations of *D. melanogaster* were virtually identical (e.g. O'Hare and Rubin, 1983; Nitasaka *et al.*, 1987), suggesting a recent common evolutionary origin. The strongest evidence for the recent invasion hypothesis is provided by the distribution of *P* elements in the subgenus *Sophophora*. There is a complete absence of *P* elements in species of the *melanogaster* species group that are the most closely related to *D. melanogaster* (Brookfield *et al.*, 1984) (Fig. 19.1). In contrast, *P* elements are commonly found in species of the *willistoni* group and their sequences are much more closely related to that of the *D. melanogaster P* element than would be predicted from the phylogenetic relationship of their host species (Daniels *et al.*, 1990). In addition, it has recently been shown (Powell and Gleason, 1996) that the *D. melanogaster P* element has a codon bias that is typical of that of other genes from *D. willistoni*, but not of genes from *D. melanogaster*.

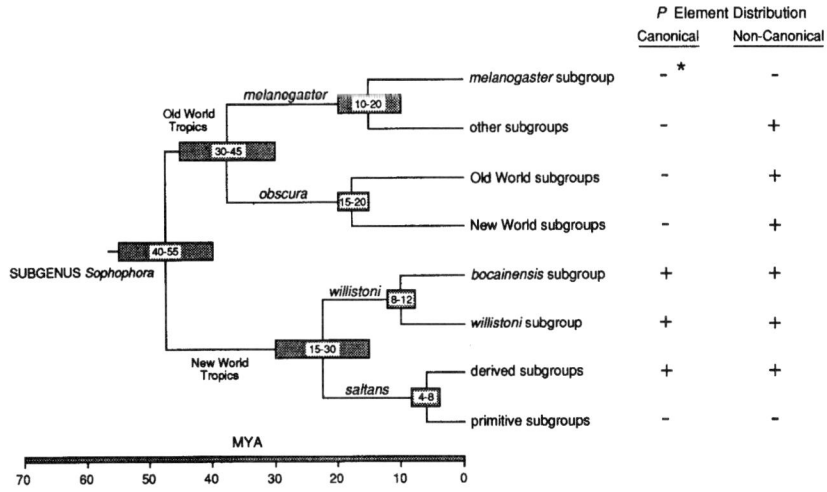

Fig. 19.1 A phylogeny of the subgenus *Sophophora*, showing estimates for divergence times of the species groups and the distribution of both canonical and non-canonical *P* elements found within these groups. The relationships among the four principal species groups is indicated and within each species group a major split among subgroups is shown. Phylogeny and divergence times of the flies were estimated from a number of morphological, cytogenetic, biogeographical and molecular studies and synthesized into this figure (see references in Clark *et al.*, 1994). Asterisk (*) identifies the *P* element transferred horizontally to *D. melanogaster*.

It is not a coincidence that the horizontal transfer that introduced the *P* element into *D. melanogaster* has occurred only relatively recently. This species is a relative newcomer to the Americas, having likely been introduced to the West Indies from Africa since the early nineteenth century (Johnson, 1913; Sturtevant, 1921). Consequently only during the last century or so has there actually been an opportunity for *D. melanogaster* to share the host range of *D. willistoni*, which is found only in Central and South America. Therefore, horizontal transfer of the *P* element from *D. willistoni* to *D. melanogaster* would not have been possible in nature until very recently.

A preliminary phylogenetic analysis of complete *P* element sequences suggested that the evolution of these mobile sequences may be complex, involving horizontal transfer between species as well as vertical transmission (Clark *et al.*, 1994). To examine the prevalence of horizontal transfer, the phylogenetic analysis was extended to other species in the subgenus *Sophophora*. Here, we report the results of a survey which spans the diversity of this subgenus, which originated between 40 and 60 million years ago (Throckmorton, 1975). This includes 183 partial

Table 19.1 Distribution of *P* element sequences among the four species groups of the subgenus *Sophophora*

Species group	Number of subgroups examined (total number)	Number of species examined	Number of sequences
melanogaster	4 (8)	9[a]	35
obscura	1 (4)	7[a]	55
saltans	5 (5)	6[b]	41
willistoni	2 (2)	10	52
Total	12 (19)	32	183

[a] In progress.
[b] Additional species examined do not have detectable *P* element sequences.

sequences isolated from 32 species distributed among the four principal species groups.

19.2 RESULTS AND DISCUSSION

Most of the species examined in this survey were obtained from the National Drosophila Species Resource Center in Bowling Green, Ohio. To date, we have sampled *P* element sequences from 32 species, as listed in Table 19.1. The relative relationships among the four species groups of *Sophophora* examined are shown in Fig. 19.1, along with time estimates for major divergence events. The distribution of *P* elements in these and other species was examined using Southern blots of genomic DNA probed with the canonical *P* element from *D. melanogaster* (Daniels *et al.*, 1990). These results have subsequently been verified by PCR in this and other studies.

For each species, genomic DNA was used as a template in PCR amplifications with two sets of degenerate primers specific for a region of exon 2 of the canonical *D. melanogaster* *P* element (the first *P* element sequenced and the standard frame of reference for subsequent sequences) (Fig. 19.2). Two alternative 5'-primers are 2015, complementary to positions 1230–1251 of the canonical *P* element from *D. melanogaster* (O'Hare and Rubin, 1983), and 2016, complementary to positions 1305–1327. A single 3'-primer, 2017, was used, complementary to positions 1758–1780. Primers 2015 and 2017 amplify a 550 bp fragment, and primers 2016 and 2017 a 450 bp fragment. The details of genomic DNA isolation and purification, PCR amplification, cloning and sequencing are found in Clark *et al.*, 1995. Also included in the analysis were several *P* element sequences obtained from the literature: *D. subobscura*

A1 and G2 (Paricio *et al.*, 1991), *D. guanche* G1 (Miller *et al.*, 1992), *D. nebulosa* N10 (Lansman *et al.*, 1987), *D. bifasciata* M-type (Hagemann *et al.*, 1992), *D. melanogaster* (O'Hare and Rubin 1983) and *Scaptomyza pallida* 2 and 18 (Simonelig and Anxolabéhère, 1991). These are used as reference sequences and are boxed in Figures 19.3, 19.4 and 19.5.

This particular region of the complete *P* element was chosen for several reasons:

- The results of phylogenetic analyses using complete *P* element sequences and this portion are consistent, indicating that this region is representative of the complete *P* element sequence.
- Because the divergence of *P* element sequences can approach 50%, and because some elements are internally deleted, primers could be designed to only certain conserved regions.
- The 500 bp region is long enough to yield phylogenetically useful information, but short enough to allow rapid sequencing in a number of species.

The primers were designed to preserve amino acid identities in these conserved regions, but codon usage preferences were also considered so that the degeneracy of the primers could be minimized. It is possible that certain more divergent subfamilies of *P* elements were missed with these primer combinations. Thus, the results presented here should be viewed not as an exhaustive survey but rather as a sample of the diversity that may exist among the *P* elements in a genome. Details of the phylogenetic analysis can be found in Clark *et al.* (1995).

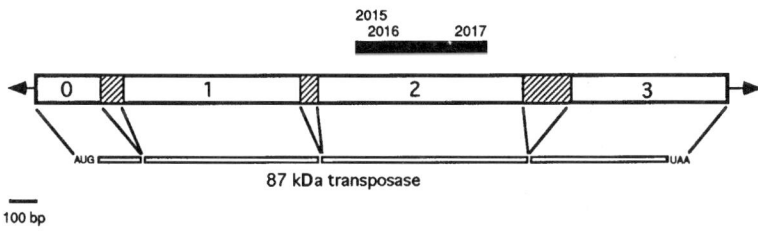

Fig. 19.2 Schematic representation of the *P* element showing location of primers used for phylogenetic analysis. The canonical *P* from *D. melanogaster* is 2907 bp in length and is flanked by perfect 31 bp inverted repeats (arrowheads). Exons 0–3 (open boxes) encode the transposase necessary for *P* element mobility. The relative locations of the primers (2015, 2016 and 2017) used to amplify the DNA fragment used for phylogenetic analysis are indicated above exon 2.

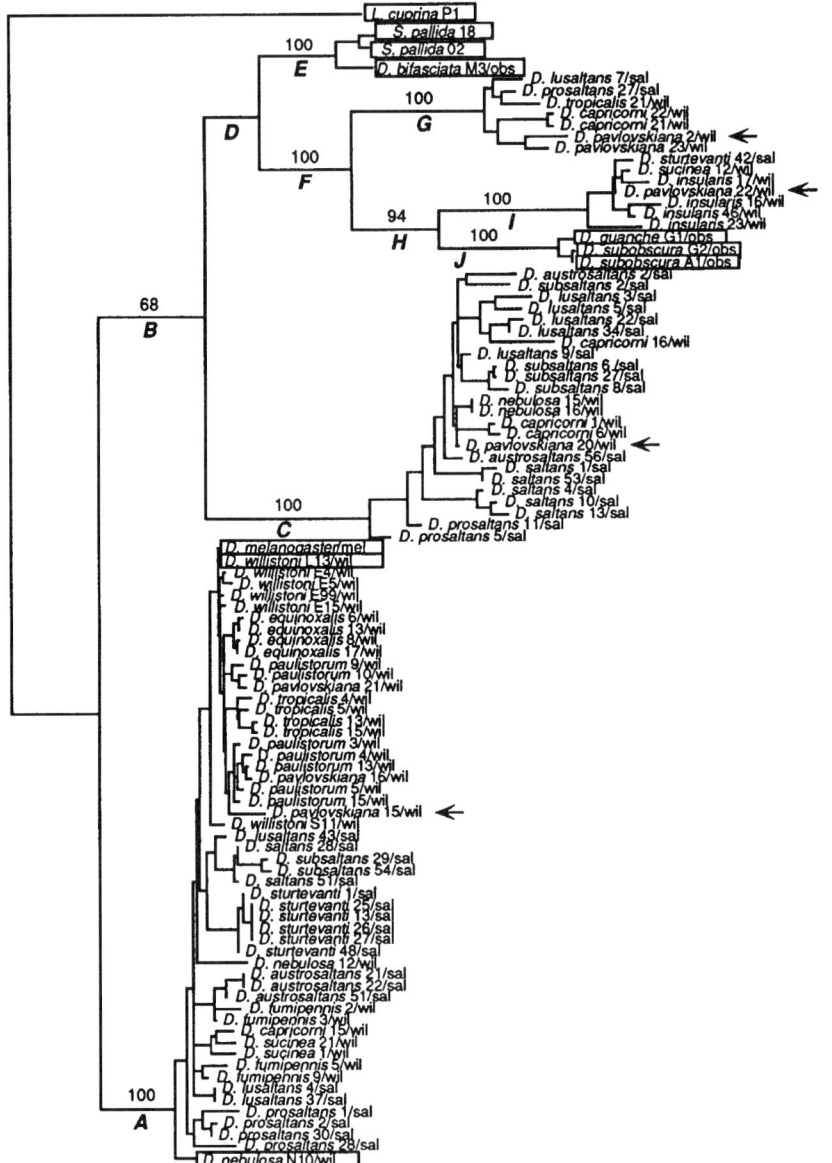

Fig. 19.3 Phylogenetic analysis of *P* element nucleotide sequences from the *D. saltans* and *D. willistoni* species groups. Comparisons were limited to 448 bp between primers 2016 and 2017 and used the sequence from *L. cuprina* as an outgroup. This cladogram was generated by parsimony analysis as implemented in PAUP 3.1.1 (Swofford 1993) using the heuristic search algorithm with TBR branch swapping and random stepwise addition of taxa. This is an arbitrarily chosen representative of 1000 equally parsimonious trees, each requiring

19.2.1   THE *SALTANS* AND *WILLISTONI* SPECIES GROUPS

P element sequences from these two species groups were analyzed together because they represent the New World lineage of the subgenus *Sophophora* (Fig. 19.1). These results have been published previously (Clark *et al.*, 1995) and so we concentrate here on only the major findings. Regardless of species, *P* elements from the *saltans* and *willistoni* species groups fall into four subfamilies, or clades (Fig. 19.3, clades *A*, *C*, *G*, *I*). The largest subfamily is clade *A*, which includes those sequences similar to the canonical element from *D. melanogaster*. In spite of the fact that these elements represent more than 50% of the sequences identified in these two groups, the nucleotide divergence within this clade is quite low (< 9%).

Most of the species sampled carry in their genomes *P* elements belonging to more than one clade. For example, *D. pavlovskiana* of the *willistoni* species group has elements that belong to each of the four major subfamilies (arrows, Fig. 19.3). What is not apparent in the figure is the absence of particular *P* element subfamilies in certain species. For example, neither *D. emarginata* nor *D. neocordata* of the *saltans* species group possesses detectable *P* elements of any kind. This result, obtained from PCR, verifies the same conclusion by Southern blots (Daniels *et al.*, 1990). Because *P* elements are found in all of the other species of the *saltans* group, this probably reflects a loss of *P* elements in these two species at some point in their diversification from the rest of the *saltans* species group. While *D. insularis* of the *willistoni* species group possesses non-canonical *P* elements (clade *I*), canonical *P* elements are absent from this species. This observation was again originally made on the basis of Southern blots. *D. insularis* almost certainly had canonical *P* elements at one time because they are prevalent in both its near and more distant relatives in the *willistoni* species group.

19.2.2   THE *OBSCURA* SPECIES GROUP

P element sequences have thus far been obtained from the *obscura* and *subobscura* complexes of the *obscura* subgroup within the *obscura* species

1340 steps. The consistency index is 0.579 and the retention index 0.846. The bootstrap values above the branches are percentages for 200 replicates. Only values of 50% or greater are shown. Letters refer to clades that are discussed in the text. Arrows denote four *D. pavlovskiana* sequences mentioned in the text. Species names are given in italics followed by a numbered clone designation. Species group affiliations are given after the slash: obs, *obscura* species group; mel, *melanogaster* species group; sal, *saltans* species group; wil, *willistoni* species group. Boxed sequences are used to provide a frame of reference in Figs 19.3, 19.4 and 19.5.

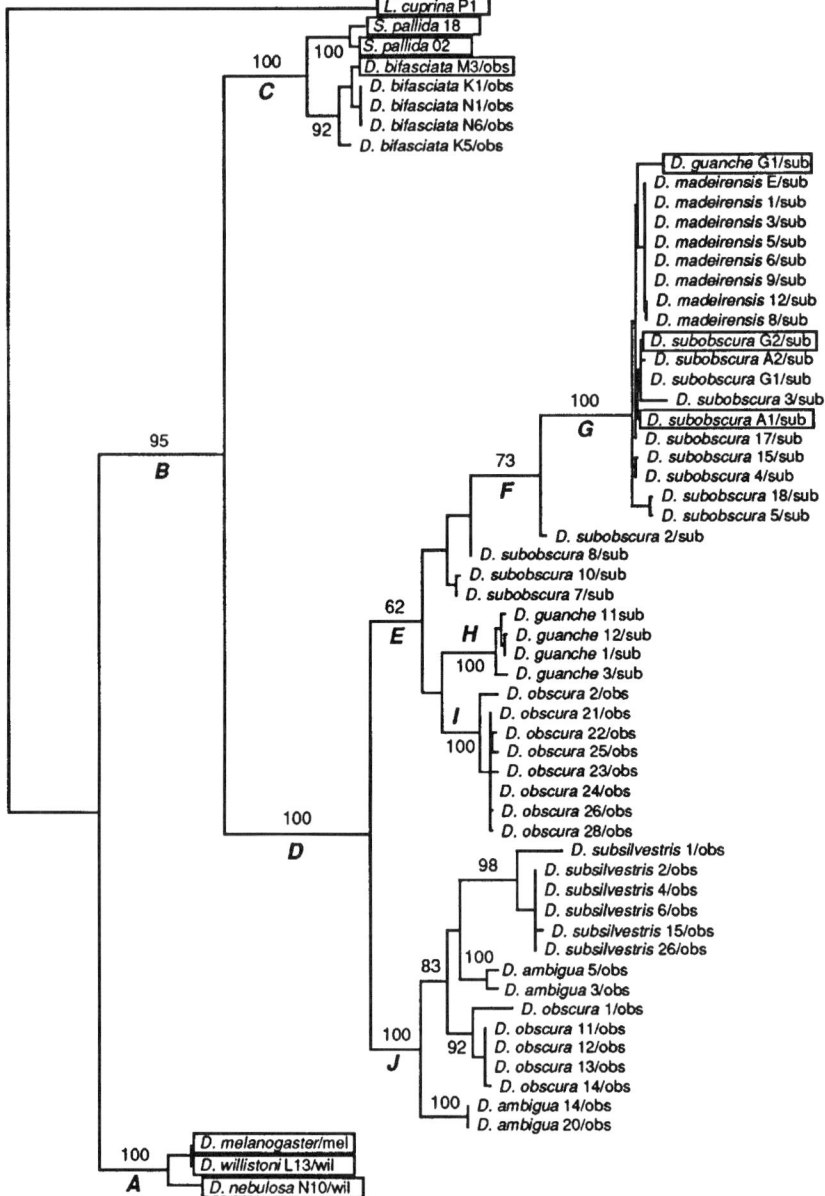

Fig. 19.4 Phylogenetic analysis of *P* element nucleotide sequences from the *D. obscura* species group. Details as for Fig. 19.3 except that this is an arbitrarily chosen representative of 1000 equally parsimonious trees, each requiring 731 steps; consistency index 0.654; retention index 0.907. Subgroup or species group affiliations given after the slash: wil, *willistoni* species group; mel, *melanogaster* species group; within the *obscura* subgroup of the *obscura* species group: obs, *obscura* complex; sub, *subobscura* complex.

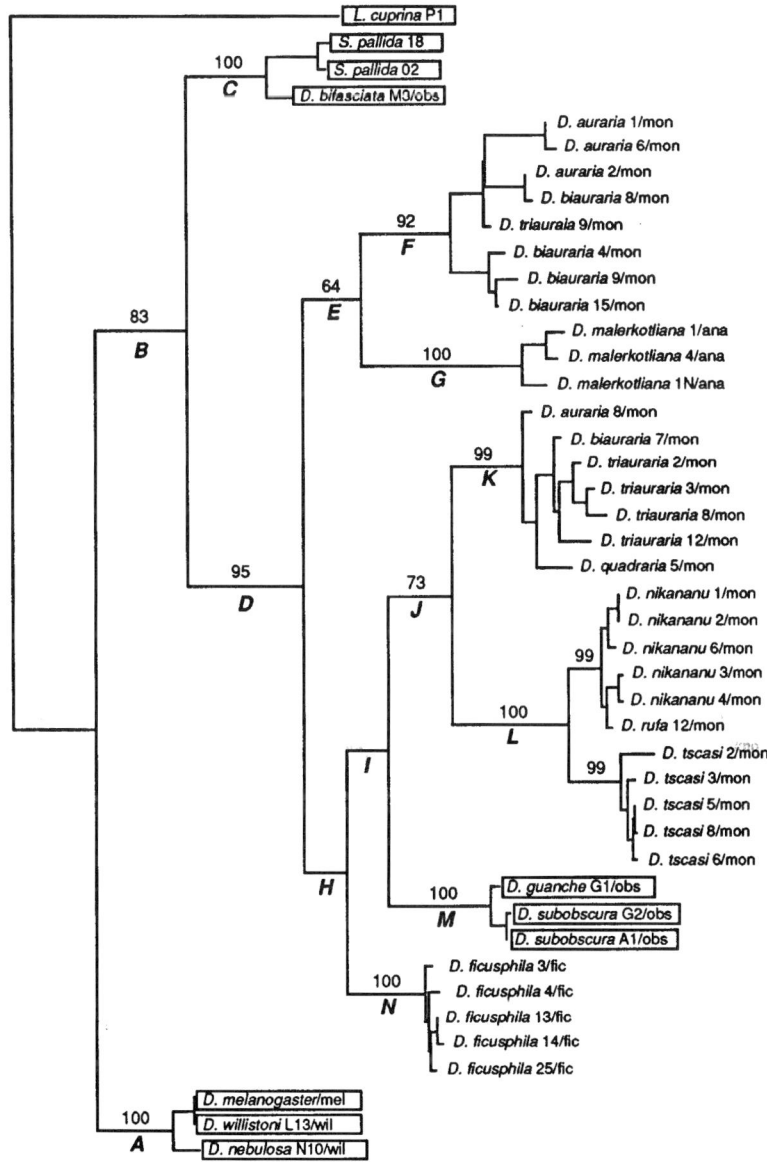

Fig. 19.5 Phylogenetic analysis of *P* element nucleotide sequences from the *melanogaster* species group. Details as for Fig. 19.3 except that this is an arbitrarily chosen representative of 256 equally parsimonious trees, each requiring 1065 steps; consistency index 0.594; retention index 0.850. Subgroup or species group affiliations given after the slash: obs, *obscura* species group; wil, *willistoni* species group; within the *melanogaster* species group: mon, *montium* subgroup; mel, *melanogaster* subgroup; ana, *ananassae* subgroup; fic, *ficusphila* subgroup.

group (J. García-Plannells, N. Paricio, J.B. Clark, R. de Frutos and M.G. Kidwell, unpublished). *P* elements from all but one species form a monophyletic group (clade *D*, Fig. 19.4). The exceptions are the M-type *P* sequences from *D. bifasciata*, which are closely related to those of *S. pallida* (clade *C*). M-type *P* elements, or their progenitors, may have been transferred horizontally from the genus *Scaptomyza* (Hagemann *et al.*, 1992; Clark *et al.*, 1994). We have sequenced four additional clones (K1, K5, N5, N6) from the *D. bifasciata* genome and all of them cluster with the M-type sequences of this species (clade *C*).

Clade *G* comprises *P* elements from three closely related species of the *subobscura* complex: *D. guanche*, *D. madeirensis* and *D. subobscura*. These sequences differ by at most 3% over the region examined here, reflecting the close relationships among these three species. A number of distinct but related elements from *D. subobscura*, that fall outside of clade *G*, have also been found (clones 2, 7, 8, 10). Also outside of this clade is a second subfamily of *P* elements from *D. guanche*, which forms clade *H*. The remainder of the sequences in Fig. 19.4 are from species belonging to the *obscura* complex: *D. obscura*, *D. subsilvestris* and *D. ambigua*. With the exception of a single subfamily of sequences from *D. obscura* (clade *I*), these fall into a single group designated as clade *J*. Overall, these preliminary results indicate that genomes of most of the species possess at least two distinct *P* element populations. This parallels the situation seen in the *saltans* and *willistoni* species groups.

19.2.3   THE *MELANOGASTER* SPECIES GROUP

The *melanogaster* species group is by far the largest in the subgenus *Sophophora*, containing over 150 described species (Lemeunier *et al.*, 1986). A preliminary examination using Southern blots led to one of the first pieces of evidence for horizontal transfer of the canonical *P* element to *D. melanogaster* (Brookfield *et al.*, 1984; Anxolabéhère *et al.*, 1988). Although *P* elements are found sporadically throughout this species group, they are apparently absent from the closest relatives of *D. melanogaster*, those species in the *melanogaster* subgroup (Daniels *et al.*, 1990) (Fig. 19.1). In addition, two other subgroups (*elegans* and *eugracilis*) lack detectable *P* elements.

The survey of the *melanogaster* species group is not yet complete, but the phylogenetic structure that is emerging is presented in Fig. 19.5. With the exception of the canonical element from *D. melanogaster*, *P* elements from the *melanogaster* species group are monophyletic (clade *D*, Fig. 19.5), showing a clear affiliation with the representative *P* elements from the *obscura* species group (clade *M*). Subgroup-specific branching of *P* elements, indicative of vertical transmission, is apparent. For example, both clades *F* and *J* comprise *P* element sequences from only the *montium* subgroup. Also, clades *G* and *N* are well-supported groups that include elements from a single species within particular subgroups. This could be a reflection of the limited sample of sequences from the *melanogaster* species group to date, and more diversity may be uncovered as the sample size increases. As was the case for the other *Sophophora* species groups, the genomes of many species in the *melanogaster* group possess more than one *P* element population (cf. clades *F* and *K*).

19.2.4   HORIZONTAL TRANSFER OF *P* ELEMENTS

As with the *obscura* species group, there are no canonical *P* elements found in the *melanogaster* species group. Thus, the canonical *P* elements, which are prevalent in the New World *saltans* and *willistoni* species groups, were probably not present at the origin of the subgenus *Sophophora*. Since canonical *P* elements have not been detected in other subgenera of *Drosophila* (Daniels *et al.*, 1990), their presence in the *saltans* and *willistoni* species groups may be explained by horizontal transfer into the ancestor of these New World lineages. Since almost all of the New World *Sophophora* have canonical *P* elements, this must have occurred before the radiation of these two species groups, between 15 and 30 million years ago (Fig. 19.1). An alternative explanation, that canonical *P* elements were lost in the Old World lineages but retained in the *saltans* and *willistoni* groups, seems unlikely when one considers the present distribution of the canonical elements and the phylogeny of their host species. It is interesting that in each of the analyses presented here, the canonical sequences form a single clade (*A* in all figures) that is a sister group to the rest of the sequences, including those from the other three *saltans–willistoni* subfamilies and from *Scaptomyza*. This is consistent with an origin of these elements outside of the subgenus *Sophophora*.

Overall, the results presented here indicate that the evolution of the *P* element in *Sophophora* has been dominated by vertical transmission from parent to offspring. However, in the recent past, there have been two unequivocal cases of horizontal transfer: from *Scaptomyza* to *D. bifasciata* and from *D. willistoni* to *D. melanogaster* (Clark *et al.*, 1994). It is interesting that both recipients (*D. bifasciata* and *D. melanogaster*) are among the small number of cosmopolitan members of their respective species groups. It may be that this expanded geographic range has increased the chance for horizontal transfer into these recipient species. It is possible that ancient horizontal transfers are indeed responsible for the overall phylogenetic structure of *P* element sequences presented here – for example, the origin of the four *saltans–willistoni* subfamilies. Identification of *P* element sequences in *Drosophila* outside of the subgenus *Sophophora* which show significant sequence similarity to the canonical elements suggests additional horizontal transfers may have occurred at more distant times in the past (Loreto *et al.*, 1996). It is clear that phylogenetic analysis is a powerful approach for detecting horizontal transfer and identifying potential donor species.

REFERENCES

Anxolabéhère, D., Kidwell, M.G. and Periquet, G. (1988) Molecular characteristics of diverse populations are consistent with the hypothesis of a recent invasion of *Drosophila melanogaster* by mobile *P* elements. *Mol. Biol. Evol.* **5**: 252–269.

Brookfield, J.F.Y., Montgomery E. and Langley, C. (1984) Apparent absence of transposable elements related to the *P* elements of *D. melanogaster* in other species of *Drosophila*. *Nature* **310**: 330–332.

Clark, J.B., Maddison, W.P. and Kidwell, M.G. (1994) Phylogenetic analysis supports horizontal transfer of *P* transposable elements. *Mol. Biol. Evol.* **11**: 40–50.

Clark, J.B., Altheide, T.K., Schlosser, M.J. and Kidwell, M.G. (1995) Molecular evolution of *P* transposable elements in the genus *Drosophila*. I. The *saltans* and *willistoni* species groups. *Mol. Biol. Evol.* **12**: 902–913.

Daniels, S.B., Peterson, K.R., Strausbaugh, L.D. *et al.* (1990) Evidence for horizontal transmission of the *P* transposable element between *Drosophila* species. *Genetics* **124**: 339–355.

Finnegan, D.J. (1989) Eukaryotic transposable elements and genome evolution. *Trends in Genetics* **5**: 103–107.

Hagemann, S., Miller, W.J. and Pinsker, W. (1992) Identification of a complete P element in the genome of *Drosophila bifasciata*. *Nucl. Acids Res.* **20**: 409–413.

Hurst, G.D.D., Hurst, L.D. and Majerus, M.E.N. (1992) Selfish genes move sideways. *Nature* **356**: 659–660.

Johnson, C.W. (1913) The distribution of some species of Drosophila. *Psyche* **20**: 202–205.

Kidwell, M.G. (1979) Hybrid dysgenesis in *Drosophila melanogaster*: the relationship between the P–M and I–R interaction systems. *Genet. Res.* **33**: 105–117.

Kidwell, M.G. (1983) Evolution of hybrid dysgenesis determinants in *Drosophila melanogaster*. *Proc. Natl Acad. Sci. USA* **80**: 1655–1659.

Kidwell, M.G. (1992) Horizontal transfer of *P* elements and other short inverted repeat transposons. *Genetica* **19**: 913–916.

Kidwell, M.G. (1993. Lateral transfer in natural populations of eukaryotes. *Ann. Rev. Genet.* **27**: 235–256.

Lansman, R.A., Shade, R.O., Grigliatti, T.A. and Brock, H.W. (1987) Evolution of *P* transposable elements: sequences of *Drosophila nebulosa P* elements. *Proc. Natl Acad. Sci. USA* **84**: 6491–6495.

Lemeunier, F., David, J.R., Tsacas, L. and Ashburner, M. (1986) The *melanogaster* species group, in *The Genetics and Biology of Drosophila*, Vol. 3e, (eds M. Ashburner, H.L. Carson and J.J.N. Thompson), Academic Press, London, pp. 147–256.

Loreto, E.L.S., Basso da Silva, L., Zaha, A. and Valente, V.L.S. (1996) A P-homologous sequence occurs in a species of the tripunctata group: *Drosophila mediopunctata*. *Brazil. J. Genet.* **19**: 267.

Miller, W.J., Hagemann, S., Reiter, E. and Pinsker, W. (1992) *P* element homologous sequences are tandemly repeated in the genome of *Drosophila guanche*. *Proc. Natl Acad. Sci.* **89**: 4018–4022.

Misra, S. and Rio, D.C. (1990) Cytotype control of *Drosophila P* element transposition: the 66 kd protein is a repressor of transposase activity. *Cell* **62**: 269–284.

Nitasaka, E., Mukai, T. and Yamazaki, T. (1987) Repressor of *P* elements in *Drosophila melanogaster*: cytotype determination by a defective *P* element with only open reading frames 0 through 2. *Proc. Natl Acad. Sci. USA* **84**: 7605–7608.

O'Hare, K. and Rubin, G.M. (1983) Structures of *P* transposable elements and their sites of insertion and excision in the *Drosophila melanogaster* genome. *Cell* **34**: 25–35.

Paricio, N.M., Perez-Alonso, M., Martínez-Sebastián, M.J. and de Frutos, R. (1991) *P* sequences of *Drosophila subobscura* lack exon 3 and may encode a 66 kd repressor-like protein. *Nucleic Acids Res.* **19**: 6713–6718.

Powell, J.R. and Gleason, J.M. (1996) Codon usage and the origin of *P* elements. *Mol. Biol. Evol.* **13**: 278–279.

Rio, D.C., Laski, F.A. and Rubin, G.M. (1986) Identification and immunochemical analysis of biologically active *Drosophila P* element transposase. *Cell* **44**: 21–32.

Simonelig, M. and Anxolabéhère, D. (1991) A *P* element of *Scaptomyza pallida* is active in *Drosophila melanogaster*. *Proc. Natl Acad. Sci. USA* **88**: 6102–6106.

Sturtevant, A. H. (1921) *The North American Species of Drosophila*, Publication 301, Carnegie Institution, Washington, DC.

Swofford, D. (1993) *PAUP: Phylogenetic Analysis Using Parsimony. Version 3.1.1*, Smithsonian Institution, Washington, DC.

Throckmorton, L.H. (1975) The phylogeny, ecology, and biogeography of *Drosophila*, in *Handbook of Genetics. Vol. 3. Invertebrates of Genetic Interest*, (ed. R.C. King), Plenum, New York, pp. 421–469.

The *mariner* transposons of animals: horizontally jumping genes

20

*Hugh M. Robertson, Felipe N. Soto-Adames,
Kimberly K.O. Walden, Rita M.P. Avancini
and David J. Lampe*

SUMMARY

The *mariner* transposons of insects and other animals are now known to comprise a large family of small (±1300 bp) transposable elements characterized by a D,D34D catalytic domain in their encoded transposases. They transpose by a DNA-mediated cut-and-paste mechanism, and form one branch of the D,D35E transposase/integrase megafamily of proteins, being most closely related to the *Tc1* family that is also widespread in animals. Within the *mariner* family at least 16 distinct subfamilies can be recognized, five of which have numerous representatives in the genomes of diverse insects, other arthropods, nematodes, flatworms, hydras and mammals, including humans. Phylogenetic analysis of the relationships of *mariners* within particular subfamilies clearly demonstrate that they have undergone multiple horizontal transfers between hosts, sometimes across animal phyla. We have studied three particular instances of relatively recent horizontal transfers in detail, involving transfer of irritans subfamily *mariners* across two orders of insects, mellifera subfamily *mariners* across four orders of insects, and cecropia subfamily *mariners* across three phyla of animals. In most cases these *mariners* evolve neutrally and accumulate incapacitating mutations within particular hosts, whereas comparisons between hosts indicate that most of the evolutionary conservation of their transposase genes occurs in conjunction with horizontal transfers between hosts. The ability of *mariner* transposase to catalyze transposition without species-specific host factors appears to allow this unusual evolutionary pattern. *Mariners* and the related *Tc1* family of transposons, which evidence many of these same characteristics, have thereby affected the composition of most animal genomes.

20.1   INTRODUCTION

The first *mariner* element was discovered in the fruit fly *Drosophila mauritiana* (Hartl, 1989) and a large body of work has characterized the distribution of this element in related drosophilids, including a likely horizontal transfer event (Maruyama and Hartl, 1991), and its functionality (e.g. Lohe and Hartl, 1996; Lohe *et al.* 1997). Following the serendipitous discovery of a distantly related *mariner* in the genome of the giant silk moth, *Hyalophora cecropia* (Lidholm *et al.*, 1991), we employed a polymerase chain reaction (PCR) assay to detect and preliminarily characterize *mariner* transposons with a widespread but patchy distribution in insects (Robertson, 1993) and related arthropods (Robertson and MacLeod, 1993). Since then we have used this assay, as well as additional primers designed from even more conserved amino acid blocks, to detect members of this family of transposons in the genomes of mites, flatworms, hydras and mammals (Robertson, 1996a, 1997; Robertson and Zumpano, 1997). Several additional *mariners* have been identified by us and others in the genomes of various insects as well as the nematode *Caenorhabditis elegans* and humans (Robertson and Asplund, 1996; Robertson, 1997).

In our initial phylogenetic analyses of *mariner* relationships it was clear that horizontal transfers across large host phylogenetic distances, such as orders of insects, must have occurred. Since then work with hosts from other phyla has shown that transfers across phyla have also occurred. Here we describe an additional dataset of PCR fragments from diverse insects, and analyze these within individual subfamilies, rather than across the entire family. For the central most conserved region of their transposase genes that are amplified with our PCR assay, sequences within subfamilies are colinear with each other and generally share at least 40% encoded amino acid identity, indicating that they diverged relatively recently (see Robertson and MacLeod, 1993, for definitions of the subfamilies). Although the rates of evolution of *mariners* are unclear, most evidence suggests that within particular hosts they evolve neutrally and hence relatively rapidly (for example, the neutral rate for DNA sequence divergence in *Drosophila* flies is about 1% per million years). Thus phylogenetic analyses of *mariners* that reveal major incongruencies with host phylogenies (for example, closely related within-subfamily *mariners* in hosts from different insect orders or animal phyla) must imply horizontal transfers because these transposons could not have evolved for 200–600 million years as separate lineages within their host genomes and remained so similar to each other.

20.2   METHODS

Since our initial reports, 267 additional species of insects have been examined for the presence of *mariners* using our PCR assay (Table 20.1). These reflect a similar diversity of insects to our previous study of over 400 species (Robertson and MacLeod, 1993), but instead of a quick DNA extraction method, DNA was extracted by conventional methods involving denaturation of proteins using heat and SDS, precipitation of

Table 20.1 Numbers of species tested and positive for *mariners* from various orders of insects (following nomenclature of Borror, Triplehorn and Johnson, 1989)

Order	Tested	Positive
Diplura	2	1
Collembola	8	0
Microcoryphia	2	2
Thysanura	2	1
Ephemeroptera	1	0
Grylloblattaria	1	1
Zoraptera	1	0
Embiidina	1	0
Isoptera	1	0
Orthoptera	5	2
Hemiptera	12	5
Mantodea	1	1
Homoptera	24	13
Blattaria	5	3
Dermaptera	4	1
Psocoptera	2	0
Neuroptera	16	8
Mecoptera	2	1
Coleoptera	46	24
Strepsiptera	2	0
Diptera	79	21
Trichoptera	2	1
Lepidoptera	23	10
Hymenoptera	39	18
Total	267	113

proteins using high salt, extraction of proteins using phenol/chloroform, and precipitation of the DNA with ethanol. Over 100 of these species tested positive using our original PCR assay, which yields a ±500 bp fragment (Robertson, 1993). Most of the 267 samples were also examined with two additional primers designed to regions encoding conserved amino acid blocks inside the original PCR fragment region (Robertson, 1997), and in each case this additional assay was also positive. A few samples only yielded amplification with the second set of primers, and one example – the grape phylloxera *Daktulosphaira vitifoliae* – was examined further and is included below. Twenty-five species (Table 20.2) were examined further by sequencing of a sample of 2-15 PCR fragment clones from each. These were conceptually translated, with judicious introduction of frameshifts to maintain aligned reading frames in some cases, and aligned with the available dataset of PCR fragments (Robertson, 1993, 1997; Robertson and MacLeod, 1993; Robertson and Lampe, 1995), as well as the equivalent regions of several full-length clones and consensus sequences now available (Robertson and Asplund, 1996; Robertson, 1997). Representative sequences, essentially all of those shown in bold in the figures, have been submitted to GenBank (accession numbers U91342-U91393).

There are many reasons to be confident that these sequences originate from the insects examined (Robertson, 1993, 1997). PCR contamination of samples was avoided by performing pre- and post-PCR work in different rooms with completely separate equipment and reagents. In any case contaminants would appear as identical sequences from different species, and while the remarkable aspect of our results is the extreme similarity of *mariners* from different hosts, no instances of identical sequences were obtained. Futhermore, as described below, we have cloned and sequenced copies of many particularly recently horizontally transferred *mariners* from genomic libraries of their hosts, thereby confirming that they originate from these genomes.

Phylogenetic analysis was conducted on the aligned ±170 amino acid sequences using maximum parsimony as implemented by PAUP, version 3.1.1 for the Macintosh (Swofford, 1993), employing the Heuristic search algorithm with random addition of sequences, 10 iterations and tree-bisection-and-reconnection branch swapping. All positions and amino acid changes were weighted equally. Where sets of multiple clones of similar kind were obtained – for example, in Robertson (1993) there were six clones representing the canonical *Drosophila mauritiana mar1* element – only one was used in the phylogenetic analysis to reduce the large numbers of equally parsimonious trees typically obtained in these analyses, and to allow bootstrap analysis. In each case the most intact clone – that is, with the fewest frameshifts and/or encoded stop codons – was employed to represent that type (an indication of the divergence of the

Table 20.2 Species from which *mariner* sequences were obtained

Common name	Scientific name	Family	Order
Surinam cockroach	*Pycnocelus surinamensis*	Blaberidae	Blattaria
Backswimmer	*Buenoa* sp.	Notonectidae	Hemiptera
Damsel bug	*Nabis* sp.	Nabidae	Hemiptera
Grape phylloxera	*Daktulosphaira vitifoliae*	Phylloxeridae	Homoptera
Hangingfly	*Bittacus strigosus*	Bittacidae	Mecoptera
Mantidfly	*Mantispa pulchella*	Mantispidae	Neuroptera
Blister beetle	*Epicauta funebris (=pestifera)*	Meloidae	Coleoptera
Locust borer	*Megacyllene robiniae*	Cerambycidae	Coleoptera
Flour bettle	*Tribolium madens*	Tenebrionidae	Coleoptera
Rusty grain beetle	*Cryptolestes ferrugineus*	Cucujidae	Coleoptera
Moquito	*Culex restuans*	Culicidae	Diptera
Tsetse fly	*Glossina palpalis*	Muscidae	Diptera
Stable fly	*Stomoxys uruma*	Muscidae	Diptera
False stable fly	*Muscina stabulans*	Muscidae	Diptera
Soybean nodule fly	*Rivellia quadrifasciata*	Platystomatidae	Diptera
Otitid fly	*Delphia picta*	Otitidae	Diptera
–	*Drosophila ananassae*	Drosophilidae	Diptera
Microcaddisfly	*Orthotrechia cf. cristata*	Hydroptilidate	Trichoptera
Parsnip webworm	*Depressaria pastinacella*	Oecophoridae	Lepidoptera
Ailanthus webworm	*Atteva punctella*	Yponomeutidae	Lepidoptera
Diamondback moth	*Plutella xylostella*	Plutellidae	Lepidoptera
Silkworm moth	*Bombyx mori*	Bombycidae	Lepidoptera
Stingless bee 1	*Plebia frontalis*	Apidae	Hymenoptera
Stingless bee 2	*Plebia jatiformis*	Apidae	Hymenoptera
Andrenid bee	*Andrena erigenia*	Andrenidae	Hymenoptera

available clones is given after the name in the figures). The consensus sequences of genomic clones were employed where possible. The four largest subfamilies (mauritiana, mellifera, cecropia and irritans) were analyzed separately, while the small capitata and lineata subfamilies were analyzed together with single sequence types representing apparently novel subfamilies. The outgroup for analyses of the four large subfamilies was chosen to be all of the sequences from Robertson and Asplund (1996) – that is, three divergent representatives from each of the mauritiana, mellifera, cecropia and irritans subfamilies, as well as two from the nematode *Caenorhabditis elegans* (for each subfamily analyzed, that subfamily's three sequences were removed from the outgroup). The basal *mariner* lineage from *Bombyx mori* (Robertson and Asplund, 1996) was not included in the outgroup because it is so divergent it only complicates the analyses: it shares just 20% amino acid identity with other *mariners*. Sequences from different subfamilies generally share 25–40% amino acid identity with each other. The outgroup for the capitata, lineata and smaller subfamilies was the three irritans subfamily sequences from Robertson and Asplund (1996), because they are a relatively basal lineage within the *mariner* family. Bootstrap analyses of at least 100 replications were performed on each dataset.

20.3   RESULTS

The overall rate of positive insect species found in this survey (42%) is considerably higher than in our earlier surveys (around 15%). In large part we ascribe this difference to the use of highly purified genomic DNA. The effect is particularly strong for certain groups, e.g. the Coleoptera, which are hard to extract using our earlier quick extraction method. Bigot *et al.* (1994) reported that the great majority of Hymenoptera they examined had evidence of mauritiana subfamily *mariners* alone, so we may still be underestimating the proportion of insects with *mariner* transposons in their genomes.

The alignment of these ±500 bp fragments encoding ±170 amino acids is largely unambiguous (e.g. Robertson, 1997) and entirely colinear within subfamilies, which is one defining feature of the subfamilies (Robertson and MacLeod, 1993). This fragment of *mariners* includes most of the D,D34D region homologous to the catalytic site of the D,D35E megafamily of transposases and integrases (Doak *et al.*, 1994; Robertson, 1995) and represents fully half of the transposase protein sequence. It therefore constitutes a good sample of each *mariner* that provides considerable phylogenetic information. A phylogenetic analysis of this region for the entire family, using the *Tc1* family as an outgroup and a sample of *mariner* sequences, is available in Robertson (1997), so here we focus on analyses of all available distinct *mariner* sequences within individual subfamilies. The 54 new types of *mariners* described here from 25 additional species reinforce and extend previous conclusions about *mariners*; for example, we again find that many insects have multiple different kinds of *mariners*, and their phylogenetic relationships indicate additional horizontal transfers across orders of insects and phyla of animals.

The mauritiana subfamily (Fig. 20.1) now includes *mariners* in three phyla of animals (arthropods, platyhelminths and cnidaria) and the new sequences reveal additional examples of horizontal transfers. For example, sequences very similar to the Deer fly 9.6 clone were found in a fruit fly, *Drosophila ananassae*, a mantidfly, *Mantispa pulchella*, and a hangingfly, *Bittacus strigosus*, the latter two species belonging to the orders Neuroptera and Mecoptera, respectively. This relationship is strongly supported by bootstrapping. These three *mariners* share 89–95% amino acid identity and, as is typical for closely related *mariners*, their percentage DNA identity is similar (91–96%). These orders are at least 265 million years old (Robertson and MacLeod, 1993). In contrast, there appears to be a basal lineage within this subfamily found to date only in various Hymenoptera, although it is not supported by bootstrapping.

The mellifera subfamily (Fig. 20.2) is now the largest subfamily of *mariners*, with over 50 different kinds. The placement of the flatworm *Dugesia tigrina* 1 and the mite *Tetranychus urticae* 1 sequences at the base of this family is tentative (Robertson, 1996a, 1997) because while they have length variants that place them in this or the cecropia subfamilies, they have sequence features that in analyses of the entire family cause them to cluster at the base of the mauritiana subfamily. The new sequences provide many additional examples of horizontal transfers in this subfamily in that almost all of their relationships are highly incongruent with those of their hosts. For example, several clones from the blister beetle *Epicauta funebris* cluster confidently with those from the European honey bee, European earwig and Mediterranean fruit fly (the first three of these *mariners* share at least 90% amino acid and DNA identity). Clones from the rusty grain beetle, *Cryptolestes ferrugineus*, cluster with the full length *mariner* described from the tsetse fly *Glossina palpalis* (Blanchetot and Gooding, 1995). Some sequences from a micro-caddisfly cluster with those from several moths, while others cluster with those of several bees, and the Hangingfly 4 clone is the closest relative of the Chloropid fly 36.4 clone.

Cecropia subfamily *mariners* (Fig. 20.3) have been found in the most diverse hosts, including many insects, two flatworms, a hydra and primates. Again a micro-caddisfly clone clusters confidently with several moth clones, perhaps suggesting that transfers among Lepidoptera and Trichoptera are common. A tsetse fly clone provides yet another insect with *mariners* closely related to the first known human *mariner*, *Hsmar1*, and *mariners* from *Hydra littoralis* and the marine flatworm *Stylochus*

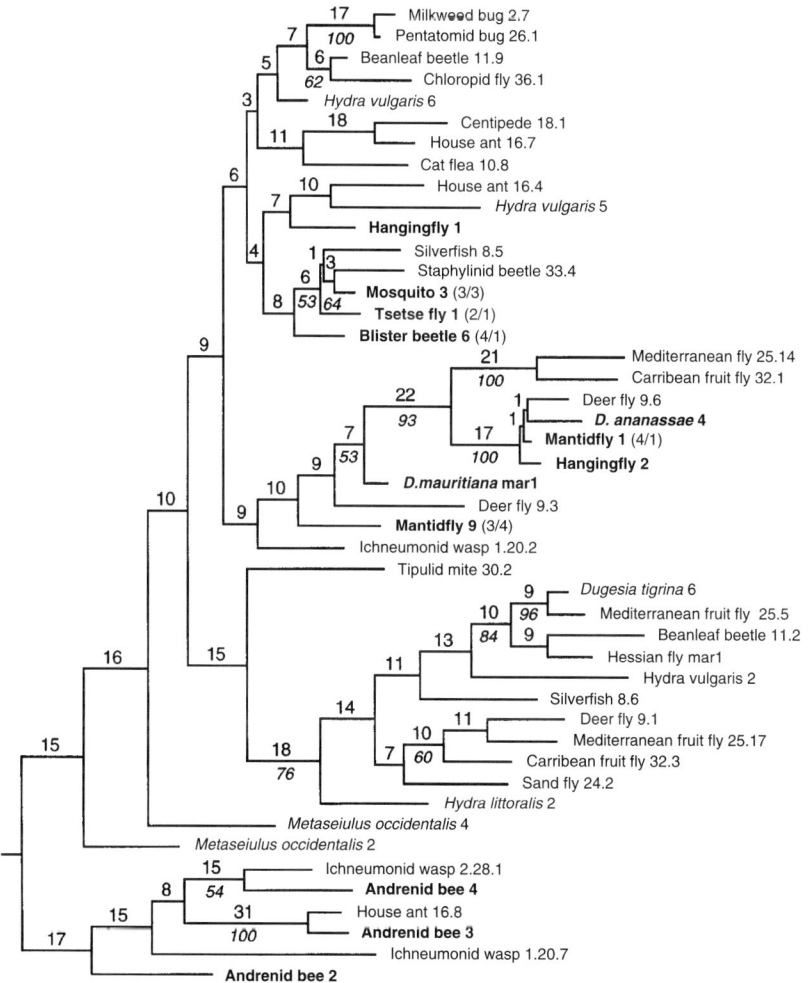

Fig. 20.1 Phylogenetic relationships of the mauritiana subfamily *mariners*. A single most parsimonious tree was obtained. Branch lengths in number of amino acid changes are shown above all branches supporting nodes. Bootstrap percentages are shown in italics below branches supporting nodes with more than 50% support. Names of clone sequences reported here for the first time are in bold. Where appropriate the average percentage amino acid divergence between the encoded transposase sequences of multiple similar clones is shown in parentheses after these bold names.

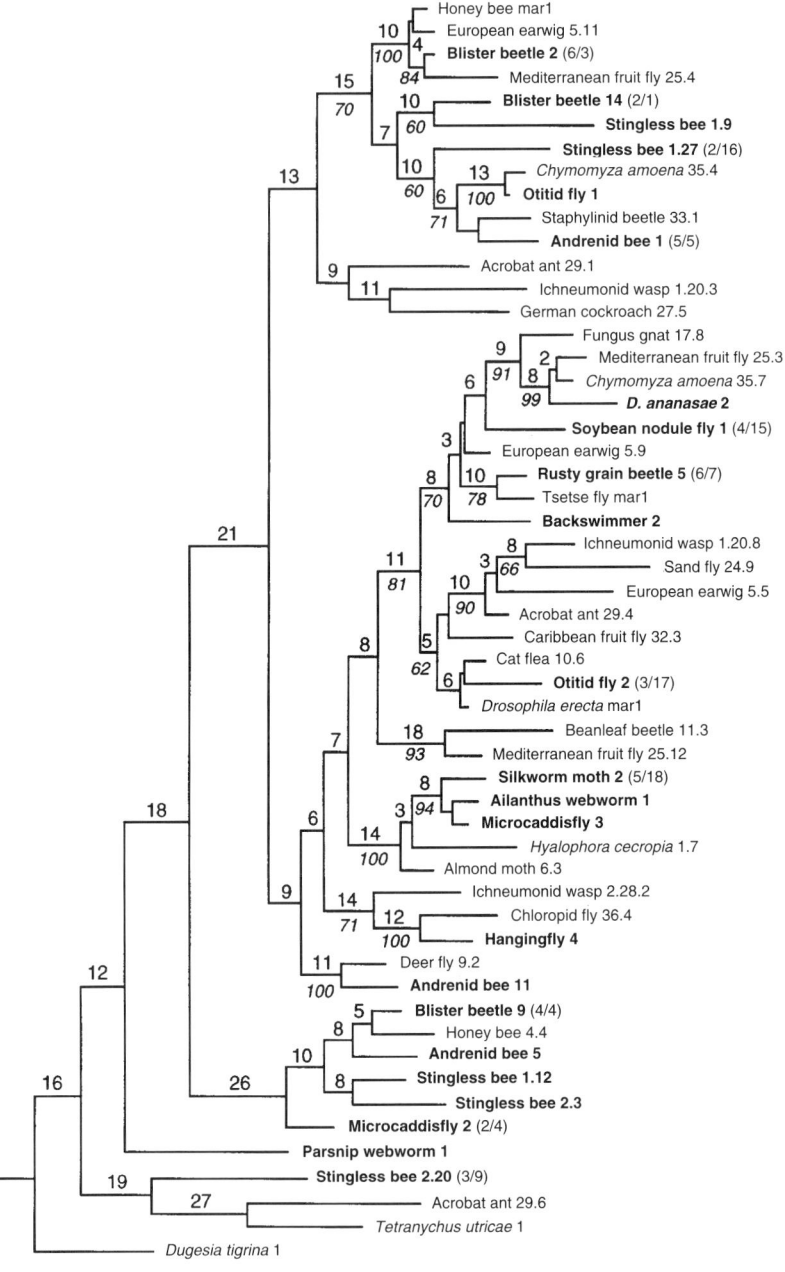

Fig. 20.2 Phylogenetic relationships of the mellifera subfamily *mariner*s. This tree is an arbitrary representative of the 72 equally parsimonious trees obtained. Branch lengths are shown only for those branches supporting nodes present in 100% of the trees, according to the semi-strict consensus option of PAUP. Other details as for Fig. 20.1.

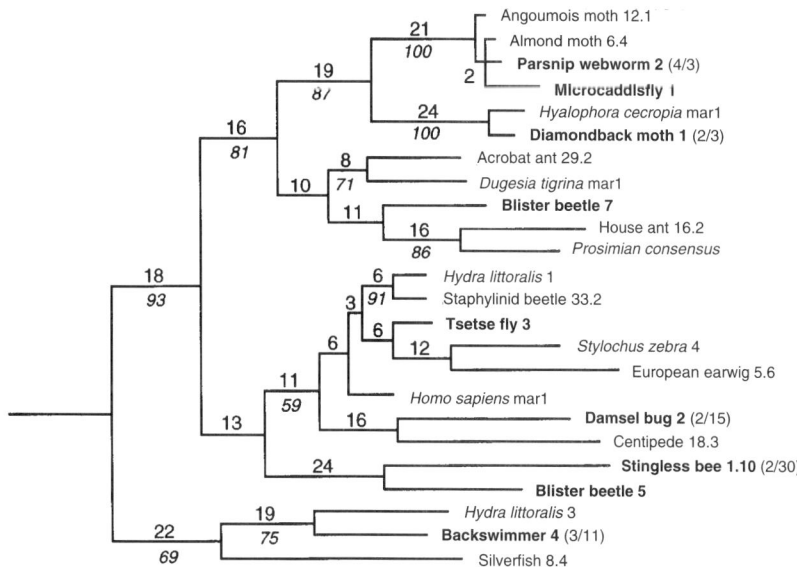

Fig. 20.3 Phylogenetic relationships of the cecropia subfamily *mariner*s. Other details as for Fig. 20.1.

zebra (Robertson, 1997; Robertson and Zumpano, 1997). The staphylinid beetle, *Hsmar1* consensus, and *Hydra littoralis mariners* share at least 85% amino acid and DNA identity, which is extraordinary for sequences from three phyla that last shared a common ancestor at least 600 million years ago (Robertson 1997). Another cluster of sequences from extremely diverse hosts is the original flatworm *Dugesia tigrina mar1* (Garcia-Fernàndez *et al.*, 1995), the insect Acrobat ant 29.2, House ant 16.2 and Blister beetle 7 sequences, and the consensus of multiple sequences found in diverse prosimians (Robertson and Zumpano, 1997).

The best-studied example of recent horizontal transfer of *mariner* elements is in the irritans subfamily (Fig. 20.4) involving a green lacewing, the horn fly *Anopheles gambiae* and *Drosophila ananassae* (Robertson and Lampe, 1995), to which sequences from a stable fly, *Stomoxys uruma*, can now be added (they all share at least 88% amino acid and DNA identity). The other new sequences in this subfamily provide additional examples of *mariner* phylogenetic relationships incongruent with those of their hosts. The second human *mariner*, *Hsmar2*, has been tentatively placed as a basal lineage of this subfamily (Oosumi *et al.*, 1995; Reiter *et al.*, 1996; Robertson *et al.*, 1996; Robertson and Asplund, 1996; Robertson, 1997) and it is not particularly closely related to any other *mariner*.

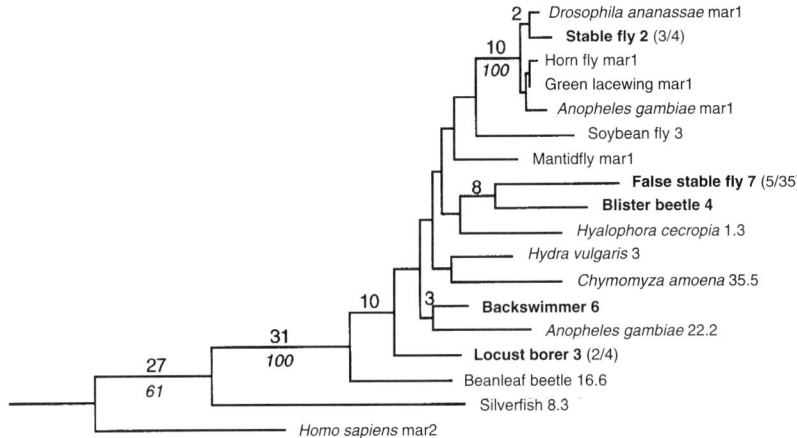

Fig. 20.4 Phylogenetic relationships of the irritans subfamily *mariner*s. This tree is an arbitrary representative of the 22 equally parsimonious trees obtained. Branch lengths are shown only for those branches supporting nodes present in 100% of the trees, according to the semi-strict consensus option of PAUP. The Soybean.fly.3 and Blister.beetle.4 sequences were reported in Lampe and Robertson (1995) and hence are not in bold. Other details as for Fig. 20.1.

Finally, in Fig. 20.5 the relationships within the small capitata and lineata subfamilies, as well as single *mariner* types that might represent novel subfamilies, are shown. These include sequences from all four phyla, as well as four from the nematode *Caenorhabditis elegans* (which also has three more *mariner*s similar to the extremely basal *Bombyx mori mariner*1: Robertson and Asplund, 1996; Robertson, 1997). Several of the new sequences might represent novel subfamilies – for example, those from the grape phylloxera, the Surinam cockroach and the flour beetle. In each case these sequences have length differences from even their closest relatives in the tree, and generally share with them and each other only 20–40% amino acid identity (values typical of subfamily relationships). It is not yet known whether these are very rare subfamilies, or are common in animals not yet sampled, or are so divergent in sequence that they do not readily amplify with the PCR primers employed to date – which might explain why the grape phylloxera sequences were only amplified with the internal primers.

20.4   DISCUSSION

The phylogenetic relationships of *mariner*s revealed by our work and that of others (e.g. Garcia-Fernàndez *et al.*, 1995; Lohe *et al.*, 1995) are clearly highly incongruent with those of their hosts. These incongruencies can

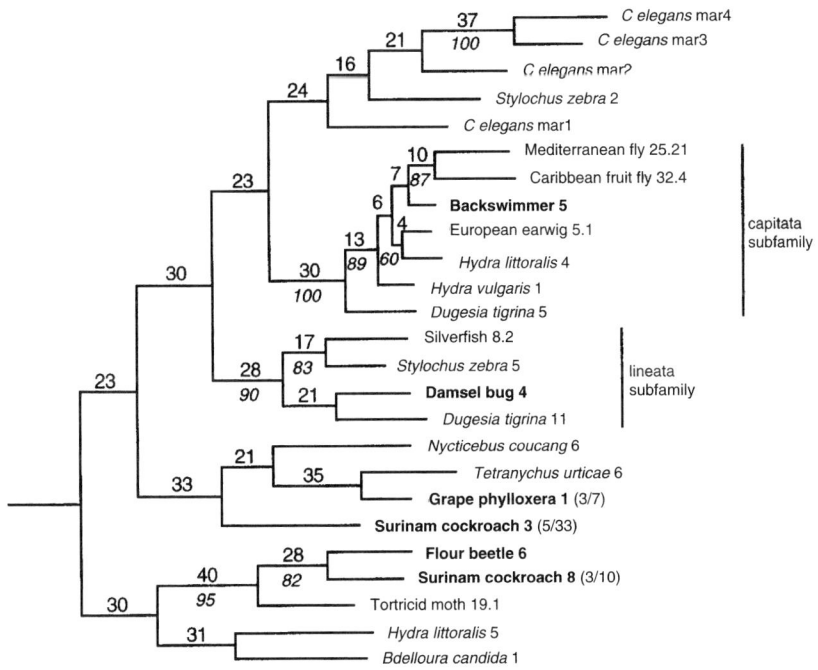

Fig. 20.5 Phylogenetic relationships of the smaller subfamilies of *mariner*s. Other details as for Fig. 20.1.

only be reasonably explained by repeated, indeed seemingly fairly frequent, horizontal transfers between hosts. The alternative explanation of vertical inheritance of particular *mariner* lineages within only certain host lineages, and their loss from all others examined, is untenable, because it is inconceivable that these *mariner*s could have remained so similar to each other (commonly at least 85% identity) for such long periods in such divergent hosts (all over 200 million years old). To provide additional data to support this claim we have examined in detail three particularly clear examples of relatively recent horizontal transfers across large host phylogenetic gaps.

Robertson and Lampe (1995) describe the remarkably similar irritans subfamily *mariner*s (Fig. 20.4) found in the green lacewing *Chrysoperla plorabunda* (order Neuroptera), the horn fly *Haematobia irritans* and a drosophilid fly *D. ananassae* (order Diptera, suborder Brachycera), and the African malaria mosquito *Anopheles gambiae* (order Diptera, suborder Nematocera). For example, the consensus DNA sequences of the 1044 bp transposase genes of the lacewing and horn fly *mariner*s differ by just

2 bp, leading to the 345 amino acid sequences of their encoded transposases differing by just one. Given the age of these insect lineages (200 million years in the case of fly suborders, and 265 million years for the orders) it is inconceivable that these *mariners* have evolved vertically within these lineages, been conserved to such an extent, including third codon positions and non-coding sequence, and have been lost from most other neuropteran and fly lineages examined. Indeed, when the evolution of the individual copies from their consensus sequences within each host lineage was examined, most copies appeared to be evolving neutrally, and therefore rapidly, because the rate of synonymous changes in the transposase genes usually equalled that of non-synonymous changes.

These results, and similar results obtained by others (e.g. Lohe *et al.*, 1995), have led to the view that horizontal transfers to new hosts are not simply an evolutionary curiosity of these transposons, but rather are central to their evolutionary persistence (Robertson and Lampe, 1995). Thus it appears that once such a transfer has occurred, presumably by a single active copy of a *mariner*, the copy number increases and the element spreads throughout the host species. The individual copies in most cases appear to evolve independently and neutrally from each other thereafter, with their transposition activity repressed either by the host, or by themselves, or by accumulation of internally deleted copies that titrate transposase (Lohe and Hartl, 1996). They will eventually be lost from the host genome or become an unrecognizable part of the DNA of the host. Occasionally a still active copy might undergo a horizontal transfer to a new host, repeating the cycle. This is probably the only stage at which significant selection for activity takes place, because while this copy may have acquired mutations, none of these can compromise activity for it to be able to achieve the horizontal transfer and establish a new population of *mariners* in its new host. From the example of the hornfly and green lacewing *mariners* above, we believe that in fact most such horizontal transfers to new hosts occur reasonably soon after a *mariner* invades a host, with at most only a few mutations being acquired, and that therefore each comparison described above of relatively closely related (> 85% identity) *mariners* in highly divergent hosts in fact reflects tens to hundreds of horizontal transfers to species we have not examined. In no case do we believe we have identified direct donor and recipient hosts, in contrast to the example of extremely recent horizontal transfer of the *P* element among *Drosophila* species (Chapter 19).

Additional support for this model of *mariner* molecular evolution is provided by our unpublished analyses of two other sets of relatively recent horizontal transfers of *mariners*. One is the set of closely related mellifera subfamily *mariners* found in insects from four different orders (Fig. 20.2); the European honey bee *Apis mellifera* (Hymenoptera), the

European earwig *Forficula auricularia* (Dermaptera), the Mediterranean fruit fly *Ceratitis capitata* (Diptera), and a blister beetle *Epicauta funebris* (= *pestifera*) (Coleoptera). Again, multiple full-length copies of these *mariners* have been cloned from genomic libraries of these species, with the exception of the blister beetle, where for reasons unexplained we cannot obtain them from a genomic library, but can by PCR. Comparison of these sequences reveals that they have generally evolved neutrally within their hosts, but show evidence of conservation when compared across hosts.

Similarly, we have cloned and sequenced multiple copies of the cecropia subfamily *mariners* (Fig. 20.3) found in the human genome, *Hydra littoralis* and a staphylinid beetle, *Carpelimus* sp. The latter two *mariners* are relatively recent invaders of their hosts, as judged by the high similarity of the copies within their genomes (e.g. 99% DNA identity for the beetle copies), but this human *mariner*, *Hsmar1*, is an old constituent of our genome, having invaded at least 30 million years ago: most copies are highly mutated and defective, differing from their consensus by ±7.5% in DNA sequence (Robertson and Zumpano, 1997).

These results from genomic libraries abundantly confirm the observations based on PCR fragments. That such horizontal transfers occur is also supported by the findings of Lohe *et al.* (1995) of a mellifera subfamily *mariner* in *Drosophila erecta* that is very similar to a PCR fragment we reported from the cat flea (Fig. 20.2). Furthermore, Garcia-Fernàndez *et al.* (1995) describe a cecropia subfamily *mariner* in the flatworm *Dugesia tigrina* that shares 75% amino acid identity with an acrobat ant *mariner* PCR fragment (Fig. 20.3). The copies within this flatworm differ from each other by less than 1% DNA divergence, and no congeners examined had this *mariner*, strongly supporting the argument for a recent horizontal transfer into the flatworm, but not of course directly from the acrobat ant. We can therefore be fairly confident that all of the unusual relationships in the phylogenetic trees of *mariners* shown herein are genuine examples of horizontal transfers. Indeed, it seems inescapable that each example in fact reflects many horizontal transfers, because we have only examined a very small sample of animals in detail.

The implications of these observations for *mariner* transposons, and likely also the closely related *Tc1* family of transposons, are two-fold. First, they must be capable of functioning in diverse host cellular environments. This inference has been confirmed for one *mariner* (*Himar1*, the horn fly irritans subfamily element) (Lampe *et al.*, 1996) and the *Tc1* element from *Caenorhabditis elegans* (Vos *et al.*, 1996). In each case purified transposase was able to catalyze transposition of a marked cognate transposon from one plasmid to another, with the products recovered in bacteria. Therefore these transposons are unlikely to require species-specific host factors for mobility, beyond the common host enzymes involved in repair of excision sites and new integrations. This transposon lifestyle probably represents one extreme on a continuum including elements like the *P* elements (Chapter 19) which reveal reasonably frequent horizontal transfers among a phylogenetically restricted set of hosts, the family Drosophilidae, to retrotransposons such as the *R* elements that are widespread in insects, yet appear to persist primarily by vertical and conserved evolution within their hosts (e.g. Lathe *et al.*, 1995).

Second, there must be some mechanism by which these transposons are able to move from one host to another. This movement has to be into the germline of the new host and across orders of insects and phyla of animals. It seems unlikely that there will be a single mechanism for such transfers (e.g. Kidwell, 1992); however, the best current candidates are perhaps various DNA viruses, some of which are known to be suitable targets for transposons from their hosts (e.g. Fraser *et al.*, 1985), including a member of the *Tc1* family (Jehle *et al.*, 1995).

Consequences of such horizontal transfers for the hosts are that essentially all animals are probably vulnerable to invasion by these transposons; indeed it appears that while the *mariner* family is fairly widespread, the *Tc1* family has multiple representatives in the genomes of most animals (Avancini *et al.*, 1996). Given the antiquity of these two transposon families, this is a process animals have been subjected to for a very long time. In some cases the hosts appear to be particularly vulnerable to such invasions; for example, the horn fly has a relatively enormous genome of 2.2×10^9 bp, nearly the size of ours and 10 times the size of *Drosophila melanogaster*. Fully 1% of this DNA consists of 17 000 copies of the particular *mariner* described above (Robertson and Lampe, 1995), while at least 17 different kinds of *Tc1* family elements constitute an undetermined portion (Avancini *et al.*, 1996).

The recent discovery of the two human *mariner*s (Auge-Gouillou *et al.*, 1995; Morgan, 1995; Oosumi *et al.*, 1995; Reiter *et al.*, 1996; Smit and Riggs, 1996; Robertson *et al.*, 1996; Robertson and Zumpano, 1997) demonstrates that these consequences have applied to the evolution of our genome as well. In each case these are very old constituents and comprise a very small portion of the human genome. There is evidence of at least two more similarly ancient DNA-mediated transposons in our genome, called *Tiggers* (both members of the *pogo* family of elements distantly related to *mariner*s and *Tc1* family) (Smit and Riggs, 1996; Robertson, 1996b), and Smit and Riggs (1996) infer that altogether remnants of DNA-mediated transposons constitute about 1% of our genome. At present it seems unlikely that any of these elements are active in our genome (but see Reiter *et al.*, 1996); however, they likely had some mutagenic and perhaps recombinational influence on its evolution.

ACKNOWLEDGMENTS

We thank the many colleagues who contributed insects and/or DNA samples for this work, particularly Alain Blanchetot, Raul Cano, Jan Conn, Thomas Coudron, Jeffrey Granett, Kostas Iatrou, Karen McClellan, Barbara Stay, Durdica Ugarkovic, Lisa Vawter and David Weaver. Matthew Sharkey, Karen Zumpano, Michelle Lepkowitz and Paul White provided technical assistance. This work was supported by NSF grant MCB 93-17586.

REFERENCES

Auge-Gouillou, C., Bigot, Y., Pollet, N. *et al.* (1995) Human and other mammalian genomes contain transposons of the *mariner* family. *FEBS Letts* **368**: 541–546.

Avancini, R.M.P., Walden, K.K.O. and Robertson, H.M. (1996) The genomes of most animals have multiple members of the *Tc1* family of transposable elements. *Genetica* **98**: 131–140.

Bigot, Y., Hamelin, M-H., Capy, P. and Periquet, G. (1994) Mariner-like elements in hymenopteran species: insertion site and distribution. *Proc. Natl Acad. Sci. USA* **91**: 3408–3412.

Blanchetot, A. and Gooding, R.H. (1995) Identification of a *mariner* element from the tsetse fly, *Glossina palpalis palpalis*. *Insect Mol. Biol.* **4**: 89–96.

Borror, D.J., Triplehorn, C.A. and Johnson, N.E. (1989) *An Introduction to the Study of Insects*, 6th edn, Saunders College Publishing, Philadelpha, PA.

Doak, T.G., Doerder, F.P., Jahn, C.L. and Herrick, G. (1994) A proposed super-family of transposase-related genes: new members in transposon-like elements of ciliated protozoa and a common 'D35E' motif. *Proc. Natl Acad. Sci. USA* **91**: 942–946.

Fraser, M.J., Brusca, J.S., Smith, G.E. and Summers, M.D. (1985) Transposon-mediated mutagenesis of a baculovirus. *Virology* **145**: 356–61.

Garcia-Fernàndez, J., Bayascas-Ramírez, J.R., Marfany, G. *et al.* (1995) High copy number of highly similar *mariner*-like transposons in planarian (Platyhelminthe): evidence for a trans-phyla horizontal transfer. *Mol. Biol. Evol.* **12**: 421–431.

Hartl, D.L. (1989) Transposable element *mariner* in *Drosophila* species, in *Mobile DNA* (eds D.E. Berg and M.M. Howe), American Society for Microbiology, Washington, DC, pp. 5531–5536.

Jehle, J.A., Fritsch, E., Nickel, A. *et al.* (1995) TCI4.7:a novel lepidopteran transposon found in *Cydia pomonella* granulosis virus. *Virology* **207**: 369–379.

Kidwell, M.G. (1992) Horizontal transfer. *Curr. Opin. Genet. Dev.* **2**: 868–873.

Lampe, D.J., Churchill, M.E.A. and Robertson, H.M. (1996) A purified *mariner* transposase is sufficient to mediate transposition *in vitro*. *EMBO J.* **15**: 5470–5479.

Lathe, W.C. III, Burke, W.D., Eickbush, D.G. and Eickbush, T.H. (1995) Evolutionary stability of the *R1* retrotransposable element in the genus *Drosophila*. *Mol. Biol. Evol.* **12**: 1094–1105.

Lidholm, D.-A., Gudmundsson, G.H. and Boman, H.G. (1991) A highly repetitive, *mariner*-like element in the genome of *Hyalophora cecropia*. *J. Biol. Chem.* **266**: 11518–11521.

Lohe, A.R. and Hartl, D.L. (1996) Autoregulation of *mariner* transposase activity by overproduction and dominant-negative complementation. *Mol. Biol. Evol.* **13**: 549–555.

Lohe, A.R., Moriyama E.N., Lidholm D.-A. and Hartl, D.L. (1995) Horizontal transmission, vertical inactivation, and stochastic loss of *mariner*-like transposable elements. *Mol. Biol. Evol.* **12**: 62–72.

Lohe, A.R., De Aguinar, D. and Hartl, D.L. (1997) Mutations in the *mariner* transposase: the D,D(35)E consensus sequence is nonfunctional. *Proc. Natl Acad. Sci. USA* **94**: 1293–1297.

Maruyama, K. and Hartl, D.L. (1991) Interspecific transfer of the transposable element *mariner* between *Drosophila* and *Zaprionus*. *J. Mol. Evol.* **33**: 514–524.

Morgan, G.T. (1995) Identification in the human genome of mobile elements spread by DNA-mediated transposition. *J. Mol. Biol.* **254**: 1–5.

Oosumi, T., Belknap, W.R. and Garlick, B. (1995) *Mariner* transposons in humans. *Nature* **378**: 672.

Reiter, L.T., Murakami, T., Koeuth, T. *et al.* (1996) A recombination hotspot responsible for two inherited peripheral neuropathies is located near a *mariner* transposon-like element. *Nature Genet.* **12**: 288–297.

Robertson, H.M. (1993) The *mariner* transposable element is widespread in insects. *Nature* **362**: 241–245.

Robertson, H.M. (1995) The *Tc1-mariner* superfamily of transposons in animals. *J. Insect Physiol.* **41**, 99–105.

Robertson, H.M. (1996a) *Mariner* transposable elements in mites. *Proc. IX Int. Cong. Acarol.* (in press).

Robertson, H.M. (1996b) Members of the pogo superfamily of DNA-mediated transposons in the human genome. *Mol. Gen. Genet.* **252**: 761–766.

Robertson, H.M. (1997) Multiple *mariner* transposons in flatworms and hydras are related to those of insects. *J. Heredity* **88**: 195–201.

Robertson, H.M. and Asplund, M.L. (1996) *Bmmar1*: a basal lineage of *mariner* family of transposable elements in the silkmoth, *Bombyx mori. Insect Biochem. Mol. Biol.* **26**: 945–954.

Robertson, H.M. and Lampe, D.J. (1995) Recent horizontal transfer of a *mariner* element between Diptera and Neuroptera. *Mol. Biol. Evol.* **12**: 850–862.

Robertson, H.M. and MacLeod, E.G. (1993) Five major subfamilies of *mariner* transposable elements in insects, including the Mediterranean fruit fly, and related arthropods. *Insect Mol. Biol.* **2**: 125–139.

Robertson, H.M. and Zumpano, K.L. (1997) Molecular evolution of an ancient *mariner* transposon, Hsmar1, in the human genome. *Gene* **205**: 203–217.

Robertson, H M., Zumpano, K.L., Lohe, A.R. and Hartl, D.L. (1996) Reconstruction of the ancient *mariners* of humans. *Nature Genet.* **12**: 360–361.

Smit, A.F.A. and Riggs, A.D. (1996) Tiggers and other DNA transposon fossils in the human genome. *Proc. Natl Acad. Sci. USA* **93**: 1443–1448.

Swofford, D.L. (1993) *PAUP: Phylogenetic Analysis Using Parsimony, Version 3.1.1,* Smithsonian Institution, Washington, DC.

Vos, J.C., De Baere, I. and Plasterk, R.H.A. (1996) Transposase is the only nematode protein required for *in vitro* transposition of *Tc1. Genes Dev.* **10**: 755–761.

Horizontal and vertical transmission of *hobo*-related sequences between *Drosophila melanogaster* and *Drosophila simulans*

21

Gail M. Simmons, Dahlia Plummer, Alex Simon, Ian A. Boussy, Julie Frantsve and Masanobu Itoh

SUMMARY

The *hobo* transposable element in *Drosophila melanogaster* and *D. simulans* is characterized by a strong 2.6 kb *Xho*I restriction fragment in a Southern blot that is found only in more recently collected strains (so-called H strains) of these species. Older strains (called E strains) typically contain faint, high molecular weight bands whose properties have heretofore been unknown. We used the polymerase chain reaction to amplify a portion of the coding region of *hobo* elements from E strains of both species, and we screened a *D. melanogaster* E strain library for *hobo*-homologous sequences. The DNA sequences suggest that the high molecular weight bands in the E strains represent ancient *hobo* elements that have degenerated, containing a great deal of indel variation as well as single base-pair changes that render them nonfunctional. A phylogenetic analysis of the sequences suggests that the modern, functional *hobo* element is most closely related to the degenerate sequences from *D. simulans*. We conclude that the modern *hobo* element may have had its origin in *D. simulans* rather than in *D. melanogaster*.

21.1   INTRODUCTION

The *hobo* transposable element of *Drosophila melanogaster* was discovered as an insertion into the *Sgs4* gene causing a visible mutation (McGinnis *et al.*, 1983). Subsequent cloning and sequencing revealed structural similarities, but no obvious homology, between *hobo* and the *P* element of *D. melanogaster* (Streck *et al.*, 1986). *Hobo* has since been placed, on the basis of the amino acid sequence of its transposase, into a family of transposable elements known as the *hAT* family, which includes the *Ac* element of maize, the *Tam3* element of snapdragon, and a number of elements found in other insect genera (Calvi *et al.*, 1991; Atkinson *et al.*, 1993; Warren *et al.*, 1994).

The full-length *hobo* element is about 3 kb long, with some variability in the number of copies of a 9 bp repeat (the S repeat) in the center of the coding region (Calvi *et al.*, 1991; Bazin and Higuet, 1996). It has 12 bp inverted terminal repeats and a single long open reading frame encoding a transposase (Fig. 21.1). This structure places it in the Class II, Type I category of transposable elements (Finnegan, 1989). The presence of full-length elements is easily detected in Southern blots of genomic DNA by the presence of a 2.6 kb *Xho*I fragment that hybridizes strongly to a *hobo* probe. Southern blotting reveals population genetic similarity between *hobo* and the *P* element – the existence of strains of *D. melanogaster* that appear to lack full-length *hobo* elements. These strains, dubbed E, or empty (as opposed to H, or *hobo*-containing) strains, are functionally similar to M strains in the P-M system, in that a cross between E strain females and H strain males produces stigmata typical of hybrid dysgene-

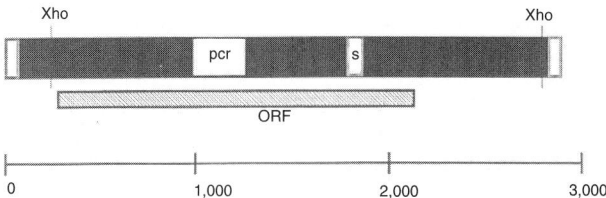

Fig. 21.1 Structure of a complete *hobo* element, based on the sequence of Hfl1 (Calvi *et al.*, 1991). The element is 2959 bp long and has 12 bp inverted terminal repeats at each end (white boxes, not to scale). The two *Xho*I sites delimit a 2.6 kb restriction fragment which is diagnostic for the presence of full-length elements in Southern blot hybridization analysis. ORF (hashed bar), putative single open reading frame; S, region in which a 9 bp sequence is repeated a variable number of times in different *hobo* elements; pcr, region of the element amplified by our primers for the present analysis.

sis. These strains, like M strains in the P-M system, are invariably ones that were collected prior to the 1960s (Boussy and Periquet, 1993; Periquet *et al.*, 1994). There are also putative E strains in *D. simulans*, although the dates of their collection and their functional equivalence to *D. melanogaster* E strains are uncertain.

More extensive Southern blot surveys (Daniels *et al.*, 1990; Boussy and Daniels 1991) have shown that full-length *hobo* elements are restricted to members of the *D. melanogaster* complex, a group of sibling species that includes *D. melanogaster*, *D. simulans*, *D. mauritiana* and *D. sechellia*. Older strains of *D. melanogaster* and *D. simulans*, as well as other species in the melanogaster subgroup, lack the 2.6 kb *Xho*I fragment but do hybridize to a *hobo* probe, producing bands that are larger than 2.6 kb and generally faint. A survey of the genus *Drosophila* found evidence of *hobo* hybridization in only the related montium subgroup – and again there is no evidence for the presence of a conserved full-length *hobo* element. *Hobo* appears to have one of the most restricted host ranges of all known *Drosophila* transposons.

The limited distribution of full-length elements within the genus, coupled with their absence from older strains of *D. melanogaster* and *D. simulans*, has led several authors to speculate that functional *hobo* elements had been horizontally transferred into and among the *D. melanogaster* species complex in the relatively recent past. The discovery of the exogenous source of the *P* element in the distantly related *D. willistoni* lends credence to the idea that *D. melanogaster* (or perhaps *D. simulans*) may have acquired *hobo* in a similar fashion and at about the same time, and then passed it to its sibling species via yet another horizontal transfer event, possibly by interspecific hybridization.

Simmons (1992) tested the hypothesis of horizontal transfer by sequencing the 2.6 kb *Xho*I fragment of cloned elements from *D. melanogaster*, *D. simulans* and *D. mauritiana*. She demonstrated that the full-length *hobo* elements in these species are essentially identical, differing by only about 0.1% of their nucleotides. Given that ordinary nuclear genes in these species typically differ at 5% or more of their nucleotides, the extremely low level of divergence supports the conclusion that the *hobo* elements in these flies have a much more recent common ancestor than do the species themselves. However, the DNA sequences reveal nothing about either the source of the active *hobo* element or the timing of its introduction into any of the species.

In an attempt to understand the horizontal transfer event in particular and the evolutionary history of *hobo* in the genus *Drosophila* in general, we have focused our attention on the identity of the 'ghost bands' seen in all Southern blots of *melanogaster* subgroup genomic DNAs – the high molecular weight *Xho*I fragments that hybridize faintly but distinctly to a *hobo* probe. We describe here the results of a two-pronged approach to identifying these sequences: a library screen of a *D. melanogaster* E strain for *hobo*-hybridizing clones and subsequent sequence analysis of one of them; and a PCR-based search of E strains in both *D. melanogaster* and *D. simulans* for sequences homologous to a conserved part of of the *hobo* coding region. Our results indicate the possibility that functional *hobo* elements are descended from 'fossil' elements dating to before the divergence of *D. melanogaster* and *D. simulans* but still found in the genomes of these species, and that the functional sequences had origin in *D. simulans*.

21.2 MATERIALS AND METHODS

We used the *D. melanogaster* E strain called Samarkand and the *D. simulans* E strain known as Peru 0251.5. The former was obtained from the Midwest Drosophila Stock Center (Bowling Green, Ohio) and the latter from the National Drosophila Species Resource Center (Bowling Green, Ohio), and both were maintained on standard *Drosophila* food and under standard laboratory conditions. Although both strains had previously been surveyed for the presence or absence of full-length *hobo* elements, we performed additional Southern hybridizations to confirm that these flies lacked any full-length *hobo* elements that might have arisen via contamination during culturing. Genomic DNA was extracted from live or frozen flies by standard methods. For the PCR experiments, single flies were used.

An *Eco*RI library of the Samarkand strain of *D. melanogaster* was constructed in the vector LambdaZap II according to the manufacturer's directions (Stratagene). The library was screened using a doubly gel-purified 2.5 kb *Ava*I-*Ava*II fragment of the *hobo* element in plasmid pRG2.6X (Blackman *et al.*, 1987). From plaques that hybridized to the probe, phages were purified and the embedded Bluescript KS(-) plasmids carrying the inserts were excised following manufacturer's directions (Stratagene). Plasmids were grown in *E. coli* XL1-Blue (Stratagene). After preliminary restriction mapping to identify sizes of *hobo*-hybridizing parts of inserts, attention focused on a single isolate, which was subcloned and sequenced with the aid of Drs Ling-wen Zheng and Marty Kreitman (University of Chicago). Sequencing of both strands was performed on an automated sequencer (ABI).

Single-fly extracts from the Samarkand and Peru strains were amplified by the polymerase chain reaction using primers AS9 (5' TTGGGCAT-CACTTTCCATTA 3') and S12 (5' AGTTCATTGGCCTCGTTAAA 3'), which were previously developed for sequencing *hobo* elements (Simmons 1992). In strains that contain complete *hobo* elements these primers amplify a 278 bp fragment that spans from position 1033 to 1310 of *hobo* element *Hfl1*, a full-length *hobo* element known to be functional (Calvi *et al.*, 1991). Following PCR amplification (which generally produced only a single band) the products were purified using Wizard columns (Promega) and

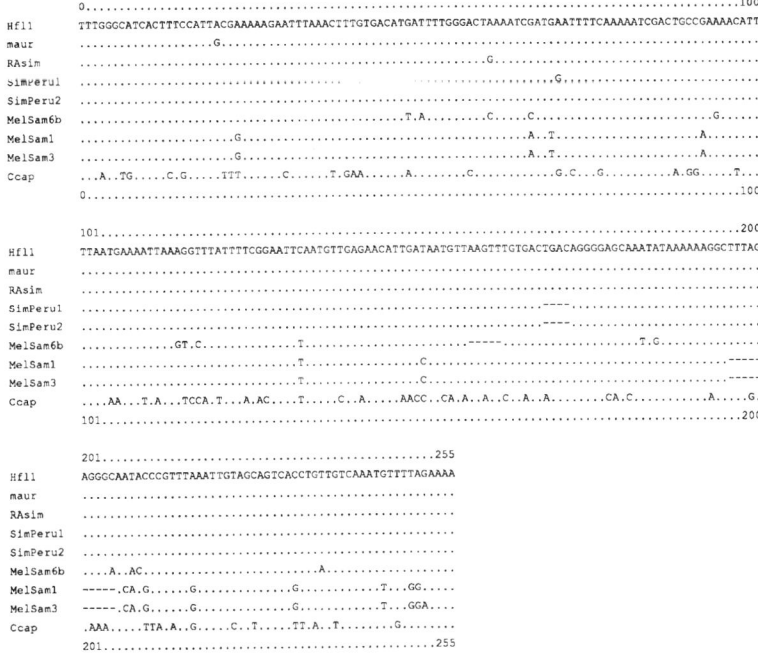

Fig. 21.2 Alignment of PCR amplification products from Samarkand *D. melanogaster* (MelSam1, MelSam3, MelSam6B) and from Peru *D. simulans* (SimPeru1, SimPeru2), along with the Nash sequence from the Samarkand *D. melanogaster* library screen and the corresponding region of *Hfl*1, a complete *hobo* element (Calvi *et al.*, 1991). Also included are putatively functional sequences of hobo elements from *D. simulans* (Rasim) and *D. mauritiana* (Maur) sequenced by Simmons (1992), and a *hobo*-like element from *C. capitata* (Handler and Gomez, 1996). Note the presence of small deletions in all the PCR-amplified sequences. These deletions were excluded from the alignment used in the cladistic analysis.

the fragments cloned into the pGEM-T vector (Promega). Several white colonies that produced plasmids with the appropriate restriction map were selected for sequencing. The cloned PCR products were sequenced manually using ^{32}P end-labeled primers and the fmol sequencing kit according to manufacturer's directions (Promega). Sequences were aligned using ClustalV and manually (Fig. 21.2), and trees were constructed by the method of maximum parsimony using PAUP 3.1.1 (Swofford).

21.3   RESULTS AND DISCUSSION

We constructed a library from a *D. melanogaster* E strain (Samarkand) and screened for clones that hybridized to a *hobo* probe, while also attempting to PCR-amplify from the same strain a segment of sequence that corresponds to the center of the coding region of an active *hobo* element (Fig. 21.1). This fragment corresponds to a conserved region of *hobo* transposases amino acid sequence found by Calvi *et al.* (1991) to be similar to that of other *hAT* family transposases. The PCR experiment was expected to favor amplification of portions of related elements that retained a full-length structure, since most internal deletion derivative elements would completely lack target sites for the primers. The library search was expected to identify any sequences with significant similarity to the canonical *hobo* sequence, regardless of length or structure.

One positive clone from the Samarkand library screen has been characterized and nearly completely sequenced. This sequence (known as *Nash* in memory of a less-than-dashing automobile of a bygone era) has substantial similarity to that of *Hfl1* but its structure is complicated. It has either lost its 5' end or contains a large insertion of unrelated material. It contains numerous deletions throughout the coding region as well as at the 5' end and insertions of sequences that align to other *Drosophila* transposons known as *Hoppel* (Kurenova *et al.*, 1990) and *1360* (Kholodilov *et al.*, 1988; Balakireva *et al.*, 1992). Across its length, *Nash* displays varying amounts of similarity to *Hfl1*, ranging from a high of 94% in the region that overlaps our PCR amplification target to a low of 64% in regions that overlap the 3' end of the *Hfl1* open reading frame. There seems to be no question that *Nash* is no longer functional; it appears to be a highly degenerate copy of *hobo* or a *hobo*-related element.

In the PCR experiments we isolated six clones from *D. melanogaster* Samarkand, which yielded three different sequences. All were easily aligned to *Hfl1* (Calvi *et al.*, 1991). In every case, the sequences have substantial similarity (> 94% identity) to *Hfl1*, but all sequences contain base pair substitutions and indel variations (typically small deletions) that create stop codons and/or destroy the reading frame. We have examined four clones from *D. simulans* Peru, yielding two different sequences. These sequences are extremely similar (> 98% identity) to those of the intact *D. simulans hobo* element *RAsim* (Simmons, 1992) and of *Hfl1* of *D. melanogaster* (as well as to each other) but contain a single indel that renders them probably non-functional.

Figure 21.2 shows an alignment of the PCR product sequences from *D. melanogaster* and *D. simulans* with the *Nash* sequence from the library screen as well as the comparable region of *Hfl1*. A *hobo*-like element from the Mediterranean fruit fly, *Ceratitis capitata* (Handler and Gomez, 1996), is included as an outgroup. We also include additional sequences of presumably functional *hobo* elements found in *D. simulans* and *D. mauritiana* (Simmons, 1992). The *Nash* sequence is missing a portion of the amplified region. The PCR products are all full length; nonetheless each has a characteristic small deletion relative to *Hfl1*. The portions of the sequences overlapping these small deletions were removed from the dataset in order to improve the resolution of the cladistic analysis.

We used this alignment to generate most parsimonious trees using PAUP 3.1.1. Because the *Nash* sequence lacks a substantial portion of the region amplified by PCR, we performed two analyses: one in which *Nash* was eliminated from the analysis, and one in which only those sequences overlapping those of *Nash* were used. The result of the analysis without *Nash* is shown in Fig. 21.3a. A heuristic search with random addition of sequences results in a single most parsimonious tree based on 19 informative characters. It has a consistency index of 0.88 and a retention index of 0.87. The analysis including *Nash* was not robust, due to the small number of base pairs involved; however, we were able to determine that *Nash* is most closely related to the MelSam6B sequence and have indicated its probable position on the tree in Fig. 21.3a.

We see two possible hypotheses consistent with the result of Simmons (1992) concerning the origin of the functional *hobo* elements seen in modern collections of *D. melanogaster* and *D. simulans*. In one model (similar to the *P* element case) the active element has been introduced recently into one or more of these species from an exogenous source. Under such a model, the defunct *hobo* fragments we amplify from the *D. melanogaster* and *D. simulans* E strains are relics of an ancient round of *hobo* invasion and extinction and are only distant cousins of the modern, functional elements. Alternatively, the modern element might be a direct descendent of an ancient element that has persisted in one species and has spread by horizontal transfer to the others in the relatively recent past. In this case the defunct *hobo* fragments we amplify are direct ancestors of the modern elements in one of the species.

These alternative hypotheses suggest different predictions for the phylogenetic relationship between the defunct elements found in the Samarkand E strain of *D. melanogaster* and the Peru E strain of *D. simulans* and the modern, functional elements represented by *Hfl1* (Fig. 21.3b). Under the 'exogenous source' hypothesis, E strain fragments (which have evolved vertically) should be more closely related to one another than to the functional element sequence, and both should be approximately equidistant from the functional sequence, which has evolved in an alternative host. Under the 'endogenous origin' hypothesis, functional elements have persisted in at least one population of one of the species. These elements should be more closely related to the defunct elements in the same species than to the defunct elements in the other species.

The sequences of PCR products from Samarkand and Peru, along with *Nash* sequences from Samarkand, support the 'endogenous origin' hypothesis. Two different sequences obtained from the Peru E strain of *D. simulans* differ from the functional *D. simulans* element by only 0.7% of their nucleotides and a small deletion. By contrast, the four sequences

(a)

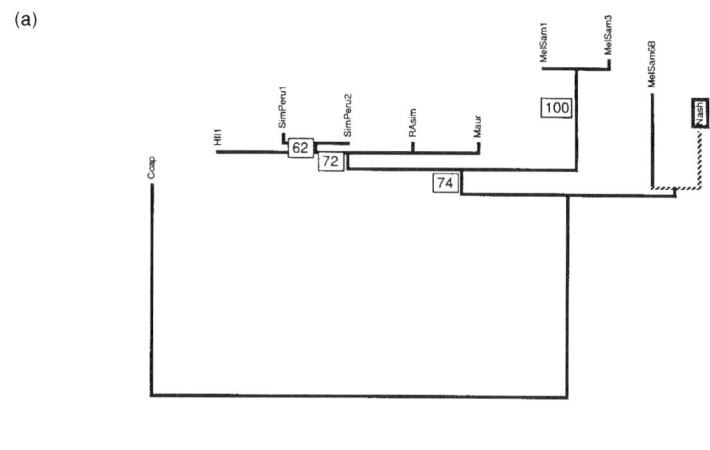

(b)

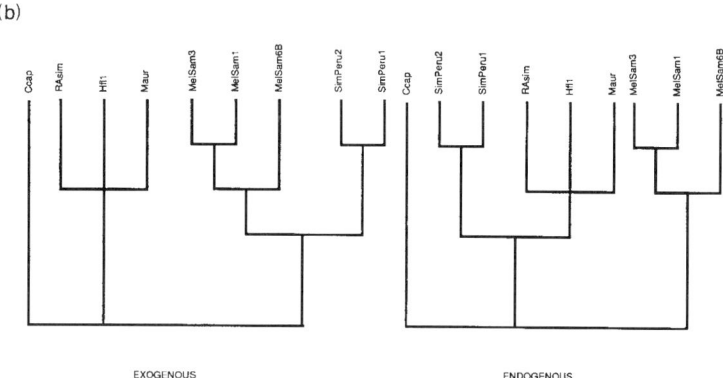

EXOGENOUS   ENDOGENOUS

Fig. 21.3 (a) The most-parsimonious tree produced by a heuristic search of the sequences shown in Fig. 21.2 using PAUP 3.1.1 (see text for details of the analysis). The analysis was performed without the *Nash* sequence; the position of *Nash* was determined in a separate analysis using a truncated dataset and is shown with hatched branches. Numbers on the branches are bootstrap values for 100 replicates. (b) Two hypotheses for the origin of the functional *hobo* element (see text for explanation). Note that the topology of the tree labeled 'endogenous' more closely matches that of the actual tree above, thus lending support to the hypothesis that the modern *hobo* element originated within one of these species, probably *D. simulans*.

obtained from the Samarkand *D. melanogaster* E strain (three by PCR, plus the *Nash* sequence) differ from the functional *D. melanogaster* element *Hfl1* by approximately 4–6% of their nucleotides across the amplified region, in addition to containing small deletions. This asymmetry, evident from a comparison of Figs 21.3a and 21.3b, supports the idea that the modern active element was retained by *D. simulans* and has spread by horizontal transfer into *D. melanogaster*, whose genome still contains relics of the earlier *hobo* invasion.

A cladistic analysis of several PCR product sequences from E strains of *D. melanogaster* and *D. simulans*, along with homologous regions of putatively functional elements from *D. melanogaster*, *D. simulans*, *D. mauritiana* and *C. capitata*, shows substantial congruence with trees based on non-mobile DNA (e.g. Jeffs *et al.*, 1994, for *Adh*), but only as long as the functional *hobo* sequences from genus *Drosophila* are not included in the analysis. Adding the functional sequences from *D. melanogaster*, *D. simulans* and *D. mauritiana* creates a polytomy with a cluster of sequences that includes all the active sequences, regardless of species of origin, and the sequences of the fragments from the *D. simulans* E strain. This also supports a recent origin for the active element in *D. simulans* or one of its sibling species.

The degree of divergence (4–6%) between the active *hobo* element and the non-functional PCR-derived and *Nash* sequences found in Samarkand is of a similar order of magnitude to the divergence between genes found in both *D. melanogaster* and *D. simulans*. These species are presumed to have diverged about 5 million years ago. Thus these sequences may have been evolving in *D. melanogaster* since approximately the time of the divergence of *D. melanogaster* and *D. simulans*. The inclusion of homologous sequences from other members of the *melanogaster* subgroup in the analysis (Simmons *et al.*, 1998) suggests, in fact, that these 'defunct' *hobo* sequences may have been present in the common ancestor of the subgroup, or possibly even earlier. The functional *hobo* element apparently represents the second time around for the *hobo* element family in these species.

ACKNOWLEDGMENTS

We thank Ling-Wen Zeng and Marty Kreitman of the University of Chicago for assistance with the cloning and sequencing of *Nash*. Portions of this work were supported by grants from the NSF and PSC-CUNY to GMS and by the NIH RCMI Center for the Study of the Cellular and Molecular Basis of Development at the City College of New York.

REFERENCES

Atkinson, P.W., Warren, W.D. and O'Brochta, D.A. (1993) The *hobo* transposable element of *Drosophila* can be cross-mobilized in houseflies and excises like the *Ac* element of maize. *Proc. Natl Acad. Sci. USA* **90**: 9693–9697.

Balakireva, M.D., Shevelyov, Y.Y., Nurminsky, D.I. *et al.* (1992) Structural organization and diversification of Y-linked sequences comprising *Su(Ste)* genes in *Drosophila melanogaster*. *Nucleic Acids Res.* **20**: 3731–3736.

Bazin, C. and Higuet, D. (1996) Lack of correlation between dysgenic traits in the *hobo* system of hybrid dysgenesis in *Drosophila melanogaster*. *Genet. Res.* **67**: 219–226.

Blackman, R.K., Grimaila, R., Macy, M. *et al.* (1987) Mobilization of *hobo* elements residing within the Decapentaplegic gene complex: suggestion of a new hybrid dysgenesis system in *Drosophila melanogaster*. *Cell* **49**: 497–505.

Boussy, I.A. and Daniels, S.B. (1991) *Hobo* transposable elements in *Drosophila melanogaster* and *D. simulans*. *Genet. Res.* **58**: 27–34.

Boussy, I.A. and Periquet, G. (1993) The transposable element *hobo* in *Drosophila melanogaster* and related species, in *Transposable Elements and Evolution*, (ed. J.F. McDonald), Kluwer Academic Publishers, Leiden, p. 347.

Calvi, B.R., Hong, T.J., Findley, S.D. and Gelbart, W.M. (1991) Evidence for a common evolutionary origin of inverted repeat transposons in *Drosophila* and plants: *hobo, Activator*, and *Tam3*. *Cell* **66**: 465–471.

Daniels, S.B., Chovnick, A. and Boussy, I.A. (1990) Distribution of *hobo* transposable elements in the genus *Drosophila*. *Mol. Biol. Evol.* **7**: 589–606.

Finnegan, D.J. (1989) Eukaryotic transposable elements and genome evolution. *Trends Genet.* **5**: 103–107.

Handler, A.M. and Gomez, S.P. (1996) The *hobo* transposable element excises and has related elements in tephritid species. *Genetics* **143**(3): 1339–1347.

Jeffs, P., Holmes, E.C. and Ashburner, M. (1994) The molecular evolution of the alcohol dehydrogenase and alcohol dehydrogenase-related genes in the *Drosophila melanogaster* species subgroup. *Molec. Biol. Evol.* **11**: 287–304.

Kholodilov, N.G., Bolshakov, V.N., Blinov, V.M. *et al.* (1988) Intercalary heterochromatin in *Drosophila*. Part III. Homology between DNA sequences from the Y chromosome, bases of polytene chromosome limbs, and chromosome *4* of *Drosophila melanogaster*. *Chromosoma* **97**: 247–253.

Kurenova, E.V., Leibovich, B.A., Bass, I.A. *et al.* (1990) Hoppel – the family of *Drosophila melanogaster* mobile elements flanked by short inverted repeats and localized preferentially within the heterochromatic genomic regions. *Genetika (Moscow)* **26**: 1701–1712.

McGinnis, W., Shermoen, A.W. and Beckendorf, S.K. (1983) A transposable element inserted just 5' to a *Drosophila* glue protein gene alters gene expression and chromatin structure. *Cell* **34**: 75–84.

Periquet, G., Lemeunier, F., Bigot, Y. *et al.* (1994) The evolutionary genetics of the *hobo* transposable element in the *Drosophila melanogaster* complex. *Genetica* **93**: 79–90.

Simmons, G.M. (1992) Horizontal transfer of *hobo* transposable elements within the *Drosophila melanogaster* species complex: evidence from DNA sequencing. *Mol. Biol. Evol.* **9**: 1050–1060.

Simmons, G.M., Tao, W., Plummer, D., Simon, A. and Wei, D. (1998) A phylogenetic hypothesis for the history of the *hobo* transposable element in the *melanogaster* subgroup of the genus *Drosophila*. *Mol. Biol. Evol.* (submitted).

Streck, R.D., MacGaffey, J.E. and Beckendorf, S.K. (1986) The structure of *hobo* transposable elements and their insertion sites. *EMBO J.* **5**: 3615–3623.

Warren, W.D., Atkinson, P.W. and O'Brochta, D.A. (1994) The *Hermes* transposable element from the house fly, *Musca domestica*, is a short inverted repeat-type element of the *hobo, Ac* and *Tam3* (*hAT*) element family. *Genet. Res.* **64**: 87–97.

The splicing of transposable elements: evolution of a nuclear defense against genomic invaders? 22

Michael D. Purugganan

SUMMARY

In the last decade, it has become apparent that many eukaryotic transposable elements possess the ability to act as introns and be spliced from pre-mRNA. The ability to act as introns has associated fitness consequences for both the transposable element as well as the host organism. Splicing may represent a mechanism to circumvent negative selection on the element that may arise from insertions into gene coding regions. Based on the details of the molecular features of transposable element splicing, it is possible that the intron-like features of many transposable element insertions may result from the coevolution of element-encoded splicing signals and the host cellular splicing machinery. The cellular splicing machinery may thus act as a general defense against foreign insertions into genes, including those elements that may be transposing at high frequencies as a result of recent horizontal transfer into new genomes.

22.1 INTRODUCTION

Nearly five decades ago, Barbara McClintock observed the genetic effects of a series of loci that changed their position in the maize genome (McClintock, 1950,1984). These mobile loci are now referred to as transposable elements, and over the last few years molecular geneticists have managed to isolate and characterize hundreds of these elements in both prokaryotic and eukaryotic systems (Berg and Howe, 1989; Wessler, 1989; Robertson and Lampe, 1995a, b). These mobile sequences are ubiquitous components of organismal genomes, and they have been shown to exist in almost every organism where their presence has been sought (Berg and Howe, 1989). The widespread distribution and mobility of these sequences have led to suggestions that transposable elements serve as agents of evolutionary change by increasing genetic variability within populations (Syvanen, 1984; McDonald, 1990, 1995). Several others have suggested, however, that transposable elements are essentially genomic parasites that confer no advantage to the host and may even be deleterious (Hickey, 1980; Doolittle and Sapienza, 1984; Orgel and Crick, 1984).

Transposable elements are not only persistent components of organismal genomes, but they also appear capable of moving between species via horizontal transfer events. *Drosophila melanogaster P* elements provide a classic example of the lateral movement between species of a transposable element (Anxolabéhère *et al.* 1988; Clark *et al.* 1994). *P* elements in the *melanogaster* genome, which were introduced in this species in the 1950s, appear to have originated from *Drosophila willistoni*. It is believed that this horizontal transfer of elements occurred through the agency of the mite *Proctolaelaps regalis* (Houck *et al.*, 1991). Phylogenetic analysis of several other transposable elements suggests that the evolution of several element families are characterized by horizontal transfer of these mobile sequences between species genomes (Flavell, 1992; Voytas *et al.*, 1992; Robertson, 1993; Robertson and Lampe, 1995a,b; Chapter 20, this volume). Among plants, the prolific ability of species to undergo interspecific hybridization presents yet another, albeit sexual, mechanism by which elements can invade the genomes of other taxa (Purugganan and Wessler, 1994).

What are the consequences of the movement of transposable elements between species genomes? The establishment of invading elements in new host genomes by horizontal transfer events may present a whole suite of challenges for both invader and new host. Unrestricted movement of transposable elements in new host genomes promotes the spread of these mobile loci. In interspecific hybrids between *Drosophila buzzatii* and *D. koepferae*, high rates of germline transposition of several elements appear to occur (Labrador and Fontdevila, 1994). Elevated rates of transposition could have severe consequences for the host organism, resulting in higher mutation rates, chromosomal aberrations and dysgenesis (Kidwell, 1985; Mackay *et al.*, 1992). The transposition of these mobile sequences has been shown to be a significant source of spontaneous insertion mutations in maize, yeast and *Drosophila*. In the maize *waxy* locus, five of 17 spontaneous mutant alleles are the result of transposable element insertions (Wessler and Varagona, 1985; Varagona *et al.*, 1992). Cytogenetic effects, such as chromosomal breakage (Weil and Wessler, 1993) and inversions, are also observed to accompany transposon movement. Finally, hybrid dysgenesis, with accompanying lost of fertility, provides another example of the effects of element activity on organismal viability (Kidwell, 1985). The dynamics of element spread in natural populations thus reflects the balance between the fitness of the transposable element and the possible negative impact the element may have towards the new host organism (Hickey, 1980; Charlesworth and Charlesworth, 1983; Charlesworth and Langley, 1989).

Transposable elements which have been resident in a genome over a protracted period of evolutionary time are likely to coevolve with their host genomes to prevent serious disruption of host gene activity that could lead to an excessive reduction in host fitness (Hickey, 1980). Charlesworth and co-workers have also suggested that transposable elements evolve mechanisms to inhibit high rates of transposition within host genomes. Indeed, under normal conditions in the wild the activity of many transposable elements is low or genetically suppressed (McDonald, 1990). There are numerous examples that illustrate a variety of mechanisms that control transposition behavior (Fedoroff, 1995; Lozovskaya *et al.*, 1995). The maize *Ac* element, for example, exhibits a negative dosage effect in which increasing copies of the element reduces the rate by which *Ac* transposes (McClintock, 1951). *P* elements introduced into *D. simulans* have also been shown to evolve towards weak P or M' types that transpose at reduced rates (Kimura and Kidwell, 1994). Finally, the *Tc1* element in *Caenorhabditis elegans* appear to be under host control in nematode somatic tissues (Emmons and Yesner, 1984).

These coevolved mechanisms, however, may not be operational initially when a new transposable element enters a host genome via a horizontal transfer event. Horizontal transfer events place transposable elements in new genomic environments where host controls on the activity of these specific invading elements are inefficient or absent (Kidwell, 1993). The maize *Ac* system again provides an illustrative example: when *Ac* is introduced into new plant species as a transgenic construct, it loses the negative dosage effect that partially controls element transposition rate in its nominal host, *Zea mays* (Scofield *et al.*, 1993). A new series of coevolutionary changes must then occur over time to control the number and activity of these invading transposable sequences if a new host organism is to remain viable after a horizontal transfer event. Until these new controls evolve, however, there remains a risk that the activities of the new invading elements are so severe as to cause widespread genetic damage to the host. In these instances, mechanisms to prevent transposable elements from damaging the genetic constitution of new hosts may operate as a short-term defense against the new genomic invaders. The ability of transposable elements to be spliced from pre-mRNA appears to provide one such mechanism that ameliorates some of the mutational impact of element insertions in eukaryotic genes (Purugganan and Wessler, 1992; Purugganan, 1993).

22.2   THE SPLICING OF TRANSPOSABLE ELEMENTS

Leaky phenotypes exhibited by transposable element insertion alleles provided early clues as to mechanisms that could mitigate the impact of element insertions into eukaryotic genes. In several model species, it has been known that not all element-induced mutant alleles result in null phenotypes (Wessler, 1989; Weil and Wessler, 1990). In the maize *waxy* locus, for example, the allele *wx-m9* displays residual levels of gene expression despite the insertion of a 4.37 kb *Ds* element in a *waxy* exon (Wessler *et al.*, 1987). In *Drosophila*, the *vermilion* allele v^k also displays appreciable levels of gene activity despite the presence of a large *412* element insertion into the *v* coding region (Fridell *et al.*, 1990).

The molecular basis for these leaky allele phenotypes was elucidated in 1987, when Wessler and her co-workers demonstrated that the partial restoration of gene activity in the *Ds* transposable element allele *wx-m9* was due to the splicing of the element insertion from the gene transcript (Wessler *et al.*, 1987). In this allele, the 4.37 kb *Ds* element insertion was located in exon 10 near the 3' end of the *waxy* coding region. The allele encodes a wild-type-sized protein and mRNA despite this large insertion in a translated exon. In *wx-m9*, it appears that the *Ds* element behaves as a nuclear intron, and the element sequence is processed out of pre-mRNA by the nuclear splicing machinery, resulting in the partial restoration of gene function. Three cryptic splice donor sites at the termini of the element are utilized by the splicing machinery, as well as one of two cryptic splice acceptor sites – one near the 3' site of the *Ds* element and downstream of the *Ds* insertion in *wx* exon 10 (Fig. 22.1). Some of these splicing combinations result in a mutant transcript in the correct reading frame, permitting the translation of a wild-type-sized protein with mutational alterations at the sequence surrounding the site of element insertion.

The splicing of transposable elements has turned out to be a fairly widespread phenomenon and is apparently an integral part of transposon biology (Table 22.1). In maize, transposable element splicing is not confined to alleles containing insertions of the *Ac/Ds* family. The *a2-m1* allele of the anthocyanin biosynthetic gene *A2*, for example, contains a *dSpm* transposon that is spliced from pre-mRNA (Menssen *et al.*, 1990). The *a2-m1* allele posseses a 2.2 kb *dSpm* insertion in the *A2* coding region, and displays low-level anthocyanin pigmentation in the kernel. The

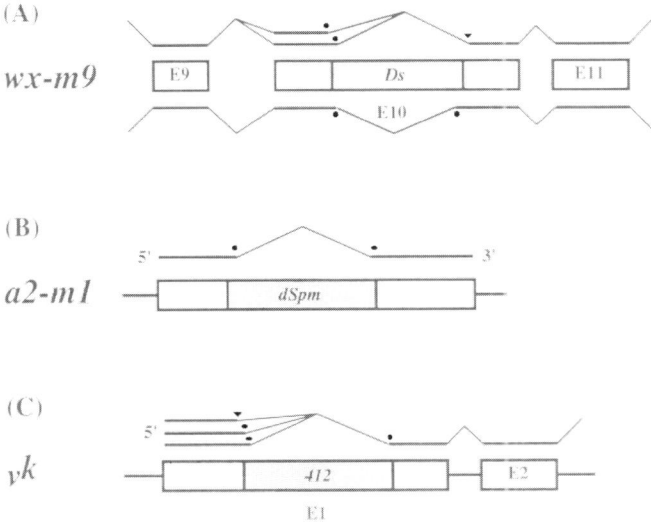

Fig. 22.1 Schematic of three examples of transposable element splicing. (A) The maize *Ds* insertion in *wx-m9*; (B) the maize *dSpm* insertion in *a2-ml*; (C) the *Drosophila 412* retrotransposon insertion in *v^k*. Boxes indicate exons; horizontal lines are introns. Shaded boxes represent transposable element insertions. Splicing events are indicated above and below by horizontal lines for exons and slanted lines for spliced introns or element sequences. Dots indicate splice signals within the element; arrows denote splice sites in host sequences used to process transposable element sequences.

dSpm element is spliced in this allele using splice sites within the element itself (Fig. 22.1). A derivative of *a2-m1*, the allele *a2-m1 (Class II)*, possesses an inserted *dSpm* element with a 900 bp deletion; this appears to increase splicing efficiency and the allele displays even higher levels of pigmentation. This allele also shows decreased responsiveness to transpose in the presence of the autonomous *Spm* element.

   Numerous cases of transposable element splicing have since been identified not only in maize but in other species as well. In *Drosophila melanogaster*, the hypomorphic *v^k* alelle of the *vermilion* gene contains a 7.5 kb *412* retrotransposon insertion in the untranslated leader region of exon 1 (Fridell *et al.*, 1990). As in *wx-m9*, the *v^k* allele conditions an intermediate phenotype and transcribes wild-type-sized mRNAs. Analysis of *v^k* shows that the *412* retrotransposon sequences are processed from the

Table 22.1 Some transposable elements that are spliced from pre-mRNA

Species	Transposable elements	Host genes
Zea mays	*Ds* family, *dSpm*	*waxy, adh, bronze, a2, shrunken*
Drosophila melanogaster	*P, 412*	*vermilion, yellow*
Caenorhabditis elegans	*Tc1*	*unc-54, hlh-1, mlc-2*

vermilion pre-mRNA through splice sites located within the element's LTR (Fig. 22.1). Three donor sites are clustered at the 5' end of the element sequence; an additional chimeric donor site containing both *vermilion* and *412* sequence is also utilized. These sites are then joined to a cryptic acceptor located just upstream of the element's 3' termini. Interestingly, an intragenic revertant of v^k, the allele v^{+37}, was found to contain an insertion of the *B104/roo* element within the original *412* insertion; the combined insertion now totals 11.1 kb in *vermilion* exon 1 (Pret and Searles, 1991). The *roo* element, located at the *412* 5' terminus, appears to provide better splice donor sites that increase the efficiency of element splicing.

Transposable element insertion alleles that display element splicing are identified in genetic screens as hypomorphic alleles with intermediate mutant phenotypes. This biases the number of insertion alleles identified – those alleles that are spliced efficiently and confer no mutant phenotype escape detection in these genetic screens. A relatively unbiased sampling of *Tc1* insertions in the *C. elegans hlh-1*, *mlc-2* and *unc-54* loci by molecular (as opposed to genetic) screening suggests that the fraction of element insertions that are spliced but give no phenotype may be high (Rushforth and Anderson, 1996). Although null alleles of these genes have clear mutant phenotypes, eight independent molecularly identified insertion alleles of these loci display a wild-type phenotype. The *Tc1* insertions in all these alleles are spliced from pre-mRNA, resulting in the production of in-frame messages. In many cases, the splicing machinery recognized cryptic splice sites in both the element and the surrounding exon.

These are just a few examples of element splicing patterns that have thus far been characterized (Kim *et al.*, 1987; Dennis *et al.*, 1988; Geyer *et al.*, 1991). Splicing of transposable elements in all these cases relies on cryptic splice sites encoded either within the element or in the host gene's exon (Purugganan and Wessler, 1992). These sites are either at the element termini or, if located in the host gene, close to the insertion junction; utilizing these splice sites thus leads to the removal of most of the inserted element sequences. The splicing is imprecise, however, since these splice sites are not usually found at the exact insertion junctions.

Splicing can delete host gene sequence, or, more often, lead to the addition of nucleotide sequences at the insertion junction in the mature transcript, resulting in translational frameshifts or altered amino acid sequences. Expressed proteins from these genes, although mutant, may still be capable of activity. Thus, spliced insertion alleles can lead to leaky or even wild-type phenotypes (Wessler, 1989; Weil and Wessler, 1990), rather than the null phenotypes commonly exhibited by transposon insertion alleles that do not exhibit splicing.

22.3   THE EVOLUTION OF TRANSPOSABLE ELEMENT SPLICING: MECHANISMS AND CONSEQUENCES

If transposable elements are viewed as genomic parasites, then splicing appears to represent a host nuclear defense against element insertions. Moreover, the ability to be spliced provides unique advantages to the new, invading mobile elements as they move throughout the genome (Kidwell, 1993; Purugganan, 1993). Splicing circumvents some of the mutational effects associated with element insertions. The data on *Ds* and *dSpm* element insertions in maize, *P* and *412* insertions in *Drosophila* and *Tc1* insertions in *C. elegans*, demonstrate that splicing of elements may result in partial to full restoration of gene expression (for a review, see Purugganan and Wessler, 1992). By permitting even limited levels of gene expression, splicing can reduce the negative fitness consequences of transposable element activity within eukaryotic genomes (Gierl, 1990; Purugganan, 1993). Transposable elements that evolve the ability to be spliced would thus be able to escape negative selection in spite of their insertional effects into host genes.

There are several reasons to believe that cryptic splice signals within the termini of transposable elements may have evolved to allow for the removal of the element from exons (Purugganan and Wessler, 1992). First, most element-encoded terminal splice sites are positioned fairly close to element ends, thus ensuring that most of the element sequences are removed by splicing. Second, the ability to be spliced can occur among differing members of specific transposable element families. The maize *Ds* elements, for example, have diverse internal sequences but all share conserved terminal ends that are required for element transposition in the presence of an autonomous *Ac* element. It also appears that the splice sites that mediate the removal of different *Ds* elements from pre-mRNA are conserved between these disparate elements, suggesting that the presence of signals for splicing are under positive selection (Wessler, 1991).

It has also been suggested that the cellular splicing machinery itself may have evolved in part to deal with transposable elements that insert themselves into eukaryotic genes. Even before the demonstration that transposable elements could be spliced, Crick (1979) speculated that splicing evolved as a defense against transposons within genomes. If this is indeed the case, then the mechanism by which this is accomplished appears to involve the ability of the eukaryotic splicing apparatus to differentiate between exon and intron sequences in a process referred to as 'exon definition' (Robberson *et al.*, 1992; Berget, 1995).

Recent results suggest that during nuclear splicing, cellular factors initially search for and recognize a pair of splice sites across an exon in the pre-mRNA (Niwa *et al.*, 1992; Berget, 1995). When such a pair is found, the binding of U1 and U2 snRNPs and other associated splicing factors occurs. These associated factors include the 3' splice site recognition factor U2AF, and the 5' splice site recognition factor ASF/SF2; together, these then define the exon as a unit in the pre-mRNA. Neighboring defined exons are then brought together and ligated to form the mature processed transcript.

Exon definition is facilitated by a group of serine/arginine-rich SR proteins that bind to pre-mRNA exons and facilitate the assembly of the spliceosome (Fu, 1995; Adams *et al.*, 1996; Reed, 1996). These SR proteins bind to exonic splicing enhancers, which contain purine-rich enhancer elements that promote the splicing of adjacent 5' and 3' splice sites (Kohtz *et al.*, 1994; Staknis and Reed, 1994) (Fig. 22.2A). SR proteins bound to exons also promote the interaction between U2 snRNP and the intron branchpoint sequence. In general, the SR proteins appear to regulate the initial steps in the recruitment of snRNPs in the formation of functional spliceosomes (Adams *et al.*, 1996; Reed, 1996).

Based on studies of transposable element splicing, it may be that the mechanism of exon definition is involved in the ability of the splicing machinery to recognize transposable element insertions as 'non-exon' sequences. One scenario could be that in alleles with transposable element insertions, the SR proteins initially recognize constitutive exon sequences by binding to exonic enhancers. The presence of non-exon sequences (the transposable element) is then detected by the splicing machinery, possibly by the lack of SR binding to the insertion sequence (see Fig. 22.2b). The splicing machinery then positions splicing factors to recognize cryptic splice sites at the boundary of the exon and transposable element insertion; these include element-encoded splice sites as well as cryptic sites in exon sequences. The assembled spliceosomes then proceed to process the element insertion from the pre-mRNA as an intron.

There are several lines of evidence that suggest that the splicing of transposable elements may be a generalized mechanism of the cell to remove insertions in pre-mRNA. First, the ability to be spliced appears to be a widespread trait, occurring in various species and types of transposons (both IR transposons and retrotransposons) (Purugganan and

(A)

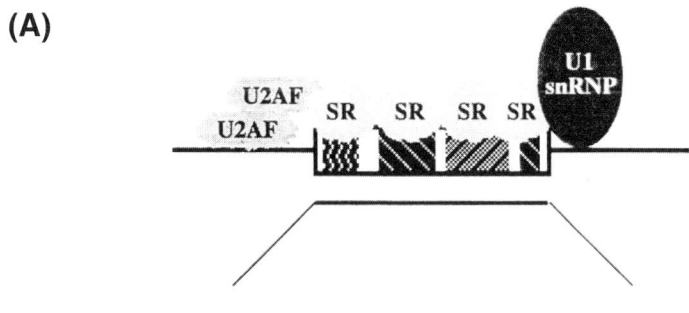

(B)

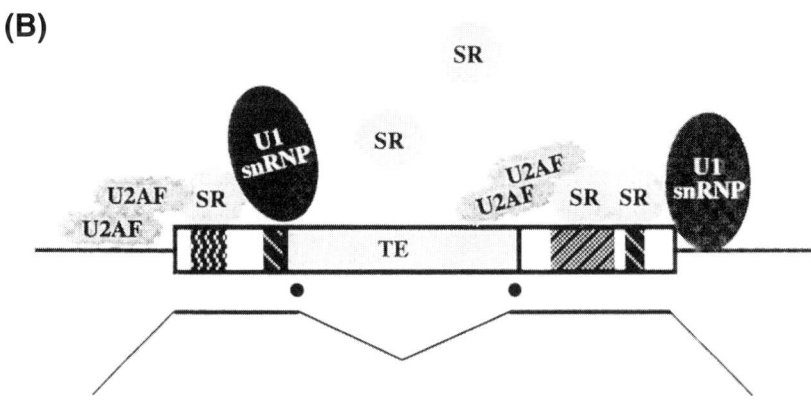

Fig. 22.2 Possible mechanism for transposable element splicing. (A) Early steps in normal spliceosome assembly. Boxes represent exons; patterned areas denote exonic splicing enhancers. The binding of SR proteins to these enhancers as well as their interactions with U1 snRNP and U2AF factors define the exon. (B) Hypothetical mechanism for element splicing. The lack of SR binding to the inserted transposable element sequence forces weak binding of U1 snRNP and U2AF factors at the junctions of the exon and element insertion. The weak binding is facilitated by strong SR protein binding to exonic enhancers not disrupted by the transposon. Positioning of the splicing factors may also be facilitated by a mechanism to recognize the presence of non-exon sequences.

Wessler, 1992; Purugganan, 1993). The elements themselves are of diverse origin, and the terminal sequences harboring the splice sites are unrelated to each other. Thus, the splice sites within these elements evolved independently several times in diverse transposable element evolutionary lineages, and it appears likely that these splice sites were responding to the general ability of the splicing machinery to remove these element insertions. Second, the use of cryptic sites not only in the element but also in adjacent host sequences lends support to the notion that the splicing machinery can recognize the presence of transposable element insertions. The use of these host exon-encoded cryptic splice sites suggests that transposable element splicing does not rely solely on element encoded splice signals (Purugganan and Wessler, 1992). Rather, the splicing machinery itself may distinguish between exon and non-exon (transposable element) sequences in the pre-mRNA, and utilize appropriate cryptic splice signals near the insertion junction to remove element sequences.

Selection may thus operate on both the transposable elements and the eukaryotic splicing machinery – the splicing of transposable elements can then be viewed as a coevolved cellular system as a genomic defense against these mobile sequences. The splicing machinery develops the ability to splice element insertions, while the element itself facilitates the process by evolving more intron-like features, such as terminal splice signals. The organism can minimize the deleterious effects of transposable element insertions and thus increase its fitness.

From the standpoint of the transposable element, there are distinct advantages in remaining camouflaged within the genome as an intron. By mitigating the negative selection pressure that accompanies deleterious insertion into host genes, the transposable element can increase its probability of survival within a genome. For the continued persistence of the transposon, however, the element must retain the ability to spread within populations via transposition. Although transposons can act as introns, they cannot evolve to be simply introns; the result would be insertions that are spliced but cannot move. Indeed, mutations within element insertions that decrease transposition rates while increasing splicing efficiency have been observed (Mennsen et al., 1990). These element insertions effectively become introns, and their ability to transpose and subsequently replicate is lost. For splicing to be effective as a strategy for increasing element fitness, it must function in concert with the continued ability of elements to move within genomes.

It is clear that there exist alternative strategies for transposable elements to escape negative selection, including the modulation of transposition activity by both host-encoded and transposon-encoded mechanisms (Emmons and Yesner, 1984; Kimura and Kidwell, 1994; Fedoroff, 1995). The importance of splicing as a strategy towards enhancing fitness, however, has yet to be explored in a systematic fashion. It would be interesting to determine the conditions in which the ability to be spliced has real selective advantages. Empirical and theoretical studies to address this question can be undertaken to assess whether transposable elements that possess the ability to be spliced can compete effectively against elements that cannot. Together, these approaches may begin to address the question of whether splicing truly represents a cellular defense against transposable element invasions, and the subsequent mutagenic insertions they create.

ACKNOWLEDGMENTS

The author wishes to thank the anonymous reviewers for critical comments that improved the manuscript. The author would also like to express his gratitude to Sue Wessler and John McDonald for numerous interesting discussions on the subject of transposable element splicing. This work was supported by the North Carolina Agricultural Service and an Alfred P. Sloan Young Investigator Award.

REFERENCES

Adams, M.D., Rudner, D.Z. and Rio, D.C. (1996. Biochemistry and regulation of pre-mRNA splicing. *Curr. Op. Cell Biol.* **8**: 331–339.

Anxolabéhère, D., Kidwell, M.G. and Periquet, G. (1988) Molecular characterization of diverse populations are consistent with the hypothesis of a recent invasion of *D. melanogaster* by mobile *P* elements. *Mol. Biol. Evol.* **5**: 252–269.

Berg, D. and Howe, M. (1989) *Mobile DNA*, American Society for Microbiology, Washington, DC.

Berget, S.M. (1995) Exon recognition in vertebrate splicing. *J. Biol. Chem.* **270**: 2411–2414.

Bingham, P.M. and Zachar, Z. (1989) in *Mobile DNA* (eds D. Berg and M. Howe), American Society for Microbiology, Washington, DC.

Charlesworth, B. and Charlesworth, D. (1983) The population dynamics of transposable elements. *Genet. Res.* **42**: 1–27.

Charlesworth, B. and Langley, C.H. (1989) The population genetics of *Drosophila* transposable elements. *Ann. Rev. Genet.* **23**: 251–287.

Clark, J.B., Maddison, W.P. and Kidwell, M.G. (1994) Phylogenetic analysis supports horizontal transfer of *P* transposable elements. *Mol. Biol. Evol.* **11**: 40–50.

Crick, F. (1979) Split genes and RNA splicing. *Science* **204**: 264–271.

Dennis, E., Sachs, M., Gehrlach, W. *et al.* (1988) The *Ds1* transposable element acts as an intron in the mutant allele *adh1-Fm335* and is spliced from the message. *Nucl. Acids Res.* **16**: 3315–3328.

Doolittle, W.F. and Sapienza, C. (1984) Selfish genes: The phenotype paradigm and genome evolution. *Nature* **284**: 601–603.

Emmons, S. and Yesner, L. (1984) High frequency excision of transposable element *Tc1* in the nematode *C. elegans*. *Cell* **36**: 599–605.

Fedoroff, N. (1995) Maize transposable element regulation. *Maydica* **40**: 7–12.

Flavell, A.J. (1992) *Ty1-copia* group retrotransposons and the evolution of retroelements in the eukaryotes. *Genetica* **86**: 203–214.

Fridell, R., Pret, A. and Searles, L. (1990) A retrotransposon *412* insertion within an exon of the *D. melanogaster vermilion* gene is spliced from precursor RNA. *Genes & Dev.* **4**: 559–566.

Fu, X.D. (1995) The superfamily of arginine/serine-rich splicing factors. *RNA* **1**: 663–680.

Geyer, P., Chien, A., Corces V. and Green, M. (1991) Mutations in the *su(s)* gene affects RNA processing in *Drosophila melanogaster*. *Proc. Natl Acad. Sci. USA* **88**: 7116–7120.

Gierl, A. (1990) How maize transposable elements escape negative selection. *Trends Genet.* **6**: 155–158.

Hickey, D. (1980) Selfish DNA, a sexually-transmitted nuclear parasite. *Genetics* **101**: 519-531.

Houck, M.A., Clark, J.B., Peterson, K.R. and Kidwell, M.G. (1991) Possible horizontal transfer of *Drosophila* genes by the mite *Proctolaelaps regalis*. *Science* **253**: 1125-1129.

Kidwell, M.G. (1985) Hybrid dysgenesis in *D. melanogaster*: nature and inheritance of *P* element regulation. *Genetics* **111**: 337–350.

Kidwell, M.G. (1993) Lateral transfer in natural populations of eukaryotes. *Ann. Rev. Genet.* **27**: 235–256.

Kim, H., Schiefelbein, J., Raboy, V. *et al.* (1987) RNA splicing permits expression of a maize gene with a defective *Spm* transposable element insertion in an exon. *Proc. Natl Acad. Sci. USA* **84**: 5863–5867.

Kimura, K. and Kidwell, M.G. (1994) Differences in *P* element population dynamics between the sibling species *D. melanogaster* and *D. simulans*. *Genet. Res.* **63**: 27–38.

Kohtz, J.D., Jamison, S.F., Will, C.L. *et al.* (1994) Protein–protein interaction and 5' splice site recognition in mammalian mRNA precursors. *Nature* **368**: 119–124.

Labrador, M. and Fontdevila, A. (1994) High transposition rates of *Osvaldo*, a new *Drosophila buzzatii* retrotransposon. *Mol. Gen. Genet.* **245**: 661- 674.

Lozovskaya, E., Hartl, D.L. and Petrov, D.A. (1995) Genomic regulation of transposable elements in *Drosophila*. *Curr. Op. Genet. Dev.* **5**: 768–773.

Mackay, T.F.C., Lyman, R.F. and Jackson, M.S. (1992) Effects of *P* element insertions on quantitative traits in *Drosophila melanogaster*. *Genetics* **130**: 315–332.

McClintock, B. (1950) The origin and behaviour of mutable loci in maize. *Proc. Natl Acad. Sci. USA.* **36**: 344–355.

McClintock, B. (1951) Chromosome organization and genic expression. *Cold Spr. Harb. Symp. Quant. Biol.* **16**: 13–47.

McClintock, B. (1984) The significance of responses of the genome to challenges. *Science* **226**: 792–801.

McDonald, J.F. (1990) Macroevolution and retroviral evolution. *BioScience* **40**: 183–191.

McDonald, J.F. (1995) Transposable elements – possible catalysts of organismic evolution. *Trends Ecol. Evol.* **10**: 123–126.

Menssen, A., Hohmann, W.M., Schnable, P *et al.* (1990) The *En/Spm* transposable element of *Z. mays* contains splice sites at the termini generating a novel intron from a *dSpm* element in the *A2* gene. *EMBO J.* **9**: 3051–3057.

Niwa, M., MacDonald, C. and Berget, S. (1992) Are vertebrate exons scanned during splice site selection? *Nature* **360**: 277–280.

Orgel, L.E. and Crick, F. (1984) Selfish DNA: the ultimate parasite. *Nature* **284**: 604–607.

Pret, A. and Searles, L. (1991) Splicing of retrotransposon insertion from transcripts of the *Drosophila melanogaster vermilion* gene in a revertant. *Genetics* **129**: 1137–1145.

Purugganan, M.D. (1993) Transposable elements as introns: evolutionary connections. *Trends Ecol. Evol.* **8**: 239–243.

Purugganan, M.D. and Wessler, S.R. (1992) The splicing of transposable elements and its role in intron evolution. *Genetica* **86**: 295–303.

Purugganan, M.D. and Wessler, S.R. (1994) Molecular evolution of *magellan*, a maize *Ty3/gypsy*-like retrotransposon. *Proc. Natl Acad. Sci. USA* **91**: 11674–11678.

Reed, R. (1996) Initial splice-site recognition and pairing during pre-mRNA splicing. *Curr. Op. Genet. Dev.* **6**: 215 -220.

Robberson, B.L., Cote, G.J. and Berget, S.M. (1992) Are vertebrate exons scanned during splice site selection? *Nature* **360**: 277–280.

Robertson, H.M. (1993) The *mariner* transposable element is widespread in insects. *Nature* **362**: 241–245.

Robertson, H.M. and Lampe, D.J. (1995) Recent horizontal transfer of a *mariner* transposable element among and between Diptera and Neuroptera. *Mol. Biol. Evol.* **12**: 850–862.

Robertson, H.M. and Lampe, D.J. (1995) Distribution of transposable elements in arthropods. *Ann. Rev. Entomol.* **40**: 333–357.

Rushforth, A.M. and Anderson, P. (1996) Splicing removes the *C. elegans* transposon *Tc1* from most mutant pre-mRNAs. *Mol. Cell. Biol.* **16**: 442–429.

Scofield, S.R., English, J.J. and Jones, J.D. (1993) High level expression of the *Ac* transposase gene inhibits the excision of *Ds* in tobacco cotyledons. *Cell* **75**: 507–517.

Staknis, D. and Reed, R. (1994) SR proteins promote the first specific recognition of pre-mRNA and are present together with the U1 small nuclear ribonucleoprotein particle in a general splicing enhancer complex. *Mol. Cell. Biol.* **14**: 7670–7682.

Syvanen, M. (1984) The evolutionary implications of mobile genetic elements. *Ann. Rev. Genet.* **18**: 271–293.

Varagona, M.J., Purugganan, M.D. and Wessler, S.R. (1992) Alternative splicing induced by insertions of retrotransposons into the maize *waxy* gene. *Plant Cell* **4**: 811–820.

Voytas, D.F., Cummings, M.P., Konieczny, A. *et al.* (1992) *Copia*-like retrotransposons are ubiquitous among plants. *Proc. Natl Acad. Sci. USA.* **86**: 7124–7128.

Weil, C.F. and Wessler, S.R. (1990) The effects of plant transposable element insertions on transcription initiation and RNA processing. *Ann. Rev. Plant. Physiol. Plant Mol. Biol.* **41**: 527–552.

Weil, C.F. and Wessler, S.R. (1993) Molecular evidence that chromosome breakage by *Ds* elements is caused by aberrant transposition. *Plant Cell* **5**: 515–522.

Wessler, S.R. (1989) Phenotypic diversity mediated by the maize transposable elements *Ac* and *Spm*. *Science* **242**: 399–405.

Wessler, S. (1991) The maize transposable *Ds1* element is alternatively spliced from exon sequences. *Mol. Cell. Biol.* **11**: 6192–6196.

Wessler, S.R., Baran, G. and Varagona, M.J. (1987) The maize transposable element *Ds* is spliced from RNA. *Science* **237**: 916–918.

Wessler, S.R. and Varagona, M.J. (1985) Molecular basis of mutations at the *waxy* locus of maize: Correlation with the fine-structure genetic map. *Proc. Natl Acad. Sci. USA.* **82**: 4177–4181.

PART FOUR

Evidence of Ancient Transfer

The case for gene transfers between very distantly related organisms

23

Russell F. Doolittle

SUMMARY

There have been many claims of individual gene transfers between prokaryotes and eukaryotes, but most of them have proved to be premature. Nonetheless, a few apparently authentic cases remain. This chapter considers the most common reasons for mistaken claims and the kinds of supporting data needed to make a valid case and, finally, discusses a few enzymes that really do seem to have been transferred between eukaryotes and prokaryotes.

23.1 INTRODUCTION

My very distant relative, Ford Doolittle, once remarked to me: 'Horizontal gene transfers are the last refuge of frustrated sequence aligners.' He may have been correct, but that is not to say that such phenomena never occur. What often happens is that the sequences of some proteins from a variety of organisms are aligned and a phylogenetic tree constructed. If the branching order is not consistent with the expected relationship, several precautionary checks are in order before one thinks about the possibility of horizontal transfer. First, the data must be robust enough that the same phylogenies result from a variety of tree-construction approaches. Serious claims cannot rest upon the small differences that may result from the use of one method or another. Second, it should be ascertained that a variety of other proteins do indeed give the expected branching order when their sequences are compared from the same organisms. Finally, great effort must be made to show that the comparisons do not involve proteins from paralogous genes, the results of gene duplications that occurred before the divergence of the organisms that appear out of order. This is the most difficult of the preliminary tests, because often the authentic orthologous protein has not yet been reported. Sequence comparers must constantly remind themselves that they are always dealing with an incomplete deck. The recent publication of complete genomes for two (small) eubacteria and an archaebacterium have eased the problem in that inspection will sometimes reveal whether paralogous genes exist. Even when all the extant sequences in the world are reported, however, the problem of the occasional gene conversion or displacement will remain. In the meantime, we must do the best we can with what we have.

There is another caveat. The computer analyst is always at the mercy of the experimentalist (the person who does the real work). Blind faith in the accuracy of every sequence and the organism listed as its source is not always warranted. It is not all that unusual for use of the polymerase chain reaction (PCR) to result in the cloning of a bacterial contaminant from a eukaryote preparation. Skepticism must be the order of the day.

On several occasions in recent years my colleagues and I have written about horizontal gene transfers between prokaryotes and eukaryotes (Doolitle et al., 1990; Smith and Doolittle, 1992a, b; Smith et al., 1992; Little et al., 1994). In this brief chapter I will review and update some of those commentaries. In at least one case, additional data have shown that our conclusions were premature. In another, we had unjustifiably questioned whether a claimed horizontal transfer was authentic. In the interim, our challenge has been correctly rebutted, and I would here like to make a graceful turnabout. In yet another case, our original notion still seems correct, in spite of a misleading sequence report; I will try to bolster its status in the light of new knowledge. In addition to there being many new sequences available, we now have a useful standard for how similar we should expect certain amino sequences to be.

23.1.1   OPPORTUNITIES FOR GENE TRANSFER

Although aberrant sequence-based phylogenies are the usual prod for considering the possibility of horizontal gene transfer, the justifications usually include a set of secondary conditions. Chief among these is the notion of opportunity. Thus, the possibility for gene transfer is often given a wider berth whenever parasitism, symbiosis or endosymbiosis is involved. Sometimes these factors actually take the lead. In the case of the Cu-Zn superoxide dismutase, horizontal gene transfer was considered for the bacterium *Photobacterium leignathii* and its ponyfish host before the sequences of the symbiont and host were even known (Martin and Fridovich, 1981). Although the sequence evidence has proved decisively to the contrary (Steffens *et al.*, 1983), the fascination associated with the idea has made it difficult to convince some that the situation is ordinary (Bannister and Parker, 1985). But opportunity **is** a factor. For example, one of the most difficult cases to decide about horizontal gene transfer has to do with a glutamine synthetase and involves root nodule bacteria and their plant hosts (more below).

23.1.2   ORGANELLAR IMPORTS

Although this chapter is not mainly concerned with organellar imports – those genes for chloroplast and mitochondrial proteins that were brought in by the prokaryotic endosymbiont and which were subsequently transferred to the eukaryotic host – we must acknowledge their existence and briefly examine the presumed mechanics. Certainly the co-mingling of genomes within the same cell presented ample opportunity for transfer. Many times both the host and the endosymbiont had genes for a particular enzyme, and many times both genes were retained in the nucleus of the host. In other cases, one or the other gene was lost, and there are instances known where the same nuclear gene provides both the cytoplasmic and organellar enzyme. In some other instances, duplication of one or the other gene has occurred after the endosymbiotic acquisition, and a new divergence between the cellular and organellar versions has begun. All of these matters are mentioned primarily as illustrations of the kinds of situation that can unfold for any horizontal gene transfer or exchange.

23.1.3   PROKARYOTES AND EUKARYOTES

It is well known that horizontal gene transfers occur between bacteria. Indeed, the observation of transfer of virulence factors between strains of *Pneumococcus* was a turning point in biology. But the transfer of very similar genes from closely related organisms is obviously difficult to detect at the sequence level. At the same time, as the sequences become significantly different through normal divergence, the less likely it will be that opportunities for gene transfer will exist (Fig. 23.1). We expect the proteins of prokaryotes and eukaryotes to be quite different, but, apart from endosymbiotic or symbiotic situations, the opportunities for transfer seem slight.

23.1.4   HOW MUCH CHANGE?

Potential horizontal gene transfers often seem apparent on the basis of sequence comparison alone, before a phylogenetic tree is even constructed. A particular prokaryote protein may just **look** more eukaryote-like than

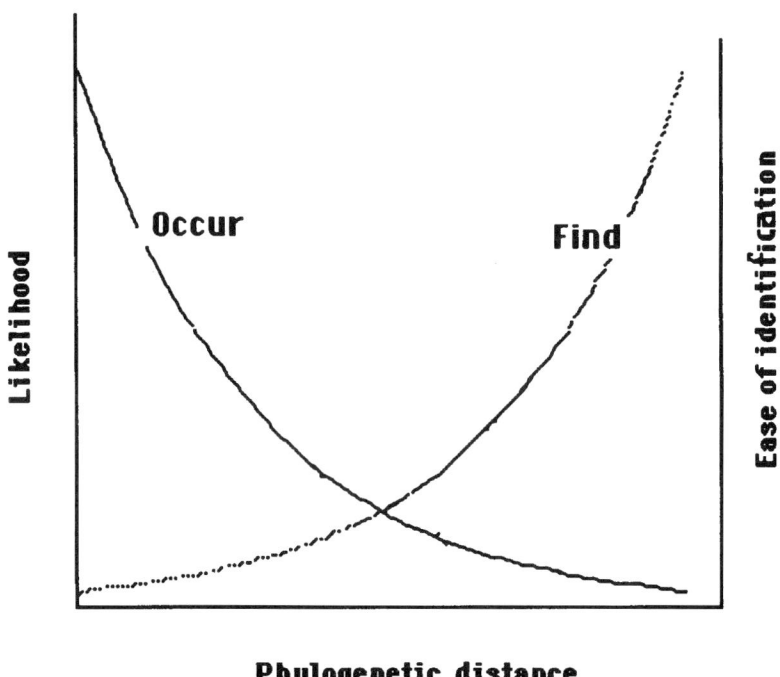

Fig. 23.1 Likelihood of horizontal gene transfers as a function of evolutionary difference between the donor and recipient organisms (solid line). Ease of detectability as realized by sequence comparison (dashed line).

expected. How similar or dissimilar **are** the proteins of prokaryotes and eukaryotes? Recently, we conducted a large-scale comparison of 57 different enzymes from many different groups of organisms (Doolittle *et al.*, 1996). On the average, the enzymes of prokaryotes and eukaryotes are 37% identical, the entire range encompassing 20–57% (Fig. 23.2). These results can serve as a compass point for considerations of possible gene transfer between prokaryotes and eukaryotes. As a rule of thumb, if an enzyme (or other protein) sequence is found to be more than 60% identical between the two groups, then horizontal gene tranfer looms as a distinct possibility.

23.2   SOME RECONSIDERATIONS

23.2.1   GAP DEHYDROGENASES

A brief review of the enzyme glyceraldehyde 3-phosphate dehydrogenase (gapdh) may serve as a lesson for the traps and pitfalls that can

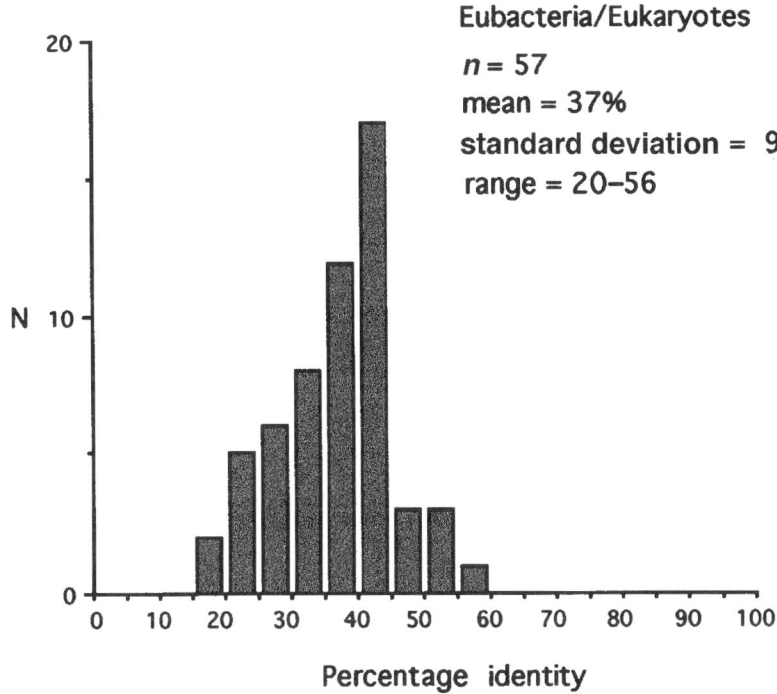

Eubacteria/Eukaryotes

$n = 57$

mean = 37%

standard deviation = 9

range = 20–56

Percentage identity

Fig. 23.2 Average resemblances (percentage identity) between multiple sequences from 57 enzymes of prokaryotes and eukaryotes as measured in blocks of five percentage points. (From Doolittle *et al.*, 1996.)

beset the sequence comparer. More than a decade ago, investigators determining the sequence of that enzyme from *E. coli* noted that it was more similar to gapdh sequences from eukaryotes than those available from prokaryotes (Branlant and Branlant, 1985). The only two prokaryote sequences known at the time were from thermophiles, however, and the authors concluded that it was the latter that were anomalous, presumably because of a faster rate of change involved in an adaptation to life at high temperatures. Not long after that, Martin and Cerff (1986) reported both cytosolic and chloroplast gapdh sequences from the mustard plant, and their analysis led them to conclude that the *E. coli* enzyme was anomalously closer to eukaryotes because of reverse transcription of an mRNA from an ancient eukaryotic host.

In 1990, we examined all available gapdh sequences and agreed that a horizontal gene transfer must have occurred from a eukaryote to a prokaryote (Doolittle *et al.*, 1990). We felt confident about the result on several grounds. For one, it had just been discovered that *E. coli* actually had two gapdh genes (Alefounder and Perham, 1989), and the newly discovered one (called B) had a sequence that placed it firmly among the prokaryotes. Moreover, this second gene was clustered chromosomally with two other glycolytic enzymes. We also showed that the anomalous eukaryotic association of the A enzyme was not an artifact of the method of tree construction, in that two fundamentally different methodologies gave the same result. Finally, the sequences of other glycolytic enzymes yielded conventional phylogenies for the same or closely related taxa.

Alas, all the data were not in. Martin *et al.* (1993) were soon to find that having two gapdh enzymes was hardly unique to *E. coli*. Indeed, cyanobacteria have three gapdh genes (and now apparently so does *E. coli*). A phylogenetic tree could be drawn that showed that there had been a long history of gene duplications, and that the alleged horizontal gene transfer was in fact the result of comparing paralogous sequences. Interestingly, the authors concluded that it looked as if the gene transfer actually went the other way and that eukaryote gapdh genes were acquired from eubacteria! They were drawn to this conclusion by the fact that the gapdh from archaebacterial methanogens was radically different in sequence from both the eubacteria and the eukaryote ones (Hensel *et al.*, 1989).

In my view, the matter is still not resolved, although recent phylogenies including the gapdh from very early diverging protists lend some credence to such a scenario (Henze *et al.*, 1995; Roger *et al.*, 1996). Still, I worry about why the gapdh from *Haloarcula vallismortis*, an archaebacterium, is 52% identical with that of *B. subtilis*, a eubacterium, but only 20% identical with the gapdh of the archaebacterial methanogens mentioned above. I also find it absolutely astonishing that the gapdh A of *E. coli*, which was the one we had previously thought had been transferred, and that of a trypansome are 81% identical; there are no gaps in the alignment. As far as I know, this is far and away the highest resemblance known for a protein from a prokaryote and eukaryote. For comparison, consider that in animals the invertebrate and vertebrate versions of this enzyme are only about 75% identical. The lesson I would urge upon the reader is that even when all the data seem to be on hand, they may not be.

23.2.2 ANTIBIOTIC SYNTHETIC GENES

The biosynthetic machinery for making penicillin-related antibiotics is found in many eubacteria but only in filamentous fungi among the eukaryotes. The possibility that those fungi had acquired the system from bacteria had long been supposed (Smith *et al.*, 1990). An early report specifically claimed that the enzyme isopenicillin N synthetase had been transferred from bacteria to fungi (Peñalva *et al.*, 1990). We worried that this conclusion was based mainly on a very limited data set and depended greatly on assumptions of very different rates of change. In our zeal to show that one can be misled by paralogous genes, we searched the data banks for possible relatives of isopenicillin N synthetase. Indeed, we found that the enzyme deacetoxycephalosporin C synthetase was a paralog that occurred in both bacteria and fungi. We used its sequence to root a phylogenetic tree that to all appearances was conventional (Smith *et al.*, 1992).

Recently, Buades and Moya (1996) have re-analyzed the situation and made the very legitimate point that a conventional phylogeny is not a sufficient condition to rule out horizontal transfer. They are correct. What convinces us that their rate considerations are on the right track is that we now have a good idea of how similar vertically transmitted enzyme sequences ought to be for prokaryotes and eukaryotes, and the resemblances of both the isopenicillin N synthetases and the deacetoxy-cephalosporin synthetases are suspiciously high. In both cases the eukaryote and prokaryote enzymes are 57% identical, at the very upper limit of what might be expected for an enzyme sequence resemblance between a prokaryote and eukaryote (Fig. 23.2). In this case the lesson to be learned is that the existence of paralogs in two distantly related taxa does not automatically make the system orthodox. Both genes may have been involved in horizontal transfers, either separately or as part of a single event involving a large cluster of genes.

23.2.3   GLUTAMINE SYNTHETASES

Glutamine synthetases come in two homologous but readily distinguishable types. Type I is dodecameric and so far has been found only in eubacteria and archaebacteria. Type II is octameric and is found in all eukaryotes and also in root nodule bacteria. Root nodule bacteria contain both forms, and the thought that they might have acquired the Type II from their plant hosts was immediately suggested (Carlson and Chelm, 1986). Careful consideration of the rates of change led others to question the need for a horizontal transfer (Shatters and Kahn, 1989). We ourselves examined the data and equivocated, although we felt overall the possibility of horizontal transfer was reasonable (Smith *et al.*, 1992). A survey by another group (Kumada *et al.*, 1993) favored an orthodox relationship in which the gene duplication leading to the two types occurred long before the divergence of eukaryotes and prokaryotes. According to this scheme, eukaryotes presumably lost the Type I entirely, and most bacteria completely gave up the Type II. If we were to judge today based on sequence resemblance alone, we would likely accept that scenario, the eukaryote enzymes being 48% identical with the root nodule bacteria sequences, on average. But the obligatory coincidences needed, combined with the opporunity factor, still make this a difficult case to resolve.

23.2.4   GLUCOSE PHOSPHATE ISOMERASES

Late in 1991, one of my postdoctoral associates, Michael W. Smith, came to me with a literature finding that the amino acid sequences of glucose phosphate isomerase (gpi) from *E. coli* and a plant (cytosolic) were 88% identical (Tait *et al.*, 1988; Froman *et al.*, 1989). I was incredulous. Clearly it was a mistake on someone's part. Smith did not have easy access to any plant DNA at the time, but we had plenty of *E. coli* material, so he promptly cloned and sequenced the gpi from that source. With the exception of a single base, it matched the published report exactly. We made a phylogenetic tree of all the gpi sequences we could find and not unexpectedly found that the *E. coli* one clustered with the plant. We published a short letter on the subject (Smith and Doolittle, 1992).

Not long after, a corrected version of the plant enzyme was reported (Thomas *et al.*, 1992), and the extraordinary resemblance with *E. coli* had disappeared. Was this the end of the case for a horizontal gene transfer? Not at all. The gpi sequences from mammals and *E. coli* are approximately 70% identical, well above the range of ordinary enzyme sequence resemblances (Fig. 23.2). Moreover, as noted above, other bacteria (e.g. *Zymomonas*) have a gpi that is at the expected level of resemblance (40% identical). Recently, Katz (1996) has conducted a study of numerous gpi sequences that in my view convincingly favors horizontal transfer and rules out any possibility of a paralogous interpretation.

23.3   A SUMMARY COMMENT

I have chosen to remark on several systems that in the past have been candidates for horizontal gene transfers between prokaryotes and eukaryotes. In one of these, glyceraldehyde 3-phosphate dehydrogenase, the matter is still confused. Although earlier considerations were off base in that paralogs were being compared, current explanations are not altogether satisfactory. In another case, the biosynthetic enzymes for antibiotics found in some bacteria and filamentous fungi, our overzealous skepticism of the claims of others was unwarranted. In still a third case, the on-again off-again transfer of a glutamine synthetase gene, current data indicate that there is no need to invoke horizontal gene transfer.

Finally, in the matter of glucose phosphate isomerase, the data still support horizontal transfer as the best explanation for why one group of bacteria have an enzyme with very high resemblance to animal versions, whereas other bacteria have an enzyme with a sequence that shows the typical eukaryote–prokaryote distance. In all of these considerations, I have been influenced by our large study that compared the amino acid sequences of 57 different enzymes from many different prokaryotes and eukaryotes (Doolittle *et al.*, 1996).

There is one aspect of this debate that I have not raised: we really do not yet understand the origin of the eukaryotic cell. The orthodox view is that cells destined to be eukaryotic branched off early, developed a cytoskeleton, and subsequently engulfed other cells, some of which became endosymbionts. The progression from eubacteria to archaebacterial to eukaryotic host is still hotly debated (Keeling and W.F. Doolittle, 1995). Some people feel that the eukaryotic cell is a chimera and that perhaps the nucleus was acquired by ensymbiosis. Until all these events are sorted out, it is very difficult to be certain about early (ancient) transfers of genetic information. Only very recent occurrences, where the candidate genes are still very similar, can be validated by sequence comparison alone. In other cases, subsidiary data – including gene arrangement, ecology and opportunity – must be taken into account.

REFERENCES

Alefounder, P.R. and Perham, R.N. (1989. Identification, molecular cloning and sequence analysis of a gene cluster encoding the class II fructose 1,6-bisphosphate aldolase, 3-phosphoglycerate kinase and a putative second glyceraldehyde 3-phosphate dehydrogenase of *Escherichia coli. Molec. Microbiol.* **3**: 723–732.

Bannister, J.V. and Parker, M.W. (1985) The presence of a copper zinc superoxide dismutase in the bacterium *Photobacterium leiognathi*: a likely case of gene transfer from eukaryotes to prokaryotes. *Proc. Natl Acad. Sci. USA* **82**: 149–152.

Branlant, G. and Branlant, C. (1985) Nucleotide sequence of the *Escherichia coli gap* gene. *Eur. J. Biochem.* **150**: 61–66.

Buades, C. and Moya, A. (1996) Phylogenetic analysis of the isopenicillin-N-synthetase horizontal gene transfer. *J. Mol. Evol.* **42**: 537–542.

Carlson, T.A. and Chelm, B.K. (1986) Apparent eukaryotic origin of glutamine synthetase II from the bacterium *Bradyrhizobium japonicum*. *Nature* **322**: 568–570.

Doolittle, R.F., Feng, D.F., Anderson, K.L. and Alberro, M.R. (1990) A naturally occurring horizontal gene transfer from a eukaryote to a prokaryote. *J. Mol. Evol.* **31**: 383–388.

Doolittle, R.F., Feng, D.F., Tsang, S. *et al.* (1996) Determining divergence times of the major kingdoms of living organisms with a protein clock. *Science* **271**: 470–477.

Froman, B.E., Tait, R.C. and Gottlieb, L.D. (1989) Isolation and characterization of the phosphoglucose isomerase gene from *Escherichia coli. Mol. Gen. Genet.* **217**: 126–131.

Hensel, R., Zwickl, P., Fabry, S. *et al.* (1989) Sequence comparison of glyceraldehyde-3-phosphate dehydrogenases from the three urkingdoms: evolutionary implication. *Can. J. Microbiol.* **35**: 81–85.

Henze, K., Badr, A., Wettern, M. *et al.* (1995) A nuclear gene of eubacterial origin in *Euglena gracilis* reflects cryptic endosymbioses during protist evolution. *Proc. Natl Acad. Sci. USA* **92**: 9122–9126.

Katz, L.A. (1996) Transkingdom transfer of the phosphoglucose isomerase gene. *J. Mol. Evol.* **43**: 453–459.

Keeling, P.J. and Doolittle, W.F. (1995) Archaea: narrowing the gap between prokaryotes and eukaryotes. *Proc. Natl Acad. Sci. USA* **92**: 5761–5764.

Kumada, Y., Benson, D.R., Hillemann, D. *et al.* (1993) Evolution of the glutamine synthetase gene, one of the oldest existing and functioning genes. *Proc. Natl Acad. Sci. USA* **90**: 3009–3013.

Little, E., Bork, P. and Doolittle, R.F. (1994) Tracing the spread of fibronectin type III domains in bacterial glycohydroleases. *J. Mol. Evol.* **39**: 631–643.

Martin, J.P. and Fridovich, I. (1981) Evidence for a natural gene transfer from the ponyfish to its bioluminescent bacterial symbiont *Photobacter leiognathi*. The close relationship between bacteriocuprein and the copper–zinc superoxide dismutase of teleost fishes. *J. Biol. Chem.* **256**: 6080–6089.

Martin, W. and Cerff, R. (1986) Prokaryotic features of a nucleus-encoded enzyme. *Eur. J. Biochem.* **159**: 323–331.

Martin, W., Brinkmann, H., Savonna, C. and Cerff, R. (1993) Evidence for a chimeric nature of nuclear genomes: eubacterial origin of eukaryotic glyceraldehyde-3-phosphate dehydrogenase genes. *Proc. Natl Acad. Sci. USA* **90**: 8692–8696.

Peñalva, M.A., Moya, A., Dopazo, J. and Ramón, D. (1990) Sequences of isopenicillin N synthetase genes suggest horizontal gene transfer from prokaryotes to eukaryotes. *Proc. Royal Soc. London* [*Biol.*] **241**: 161–169.

Roger, A.J., Smith, M.W., Doolittle, R.F. and Doolittle, W.F. (1996) Evidence for the heterolobosea from phylogenetic analysis of genes encoding glyceraldehyde-3-phosphate dehydrogenase. *J. Eukaryotic Microbiol.* **43**: 475–485.

Shatters, R.G. and Kahn, M.L. (1989) Glutamine synthetase II in Rhizobium: reexamination of the proposed horizontal transfer of DNA from eukaryotes to prokaryotes. *J. Mol. Evol.* **29**: 422–428.

Smith, D.J. *et al.* (1990) β-Lactam antibiotic biosynthetic genes have been conserved in clusters in prokaryotes and eukaryotes. *EMBO J.* **9**: 741–747.

Smith, M.W. and Doolittle, R.F. (1992a) A comparison of evolutionary rates of the two major kinds of superoxide dismutase. *J. Mol. Evol.* **34**: 175–184.

Smith, M.W. and Doolittle, R.F. (1992b) Anomalous phylogeny involving the enzyme glucose-6-phosphate isomerase. *J. Mol. Evol.* **34**: 544–545.

Smith, M.W., Feng, D.F. and Doolittle, R.F. (1992) Evolution by acquisition: the case for horizontal gene transfers. *TIBS* **17**: 489–493.

Steffens, G.J., Bannister, J.V., Bannister, W.H. *et al.* (1983) The primary structure of Cu–Zn superoxide dismutase from *Photobacterium leiognathi*. *Hoppe-Seyler's Z. Physiol. Chem.* **364**: 675–690.

Tait, R.C., Froman, B.E., Laudencia-Chingcuanco, D.L. and Gottlieb, L.D. (1988) Plant phosphoglucose isomerase genes lack introns and are expressed in *Escherichia coli*. *Plant Molec. Biol.* **11**: 381–388.

Thomas, B.R., Laudencia-Chingcuanco, D. and Gottlieb, L.D. (1992) Molecular analysis of the plant gene encoding cytosolic phosphoglucose isomerase. *Plant Molec. Biol.* **19**: 745–757.

Fungal denitrification, a respiratory system possibly acquired by horizontal gene transfer from prokaryotes

24

Naoki Takaya, Michiyoshi Kobayashi and Hirofumi Shoun

SUMMARY

The amino acid sequence of nitrite reductase purified from the denitrifying fungus *Fusarium oxysporum* was partially determined and compared with the sequences of dissimilatory nitrite reductase of bacteria. The sequence could be well aligned between the fungal and bacterial enzymes, which revealed as much as 80–90% sequence identities between them. The molecular identity between them was further confirmed immunologically. The amino-terminal amino acid sequence of another component of the fungal denitrifying system, azurin, was also compared with those of bacterial counterparts, which again showed surprising agreement between them. These results are indicative of horizontal gene transfer of the genes from prokaryote to eukaryote that occurred at a rather recent evolutionary stage.

24.1 INTRODUCTION

Biological denitrification is a process whereby fixed nitrogen (nitrate or nitrite) is reduced to gaseous forms of nitrogen such as N_2 or N_2O (nitrous oxide). It plays an important role in the global nitrogen cycle as the reverse reaction of nitrogen fixation. The process consists of four reducing steps, catalyzed by nitrate reductase (Nar), nitrite reductase (Nir), nitric oxide reductase (Nor), and nitrous oxide reductase, respectively (Zumft, 1992; Berks *et al.*, 1995), as follows:

$$NO_3^- \xrightarrow{\text{Nar}} NO_2^- \xrightarrow{\text{Nir}} NO \xrightarrow{\text{Nor}} N_2O \longrightarrow N_2$$

Denitrification physiologically acts as anaerobic respiration (nitrate respiration). Biological denitrifying activities were believed to be uniquely characteristic of prokaryotes (Berks *et al.*, 1995) until we discovered their presence in several fungi such as *Fusarium oxysporum* and *Cylindrocarpon tonkinense* (Shoun and Tanimoto, 1991; Shoun *et al.*, 1992; Usuda *et al.*, 1995). The fungal denitrifyng system first seemed to be quite different in properties from the bacterial systems, judging from the unique involvement of cytochrome P450 (P450nor) only in the fungal system (Shoun and Tanimoto, 1991; Nakahara *et al.*, 1993). However, we later purified from *F. oxysporum* other components of the denitrifying system: copper-containing Nir and its physiological electron donor (azurin), which bore a close resemblance to the bacterial counterparts (Kobayashi and Shoun, 1995). Further, we showed that the denitrifying system is localized to the respiring organelle (the mitochondrion) coupled with ATP synthesis (Kobayashi *et al.*, 1996). Thus the fungal system acts as an anaerobic respiration (nitrate-respiration) system, like bacterial denitrifying systems.

It has become generally accepted that the mitochondrion of eukaryotic cells evolved by the endosymbiosis of an aerobic bacterium with the host and that the present-day denitrifying bacteria such as *Paracoccus* and *Rhodobacter* are the direct descendants of the endosymbiont protomitochondria (Jhon and Whatley, 1975; Yang *et al.*, 1985; Gupta and Golding, 1996). It therefore seems that mitochondria lost the ability to denitrify during their evolution (Jhon and Whatley, 1975). Now that denitrifying activities have been found in fungal mitochondria, it is of evolutional interest when the fungi acquired this denitrifying activity: is it the remnant of the symbiotic protomitochondria, or was the system acquired afterwards during evolution?

Here we briefly report the partial amino acid sequences of dissimilatory nitrite reductase and its physiological electron donor, azurin, of the denitrifying fungus *F. oxysporum*, which bear a striking resemblance to those of bacterial counterparts.

24.2   RESULTS AND DISCUSSION

Nir and azurin of *F. oxysporum* were purified as reported (Kobayashi and Shoun, 1995). Tryptic peptides were obtained and sequenced in the same manner as in the case of P450nor (Nakahara and Shoun, 1996). The determined partial amino acid sequence of the fungal Nir protein could be easily aligned to those of the bacterial enzymes; we found greater than 70% amino acid identity. For example, the alignment with the Nir protein of *Pseudomonas aureofaciens*, which showed the highest sequence identity to the fungal Nir protein, is shown in Fig. 24.1a. Surprisingly, the sequence of as many as 91 amino acid residues among 104 residues in total was identical between them. In the case of the fungal azurin, the amino-terminal sequence (the first 20 residues) also exhibited unusually high identities to those of the bacterial counterparts (Fig. 24.1b). In particular, the sequences of azurins of *F. oxysporum* and *Alcaligenes* sp. were completely identical. In summary, the fungal Nir protein is 88% identical to the *Pseudomonas* counterpart, while the fungal azurin is > 86% identical ($P = 0.95$) to its *Alcaligenes* homolog. The results showed close relationships between the fungal and bacterial copper proteins.

In spite of its unique function (nitric oxide reductase) P450nor of denitrifying fungi exhibits identities in amino acid sequence to other mono-oxygenase P450s and thus belongs to the P450 superfamily (Kizawa *et al.*, 1991; Nevert *et al.*, 1991). P450nor is unique not only in its function but also in that it is the only exception among eukaryotic P450 known so far that is phylogenetically classified with bacterial P450s (Kizawa *et al.*, 1991; Nevert *et al.*, 1991). In particular, P450nor shows sequence identities up to 40% to P450s of a unique group of Gram-positive bacteria: actinomycete (Kizawa *et al.*, 1991). This fact implies that actinomycetes may also contain a similar denitrifying activity to the fungal system, although they have been thought to be aerophilic organisms like fungi. We searched and found several denitrifiers in actinomycetes (unpublished). Among them we are now characterizing the denitrifying system of *Streptomyces thioluteus*. The system seems to resemble the fungal system and to include a copper-containing Nir and azurin. Western blot analysis of the cell-free extract obtained from denitrifying cells showed that *S. thioluteus* contains a protein that reacts with the antibodies to the fungal Nir and whose molecular mass is almost identical to that of the fungal Nir (Fig. 24.2).

Although the present data are preliminary, they may be sufficient to show that Nir and azurin of *F. oxysporum* are very closely related to those of bacterial origin. Both fungal copper-proteins are too similar to each bacterial counterpart to be regarded as remnants of the protomitochondrion that have been kept in the fungal cells for as long as 1.5–2.0 billion years since the emergence of eukaryotic cells. For example, Doolittle (Chapter 23) conducted a large-scale comparison of 57 different enzymes and found that on average the enzymes of eukaryotes and prokaryotes are 32% identical, with values ranging from 20 to 57%. According to Doolittle (section 22.1.4): 'As a rule of thumb, if an enzyme sequence is

A

```
FUSOX                                           VALVAPPQVHPHE
                                                * ***** *****
PSEAR     1 MSVFRSVLGACVLLGSCASSLALAGGAEGLQRVKVDLVAPPLVHPHEQVV


FUSOX                           MSIEEKKMVIDDKGTTLQAMTFDGSMPGPTLVVHEGDYVE
                                ************ ********* **************** *
PSEAR    51 SGPPKVVQFRMSIEEKKMVIDDQGTTLQAMTFNGSMPGPTLVVHEGDYIE


FUSOX       LT
            **
PSEAR   101 LTLVNPATNSMPHNVDFHAATGALGGAGLTQVVPGQEVVLRFKADRSGTF


FUSOX                   MVPWHVVSGM    MVLPRDGLKDPDGTILRYD VYTIGEFD
                        **********    ******** ** *  * ** ****** *
PSEAR   151 VYHCAPQGMVPWHVVSGMNGALMVLPRDGLRDPQGKLLHYDRVYTIGESD


FUSOX       LYIP
            ****
PSEAR   201 LYIPKDKDGHYKDYPDLASSYQDTRAVMRTLTPSHVVFNGRVGALTGANA


FUSOX                                                       NLETWF
                                                           * ****
PSEAR   251 LTSKVGESVLFIHSQANRDSRPHLIGGHGDWVWTTGKFANPPQRNMETWF


FUSOX       IR
            *
PSEAR   301 IPGGSAVAALYTFKQPGTYVYLSHNLIEAMELGALAQIKVEGQWDDDLMT


PSEAR   351 QVKAPGPIVEPKQ
```

B

```
FUSOX     1 AECSVDIAGNDQMQFDKKEI 20
ALCSP     1 AECSVDIAGNDQMQFDKKEI 20
BORBR     1 AECSVDIAGTDQMQFDKKAI 20
PSEAE    21 AECSVDIQGNDQMQFNTNAI 40
PSEDE     1 AECSVDIQGNDQMQFSTNAI 20
PSEFD     1 AECKVDVDSTDQMSFNTKEI 20
METJ      1 AECKVDVDSTDQMSFNTKEI 20
            *** ** *** *      *
```

Fig. 24.1 (A) Comparison of the amino acid sequences of the copper-containing nitrite reductases from *Fusarium oxysporum* (FUSOX) and *Pseudomonas aureofaciens* (PSEAR; accession code, NIR_PSEAR) (Glockner *et al.*, 1993). (B) Comparison of the amino-terminal sequences of azurins from *F. oxysporum*, FUSOX; *Alcaligenes* sp., AZUR_ALCSP (accession code); *Bordetella bromchisepticca* AZUR_BORBR; *Pseudomonas aeruginasa*, AZUR_PSEAE; *Pseudomonas denitrificans*, AZUR_PSEDE; *Pseudomonas fluorescens* biotype D, AZUR_PSEFD; *Methylomonas* J., AZU1_METJ.

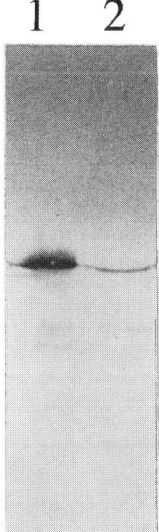

Fig. 24.2 Western blot analysis of the crude cell extract from *S. thioluteus* (lane 1; 10 mg protein) and *F. oxysporum* MT-811 (lane 2; 30 mg protein). Proteins were separated by SDS-PAGE and subjected to immunoblotting using the specific antiserum against *F. oxysporum* Nir (Kobayashi *et al.*, 1996). Nir seems to be expressed much more in the actinomycete, *S. thioluteus*, than in the fungus.

found to be more than 60% identical between two groups, then horizontal gene transfer looms as a distinct possibility.' It is therefore likely that the fungi acquired the denitrifying system by horizontal gene transfer well after the endosymbiosis that established the mitochondrion.

It is of particular interest to compare the fungal denitrifying (nitrate-respiring) systems with bacterial systems. From the knowledge accumulated to date, bacterial nitrate-respiring systems are mainly classified into two types: denitrifying and non-denitrifying (Zumft, 1992; Berks *et al.*, 1995). The denitrifying type forms N_2 or N_2O from nitrate or nitrite. The typical non-denitrifying, nitrate-respiring system forms ammonia from nitrate via nitrite, and is well characterized in *Esherichia coli* (Berks *et al.*, 1995). In the *E. coli* system, nitrite is reduced to ammonia by Nir of oligomeric cytochrome *c* type, which is in contrast to Nir of cytochrome *cd1* or the copper-containing type of denitrifying bacteria that forms nitric oxide (NO) from nitrite. A unique formate dehydrogenase (Fdh) is known to participate in the *E. coli* system that donates electrons to Nar via the ubiquinon/ubiquinol system (Berks *et al.*, 1995). As far as has been clarified to date, Nar of the membrane-bound type (this term is to discriminate it from nitrate reductase of the periplasmic type, Nap, which is found in several denitrifying bacteria) is the only reductase that is universally found in nitrate-respiring bacteria of both denitrifying and non-denitrifying types.

Although the system of *F. oxysporum* is of a denitrifying type, which forms N_2O from nitrate, we could detect Fdh activity in the fungal system (Kobayashi *et al.*, 1996) that is usually not found in denitrifying bacteria. Further, its Nar also bears resemblance to that of *E. coli* (unpublished data). So the constituents of the fungal system are quite unique. The fungal system resembles the non-denitrifying *E. coli* system in that it contains Fdh, whereas its Nir and azurin are closely related to those of denitrifying bacteria. Further, Nor of the fungal systems (P450nor) is quite unique. Nor of the cytochrome P450 type has not been known so far among bacteria, although it was suggested to be of bacterial origin (Kizawa *et al.*, 1991). So the fungal system seems to have arisen from a medley of constituents from several types of nitrate-respiring bacteria. The relevance of the fungal system to the bacterial systems will become clearer when the complete sequence of the associated genes, in addition to that of P450nor are obtained.

REFERENCES

Berks, B.C., Ferguson, S.J., Moir, J.W.B. and Richardson, D.J. (1995) Enzymes and associated electron transport systems that catalyse the respiratory reduction of nitrogen oxides and oxyanions. *Biochim. Biophys. Acta* **1232**: 97–173.

Glockner, A.B., Jungest, A. and Zumft, W.G. (1993) Copper-containing nitrite reductase from *Pseudomonas aureofaciens* is functional in a mutationally cytochrome *cd1*-free background (*NirS⁻*) of *Pseudomonas stutzeri*. *Arch. Microbiol.* **160**: 18–26.

Gupta, R.S. and Golding, G.B. (1996) The origin of the eukaryotic cell. *Trends Biol. Sci.* **21**: 166–171.

Jhon, P. and Whatley, F.R. (1975) *Paraccocus denitrificans* and the evolutionally origin of the mitochondrion. *Nature* **254**: 495–498.

Kizawa, H., Tomura, D., Oda, M. *et al.* (1991) Nucleotide sequence of the unique nitrate/nitrite-inducible cytochrome P-450 cDNA from *Fusarium oxysporum*. *J. Biol. Chem.* **266**: 10632–10637.

Kobayashi, M. and Shoun, H. (1995) The copper-containing dissimilatory nitrite reductase involved in the denitrifying system of the fungus *Fusarium oxysporum*. *J. Biol. Chem.* 270: 4146–4151.

Kobayashi, M., Matsuo, Y., Takimoto, A. *et al.* (1996) Denitrification, a novel type of respiratory metabolism in fungal mitochondrion. *J. Biol. Chem.* **271**: 16263–16267.

Nakahara, K. and Shoun, H. (1996) N-terminal processing and amino acid sequence of two isoforms of nitric oxide reductase cytochrome P450nor from *Fusarium oxysporum*. *J. Biochem.* **120**: 1082–1087.

Nakahara, K., Tanimoto, T., Hatano, K. *et al.* (1993) Cytochrome P-450 55A1 (P-450dNIR) acts as nitric oxide reductase employing NADH as the direct electron donor. *J. Biol. Chem.* **268**: 8350–8355.

Nevert, D.W., Nelson, D.R., Coon, M.J. *et al.* (1991) The P450 superfamily: update on new sequences, gene mapping, and recommended nomenclature. *DNA Cell Biol.* **10**: 1–14.

Shoun, H. and Tanimoto, T. (1991) Denitrification by the fungus *Fusarium oxysporum* and its involvement of cytochrome P-450 in the respiratory nitrite reduction. *J. Biol. Chem.* **266**: 11078–11082.

Shoun, H., Kim, D.-H., Uchiyama, H. and Sugiyama, J. (1992) Denitrification by fungi. *FEMS Micribiol. Lett.* **94**: 277–282.

Usuda, K., Toritsuka, N., Matsuo, Y. *et al.* (1995) Denitrification by the fungus *Cylindrocarpon tonkinense*: anaerobic cell growth and two isozyme forms of cytochrome P-450nor. *Appl. Environ. Micribiol.* **61**: 883–889.

Yang, D., Oyaizu, Y., Oyaizu, H. *et al.* (1985) Mitochondrial origins. *Proc. Natl Acd. Sci. USA* **82**: 4443–4447.

Zumft, W.G. (1992) The denitrifying prokaryotes, in *The Prokaryotes* (eds A. Balows, H.G. Truper, M. Dworkin *et al.*), 2nd edn, Vol. 1, Springer-Verlag, New York, pp. 554–582.

Cytochrome-*c* from *Stellaria longipes* and *Arabidopsis thaliana* was likely transferred from fungi since the radiation of terrestrial plants

25

Michael Syvanen

SUMMARY

The cytochrome-*c* gene sequences reported for *Stellaria longipes* and *Arabidopsis thaliana* show closer affinity to those from the fungi than to those from 25 other plants. This work analyzes the cytochrome-*c* molecular clock in plants and fungi and compares this with the rate of evolution of those genes from *S. longipes* and *A. thaliana*. It examines three hypotheses: that these idiosyncratic associations result from (1) a horizontal gene transfer event from fungi or (2) convergent evolution within isolated plant lineages, or (3) that the majority of plant and fungal cytochrome-*c* genes are paralogous to each other. The current analysis renders hypotheses (2) and (3) highly unlikely. The most likely explanation is that on two separate occasions, respectively, relatively recent ancestors to *A. thaliana* and *S. longipes* inherited their cytochrome-*c* from some fungal donor, presumably a fungal endophyte.

25.1 INTRODUCTION

The amino acid sequences from over 20 seed plant cytochrome-*c* proteins (cyt-*c*) have been determined. Most cluster in a single group and define

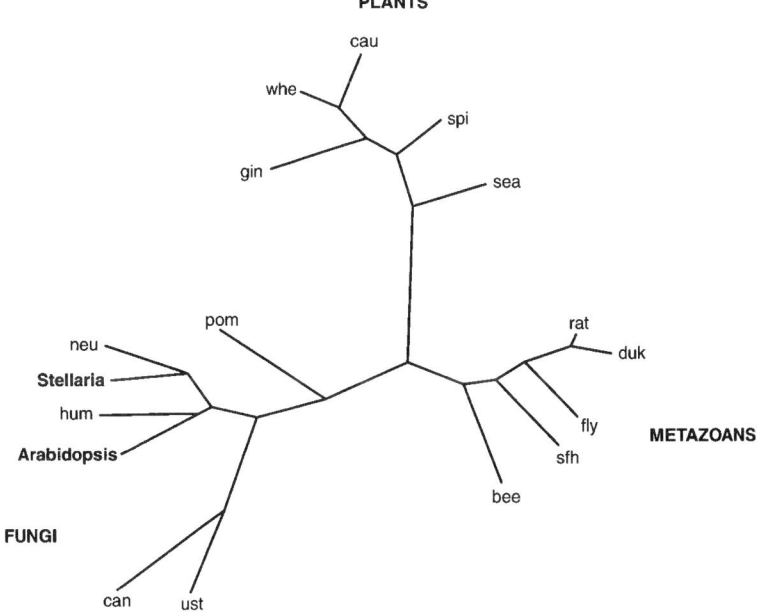

PLANTS

Fig. 25.1 Minimal replacement tree for the 15 eukaryotic cyt-*c*s. There were 32 different parsimonious trees of equal length. The others are produced by combinations of different topologies within the fungi and the plants. The one shown is representative. The length of each branch is proportional to the number of replacements, i.e. shown is a phylogram. Trees based on distances give similar results. Taxa abbreviations for Figs 25.1, 25.2 and 25.3: bee, *Apis mellifera* (honey bee); can, *Candida krusei*; cau, *Brassica oleracea* (cauliflower); duk, *Anas platyrhynchos* (duck); sea *Enteromorpha intestinalis* (seaweed); fly, *Drosophila melanogaster*; gin, *Ginkgo biloba*; hum, *Humicola lanuginosa*; neu, *Neurospora crassa*; pom, *Schizosaccharomyces pombe*; rat, *Rattus norvegicus*; sfh, *Asterias rubens* (starfish); spi, *Spinacia oleracea* (spinach); ust, *Ustilago sphaerogena*; whe, *Triticum aestivum* (wheat).

the plants as belonging to a monophyletic group that diverged from the eukaryotic lines leading to fungi and metazoans. However, two of the plant sequences, based on a genomic clone from *Arabidopsis thaliana* (Kemmerer *et al.*, 1991) and a cDNA clone from *Stellaria longipes* (Zhang and Chinnappa, 1994), are more closely related to the fungi than they are to the other plants. Figure 25.1 gives a minimal replacement tree that illustrates these relationships. Shown is the most parsimonious tree with a subset of the data. The same conclusions are reached using larger data sets and/or distance trees using the UPGMA program or nearest-neighbor algorithms (data not shown).

Could these idiosyncratic genes be the result of an inadvertent cloning of DNA derived from a fungal contaminant of the plant tissue? This possibility is weakened since in the case of *Arabidopsis* the presence of the cloned sequence in genomic DNA was confirmed by direct probing and in the case of *Stellaria* it was shown that correct transcripts were found in plant tissue using the Northern probing technique. If fungal contamination is the source then we would have to conclude that an undiscovered fungal endophyte or intimate symbiote is present in both *Arabidopsis* and *Stellaria* that contributes significant quantities of DNA and RNA to the plant tissue extracts. Artifactual association can also be deduced when genes with highly unequal rates are being compared. As shown below, rates of divergence are constant among the groups being compared, which is interpreted as evidence for horizontal gene transfer (Smith *et al.*, 1992; Syvanen, 1994).

In recent years, major incongruities between a species tree and a gene tree have been increasingly interpreted as evidence for horizontal gene transfer. However, two alternative explanations have been offered for the similarity of the *Arabidopsis* and *Stellaria* cyt-*c* to those of the fungi:

1. Their evolution converged toward the fungi from a plant ancestral form.
2. These two idiosyncratic cyt-*c* are orthologous to the fungi but paralogous to those of most plants.

Hypotheses (1) and (2) lead to testable molecular clock predictions. It will be shown here that a consideration of the rates of change is a useful tool in distinguishing horizontal transfer from the two alternative explanations.

25.2 CONVERGENT EVOLUTION

In the first hypothesis, all seed plants shared an ancestral cytochrome-*c* that subsequently diverged. At some point the uniquely shared residues that characterize vascular plants in lines leading to *Arabidopsis* and *Stellaria* began converging toward those found in fungi. The resulting phylogram of this scenario is shown in Fig. 25.2. The topology of this tree was determined by parsimony, but *Arabidopsis* and *Stellaria* were placed with the angiosperms and *Arabidopsis* and cauliflower were made monophyletic because both are in the cruciferae family. The distance along each branch was the result of determining the minimal number of replacements required to satisfy that topology. As can be seen, the hypothesis of convergent evolution leads to further postulates. The most apparent is that the molecular clock in these two lineages must have also become greatly accelerated. If this were the case, then this accelerated rate should be apparent in a relative-rate test (Sarrich and Wilson, 1969)

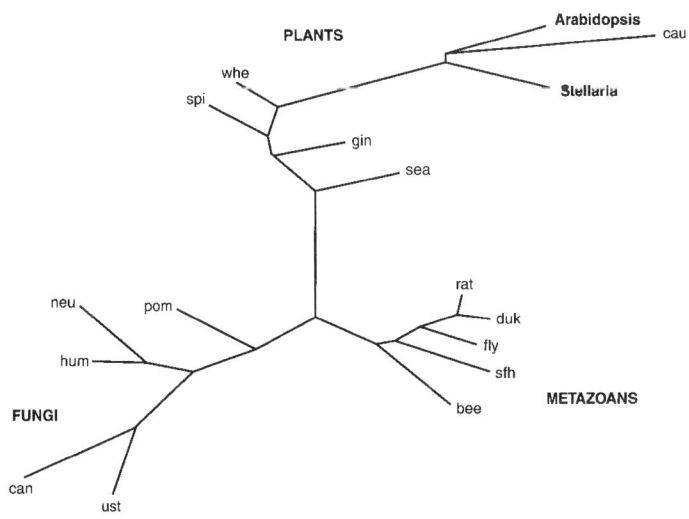

Fig. 25.2 Phylogram calculated from the topology where it is assumed that *Stellaria* and *Arabidopsis* cyt-*c* have diverged from those of the other plants. *Arabidopsis* and cauliflower are monophyletic because both are in the *Cruciferae* family. Taxa abbreviations as in Fig. 25.1.

where we compare the measured distance (d_{ij}) between these idiosyncratic plant sequences and some outgroup taxa, such as metazoan. This result is shown in Table 25.1. As can be seen, each of the five plants, along with *Arabidopsis* and *Stellaria*, has basically the same value of d_{ij} and is comparable to the five fungi. If we measure the distances through the tree in Fig. 25.2, the so-called patristic distances, we see major discrepancies with the values of d_{ij} (line 2, Table 25.1). This can be compared with the patristic distances through the minimal replacement tree in Fig. 25.1 (line 3, Table 25.1). The systematic differences between the d_{ij} and patristic distances in the minimal tree is due to homoplasy in the cyt-*c* data set and is to be expected. This homoplasy is found throughout the entire data set and is not restricted to *Stellaria* and *Arabidopsis*.

There is a second line of reasoning that argues against the hypothesis that the gene in *Arabidopsis* and *Stellaria* has evolved from the ancestral plant gene. This argument is based on the earlier analysis of the cyt-*c* data by Fitch (1971a,b, 1976) – an analysis that led to the covarion hypothesis. This began with a comparison of a very large number of cyt-*c* sequences in which it was noted that there were about 80 variable sites. However, for any given cyt-*c* it was estimated that only 12 residues were free to vary without inactivating the protein. The explanation for this is

Table 25.1 Relative rate test: expectations of the convergence explanation, comparing the outgroup taxa (average of the five metazoans) with the averages of, respectively, the five plant taxa, the five fungi taxa, and *Arabidopsis* and *Stellaria*

	Plants	Fungi	*Arabidopsis*	*Stellaria*
d_{ij}	38 ± 4	36 ± 6	36 ± 3	36 ± 3
d_{pat} (Fig. 25.2)	–	61 ± 20	104	100
d_{pat} (Fig. 25.1)	63 ± 11	64 ± 12	62 ± 3	65 ± 3

d_{ij}, uncorrected distance computed from the distance matrix; d_{pat}, patristic distance, i.e. number of replacements needed to produce branch lengths in respective phylograms (Figs. 25.1, 25.2).

that as a protein evolves, formerly functionally constrained residues become unconstrained while some variable residues will freeze. The group of residues free to vary is called a covarion. This means that certain residues in, for example, plants are invariable because of protein functional constraint, while in other taxonomic groups the homologous positions may vary.

I have looked for residues that are conserved within plants but are different and may be variable in the fungi. Eighteen residues were identified (Table 25.2). In all cases, the homologous position in *Arabidopsis* and *Stellaria* shares features with the proteins from the fungi, but in no case with the plants. Thus, the convergent evolution explanation requires the additional postulate that the presumably functionally constrained residues unique to plants must vary in the lines leading to *Arabidopsis* and *Stellaria* and, furthermore, in each case assume the fungal identity. Finally, these convergent replacements must have also occurred in parallel in the line leading to *Stellaria* and the one leading to *Arabidopsis*.

In summary, the hypothesis of convergent replacements in the lines leading to *Arabidopsis* and *Stellaria* requires in its support multiple ad hoc assumptions that quite simply argue against much of what we know about protein evolution. The hypothesis of convergence is not supported by the data.

25.3   ORTHOLOGY VERSUS PAROLOGY

The second hypothesis posits that the ancestor which gave rise to the seed plants and fungi had two copies of a cyt-*c* that diverged after an even earlier gene duplication event; and that most seed plants inherited one copy whereas the fungi and the two idiosyncratic angiosperms inherited the other. Thus, one would conclude that the seed plant cyt-*c* are paralogous to those from the fungi.

Table 25.2 Comparison of conserved residues, suggesting that *Arabidopsis* and *Stellaria* cyt-*c* belong to the fungal covarion

Text/plant	Amino acid residue number																	
	9	18	22	33	37	42	48	51	52	58	59	63	78	90	92	98	101	105
Plant	A,P	I	K	A	Q,E	N	Q	T	T	S	A	N	L	V	P	Q	A	A
Fungi	X	L	R	G,E	Y,I	H	X	Q,S	A,V	T	D,E	X	E	A	G,A	K,A	N	T
Arabidopsis	Q	L	R	E	I	H	K	S	V	T	D	Q	E	A	G	K	N	T
Stellaria	E	L	R	E	I	H	H	S	V	T	D	A	E	A	G	K	N	T

The amino acid residue numbering begins at the methionine in the *Arabidopsis* cyt-*c*. X indicates that more than two different residues are found at that position. The conserved residues in the plants were identified by comparing sequences from the following: tomato, potato, cotton, castor bean, sesame, cauliflower, pumpkin, mung bean, hemp, cuckoo-pint, maize, rice, niger, nasturtium, sunflower, leek, wheat and parsnip. The sequences from nine fungi included: *Schwanniomyces occidentalis, Candida famata, Kluyvoromyces lactis, Candida krusei, Schizosaccharomyces pombe, Neurospora crassa, Humicola lanoginosa, Ustilago sphaerogena* and *Aspergillus nidulans.* In addition to the 20 sites that are conserved either in plants or in fungi, there are 41 residues conserved in both. The only residues listed are those that are totally conserved in one group, while that residue is not present in any member in the other group. After compilation, these homologous positions were compared with those from *Arabidopsis* and *Stellaria*.

If the gene duplication event occurred prior to the fungi–plant divergence, then this would argue that the evolutionary distance between the cyt-c from *Stellaria* and *Arabidopsis*, on the one hand, and the fungi , on the other, should go back at least as far as the fungi–plant split, which is a point before the diversification of the fungi. However, the distance data in Table 25.3 just do not support this view. For example, the distance of 20 replacements per 100 amino acids between *Stellaria* and neurospora is comparable to the number observed between metazoan phyla, which indicates a divergence time 400–600 million years ago, assuming even a roughly constant molecular clock. To hold that the gene from *Stellaria* and *Arabidopsis* split from the rest of the fungi at a time significantly prior to the fungi–plant radiation would mean that the rate of evolution of the *Stellaria*/*Arabidopsis* cyt-c gene would have to be considerably lower than that of the other cyt-c genes. Since the hypothesized gene duplication event would have to have occurred more than 1.2 billion years ago (i.e. before the fungi–plant divergence) there has to be a further postulation that the two idiosyncratic plant proteins experience a rate of evolution of about one-sixth that of the other eukaryotes. Again, this can be refuted from the relative rate test shown in Table 25.3: the distance from the metazoans to *Arabidopsis* and *Stellaria* is 49, not the 20–25 needed to make these two taxa closer to the fungi.

The phylogram that results from placing the two plant cytochromes as an outgroup to those of the fungi (Fig. 25.3) also does not support retarded rates of evolution. First, as is apparent from Fig. 25.3, the branches leading to the aberrant plants are not shortened. In Table 25.3 the distances through this tree (line 2) are again compared with d_{ij} (line 1). As can be seen, placing these two plant genes as the outgroup produces patristic distances that are significantly longer than are the d_{ij}, or even the patristic distances in the minimal tree from Fig. 25.1 (line 3 of Table 25.3). For example, the distance from *Arabidopsis* to *Neurospora* in

Table 25.3 Relative rate test: expectations of the parology/orthology explanation, comparing *Stellaria* and *Arabidopsis* with each of the five fungi

Distance	neu	hum	ust	can	pom
D_{ij}	20	21	24	33	33
d_{pat} (Fig. 25.3)	49.5	49.5	53.5	61.5	45.5
d_{pat} (Fig. 25.1)	27	32	34	44	47

d_{ij}, d_{pat}, as defined in Table 25.1.
The differences between the patristic distances from Fig. 25.1 and d_{ij} from the distance matrix are predictable from the consistency index calculated for the tree in Fig 25.1.
neu, *Neurospora crassa*; hum, *Humicola lanuginosa*; ust, *Ustilago sphaerogena*; can, *Candida krusei*; pom, *Schizosaccharomyces pombe*.

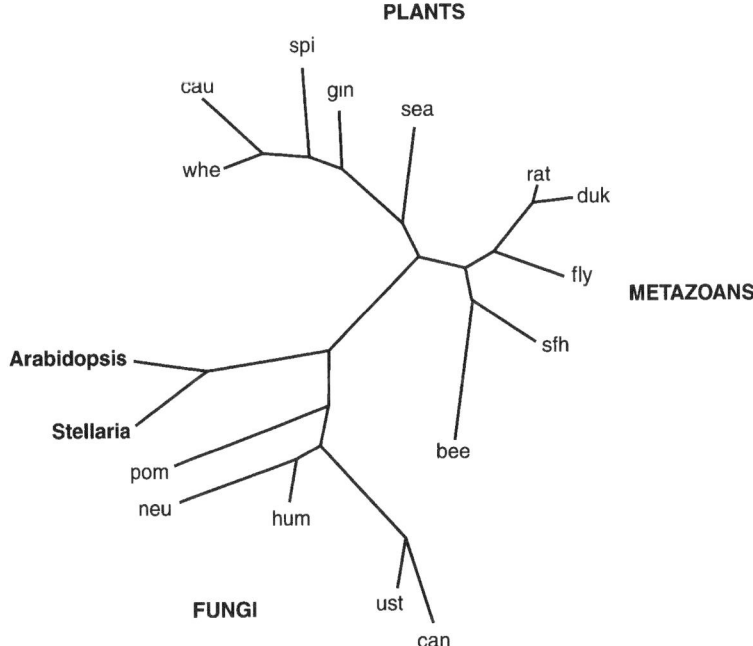

Fig. 25.3 Phylogram calculated from the topology where it is assumed that *Stellaria* and *Arabidopsis* cyt-*c* are orthologous to those from the fungi but paralogous to those from the plants. Taxa abbreviations as in Fig. 25.1.

the tree in Fig. 25.3 is 49 amino acid replacements, whereas the d_{ij} is 22 and the distance in the minimal tree is 29. Moreover, the trend in distances (increasing from *Neurospora* to *Schistosomonas*) is also observed for the patristic distances from the minimal tree (line 3) but not the patristic distances for the tree in Fig. 25.3 (line 2 of Table 25.3). Again, as was seen in Table 25.1, the systematic differences between the patristic distances from the minimal tree and d_{ij} are expected because of homoplasy in the cyt-*c* data set.

25.4   CONCLUSION

The fungal affinity of the *Arabidopsis* and *Stellaria* cyt-*c* is most obvious from the phylogenetic incongruity between the species and gene trees. This is the first indication that a horizontal gene transfer event might have occurred. There are other mechanisms that can create these incongruities. These include cloning artifacts, aberrant rates, parology/orthology confusions and convergent replacements. Clearly, it has been shown here that there is no evidence for aberrant rates. The results in this chapter show that the relative-rate test (using the metazoans as the unrelated outgroup) is satisfied. This fact also seriously weakens the parology/orthology explanation and, especially, the notion of convergent replacements. As shown previously (Syvanen, 1987), the relative rate of divergence between an unaffected outgroup and a group of genes experiencing horizontal transfer will not be affected by the consequent recombination or gene conversion events.

The finding of two unrelated dicots carrying fungal cyt-*c* means that two horizontal gene transfer events must have occurred – perhaps once from the fungi to plants and possibly from plant to plant. This scenario seems more likely (as opposed to two fungi-to-plant transfers) since there are indications that horizontal transfer among the land plants is possibly so common as to disguise natural taxonomic boundaries (Syvanen *et al.*, 1989).

In earlier work, I conjectured that the mechanism of transferring DNA involved DNA viruses or perhaps retroviruses. For a fungi-to-plant gene transfer, such a mechanism seems improbable given the paucity of known viral vectors. Direct transfer of chromosomes upon cell-to-cell contact is a more appealing idea, since there is some empirical support for such transfer events as is demonstrated by the work described by Kellner *et al.* (1993), Goff and Coleman (1995) and Wöstemeyer (Chapter 10).

REFERENCES

Doolittle, R.F. (1994) Convergent evolution: the need to be explicit. *Trends Biochem. Sci.* **19**: 15–18.

Fitch, W.M. (1971a) Rate of change of concomitantly variable codons. *J. Molec. Evol.* **1**: 84–96.

Fitch, W.M. (1971b) The nonidentity of invariable positions in the cytochromes *c* of different species. *Biochem. Genet.* **5**: 231-2-41.

Fitch, W.M. (1976) The molecular evolution of cytochrome *c* in eukaryotes. *J. Molec. Evol.* **8**: 13–40.

Goff, L.J. and Coleman, A.W. (1995) Fate of parasite and host organelle DNA during cellular transformation of red algae by their parasites. *Plant Cell* **7**: 1899–1911.

Kellner, M., Burmester, A., Wostemeyer, A. and Wostemeyer, J. (1993) Transfer of genetic information from the mycoparasite *Parasitella parasitica* to its host *Absidia glauca*. *Curr. Genet.* **23**: 334–337.

Kemmerer, E.C., Lei, M. and Wu, R. (1991) Structure and molecular evolutionary analysis of a plant cytochrome *c* gene: surprising implications for *Arabidopsis thaliana*. *J. Molec. Evol.* **32**: 227–237.

Sarrich, V.M. and Wilson, A.C. (1967) Immunological time scale for hominid evolution. *Science* **158**: 1200–1203.

Smith, M.W., Feng, D.F. and Doolittle, R.F. (1992) Evolution by acquisition: the case for horizontal gene transfers. *Trends Biochem. Sci.* **17**: 489–493.

Syvanen, M. (1987) Molecular clocks and evolutionary relationships: possible distortions due to horizontal gene flow. *J. Molec. Evol.* **26**: 16–23.

Syvanen, M. (1994) Horizontal gene transfer: evidence and possible consequences. *Annu. Rev. Genet.* **28**: 237–261.

Syvanen, M., Hartman, H. and Stevens, P.F. (1989) Classical plant taxonomic ambiguities extend to the molecular level. *J. Molec. Evol.* **28**: 536–544.

Zhang, X.H. and Chinnappa, C.C. (1994) Molecular cloning of a cDNA encoding cytochrome *c* of *Stellaria longipes* (Caryophyllaceae) – and the evolutionary implications. *Molec. Biol. Evol.* **11**: 365–375.

Phylogenetic relationship of volvocine algae to plants and animals

26

Rüdiger Schmitt, Klaus Stark, Stefan Fabry and David L. Kirk

SUMMARY

Molecular clock estimates indicate that lines leading to modern green algae and higher plants diverged about 700–750 million years ago, about 500 million years after plants and animals had last shared a common ancestor. Although placed at the apex of one line of green algal evolution, far from the plant–animal and the plant–algal branch points, *Volvox* exhibits a combination of plant-like and animal-like features (photoautotrophy vs. complete germ/soma dichotomy). This mixture of features is also reflected in the histone genes that combine plant-like promoter signals and peptide-coding sequences with animal-like introns and 3' palindromes. Possible explanations by vertical vs. horizontal gene transfer are discussed.

The volvocine lineage comprises a series of green flagellates varying in complexity from unicellular *Chlamydomonas* through colonial *Gonium*, *Pandorina*, *Eudorina* and *Pleodorina*, to multicellular *Volvox* with its fully differentiated cell types, and has been used as a paradigm for studying the way evolution 'invented' multicellularity and the division of labor between specialized cells. This group is also suited for investigating routes of gene transfer. A conserved portion of the *regA* gene that is essential for germ/soma differentiation in *Volvox carteri* has been employed to examine gene distribution among the volvocine algae. The pattern emerging among closely and distantly related volvocalean algae is evaluated.

26.1   INTRODUCTION

Because the multicellular green alga *Volvox* combines certain plant-like features (such as photoautotrophy) with certain animal-like features (such as motility and a germ–soma dichotomy) some biologists once speculated that *Volvox* might provide the 'missing link' between plants and animals. Sequence comparisons for 18S rRNA gene loci support the view that, although the green algae in the order Volvocales shared a common ancestor with higher plants, they constitute a separate lineage – with *Volvox* at its apex – that has long been separated from the lines leading to vascular plants and animals (Rausch *et al.*, 1989). This type of relationship is supported by an analysis of volvocine histone genes which reveals that they combine certain plant and metazoan features. The acquisition of these gene features is subject to three alternative hypotheses, which will be examined.

The volvocine lineage with its extant members – *Chlamydomonas, Gonium, Pandorina, Eudorina, Pleodorina* and *Volvox* – has been used as a paradigm for an evolutionary progression in size and developmental complexity, as illustrated in Fig. 26.1. Although the volvocine algae have been shown to constitute a coherent group of closely related organisms, an rRNA sequence-based tree does not support a monophyletic origin for the genus *Volvox*, but rather indicates that its members fall in at least

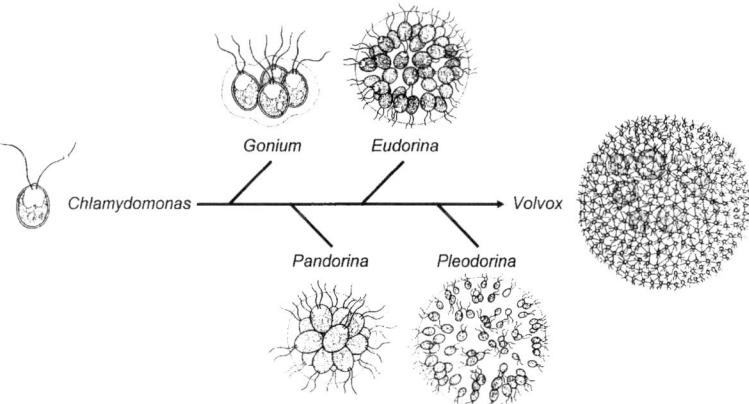

Fig. 26.1 A 'textbook view' of the genealogical progression of the volvocine lineage from single-celled *Chlamydomonas*, through increasingly more complex colonial forms toward the multicellular genus, *Volvox*, with its fully differentiated somatic and reproductive cells.

three (and possibly more) separate lineages, each of which contains also colonial members of the family (Larson *et al.*, 1992; A. Larson and D.L. Kirk, unpublished). The recently isolated regenerator gene (*regA*) of *Volvox carteri* provides a probe for independent assessment of this phylogenetic configuration.

26.2 RESULTS AND DISCUSSION

26.2.1 VOLVOCALEAN HISTONE GENES AND THE INHERITANCE OF MIXED PLANT-LIKE AND ANIMAL-LIKE FEATURES

A phylogenetic tree constructed from 18S rRNA sequences of 18 eukaryotic taxa using distance-matrix methods (Rausch *et al.*, 1989) placed *Volvox carteri* and *Chlamydomonas reinhardtii* at the tip of a green algal lineage that diverged from the main branch of eukaryotic evolution between the lines leading to the higher plants and to the animals, but closer to the former (Fig. 26.2). On the basis of cytochrome *c* sequence comparisons, Amati *et al.* (1988) estimated that lines leading to the modern green algae and to the higher plants diverged about 700–750 million years ago, about 500 million years after plants and animals had last shared a common ancestor. 'Molecular clock' methods applied to the rRNA data, and the number of silent exchanges observed in tubulin genes, led to an estimate that the evolution of *Volvox*, a multicellular organism with a complete division of labor between soma and germ line, from a *Chlamydomonas*-like unicellular ancestor may have occurred in an interval of 50–75 million years, or even less (Rausch *et al.*, 1989). Taken together, these data support the concept that modern green flagellates last shared a common ancestor with the higher plants in deep antiquity and that their relationship to animals is even more archaic, but that the radiation of the volvocine algae was a relatively recent event on an evolutionary time scale (Schmitt *et al.*, 1992). Although these results certainly do not support the earlier view that *Volvox* represents an evolutionary 'missing link' between plants and animals, it is *prima vista* consistent with the combination of plant-like and animal-like features that *Volvox* exhibits.

The first algal histone genes were characterized in our laboratory and shown to exhibit a combination of features not previously described (Müller and Schmitt, 1988; Müller *et al.*, 1990; Lindauer *et al.*, 1993; Fabry *et al.*, 1995):

1. The *V. carteri* and *C. reinhardtii* genomes both contain about 15 copies of each type of nucleosomal histone gene; their derived peptide sequences more closely resemble those of plant than animal histones (Fabry *et al.*, 1995). However, volvocine histone genes are arranged

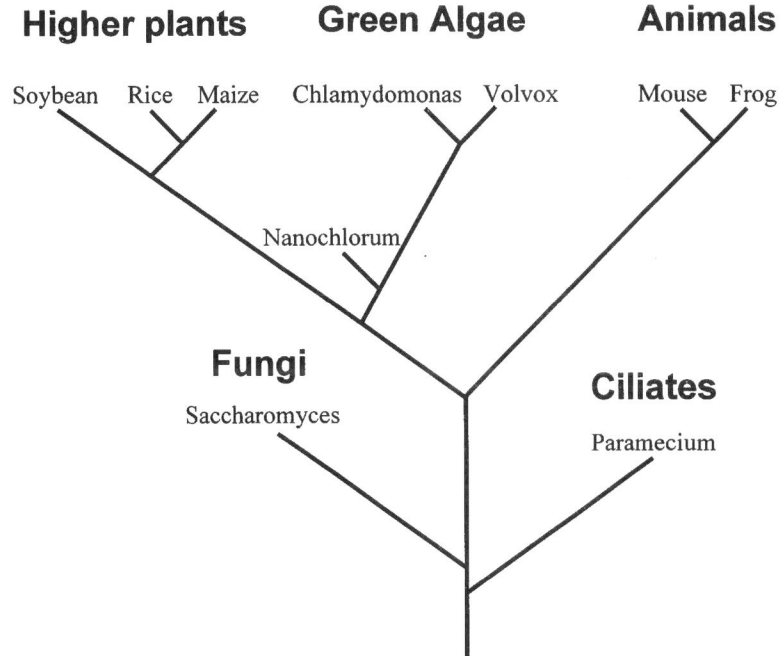

Fig. 26.2 Eukaryotic phylogenetic tree based on 18S rRNA sequence comparisons of selected organisms (modified from Rausch *et al.*, 1989). The volvocalean algae, *Chlamydomonas* and *Volvox*, appear at the tip of an algal branch that diverged long ago from the higher plant lineage.

     in clusters of H2A–H2B and H3–H4 pairs, a situation resembling that in some invertebrates and lower vertebrates. Each pair shows outwardly divergent transcription from a short intercistronic region that contains a promoter with plant-like enhancer elements (Fig. 26.3).

2.   Volvocine histone mRNAs, like the 'classical', replication-type of vertebrate histone mRNAs, but unlike those of plants (Chaboute *et al.*, 1988), are not polyadenylated; instead, their transcription is terminated by a 3' palindrome very similar to the palindromes seen in metazoan histone mRNAs.

3.   All *Volvox* and many *Chlamydomonas* H3 genes have an intron, a feature shared by fungal and protist H3 genes and by the vertebrate replacement-type H3.3 genes (Wells and Kedes, 1985; Wu *et al.*, 1986), but not by the replication-type H3 genes of either plants or animals.

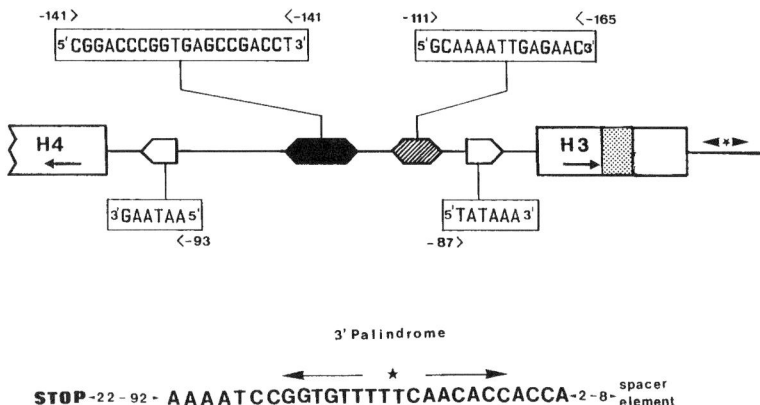

Fig. 26.3 Histone genes H3 and H4 of *Volvox* as example of the typical arrangement of volvocalean histone genes as outwardedly divergent pairs with their promoters (open arrowheads) and common enhancer elements (shaded hexagons with consensus sequences shown above), placed in the intercistronic region. The 3' palindrome consensus (symmetry indicated by divergent arrows), with distances (in bp) to the translation stop and to the spacer element (Müller *et al.*, (1990), is shown below.

This unusual combination of structural features reflects an interesting phylogenetic situation (Fig. 26.4). Although the coding sequences and enhancer elements of volvocine histone genes are more similar to plant than to animal histone genes, the algal genes resemble those of animals by possessing a 3' palindrome rather than a polyadenylation signal. On the other hand, by virtue of their split-gene organization, the volvocine H3 genes resemble the H3 genes of protists and the H3.3 genes of vertebrates (all of which are, however, polyadenylated). Three alternative hypotheses are advanced here to explain these observations:

1. The last common ancestor of plants and algae possessed only H3 genes like those now seen in the volvocine algae. On the evolutionary route leading to higher plants, introns and 3' palindromes were lost and polyadenylation was readapted from non-histone genes.
2. The last common ancestor of plants and algae possessed only histone genes like those now seen in the higher plants. On the evolutionary route leading to the volvocine algae, polyadenylation signals were lost and 3' palindromes (and introns?) were gained by horizontal transfer from animals.
3. Two types of H3 genes (some producing polyadenylated mRNAs and others producing palindrome-terminated mRNAs) were present in the last common ancestor of animals, plants and algae. Differential

Fig. 26.4 Comparison of histone H3 gene structures (bottom) from plants, *Volvox* and vertebrates with introns (black segments) and 3' signals (polyA or a palindrome, symbolized by divergent arrow heads). Their assigments to the plant or animal lineages are indicated above.

loss occurred in these three lineages: polyadenylated versions were lost in the algal lineage, palindrome-terminated versions were lost in the plant lineage, and both types were retained in the lineage leading to the vertebrates (Müller and Schmitt, 1988). This hypothesis also requires that introns were gained and/or lost in certain histone lineages, but since intron mobility appears to be a regular occurrence in evolution, this poses no major conceptual obstacle.

Although there are as yet no specific data that permit a clear choice among these hypotheses, we consider the third to be the most likely. Presumably, analysis of histone genes of other extant green algae, and particularly those that are thought to resemble most closely the algae that were ancestral to the higher plants (Devereux *et al.*, 1990), may provide evidence bearing on this question.

Assuming that the third hypothesis is correct, why might palindrome-terminated histone genes have been selected for in the volvocine lineage, but not in the higher plant lineage? We believe that it may be related to the unusual cell cycles that these algae exhibit. Whereas most eukaryotic cells, including those of higher plants, divide each time they have doubled in size, all volvocine algae alternate relatively long periods of growth in the absence of division with periods of rapid sequential divisions in the absence of growth. To be able to alternate periods in which they double their histones in the brief intervals (often << 1 hour) between successive divisions with long periods in which no histone synthesis occurs, they must be capable of rapid and efficient regulation of histone gene expression. It has been shown that palindrome-terminated histone mRNAs are well suited for such purposes, because they turn over much more rapidly than polyadenylated histone messages (Schümperli, 1988).

26.2.2   PHYLOGENETIC RELATIONSHIPS WITHIN THE VOLVOCINE LINEAGE

In an attempt to evaluate the volvocine lineage hypothesis (Fig. 26.1), a composite dendrogram of 20 volvocine algae, based on partial 18S and 28S rRNA sequences (Larson *et al.*, 1992, and unpublished data) and on the sequences of certain class II introns (Liss *et al.*, 1997), was derived (Fig. 26.5). These data indicate that the family Volvocaceae (the volvocine algae minus *Chlamydomonas*) is a coherent group that is closely related to *C. reinhardtii*, but they do not support any simple, monophyletic relationship either within or among its various genera. In all three cases in which more than one representative of a volvocacean genus has been analyzed (namely, *Eudorina*, *Pleodorina* and *Volvox*), maximum-parsimony analysis fails to cluster the members of that genus as would be expected if the genus were monophyletic. Instead, it locates representatives of each of them at quite disparate locations within the tree, interspersed with members of other genera. Based on such observations, Larson *et al.* (1992) suggested that the colonial life style and the multicellular, differentiated life style 'may represent different stable states, among which there may have been multiple transitions during the phylogeny of the group'. These authors went on to speculate that 'only a small number of genetic changes may be required to effect a transition in either direction,

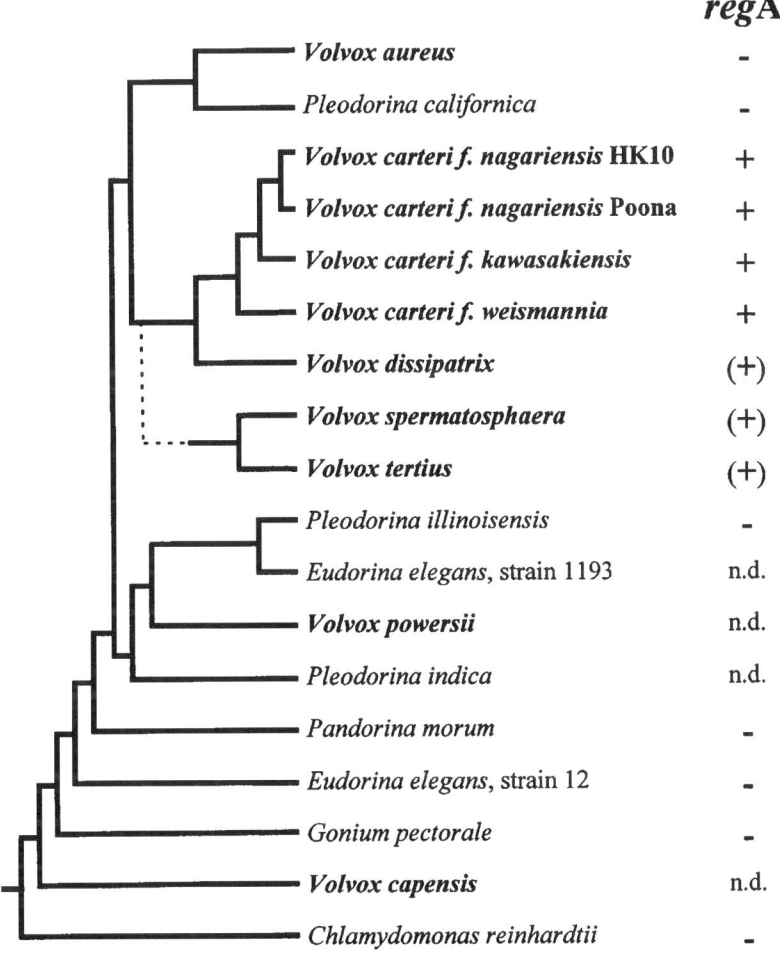

regA

Volvox aureus	-
Pleodorina californica	-
Volvox carteri f. nagariensis **HK10**	+
Volvox carteri f. nagariensis **Poona**	+
Volvox carteri f. kawasakiensis	+
Volvox carteri f. weismannia	+
Volvox dissipatrix	(+)
Volvox spermatosphaera	(+)
Volvox tertius	(+)
Pleodorina illinoisensis	-
Eudorina elegans, strain 1193	n.d.
Volvox powersii	n.d.
Pleodorina indica	n.d.
Pandorina morum	-
Eudorina elegans, strain 12	-
Gonium pectorale	-
Volvox capensis	n.d.
Chlamydomonas reinhardtii	-

Fig. 26.5 A composite dendrogram indicating sister-taxon relationships among 18 volvocine algae deduced from rRNA sequence comparisons (modified from Larson *et al.*, 1992, on the basis of additional unpublished data of D.L. Kirk and A. Larson), and on class II intron sequence comparisons (M. Liss *et al.*, 1997). Dashed line indicates provisional placement based on morphological criteria (Smith, 1944). Results from a *regA* gene screen are added: +, positive signal; (+) weak signal; -, no signal; n.d., not determined.

between the colonial and multicellular . . . life history'. Thus, in addition to falsifying the simple, traditional volvocine lineage hypothesis, this study suggests that the genetic origins of the *Volvox* cellular differentiation program might be simple enough to be amenable to detailed molecular analysis.

26.2.3   THE *REGA* GENE

The diagnostic feature linking all of the species in the genus *Volvox* is a germ–soma dichotomy, in which only a few cells (called gonidia) exhibit any reproductive potential, whereas the rest (the somatic cells) are terminally differentiated and eventually die. Mutational analysis has shown that in *V. carteri* the terminal differentiation of somatic cells is controlled by a single gene, called *regA*, which functions to repress (directly or indirectly) all of the genes required for reproduction (Kirk, 1988; Schmitt *et al.*, 1992). Recently the *Volvox* transposon, *Jordan* (Miller *et al.*, 1994), has been used to tag and recover the *regA* gene of *V. carteri* f. *nagariensis* strain HK 10. Saturation transposon mutagenesis has been used to identify an ~ 10 kb region of the genome within which the *regA* gene is located (Fig. 26.6). The sequencing of genomic and cDNA clones is in progress (Kirk *et al.*, 1998).

The cloning and characterization of the *regA* locus has opened up the possibility of asking two interesting evolutionary questions:

* Is the same molecular genetic mechanism (repression of gonidial genes by a *regA* homologue) used to generate terminally differentiated somatic cells in all species of *Volvox* or are different genetic mechanisms used to reach the same end in different *Volvox* lineages?
* Can precursors of the *regA* gene be identified in the simpler volvocacean relatives of *V. carteri* that lack its characteristic germ–soma differentiation?

Regenerator (*regA*) gene locus of *Volvox carteri*

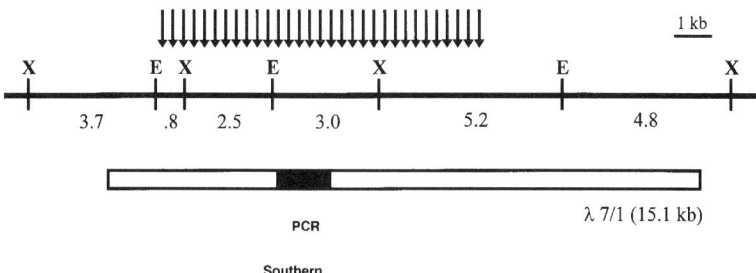

Fig. 26.6 **Top:** the *regA* locus of *Volvox* as defined by multiple *Jordan* transposon insertions identified in *regA* mutants (vertical arrows). **Center:** physical map with restriction sites (E, *Eco*RI; X, *Xho*I) and fragment sizes (in kb). **Bottom:** genomic λEMBL3-based *regA* clone with a conserved exon (black). The 0.7-kb and 3.1-kb fragments used for probing by PCR and Southern technics are shown.

As an approach toward answering these questions, we have recently initiated efforts to screen for the presence of *regA* analogues in other volvocaceans. In the first such study we used PCR in an attempt to amplify from the genomes of certain closely related volvocaceans a 700 bp fragment of the *regA* locus that encodes a deduced peptide with strong homology to a DNA-binding motif present in many families of eukaryotic transcription factors. An interfertile isolate of *V. carteri* f. *nagariensis* (the Poona strain) yielded a PCR product that was shown (by nucleotide sequence data) to encode the same polypeptide (although with three synonymous nucleotide differences), whereas the nearest non-interfertile relative, *V. carteri* f. *kawasakiensis*, exhibited 78 nucleotide differences resulting in 12 amino acid substitutions. Screening for a *regA* analogue in other volvocine algae by Southern hybridization (using a 3.1 kb *regA* fragment from HK10) (Fig. 26.6) produced a positive signal in the closely related *V. carteri* f. *weismannia*, and weak signals in *V. dissipatrix*, *V. spermatophora* and *V. tertius*, but not in *V. aureus* or any of the other volvocine algae tested (Fig. 26.5). While these results suggest that the *regA* gene, as a key developmental determinant, is common to all immediate relatives of *Volvox carteri*, either the presumptive *regA* analogue of *V. aureus* is too dissimilar for detection by Southern blotting or this species uses a different molecular mechanism of cytodifferentiation. Moreover, these data support the assignment of various volvocaceans to the *V. carteri* lineage, as is shown in Fig. 26.5. Degenerate primers, different PCR conditions and primers directed towards other regions of the *regA* gene locus are now being tested in an effort to generate positive signals in other volvocaceans and extend the analysis.

ACKNOWLEDGMENTS

This study was supported by grants from the Deutsche Forschungsgemeinschaft to R.S. (SFB 521/B1) and to S.F. (Fa232/6-1), and from the National Science Foundation (MCB9304447) and the US Department of Agriculture (95-37304-2229) to D.L.K.

REFERENCES

Amati, B.B., Goldschmidt-Clermont, M., Wallace, C.J.A. and Rochaix, J.-D. (1988). cDNA and deduced amino acid sequences of cytochrome *c* from *Chlamydomonas reinhardtii*: unexpected functional and phylogenetic implications. *J. Mol. Evol.* **28**: 151–160.

Chaboute, M.E., Chaubet, N., Clement, B. *et al.* (1988) Polyadenylation of histone H3 and H4 mRNA in dicotyledonous plants. *Gene* **71**: 217–233.

Devereux, R., Löblich, A.R. III and Fox, G.E. (1990) Higher plant origins and the phylogeny of green algae. *J. Mol. Evol.* **31**: 18–24.

Fabry, S., Müller, K., Lindauer, A. *et al.* (1995) The organization, structure and regulatory elements of *Chlamydomonas* histone genes reveal features linking plant and animal genes. *Curr. Genet.* **28**: 333–345.

Kirk, D.L. (1988) The ontogeny and phylogeny of cellular differentiation in *Volvox*. *Trends Genet.* **4**: 32–36.

Kirk, M.M., Müller, W., Stark, K. *et al.* (1998) *regA*, the gene repressing reproductive development in *Volvox*, encodes an unusual putative transcription factor. *Cell* (submitted).

Larson, A., Kirk, M.M. and Kirk, D.L. (1992) Molecular phylogeny of the volvocine flagellates. *Mol. Biol. Evol.* **9**: 85–102.

Lindauer, A., Müller, K., and Schmitt, R. (1993) Two histone H1-encoding genes of the green alga *Volvox carteri* with features intermediate between plant and animal genes. *Gene* **129**: 59–68.

Liss, M., Kirk, D.L., Beyser, K. and Fabry, S. (1997) Intron sequences provide a tool for high-resolution phylogenetic analysis of volvocine algae. *Curr. Genet.* **31**: 214–227.

Miller, S.M., Schmitt, R. and Kirk, D.L. (1993) *Jordan*, an active *Volvox* transposable element similar to higher plant transposons. *Plant Cell* **5**: 1125–1138.

Müller, K. and Schmitt, R. (1988) Histone genes of *Volvox carteri*: DNA sequence and organization of two H3–H4 gene loci. *Nucleic Acids Res.* **16**: 4121–4136.

Müller, K., Lindauer, A., Brüderlein, M. and Schmitt, R. (1990) Organization and transcription of *Volvox* histone-encoding genes: similarities between algal and animal genes. *Gene* **93**: 167–175.

Rausch, H., Larsen, N. and Schmitt, R. (1989) Phylogenetic relationships of the green alga *Volvox carteri* deduced from small-subunit ribosomal RNA comparisons. *J. Mol. Evol.* **29**: 255–265.

Schmitt, R., Fabry, S. and Kirk, D.L. (1992) In search of molecular origins of cellular differentiation in *Volvox* and its relatives. *Int. Rev. Cytol.* **139**: 189–265.

Schümperli, D. (1988) Multilevel regulation of replication-dependent histone genes. *Trends Genet.* **4**: 187–191.

Smith, G.M. (1944) A comparative study of the species of *Volvox. Trans. Amer. Micros. Soc.* **63**: 265–310.

Wells, D. and Kedes, L. (1985) Structure of a human histone cDNA: evidence that basally expressed histone genes have intervening sequences and encode polyadenylated mRNAs. *Proc. Natl Acad. Sci. USA* **82**: 2834–2838.

Wu, M., Allis, C.D., Richman, R. *et al.* (1986) An intervening sequence in an unusual histone H1 gene of *Tetrahymena thermophila*. *Proc. Natl Acad. Sci. USA* **83**: 8674–8678.

Horizontal gene transfer and fusing lines of descent: the Archaebacteria – a chimera?

27

Lorraine Olendzenski, Elena Hilario and J.Peter Gogarten

SUMMARY

Conflicting molecular phylogenies provide evidence for horizontal gene transfer events that have led to the evolution of chimeric lineages. For example, the endosymbiont theory explains the origin of mitochondria and plastids from eubacteria that lived as endosymbionts inside the eukaryotic cell. Molecular evolution studies of proteins and nucleic acids that function inside these organelles have thoroughly corroborated the proposed fusion of formerly independent lines of descent which led to the origin of these organelles. Since the recognition, based on 16S rRNA phylogenetic analysis, that archaebacteria form a domain separate from other prokaryotes, a growing list of characteristics has forced even the most reluctant biologists to accept the archaebacteria (domain Archaea) as a distinct group separate from the eubacteria (domain Bacteria). Just as the unity of this domain is being accepted, however, new trouble is brewing fueled by the increased availability of molecular sequences. While many genes in the Archaea have been found to be very similar to their eukaryotic counterparts, a number of well resolved molecular phylogenies reveal archaeal genes that are best described as prokaryotic in character. We suggest that horizontal gene transfer is the best explanation for the presence of abundant eubacterial genes in archaeal genomes.

27.1   INTRODUCTION

The phylogenies of many gene families are in striking agreement with 16S rRNA phylogenies; cases of horizontal gene transfer are so infrequent that they show up as exceptions in an otherwise coherent picture. Rapidly accumulating data force us to take a closer look at these exceptions. The genome projects currently under way provide a wealth of data that is useful in studying these ancient gene transfer events. In particular, it is possible to classify gene families according to the type of phylogeny they represent: genes that reflect the archaea as distinct from the eubacteria; genes that group the archaea among the eubacteria; and genes that did not retain sufficient phylogenetic information to address this relationship (e.g. Doolittle and Brown, 1994).

One of the best corroborated examples for the fusion of two lines of descent is the evolution of mitochondria and plastids from formerly free-living bacteria. Additional examples of the transfer of genetic information between distantly related species continue to be discovered (Smith *et al.*, 1992; for reviews, see Margulis, 1993; Margulis and Fester, 1991). The fusion of lineages, via the transfer of genes horizontally amongst distantly related organisms therefore, has played an integral role in early evolution (Gogarten, 1995).

27.2   COMPARISON OF MOLECULAR MARKERS CAN REVEAL EPISODES OF HORIZONTAL GENE TRANSFER

Phylogenies based on single molecules (single gene trees) cannot readily be used to determine a true species phylogeny since the reconstruction of phylogenies from different molecular markers often results in phylogenetic trees with differing topologies. A phylogenetic tree based on a single molecule can only reconstruct the history of that molecule, and may or may not reflect true species evolution. Rather, species phylogeny must be determined using a consensus of characters, including a variety of molecular sequence data, distribution of biochemical pathways and, when available, morphological and life history data.

Conflicting molecular phylogenies can occur for a number of reasons. For example, the phylogenies under consideration may not be sufficiently resolved. The data being used might not contain sufficient information because the molecule being analyzed is too short, or because regions within the molecule are saturated with substitutions or are not variable enough to contain useful phylogenetic information for the algorithm used to calculate the tree. Trees with identical topology can also differ drastically with respect to branch lengths because of varying substitution rates for each molecule. There is no reason to assume that the selection pressure on any given molecule is constant throughout evolution, nor is it reasonable to assume that the selection pressure varied identically for different molecular markers.

Phylogenies involving comparison of paralogous genes can also give results that differ from the species tree. Gene duplications can give rise to two diverging forms of a gene within a single organism (gene A and gene B). Each of the two forms is paralogous to each other (Fitch, 1970). In attempting to reconstruct molecular phylogeny it is mandatory that analyses be done using the appropriate orthologous genes (all gene As or all gene Bs) from different organisms. Comparing unrecognized paralogues from different organisms results in tree topologies that differ from the species phylogeny and that are misleading, since divergence between A and B reflects a gene duplication event and not a speciation event.

Comparisons of ancient gene duplications can be useful in rooting the universal tree of life (Schwartz and Dayhoff, 1978). When comparing a large number of examples (orthologs) of each of two paralogous or duplicated genes, one set of the paralogs can be used as an outgroup for the other. This approach has been applied to the catalytic and non-catalytic subunits of H^+-pumping ATPases (section 27.3) (Gogarten et al., 1989), elongation factors EF-Tu and EF-G (Iwabe et al., 1989; Cammarano et al., 1992, Baldauf et al., 1996) and aminoacyl tRNA synthetases (Brown and Doolittle, 1995). All support the position of the root of the universal tree of life as being between eubacteria on one side and the archaea and eukaryotes on the other.

Conflicting, well-resolved molecular phylogenies that are not based on unrecognized paralogs are evidence for horizontal gene transfer. Two types of transfer can be recognized; if only a single molecule deviates from a majority consensus of other markers, this represents the transfer of only a single gene. Homology that arises from the horizontal transfer of a single gene has been termed xenology (Gray and Fitch, 1983). However, if a number of trees constructed from different molecular markers all yield the same aberrant topology when compared with the species consensus, these conflicting topologies can be explained by the transfer of a large number of genes from one lineage into another. This type of homology, which represents the transfer of a significant portion of a genome (and consequently, lineage fusion), is termed synology (Gogarten, 1994), and may have contributed to major events in the history of life. These two types of homology, derived from horizontal gene transfer, can be distinguished by the comparison of phylogenies constructed from different molecular markers. Synology is revealed by two incompatible patterns, each of which is strongly supported by many molecular markers. In contrast, xenology is revealed as a deviation of a single molecular marker from the majority consensus of other molecular markers. Both cases of horizontal gene transfer reconcile conflicting rooted phylogenies by explaining the more recent of the conflicting bifurcations as a case of horizontal transfer.

27.3   TRANSFER FROM ARCHAEA TO EUBACTERIA: THE EXAMPLE OF H⁺-PUMPING ATPASES

Proton-pumping ATPases/ATP synthases are found in all groups of extant organisms. Traditionally these ATPases are divided into three categories: vacuolar (V), bacterial (F) and P-types. F-ATPases occur as the ATP synthase in mitochondria and chloroplasts of eukaryotes and as an ATP synthase or ATP-driven proton pump in eubacteria. F-ATPases are homologous to the V-ATPases found in the endomembrane system of eukaryotes. In all cases studied, the eukaryotic vacuolar-type ATPase *in vivo* functions exclusively as a proton-pumping ATPase, not as an ATP synthase (Harvey and Nelson, 1992). V-ATPases energize membranes for a multitude of secondary active transport systems and also provide compartments with an acidic interior. Archaea possess H⁺-ATPases called A-ATPases, which are similar in molecular sequence to V-ATPases but whose function and quarternary structure are more similar to the F-ATPases in that they can synthesize ATP. P-type ATPases function as cation pumps and are not homologous to the V/F/A-type ATPases.

The vacuolar (V), bacterial (F) and archaeal (A) ATPases exhibit a number of structural and functional similarities. All are large multisubunit enzymes composed of a water soluble ($V/F/A_1$) portion and a proton-conducting membrane-spanning complex ($V/F/A_0$) portion. Each hydrophilic sector contains three copies of a catalytic (β-subunit in F-ATPases, *A*-subunit in A/V-ATPases) and three copies of a non-catalytic subunit (α-subunit in F-ATPases, *B*-subunit in A/V-ATPases). The catalytic subunits bind and hydrolyze ATP. The non-catalytic or regulatory subunits also bind ATP, but they do not hydrolyze it.

The A-, V- and F-ATPases are homologous enzymes, i.e. they evolved from the same ancestral enzyme (Zimniak, 1988; Gogarten *et al.*, 1992). In addition, the catalytic and the non-catalytic subunits have evolved from the same ancestral gene after an ancient gene duplication and are therefore paralogous subunits. The gene duplication which gave rise to these subunits had already occurred in the last common ancestor of the eubacteria, archaea and eukaryotes and comparison of these paralogous subunits with each other has been used to root the tree of life (Gogarten *et al.*, 1989, Gogarten and Taiz, 1992). Such analyses support the sister group relationship of the archaea and eukaryotes and place the root of the universal tree of life between eubacteria on one side and eukaryotes and archaea on the other.

The phylogenies calculated from ATPase sequences (Fig. 27.1) agree well with phylogenies calculated from other markers (e.g. 16S-like

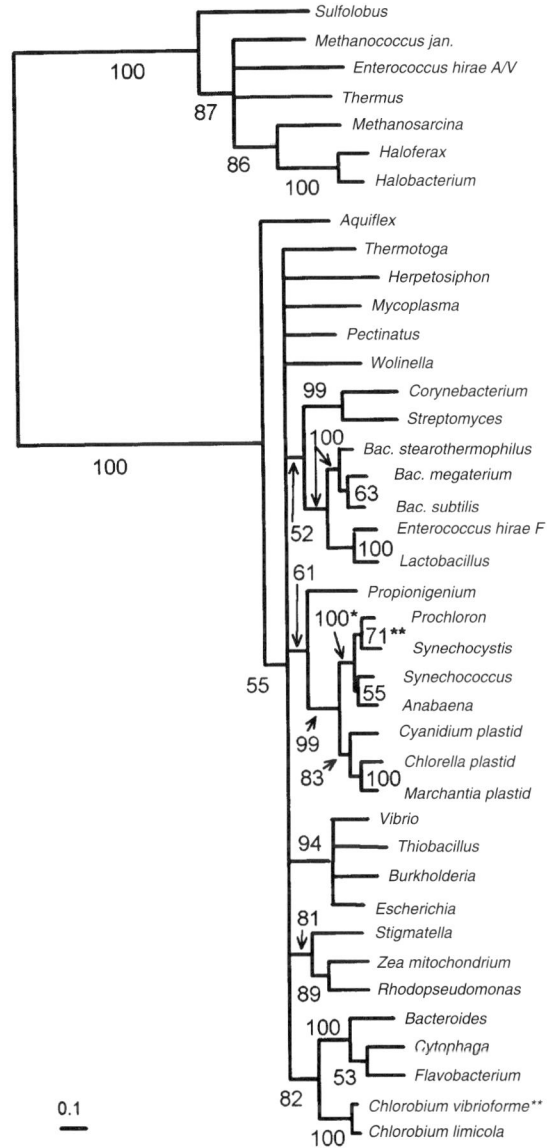

Fig. 27.1 Distance matrix analysis of catalytic subunits of F-ATPases from eubacteria (β subunit) and A-ATPases from archaea (A subunit). Amino acid sequences were aligned and bootstrapped samples were calculated using ClustalW v.1.6. Positions with gaps were excluded. All branches which were recovered in less than 50% of the bootstrapped samples were collapsed. Branch lengths were calculated using maximum likelihood as implemented in PUZZLE (Strimmer and von Haeseler, 1996) using the JTT model and the user tree option. Asterisks denote branches that were found to be only one (*) or two (**) times larger than the determined standard deviation. All other branches were at least three times as long as their standard deviation. Sequences for *Aquiflex* and *Thermotoga* were kindly provided by Drs H.-K. Schleifer and W. Ludwig, Univerisity of Munich; all others obtained from GenBank.

rRNA: Gogarten *et al.*, 1996); therefore, it appears that the coupling factor ATPases have evolved together with the organisms in which they reside and can be used as reliable phylogenetic markers. The majority of eubacteria have an F-ATPase and the archaea in which the ATPase has been studied possess the homologous A-ATPase.

Two exceptions exist: *Thermus thermophilus* and *Enterococcus hirae* possess A/V-ATPases (Hilario and Gogarten, 1993). Based on rRNAs and other biochemical characters (cell wall, lipids, antibiotic resistance) *Thermus* is classified as bacterial genus, with closest similarities to the genus *Deinococcus* (Woese, 1987; Hensel *et al.*, 1986). However, *Thermus thermophilus* does not have an F-ATPase, like other eubacteria, but an A/V-ATPase (Yokoyama *et al.*, 1990; Tsutsumi *et al.*, 1991). Both major subunits of the *Thermus* ATPase are clearly archaeal in character (Gogarten *et al.*, 1992). The eubacterium *Enterococcus hirae* has a normal bacterial F-ATPase that groups, as expected, with other low GC Gram-positives (Fig. 27.1); in addition *E. hirae* has a sodium-pumping ATPase that, based on its sequence, is classified as an archaeal type ATPase (Takase *et al.*, 1994; Hilario and Gogarten, 1993). One possible explanation is that the last common ancestor already had both a V- and an F-type ATPase, and that one or the other ATPase type would have been lost in most of the lineages leading to extant organisms. The conflicting topology of the tree in Fig. 27.1, then, would be a result of the comparison of unrecognized paralogs (Forterre *et al.*, 1993). However, the sequences of the *Thermus* and the *Enterococcus* A/V-ATPase are clearly archaeal. These sequences group **within** the archaeal domain, together with the subunits of *Halobacterium* and *Methanosarcina*. If these were eubacterial paralogs of the A-ATPases, they would be expected to form a clade separate from the archaeal A-ATPases. Additionally, the complete genome sequence of *Methanococcus jannashcii* indicates that it only possesses one A-ATPase, with no F-ATPase present. Similarly, the complete genomes of *Mycoplasma* and *Haemophilus* reveal the presence of only one complete F-ATPase, with no corresponding V-ATPase (data from the Institute for Genomic Research, http://www.tigr.org). Therefore, we interpret the grouping of the *Thermus* and *Enterococcus* ATPase within the Archaea as an indication of a horizontal gene transfer across domain boundaries (Hilario and Gogarten, 1993). If these ATPases were acquired by horizontal gene transfer, their position in the phylogeny suggests that they were transferred from the archaeal lineage (in two separate events) to these eubacteria.

27.4   THE ARCHAEA: A CHIMERIC LINEAGE?

Since the recognition that the Archaea form a cohesive group (Woese and Fox, 1977), a growing list of characteristics has forced even the most reluctant biologists to accept the archaebacteria (Archaea) as a distinct group separate from the eubacteria (Bacteria) (Table 27.1). There continues to be debate concerning the possibility that the archaea might constitute a paraphyletic group, because some or all of them share a common ancestor with a lineage that gave rise to the eukaryotic nucleocytoplasm (Rivera and Lake, 1992; Lake and Rivera, 1996; Olsen *et al.*, 1996). However, at the same time that the unity of the domain Archaea is being accepted, increased availability of molecular sequences is forcing us to take a closer look at the evolution of this group. Many genes in the Archaea have been found to be very similar to their eukaryotic counterparts, especially those involved in genome structure, transcription and translation (Marsh *et al.*, 1994; Langer *et al.*, 1995; Keeling and Doolittle, 1995). These findings support the identification of the Archaea as the group giving rise to the eukaryotic nucleocytoplasmic component, a relationship first indicated through studies of ancient duplicated genes (Gogarten *et al.*, 1989; Iwabe *et al.*, 1989).

Many other archaebacterial genes are best described as prokaryotic in character. Molecular phylogenies of these genes suggest a fundamental division between Prokaryotes and Eukaryotes. Phylogenetic trees constructed from these markers, many of which are enzymes in core metabolic pathways, do not yield branches that group archaebacteria separately from eubacteria. Rather, they yield trees that show archaebacterial branches interspersed throughout their eubacterial counterparts (Table 27.1). These include homologs of the heat shock proteins (HSP70) (Gupta and Singh, 1992; Gupta and Golding, 1993), glutamine synthetases (Brown *et al.*, 1994; Kumada *et al.*, 1993; Tiboni *et al.*, 1993), nitrogenases (Souillard *et al.*, 1988; Chien and Zinder, 1994), glutamate dehydrogenases (Benachenhou-Lafha *et al.*, 1993; Hilario and Gogarten, 1993) and an F-ATPase subunit fragment from *Methanosarcina* (Sumi *et al.*, 1992; Hilario and Gogarten, 1993). Distance and parsimony analyses of enolase and phosphoglycerate kinase (glycolytic enzymes), dihydrofolate reductase (pterine biosynthesis pathway) and dihydrolipoamide dehydrogenase also give tree topologies in which the archaeal sequences group among the eubacteria (Gogarten, unpublished data).

Distance trees constructed using portions of carbamylphosphate synthetase – an enzyme involved in pyrimidine biosynthesis, arginine biosynthesis and the urea cycle – give an even more intriguing topology (Fig. 27.2). The large subunit of this enzyme contains a duplicated region, each element of which can be compared with the other to create a rooted phylogeny (Schofield, 1993; van den Hoff *et al.*, 1995). In this tree, *Sulfolobus*, an archaebacterium, shares a sister group relationship with the eukaryotic orthologs (Fig. 27.2) (Lawson *et al.*, 1996). Two other archaeal sequences – those from *Methanococcus* and *Methanosarcina* – group within the eubacteria (Fig. 27.2) (A. Lazcano, V. Puente and J.

Table 27.1 Characters supporting the division or close association of Archaea and Bacteria

Characters supporting division	Characters supporting close association
16S-like rRNA	nitrogenases
other ribosomal RNAs (23S and 5S)	phosphoglycerate kinases
ribosomal proteins	enolases
RNA polymerases	dihydrolipoamide dehydrogenases
promoter organization	dihydrofolate reductases
TATA binding protein, transcription factors	F-ATPase subunit encoding DNA fragment isolated from *Methanosarcina*
membrane lipids	homologues of the 70 kDa heat shock proteins (HSP70)[2]
(ether lipids characterize archaebacteria)	glutamine synthetases[2]
fatty acid synthetase	glutamate dehydrogenases[2]
cell wall structure (murein in eubacteria)	carbamoylphosphate synthetases[2]
flagella (eubacterial or archaebacterial flagellin)	
N-linked glycosylation in archaebacteria	
lack of formylmethionine in archaebacteria	
sensitivity towards antibiotics	
ribosome binding motif on mRNAs	
tRNAs[1]	
dehydrogenases[1]	
proton pumping ATPases[1]	
elongation factors 1α/TU and 2/G[1]	
aminoacyl tRNA synthetases[1]	

[1] These characters underwent an ancient gene duplication before the three domains diverged from each other. All of them support the archaea as the sister group to the eukaryotic nucleocytoplasm. (Modified from Zillig *et al.*, 1992; Keeling and Doolittle, 1995.)
[2] These characters underwent an ancient gene duplication. At least some archaeal sequences were found to group among their eubacterial homologues.

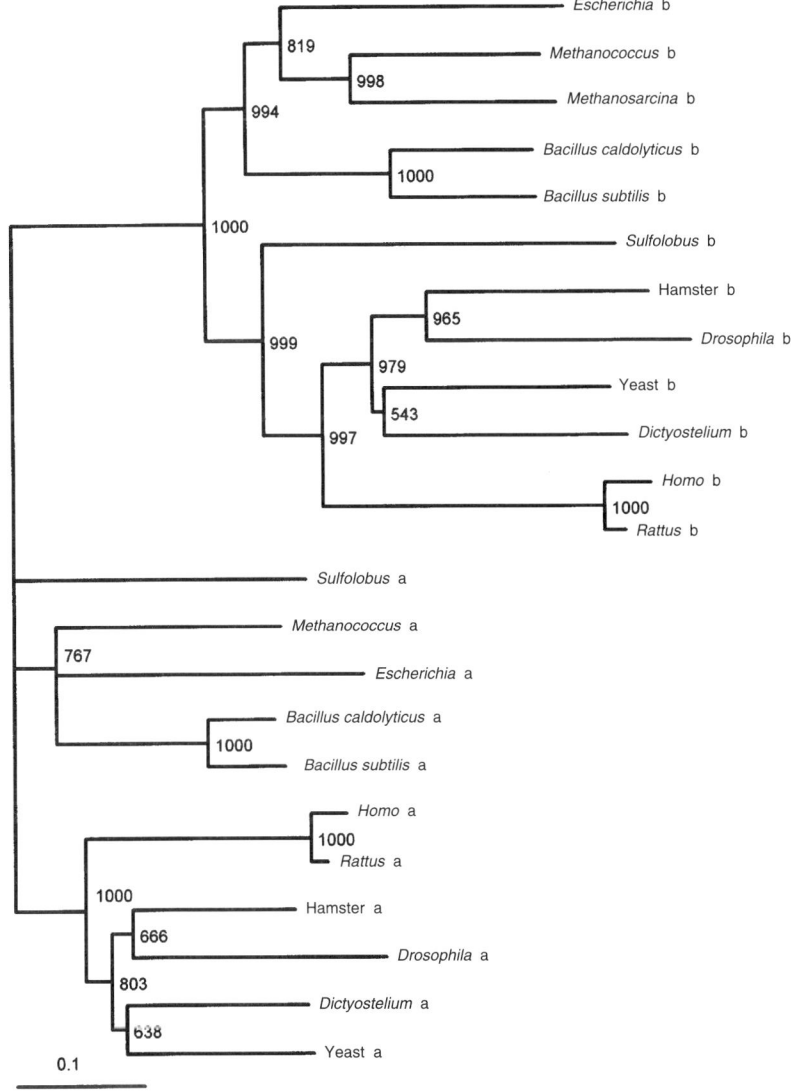

Fig. 27.2 Distance tree constructed using 273 amino acid positions within each half of the internal duplication within the large subunit of carbamylphosphate synthetase (carB). Each half of the internal duplication was determined by dot-matrix plot analyses of the yeast carB gene. Positions corresponding to homologous domains in **(a)** the amino-terminal half and **(b)** the carboxy-terminal half were aligned using ClustalW v.1.6 (Thompson *et al.*, 1994) and truncated to correspond to the homologous 273 amino acid positions of the available b fragment of *Methanosarcina barkeri* (complete sequence not available). 1000 bootstrapped replicates were performed using the phylogenetic tree option of ClustalW v.1.6 with gap positions excluded. The tree was rooted by using all the amino-terminal sequences as the outgroup. Trees with the same topology among the carboxy-terminal segment were obtained using parsimony (PROTPARS program of PHYLIP; Felsenstein, 1993) and maximum likelihood analyses (PUZZLE). Amino acid sequence alignments are available from the authors.

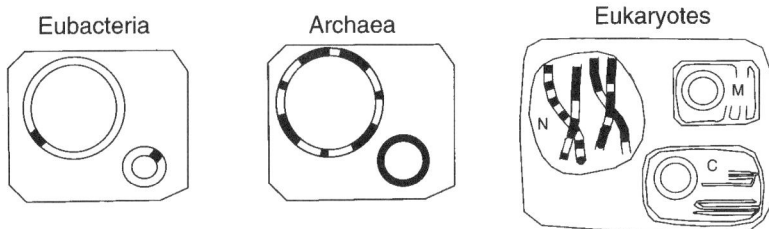

Fig. 27.3 Distribution of genes among eubacteria, archaea and eukaryotes based on conflicting molecular phylogenies. Eubacteria: few archaeal genes (dark segments); e.g. A-ATPases in *Thermus* and *Enterococcus*. Archaea: several eubacterial genes (white segments), in particular genes involving biosynthesis, e.g. glutamate dehydrogenase, glutamine synthase, carbamylphosphate synthetase, also HSP70 homologues. Eukaryotes: many archaeal genes involved in transcription and translation, also V/A-ATPases; eubacterial genes in mitochondria and chloroplasts; many genes that encode biosynthetic enzymes reflect Prokaryote/Eukaryote dichotomy.

Leguina, personal communication). Taken together, these examples suggest that there are abundant occurrences of eubacterial genes in archaebacterial genomes. The genome composition that emerges from the characterization of genes within the three cell lineages is summarized in Fig. 27.3.

Initially, gene families that grouped the archaebacteria among the eubacteria were considered as examples of bad molecular markers with ill-resolved phylogenies. However, more of these molecular phylogenies are being discovered and many of them appear to be well resolved. Horizontal gene transfer is the best explanation for the presence of eubacterial genes in archaeal genomes (Fig. 27.4) (Gogarten *et al.*, 1996). At present, insufficient information is available to pinpoint the phylogenetic positions of the donor and recipient of this (or these) horizontal gene transfer event(s). It is not even possible to decide whether these eubacterial genes were contributed to the archaeal lineage in one massive horizontal transfer (resulting from a fusion of the archaeal ancestor with a eubacterium), or if they were transferred in a trickle-down fashion (as important innovations evolved in the eubacterial domain these were passed horizontally to other organisms that could make use of them). The latter scenario is supported by the data of carbamylphosphate synthetase, which places the archaeal sequence branches into two domains (Eukaryotes and Bacteria) and the glutamine synthetases (Brown *et al.*, 1994) in which sequences from *Sulfolobus* consistently group at the base of the eubacteria, while those from other archaea group within the eubacteria.

Although current genome projects will help to identify gene families

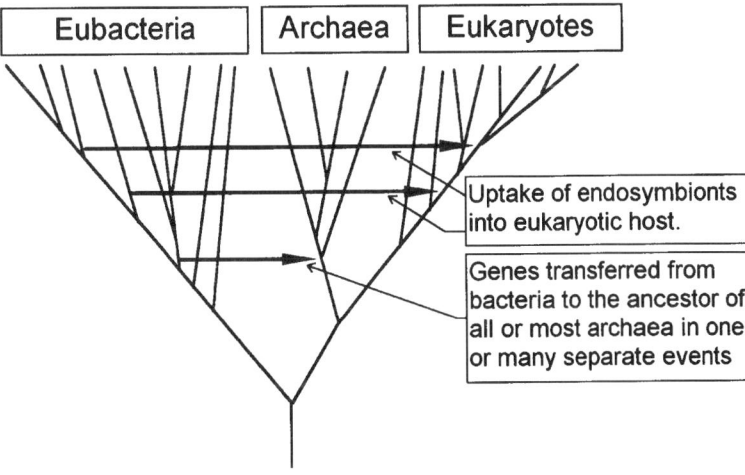

Fig. 27.4 Proposed transfer events from eubacteria to the archaea and eukaryotes.

that have undergone horizontal transfer, a greater sampling of species is needed to address the following questions. Did the formation of a chimera occur at the base of the archaeal domain? If so, what was the phylogenetic position of the eubacterial contributor? Might phylogenetic positions of donor and recipient provide hints as to the mechanisms of these transfers? If these transfers occurred after the diversification of archaea, how many independent events occurred during early evolution, and how quickly did newly evolved biosynthetic pathways spread through the different domains? The importance of gene transfer and fusion events and the molecular evidence supporting them should not be overlooked in our attempts to reconstruct the early evolutionary history of life on Earth.

ACKNOWLEDGMENTS

Research in our laboratory was supported through the NASA Exobiology program. E.H. was supported in part by a fellowship from the DGAPA-Universidad Nacional Autónoma de México.

REFERENCES

Baldauf, S.L., Palmer, J.D. and Doolittle, W.F. (1996) The root of the universal tree and the origin of eukaryotes based on elongation factor phylogeny. *Proc. Natl Acad. Sci. USA* **93**: 7749–7754.
Benachenhou-Lafha, N., Forterre, P. and Labedan, B. (1993) Evolution of gluta-

mate dehydrogenase genes: evidence for two paralogous protein families and unusual branching patterns of the archaebacteria in the universal tree of life. *J. Mol. Evol.* **36**: 335–346.

Brown, J.R. and Doolittle, W.F. (1995) Root of the universal tree of life based on ancient aminoacyl-tRNA synthetase. *Proc. Natl Acad. Sci. USA* **92**: 2441–2445.

Brown, J.R., Masuchi, Y., Robb, F.T. and Doolittle, W.F. (1994) Evolutionary relationships of bacterial and archaeal glutamine synthetase genes. *J. Mol. Evol.* **38**: 566–576.

Cammarano, P., Palm, P., Creti, R. *et al.* (1992) Early evolutionary relationships among known life forms inferred from elongation factor ef-2 (ef-g) sequences. Phylogenetic coherence and structure of the archaeal domain. *J. Mol. Evol.* 34: 396–405.

Chien, Y.T. and Zinder, S.H. (1994) Cloning, DNA sequencing, and characterization of a nifD-homologous gene from the archaeon *Methanosarcina barkeri* 227 which resembles nifD1 from the eubacterium *Clostridium pasteurianum*. *J. Bact.* **176**: 6590–6598.

Doolittle, W.F. and Brown, J.R. (1994) Tempo, mode, the progenote, and the universal root. *Proc. Natl Acad. Sci. USA* **91**: 6721–6728.

Felsenstein, J. (1993) Phylogeny Inference Package, version 3.5c. Distributed by the author. Department of Genetics, University of Washington, Seattle.

Fitch, W.S. (1970) Distinguishing homologous from analogous proteins. *Syst. Zool.* **19**: 99–113.

Forterre, P., Benachenhou-Lafha, N., Confalonieri, F. *et al.* (1993) The nature of the last universal ancestor and the root of the tree of life, still open questions. *BioSystems* **28**: 15–32.

Gogarten, J.P. (1994) Which is the most conserved group of proteins? Homology–orthology, paralogy, xenology and the fusion of independent lineages. *J. Mol. Evol.* **39**: 541–543.

Gogarten, J.P. (1995) The early evolution of cellular life. *TREE* **10**: 147–151.

Gogarten, J.P. and Taiz, L. (1992) Evolution of proton pumping ATPases: rooting the tree of life. *Photosyn. Res.* **33**: 137–146.

Gogarten, J.P., Kibak, H., Dittrich, P. *et al.* (1989) Evolution of the vacuolar H+-ATPase: implications for the origin of eukaryotes. *Proc. Natl Acad. Sci. USA.* **86**: 6661–6665.

Gogarten, J.P., Starke, T., Kibak, H. *et al.* (1992) Evolution and isoforms of V-ATPase subunits. *J. Exp. Biol.* **172**: 137–147.

Gogarten, J.P., Hilario, E. and Olendzenski, L. (1996) Gene duplications and horizontal gene transfer during early evolution, in *Evolution of Microbial Life*, (eds D.McL. Roberts, P. Sharp, G. Alderson and M. Collins), Society for General Microbiology 54, Cambridge University Press, Cambridge, UK, pp. 567–592.

Gray, G.S. and Fitch, W.S. (1983) Evolution of antibiotic resistance genes: the DNA sequence of a kanamycin resistance gene from *Staphylococcus aureus*. *Mol. Biol. Evol.* **1**: 57–66.

Gupta, R.S. and Golding, G.B. (1993) Evolution of HSP70 gene and its implications regarding relationships between archaebacteria, eubacteria and eukaryotes. *J. Mol. Evol.* **37**: 573–582.

Gupta, R.S. and Singh, B. (1992) Cloning of the HSP70 gene from *Halobacterium marismortui*: relatedness of archaebacterial HSP70 to its eubacterial homologs and a model of the evolution of the HSP70 gene. *J. Bact.* **174**: 4594–4605.

Harvey, W.R. (1992) Physiology of V-ATPases. *J. Exp. Biol.* **172**: 1–17.

Hensel, R., Demharter, W., Kandler, O. *et al.* (1986) Chemotaxonomic and molecular-genetic studies of the genus *Thermus*: evidence for a phylogenetic relationship of *Thermus aquaticus* and *Thermus ruber* to the genus *Deinococcus*. *Int. J. Syst. Bact.* **36**: 444–453.

Hilario E. and Gogarten, J.P. (1993) Horizontal transfer of ATPase genes – the tree of life becomes a net of life. *BioSystems* **31**: 111–119.

Iwabe, N., Kuman, K.-I., Hasegawa, M. *et al.* (1989) Evolutionary relationships of archaebacteria, eubacteria and eukaryotes inferred from phylogenetic trees of duplicated genes. *Proc. Natl Acad. Sci. USA.* **86**: 9355–9359.

Keeling, P.J. and Doolittle, W.F. (1995) Archaea: narrowing the gap between prokaryotes and eukaryotes. *Proc. Natl Acad. Sci. USA.* **92**: 5761–5764.

Kumada, Y., Benson, D.R., Hillemann, D. *et al.* (1993) Evolution of the glutamine synthase gene, one of the oldest existing and functioning genes. *Proc. Natl Acad. Sci. USA* **90**: 3009–3013.

Lake, J.A. and Rivera, M. (1996) Letter. *Science* **274**: 900.

Langer, D., Hain, J., Thuriaux, P. and Zillig, W. (1995) Transcription in archaea: similarity to that in eucarya. *Proc. Natl Acad. Sci. USA.* **92**: 5768–5772.

Lawson, F.S., Charlebois, R.L. and Dillon, J.R. (1996) Phylogenetic analysis of carbamoylphosphate synthetase genes: complex evolutionary history includes an internal duplication within a gene which can root the tree of life. *Mol. Biol. Evol.* **13**: 970–977.

Ludwig, W., Neumaier, J., Klugbauer, N. *et al.* (1993) Phylogenetic relationships of bacteria based on comparative sequence analysis of elongation factor Tu and ATP-synthase beta-subunit genes. *Antonie Van Leeuwenhoek* **64**: 285–305.

Margulis, L. (1993) *Symbiosis in Cell Evolution*, 2nd edn, W.H. Freeman and Company, New York.

Margulis, L. and Fester, R. (eds) (1991) *Symbiosis as a Source of Evolutionary Innovation: Speciation and Morphogenesis*. MIT Press, Cambridge, USA.

Marsh, T.L., Reich, C.I., Whitelock, R.B. and Olsen, G.J. (1994) Transcription factor IID in the Archaea: sequences in the *Thermococcus celer* genome would encode a product closely related to the TATA-binding protein of eukaryotes. *Proc. Natl Acad. Sci. USA* **91**: 4180–4184.

Olsen, G.J., Woese, C.R., White, O. and Venter, J.C. (1996) Letter. *Science* **274**: 902.

Pesole, G., Gissi, C., Lanave, C. and Saccone, C. (1995) Glutamine synthetase gene evolution in bacteria. *Mol. Biol. Evol.* **12**: 189–197.

Puhler, G., Leffers, H., Gropp, F. *et al.* (1989) Archaebacterial DNA-dependent RNA polymerases testify to the evolution of the eukaryotic nuclear genome. *Proc. Natl Acad. Sci. USA.* **86**: 4569–4573.

Rivera, M.C. and Lake, J.A. (1992) Evidence that eukaryotes and eocyte prokaryotes are immediate relatives. *Science* **257**: 74–76.

Schofield, J.P. (1993) Molecular studies on an ancient gene encoding for carbamoyl-phosphate synthetase. *Clin. Sci.* **84**: 119–128.

Schwartz, R.M. and Dayhoff, M.O. (1978) Origins of prokaryotes, mitochondria and chloroplasts. *Science* **199**: 395–403.

Smith, M.W., Feng, D.F. and Doolittle, R.F. (1992) Evolution by acquisition: the case for horizontal gene transfer. *TIBS* **17**: 489–493.

Souillard, N., Magot, M., Possot, O. and Sibold, L. (1988) Nucleotide sequence of regions homologous to nifH (nitrogenase Fe protein) from the nitrogen-fixing archaebacteria *Methanococcus thermolithotrophicus* and *Methanobacterium ivanovii*: evolutionary implications. *J. Mol. Evol.* **27**: 65–76.

Strimmer, K. and von Haeseler, A. (1996) Quartet puzzling: a quartet maximum likelihood method for reconstructing tree topologies. *Mol. Biol. Evol.* **13**: 964–969.

Sumi, M., Sato, M.H., Denda, K. *et al.* (1992) A DNA fragment homologous to F1-ATPase beta subunit was amplified from genomic DNA of *Methanosarcina barkeri*. Indication of an archaebacterial F-type ATPase. *FEBS Lett.* **314**: 207–210.

Takase, K., Kakinuma, S., Yamato, I. *et al.* (1994) Sequencing and characterization of the *ntp* gene cluster for vacuolar-type Na(+)-translocating ATPase of *Enterococcus hirae*. *J. Biol. Chem.* **269**: 11037–11044.

Thompson, J.D., Higgins, D.G. and Gibson, T.J. (1994) CLUSTAL W: improving the sensitivity of progressive multiple sequence alignment through sequence weighting, positions-specific gap penalties and weight matrix choice. *Nuc. Acids Res.* **22**: 4673–4680.

Tiboni, O., Cammarano, P. and Sanangelantoni, A.M. (1993) Cloning and sequencing of the gene encoding glutamine synthase I from the Archaeum *Pyrococcus woesei*: anomalous phylogenies inferred from analysis of archaeal and bacterial glutamine synthase I sequences. *J. Bact.* **175**: 2961–2969.

Tsutsumi, S., Denda, K., Yokoyama, K. *et al.* (1991) Molecular cloning of genes encoding major two subunits of a eubacterial V-Type ATPase from *Thermus thermophilus*. *Biochim. et Biophys. Acta* **1098**: 13–20.

van den Hoff, M.J.B., Jonker, A., Beintema, J.J. and Lamers, W.H. (1995) Evolutionary relationships of the carbamoylphosphate synthetase genes. *J. Mol. Evol.* **41**: 813–832.

Woese, C.R. (1987) Bacterial evolution. *Microbiol. Rev.* **51**: 221–271.

Woese, C.R. and Fox, G.E. (1977) Phylogenetic structure of the prokaryotic domain: the primary kingdoms. *Proc. Natl Acad. Sci. USA.* **74**: 5088–5090.

Yokoyama, K., Oshima, T. and Yoshida, M. (1990) *Thermus thermophilus* membrane-associated ATPase; indication of a eubacterial V-type ATPase. *J. Biol. Chem.* **265**: 21946–2(1950.

Zillig, W., Palm, P. and Klenk, H.-P. (1992) A model of the early evolution of organisms: the arisal of the three domains of life from the common ancestor, in *The Origin and Evolution of the Cell*, (eds H. Hartman and K. Matsuno), World Scientific, Singapore, pp. 163–182.

Zimniak, L., Dittrich, P., Gogarten, J.P. *et al.* (1988) The cDNA sequence of the 69kDa subunit of the carrot vacuolar H+-ATPase. Homology to the beta-chain of F_oF_1-ATPases. *J. Biol. Chem.* **263**: 9102–9112.

Endosymbiosis and the origins of chloroplast–cytosol isoenzymes: a revision of the gene transfer corollary

28

William F. Martin

SUMMARY

Endosymbiotic gene transfer describes the process through which chloroplasts and mitochondria relinquished the majority of their genes to the nucleus while not having surrendered the majority of proteins integral to the eubacterial nature of their metabolism. It is a special case of lateral gene transfer that was very important for the establishment of eukaryotic cells. Five examples of endosymbiotic gene transfer in the history of higher plant genes for chloroplast-cytosolic isoenzymes of the Calvin cycle and glycolysis are considered here. The data indicate that nuclear genes for several glycolytic enzymes of the eukaryotic cytosol were acquired from eubacteria early in eukaryotic evolution, probably through endosymbiosis. For those enzymes common to symbionts and host, the pre-existing nuclear homologues were replaced in many cases, suggesting that endosymbiotic events had a high likelihood of successful gene transfer. During eukaryotic evolution, the products of transferred genes have been surprisingly often rerouted to compartments other than that from which the genes were donated. A minor but not insignificant revision of endosymbiotic theory to account for the origins of enzymes of compartmentalized metabolism is proposed.

28.1 INTRODUCTION

Gray and Doolittle (1982) asked: 'Has the endosymbiont hypothesis been proven?' Today we know that the answer is a resounding 'yes' – genes encoded in organellar genomes attest beyond all reasonable doubt to the eubacterial ancestry of chloroplasts and mitochondria (Gray, 1989; De Rijk *et al.*, 1995; Van de Peer *et al.*, 1996; Martin and Schnarrenberger, 1997). But how does endosymbiotic theory currently account for the origin of organellar proteins that are not encoded in organellar DNA? Their fate is commonly explained with the help of a scenario elegantly argued by Weeden (1981) that we have come to accept as the 'gene transfer corollary'. It posits that during the course of eukaryotic history, the majority of genes for proteins integral to organellar metabolism were transferred to the nucleus, where they became integrated into the regulatory hierarchy of the nucleus and acquired a transit peptide, so that the functional gene products could be reimported into the organelle of their genetic origin on a daily basis since. This is a reasonable and logical scenario that satisfyingly explains why organelles (particularly photosynthetic plastids) have retained so much of their biochemically eubacterial heritage while having relinquished to the nucleus the majority of genes necessary to have done so. It also accounts for evolutionarily recent organelle-to-nucleus gene transfer events (Baldauf and Palmer, 1990; Gannt *et al.*, 1991; Brennicke *et al.*, 1993).

However, the simple phrase 'organelle-to-nucleus gene transfer' raises a different question. What is the origin of the nucleus under endosymbiotic theory? Discovery of the archaebacterial nature of components of the nuclear genetic apparatus (Ouzounis and Sander, 1992; Rowlands *et al.*,1994; Langer *et al.*, 1995) and the results of extensive molecular phylogenetic work have slowly unveiled the nucleus as a descendant – of sorts – of the archaebacterial domain (Doolittle and Brown, 1994; Brown and Doolittle, 1995; Doolittle, 1995; Baldauf *et al.*, 1996). This means that the host cell under endosymbiotic theory *was* an archaebacterium, though there is still some controversy as to whether the endosymbiotic host was a 'pure' archaebacterium, or whether it might have contained a eubacterial component prior to the acquisition of mitochondria and chloroplasts *via* fusion scenarios with eubacteria as envisaged by some (Zillig *et al.*, 1989; Lake and Rivera, 1994; Golding and Gupta, 1995). This is a complicated issue that is interwoven with the question of whether amitochondriate protists are primitively so, or whether they initially had mitochondria but secondarily lost them (for discussion, see Henze *et al.*, 1995; Bui *et al.*, 1996; Doolittle, 1996; Müller, 1996; Martin and Schnarrenberger, 1997). At the time of writing, that issue has not been resolved to the satisfaction of all, but it is safe to assert that the archaebacterial component of eukaryotes – the eocyte (Rivera and Lake, 1992), possibly a member of the Crenarchaeota (Woese *et al.*, 1990; Baldauf *et al.*, 1996) – had more than a genetic apparatus; it had some form of metabolism and thus it possessed genes for the enzymes of that metabolism. What has happenend to those archaebacterial genes? Are archaebacterial metabolic enzymes preserved in the eukaryotic cytosol?

Few explicit hypotheses, other than that of Weeden, have been put forth for the origin of cytosolic enzymes of metabolism. He suggested that they should reflect the evolutionary history of the host – a very reasonable idea. Extending his logic to non-phyotosynthetic eukaryotes, and taking into account what we now know about the host, that would mean that metabolic enzymes of the eukaryotic cytosol should, on the whole, tend to reflect an archaebacterial (eocyte) ancestry. Applying these thoughts to the evolutionary history of enzymes that were present in each of the cellular lineages that contributed to eukaryotic cells, we obtain a simple (and occasionally implicit) null hypothesis (Fig. 28.1). That model depicts a eukaryote with archaebacterial and eubacterial genes in the nucleus, and with compartmentalized gene products in the compartments from which the genes descend. Chloroplast–cytosol isoenzymes have been used to forge this model; they can also be used to test it.

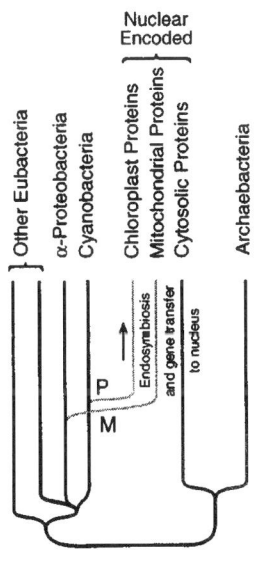

The simplest hypothesis for the origin
of genes for nuclear encoded proteins

Fig. 28.1 Explicit depiction of an often implicitly assumed working hypothesis for the evolutionary history and expected schematic phylogenetic tree of compartmentalized nuclear encoded isoenzymes. P, plastids; M, mitochondria.

Enzymes common to both the Calvin cycle of higher plant chloro-plasts and the glycolytic/gluconeogenetic pathway of the cytosol (Table 28.1) have homologs in contemporary eubacteria (Fothergill-Gilmore and Michels, 1993) and archaebacteria (Schönheit and Schäfer, 1995). They existed in both the nuclear and the cyanobacterial lineages. All are nuclear-encoded in eukaryotes. They thus permit us to study correlations between the organellar localization of nuclear gene products and evolu-tionary origin of organelles. In several recent papers, the evolution of individual Calvin cycle–glycolytic isoenzymes has been specifically addressed; a brief review has also appeared on the topic (Martin and Schnarrenberger, 1997). Some of those findings will be briefly summa-rized here with case examples of five enzymes. The results of those stud-ies will be contrasted with the simple prediction of Fig. 28.1 that Calvin cycle enzymes of chloroplasts should reveal a eubacterial (cyanobacter-ial) origin whereas their glycolytic homologs of the cytosol in higher plants and non-photosynthetic eukaryotes should reflect an archaebacte-rial ancestry.

28.2 RESULTS AND DISCUSSION

28.2.1 TRANSKETOLASE

Transketolase (TKL) (EC 2.2.1.1) is essential to both the Calvin cycle of higher plant chloroplasts and to the oxidative pentose phosphate path-way (OPPP). In spinach, both the Calvin cycle and the regenerative segment of the OPPP are localized in plastids and only one TKL enzyme is detectable which functions in both pathways (Schnarrenberger *et al.*, 1995; Flechner *et al.*, 1996). But in the dehydratable angiosperm *Craterostigma*, two *TKL* genes are expressed, the products of which lack a transit peptide and, like their homologs of the OPPP in the cytosol of non-photosynthetic eukaryotes, are evidently cytosolic enzymes (Bernacchia *et al.*, 1995). Several eubacterial TKL sequences have been reported, including that from the cyanobacterium *Synechocystis* (Kaneko *et al.*, 1996). The genome of the archaebacterium *Methanococcus jannaschii* (Bult *et al.*, 1996) encodes a TKL homolog, but this is split across two genes that roughly correspond to the N- and C-terminal halves, respec-tively, of the single-chain eukaryotic and eubacterial TKL subunit.

With this modest sample of reference sequences for comparison, one can address the ancestry of chloroplast and cytosolic TKL in plants. A phylogenetic tree was inferred from TKL data (Fig. 28.2) and several aspects of the result are of interest. Firstly, TKL enzymes from the mammalian cytosol are more similar to the archaebacterial sequence (aver-age 60% sequence identity) than they are to homologs from other sources, including the yeast and higher plant cytosol (average 50% identity). This

Table 28.1 Condensed overview of enzymes of the three central pathways of carbohydrate metabolism in higher plants

Calvin cycle	Glycolysis/ gluconeogenesis	OPPP	Symbol	Enzyme
	●		HK	Hexokinase
	●		PFK	Phosphofructokinase
	●		PGM	Phosphoglycerate mutase
	●		ENO	Enolase
	●		PYK	Pyruvate kinase*
●	●		PGK	3-Phosphoglycerate-kinase*
●	●		GAPDH	Glyceraldehyde-3-phosphate dehydrogenase*
●	●	●	FBA	Fructose-1,6-bisphosphate aldolase*
●	●	●	TPI	Triosephosphate isomerase*
●	●	●	FBP	Fructose-1,6-bisphosphatase*
●		●	TKL	Transketolase*
●		●	RPE	Ribulose-5-phosphate 3-epimerase
●		●	PRI	Ribose-5-phosphate isomerase
●			RBC	Ribulose-1,5-P2 carboxylase/oxygenase
●			PRK	Phosphoribulokinase
●			SBP	Sedoheptulose-1,7-bisphosphatase
	●	●	PGI	Glucose-6-phosphate isomerase
		●	G6PDH	Glucose-6-phosphate dehydrogenase*
		●	6GPDH	6-Phosphogluconate dehydrogenase
		●	TAL	Transaldolase

OPPP, oxidative pentose phosphate pathway; *, sequences are available in GenBank for chloroplast and cytosolic forms of the enzyme in plants; ● function of the enzyme in the pathway.

similarity is underscored by the presence of a ~40 amino acid insertion found in the region corresponding to the break in the N-terminal TKLα and C-terminal TKLβ chains of the *Methanococcus* enzyme that is shared by the eubacterial-type TKL sequences and is lacking in the mammalian homologs.

Also, the fungal and plant (both chloroplast and cytosol) sequences share much more similarity to eubacterial homologues (65–70% identity) than they do with either the archaebacterial or the mammalian cytosolic enzymes, whereby plant TKL sequences are robustly more similar to cyanobacterial TKL than they are to other eubacterial TKL homologs. Finally, the higher plant chloroplast and cytosolic TKL isoenzymes can clearly be traced to a gene duplication event which appears to have occurred specifically in the plant lineage.

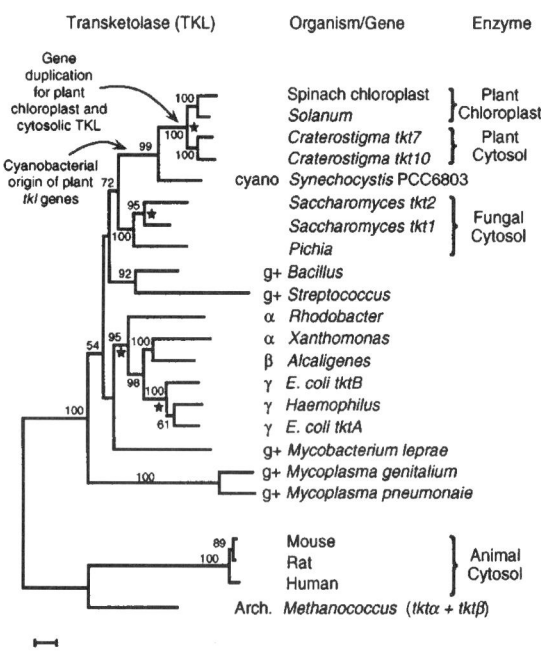

Fig. 28.2 Neighbor-joining (Saitou and Nei, 1987) tree of TKL protein sequences were constructed from a matrix of Dayhoff distances (Felsenstein, 1993). Sequences were retrieved from the database and aligned with ClustalW. Uncertain regions of the alignment were excluded from analysis, yielding 657 positions for phylogenetic inference. g+, Gram-positive; α, β and γ, proteobacteria. Only those bootstrap proportions greater than 50/100 are indicated. Stars indicate recognizable gene duplications. Scale bar indicates 0.1 substitutions per site.

Despite the lack of sequences from a large eubacterial and eukaryotic sample, this general pattern of similarity can be explained in a reasonably straightforward manner in the context of symbiotic theory, although numerous details remained to be clarified. The most simple explanation would be that both archaebacterial and eubacterial TKL genes coexisted during the early evolution of the eukaryotic nucleus. In the subsequent course of crown group radiation, at least the mammalian portion of the metazoan lineage retained the archaebacterial gene, whereas the fungal and photosynthetic lineages sampled here have retained a eubacterial copy. It is evident that plants sampled have retained a cyanobacterial gene for TKL. The eubacterial donor of fungal TKL genes cannot be identified with certainty. That is, the gene tree does not indicate whether the eubacterial TKL gene of fungi sampled was acquired from mitochondria or from an earlier 'fusion' event of some type in the history of eukaryotic cells involving an unknown eubacterium 'Z' for the origin of the nucleus (Zillig *et al*, 1989; Lake and Rivera, 1994; Golding and Gupta, 1995). Despite these vagaries, the origin of higher plant chloroplast–cytosol TKL isoenzymes *via* comparatively recent gene duplication provides clear evidence against the view that contemporary subcellular enzyme compartmentalization corresponds to the evolutionary origin of the compartment. The working hypothesis with the fewest corollary assumptions would be that the genes for both chloroplast and cytosolic TKL of higher plants descend from cyanobacteria, and that the fungi sampled acquired their TKL gene from a different eubacterial donor. Chloroplast–cytosol TKL isoenzymes of plants arose recently in evolution through gene duplication; plant and fungal cytosolic TKL are clearly not archaebacterial, whereas TKL of the mammalian cytosol apparently is.

28.2.2   TRIOSEPHOSPHATE ISOMERASE

The genes for higher plant chloroplast and cytosolic triosephosphate isomerase (TPI) (EC 5.3.1.1) were previously shown to have arisen through duplication of a pre-existing nuclear counterpart for the cytosolic enzyme during the course of eukaryotic evolution (Henze *et al*, 1994; Schmidt *et al.*, 1995). The product of the TPI gene once encoded by the antecedants of plastids was functionally replaced as a result of that duplication event. Thus, although the higher plant Calvin cycle enzyme is localized in chloroplasts, its nuclear gene is derived from a pre-existing eukaryotic copy, not from cyanobacteria, and again no correlation between chloroplast compartmentalization and cyanobacterial origin can be observed.

Does the pre-existing nuclear gene for cytosolic TPI of eukaryotes reflect an archaebacterial origin of the nucleocytoplasm? Two sequences for TPI from archaebacteria recently became available (Bult *et al.*, 1996; Kohlhoff *et al.*, 1996) and a few eubacterial sequences exist in the data base that permit this question to be addressed, as Keeling and Doolittle (1997) have recently done. As shown in Fig. 28.3, cytosolic TPI from eukaryotes is apparently a eubacterial enzyme. Whereas TPI from eukaryotic and eubacterial sources share roughly 40–50% amino acid identity, their identity with the archaebacterial homologues is only in the range of 20% (Kohlhoff *et al.*, 1996). Notably, the TPI sequence from the amitochondriate protist *Giardia lamblia* (Mowatt *et al.*, 1994) is also of the eubacterial type (Fig. 28.3). Does this suggest that *Giardia* is secondarily amitochondriate? Not directly, because if some variant of a fusion hypothesis with eubacterium 'Z' is correct, amitochondriate protists would be expected to contain some genes of eubacterial origin, without these necessarily having derived from disappearing mitochondrial symbionts. But the finding that another amitochondriate protist, *Entamoeba histolytica*, possesses clearly eubacterial genes for two typically mitochondrial proteins (Clark and Roger, 1995) should make us wary.

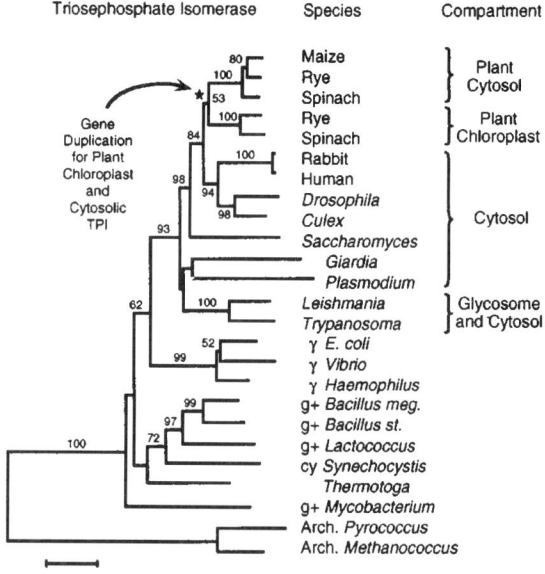

Fig. 28.3 Neighbor-joining tree for TPI protein sequences constructed as described for Fig. 28.2. Uncertain regions of the alignment were excluded from analysis, yielding 220 positions for phylogenetic inference. g+, Gram-positive; γ, proteobacteria; cy, cyanobacteria. Only those bootstrap proportions greater than 50/100 are indicated. Stars indicate recognizable gene duplications. Scale bar indicates 0.1 substitutions per site.

So although the result in Fig. 28.3 does not reveal the specific eubacterial origin (mitochodrial vs. 'fusion') of nuclear genes for cytosolic TPI, we can at least answer the question 'Does enzyme localization correlate to evolutionary origin?' with a simple 'no' for chloroplast and cytosolic TPI of higher plants. At the minimum, the chloroplast enzyme has been rerouted during evolution to an organelle from which its gene was not donated, having functionally replaced the cyanobacterial protein. The cytosolic enzyme of eukaryotes surveyed to date (including plants) is not of the type found in archaebacteria, which indicates that a eubacterial gene for TPI functionally replaced that contributed by the archaebacterial component of eukaryotes.

28.2.3 PHOSPHOGLYCERATE KINASE

Higher plants possess chloroplast (Calvin cycle) and cytosolic (glycolytic) isoenzymes for phosphoglycerate kinase (PGK) (EC 2.7.2.3), whereas *Chlamydomonas* possesses solely a chloroplast form of the enzyme (Schnarrenberger *et al.*, 1990). It was recently shown that the ancestral gene, from which contemporary homologs for the Calvin cycle and glycolytic PGK isoenzymes in higher plants descend, was acquired by the nucleus from endosymbiotic eubacteria (for details, see Brinkmann and Martin, 1996). The cyanobacterial origin suggested for plant PGK genes (Brinkmann and Martin, 1996) is strongly substantiated by phylogenetic analysis of the PGK sequence from *Synechocytis* PCC6803 (Kaneko *et al.*, 1996), which branches very robustly with plant PGK seqences (bootstrap proportion of 100% using the Kimura distance or parsimony; data not shown). That nuclear gene of cyanobacterial origin underwent duplication subsequent to separation of *Chlamydomonas* and land plant lineages, giving rise to the contemporary genes for chloroplast and cytosolic PGK isoenzymes in higher plants. This resulted in functional replacement of the pre-existing nuclear gene for cytosolic PGK of the eukaryotic cell that hosted plastids. But the pre-existing nuclear gene that was replaced in the plant lineage was itself eubacterial, since the PGK tree reveals the same general pattern observed for TPI and GAPDH (section 28.2.4), i.e. cytosolic PGK of eukaryotes shares ~50% identity with eubacterial and only ~30% identity with archaebacterial homologs.

As for TPI and TKL, the eubacterial donor for cytosolic PGK of non-photosynthesizers remains elusive, since the available data do not distinguish between the mitochondrial vs. fusion (M and Z, respectively, in the schematic PGK tree in Fig. 28.4) alternatives, as indicated by question marks in the figure. The example of PGK thus suggests that along the evolutionary route to plants, the archaebacterial copy of the nucleus was replaced by a eubacterial (mitochondrial? Z?) copy, which in turn was

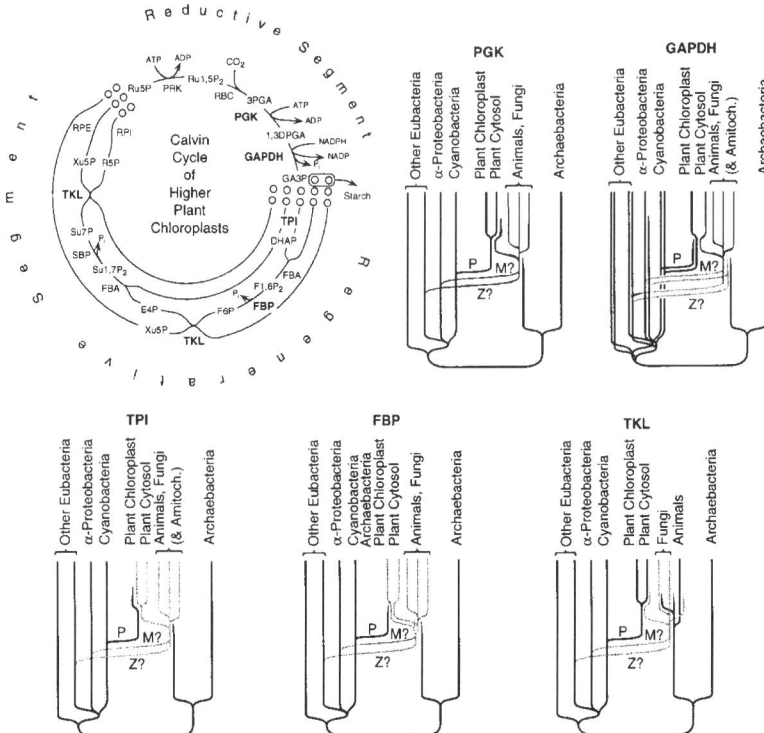

Fig. 28.4 Schematic summary of the origins of genes for chloroplast—cytosol isoenzymes of higher plants considered in this chapter (see text). Branches of nuclear genes argued here to derive from endosymbiots are shown in gray. Genes that did not persist to the present (or have not been characterized to date) are tapered. See Table 28.1 for enzyme abbreviations.

replaced with a cyanobacterial gene involving the origins of plastids. Regardless of the specific eubacterial origin of cytosolic PGK in non-photosynthetic eukaryotes, the origin of the cytosolic PGK enzyme of higher plants can be confidently traced to nuclear duplication of a trans-ferred cyanobacterial gene, providing strong evidence against the view that the compartment within which an enzyme resides corresponds to the evolutionary history of that compartment.

28.2.4   GLYCERALDEHYDE-3-PHOSPHATE DEHYDROGENASE

Glyceraldehyde-3-phosphate dehydrogenase (GAPDH, phosphorylat-ing) (EC 1.2.1.12 and 1.2.1.13) represents one of the more complicated cases of gene evolution for chloroplast–cytosol isoenzymes (for details, see Martin *et al.*, 1993; Henze *et al.*, 1995). This is because at least one cyanobacterium (*Anabaena variabilis*) and at least one proteobacterium (*Escherichia coli*) possess small gene families of three members each for this enzyme. The age of the eubacterial gene family appears to predate cyanobacterial–proteobacterial divergence. Although several gene sequences are available from archaebacteria, those from methanogens encode what is termed Class II GAPDH, because the proteins share only minimal (~15%) amino acid identity with the Class I GAPDH enzymes found in eubacteria, eukaryotes and the halophilic archaebacterium *Haloarcula vallismortis* (Brinkmann and Martin, 1996). The sequence of the nuclear encoded chloroplast enzyme in plants is significantly more similar to its homologs from cyanobacteria than it is to the enzyme from any other source, indicating that the gene product is reimported into the organelle that donated the gene (Martin *et al.*, 1993). The origin of the gene for the cytosolic (glycolytic) enzyme of higher plants and non-photosynthetic eukaryotes has been a debated matter over the years (Doolittle *et al.*, 1990; Wiemer *et al.*, 1995). Since it has gradually become apparent that the nucleus can be expected to contain either eubacterial or archaebacterial genes, and since GAPDH of the eukaryotic cytosol is significantly more similar to proteobacterial orthologues (~60–70% identity) than it is to the archaebacterial class II enzyme from methanogens or the class I (~40–50%) enzyme from *Haloarcula*, a case can be made from study of the plant enzymes (Martin *et al.*, 1993; Henze *et al.*, 1995; Martin and Schnarrenberger, 1997) that eubacterial GAPDH genes have persisted in the eukaryotic nucleus, whereas homologs contributed by archaebacterial component of the nucleus have not. Under that view, the chloroplast–cytosol GAPDH isoenzymes of higher plants are related through ancient gene duplications that occurred in eubacterial genomes (Fig. 28.4). Notably, the amitochondriate protist *Trichomonas vaginalis* possesses a cytosolic GAPDH enzyme with strikingly eubacterial features (Markos *et al.*, 1993) and only 45% identity to the *Haloarcula* homologue. Two other amitochondriate protists (*Giardia lamblia* and *Entamoeba histolytica*) possess GAPDH genes of the eubacterial type found in metazoa and fungi (as in the case of TPI described above) but phylogenetic resolution and eubacterial sampling are insufficient to discern between a mitochondrial or other eubacterial origin (Henze *et al.*, 1995). Thus, for GAPDH it appears once again that the cytosolic enzyme of eukaryotes does not betray an archaebacterial heritage of the nucleocytoplasm, but rather that the GAPDH gene possessed by the archaebacterial component of eukaryotes has been replaced by a descendant member of an ancient and diverse eubacterial gene family.

28.2.5   FRUCTOSE-1,6-BISPHOSPHATASE

As a last example, one can consider higher plant chloroplast and cytosolic isoenzymes for fructose-1,6-bisphosphatase (FBP) (EC 3.1.3.11) involved in the Calvin cycle and gluconeogenesis, respectively. It has recently been shown that these isoenzymes arose through gene duplication early in eukaryotic evolution, apparently prior to crown group radiation (for details, see Martin *et al.*, 1996). Although several archaebacteria (including methanogens) clearly possess FBP enzyme activity (Schönheit and Schäfer, 1995), no sequence with detectable similarity to FBP has yet been identified in an archaebacterial genome. Nonetheless, the FBP gene tree is characterized by two findings typical of other cases of eubacterial origin:

* Sequence divergence within eubacterial sequences (in the order of 40% identity) is greater than between eukaryotic and eubacterial homologues (in the order of 50% identity).
* Eukaryotic FBP sequences appear on a branch that emerges from a 'trunk' of more highly diverse eubacterial sequences, strikingly similar to the pattern observed for eukaryotic PGK, GAPDH, TPI, and plant TKL.

Notably, the genes for higher plant chloroplast and cytosolic FBP appear to descend from the same eubacterial source as their homologues from non-photosynthetic eukaryotes surveyed and in contrast to GAPDH, none of the eukaryotic nuclear genes reveals a significant similarity to FBP sequences from three cyanobacteria available for comparison. This suggests that, again, the gene for the Calvin cycle enzyme of higher plant chloroplasts was not donated to the nucleus by the eubacterial antecedent of the organelle into which the enzyme is now imported.

28.3   CONCLUSIONS

As summarized in Fig. 28.4, the majority of eukaryotic genes considered here come from eubacteria, and quite a bit of endosymbiotic gene replacement has taken place in evolution. As a gratuitous aspect of these studies, two cases (GAPDH and TPI) were found of eubacterial genes in the nuclei of protists that lack mitochondria. A general answer to the question of whence such eubacterial genes in amitochondriate protists ultimately derive may turn out to be very difficult to come by, because it hinges upon our ability to assign such genes to contemporary eubacterial genomes (i.e. α-proteobacteria or an unknown 'Z'-fusion type source). This is impaired by eubacterial ancient gene duplications followed by differential loss (Martin and Schnarrenberger, 1997), and by the limited amount of phylogenetic resolution contained in a single protein of only a few hundred amino acids in length. Furthermore, if eubacteria exchanged their genes (at the rate of once every billion years or so per gene) across broad taxonomic boundaries either prior or subsequent to the endosymbiotic origins of organelles, as other contributions in this volume might suggest to be a tenable view, then eubacterial genes in the nucleus will be subject to an ancient allele-sampling problem that will be extremely tedious to sort out finally. Finding a few genes of eubacterial origin (Henze *et al.*, 1995), or of specifically α-proteobacterial origin (Doolittle, 1996), in archaezoan genomes is evidence that should weigh heavily in this issue.

With regard to the question of the gene transfer corollary as outlined in section 28.1, we have obtained a few reasonably solid answers. For five enzymes common to each of the antecedents of the chloroplast, the mitochondrion and the nucleocytoplasm, we see that the cellular local-ization of at least one of the isoenzymes in plants – and in some cases both – does not correlate to the endosymbiotic ancestry of the organelle. Rather, the products of genes transferred from eubacteria have often taken hold in the cytosol, replacing pre-existing functions of the nuclear lineage, and in some cases mitochondrial enzymes have replaced the chloroplast homolog.

From the standpoint of enzymes considered here, a very complex picture of the eukaryotic cell is emerging indicating that, in contrast to components of the nuclear genetic apparatus, the archaebacterial identity of the cytosol has been largely obscured, having been replaced with eubacterial enzymes, the nuclear genes for which were evidently acquired from (endosymbiotic) eubacteria. The reasons for the preferen-tial fixation of the eubacterial enzymes in the eukaryotic cytosol are unknown, but it is not unreasonable to speculate that some form of selec-tion for enzymatic properties may have been at work.

The nuclear genes for enzymes active in the Calvin cycle of higher plant chloroplasts can in some cases be traced to cyanobacterial donors (GAPDII, PGK), but in other cases they arose through nuclear duplica-tions of pre-existing genes that can be traced to eubacteria, but not specifically to cyanobacteria (TPI, TKL, FBP). The origin of other chloro-plast–cytosol isoenzymes, for example fructose-1,6-bisphosphate aldolase (Schnarrenberger *et al.*, 1994) and glutamine synthase (Kumada *et al.*, 1993), can also be traced to similar gene duplications.

The evolutionary patterns revealed by chloroplast–cytosol isoenzymes of the Calvin cycle and glycolysis would be altogether impossible to interpret in the absence of endosymbiotic theory. Yet at the same time, they require a minor modification of the theory. The prevailing view that the contemporary cellular compartmentation of a given enzyme should reflect its evolutionary origin is incorrect in many cases. Case examples from carbohydrate metabolism indicate that endosymbiotic events had a very high likelihood of successful endosymbiotic gene transfer for a given protein, and in the ensuing struggle for survival between genes for functionally equivalent products, pre-existing nuclear genes have repeatedly been replaced by eubacterial intruders. The latter are, in turn, highly prone to undergoing gene duplication, giving rise to products that may find their way to new target compartments and may there ursurp the function of pre-existing compartment-specific enzymes. We may free ourselves from the confines of anticipation that the products of genes subject to endosymbiotic gene transfer return, as a rule, to the organelle from which they came.

ACKNOWLEDGMENTS

I thank Claus Schnarrenberger and Miklos Müller for numerous stimulating discussions, and Patrick Keeling and Ford Doolittle for communicating data prior to publication. Generous financial support from the Deutsche Forschungsgemeinschaft is gratefully acknowledged.

REFERENCES

Baldauf, S. and Palmer, J.D. (1990) Evolutionary transfer of the chloroplast *tuf*A gene to the nucleus. *Nature* **344**: 262–265.

Baldauf, S., Palmer, J.D. and Doolittle, W.F. (1996) The root of the universal tree and the origin of eukaryotes based on elongation factor phylogeny. *Proc. Natl Acad. Sci. USA* **93**: 7749–7754.

Bernacchia, G., Schwall, G., Lottspeich, F. *et al.* (1995) The transketolase gene family of the resurrection plant *Craterostigma plantagineum*: differential expression during the rehydration phase. *EMBO J.* **14**: 610–618.

Brennicke, A., Grohmann, L., Hiesel, R. *et al.* (1993) The mitochondrial genome on its way to the nucleus: Different stages of gene transfer in higher plants. *FEBS Letts* **325**: 140–145.

Brinkmann, H. and Martin, W. (1996) Higher plant chloroplast and cytosolic 3-phosphoglycerate kinases: a case of endosymbiotic gene replacement. *Plant Mol. Biol.* **30**: 65–75.

Brown, J.R. and Doolittle, W.F. (1995) Root of the universal tree of life based on ancient aminoacyl-tRNA synthtase gene duplications. *Proc. Natl Acad. Sci. USA* **92**: 2441–2445.

Bui, E.T.N., Bradley, P.J. and Johnson, P.J. (1996) A common evolutionary origin for mitochondria and hydrogenosomes. *Proc. Natl Acad. Sci. USA* **93**: 9651–9656.

Bult, C.J., White, O., Olsen, G.J. *et al.* (1996) Complete genome sequence of the methanogenic Archeon, *Methanococcus jannaschii. Science* **273**: 1058–1073.

Clark, C.G. and Roger, A.J. (1995) Direct evidence for secondary loss of mitochondria in *Entamoeba histolytica. Proc. Natl Acad. Sci. USA* **92**: 6518–6521.

De Rijk, P., Van de Peer, Y., Van den Brock, I. and De Wachter, R. (1995) Evolution according to large ribosomal subunit RNA. *J. Mol. Evol.* **41**: 366–375.

Doolittle, R.F. (1995) Of Archae and Eo: what's in a name? *Proc. Natl Acad. Sci. USA* **92**: 2421–2423.

Doolittle, R., Feng, D., Anderson, K. and Alberro, M. (1990) A naturally occurring horizontal gene transfer from a eukaryote to a prokaryote. *J. Mol. Evol.* **31**: 383–388.

Doolittle, W.F. (1996) Some aspects of the biology of cells and their possible evolutionary significance, in *Evolution of Microbial Life*, (eds D. Roberts, P. Sharp, G. Alserson and M. Collins), Society for General Microbiology Symposium 54, Cambridge University Press, Cambridge, pp. 1–21.

Doolittle, W.F. and Brown, J.R. (1994) Tempo, mode, the progenote, and the universal root. *Proc. Natl Acad. Sci. USA* **91**: 6721–6728.

Felsenstein, J. (1993. *PHYLIP (Phylogeny Inference Package) manual, Version 3.5c*, Department of Genetics, University of Washington, Seattle.

Flechner, A., Dressen, U., Westhoff, P. *et al.* (1996) Molecular characterization of transketolase (EC 2.2.1.1) active in the Calvin cycle of spinach chloroplasts. *Plant Mol. Biol.* **32**: 475–484.

Fothergill-Gilmore, L.A. and Michels, P.A.M. (1993) Evolution of glycolysis. *Progr. Biophys. Mol. Biol.* **59**: 105–238.

Gannt, J.S., Baldauf, S., Calie, P.J. *et al.* (1991) Transfer of *rpl22* to the nucleus greatly preceded its loss from the chloroplast and involved the gain of an intron. *EMBO J.* **10**: 3073–3078.

Golding, G.B. and Gupta, R.S. (1995) Protein-based phylogenies support a chimaeric origin for the eukaryotic genome. *Mol. Biol. Evol.* **12**: 1–6.

Gray, M.W. (1989) The evolutionary origin of organelles. *Trends Genet.* **5**: 294–299.

Gray, M.W. and Doolittle, W.F. (1982) Has the endosymbiont hypothesis been proven? *Microbol. Rev.* **46**: 1–42.

Henze, K., Schnarrenberger, C., Kellermann, J. and Martin, W. (1994) Chloroplast and cytosolic triosephosphate isomerase from spinach: purification, microsequencing and cDNA sequence of the chloroplast enzyme. *Plant Mol. Biol.* **26**: 1961–1973.

Henze, K., Badr, A., Wettern, M. *et al.* (1995) A nuclear gene of eubacterial origin in *Euglena gracilis* reflects cryptic endosymbioses during protist evolution. *Proc. Natl Acad. Sci. USA* **92**: 9122–9126.

Kaneko, T., Sato, S., Kotani, H. *et al.* (1996) Sequence analysis of the genome of the unicellular cyanobacterium *Synechocystis* sp. strain PCC6803. II. Sequence determination of the entire genome and assignment of potential protein-coding regions. *DNA Res.* **3**: 109–136.

Keeling, P.W. and Doolittle, W.F. (1997) Evidence that eukaryotic triosephosphate isomerase is of alpha-proteobacterial origin. *Proc. Natl Acad. Sci. USA* **94**: 1270–1275.

Kohlhoff, M., Dahm, A. and Hensel, R. (1996) Tetrameric triosephosphate isomerase from hyperthermophilic Archaea. *FEBS Letts* **383**: 245–250.

Kumada, Y., Benson, D.R., Hillemann, D. *et al.* (1993) Evolution of the glutamin synthase gene, one of the oldest existing and functioning genes. *Proc. Natl Acad. Sci. USA* **90**: 3009–3013.

Lake, J.A. and Rivera, M.C. (1994) Was the nucleus the first endosymbiont? *Proc. Natl Acad. Sci. USA* **91**: 2880–2881.

Langer, D., Hain. J., Thuriaux. P. and Zillig, W. (1995) Transcription in Archaea: similarity to that in Eukarya. *Proc. Natl Acad. Sci. USA* **92**: 5768–5772.

Markos, A., Miretsky, A. and Müller, M. (1993) A glyceraldehyde-3-phosphate dehydrogenase with eubacterial features in the amitochondriate eukaryote *Trichomonas vaginalis. J. Mol. Evol.* **37**: 631–643.

Martin, W. and Schnarrenberger, C. (1997) The evolution of the Calvin cycle from prokaryotic to eukaryotic chromosomes: A case study of functional redundancy through endosymbiosis. *Curr. Genet.* **32**: 1–18.

Martin, W., Brinkmann, H., Savona, C. and Cerff, R. (1993) Evidence for a chimaeric nature of nuclear genomes: eubacterial origin of eukaryotic glyceraldehyde-3-phosphate dehydrogenase genes. *Proc. Natl Acad. Sci. USA* **90**: 8692–8696.

Martin, W., Mustafa, A.-Z., Henze, K. and Schnarrenberger, C. (1996) Higher plant chloroplast and cytosolic fructose-1,6-bisphophosphatase isoenzymes: origins via duplication rather than prokaryote–eukaryote divergence. *Plant Mol. Biol.* **32**: 485–491.

Mowatt, M.R., Weinbach, E.C., Howard, T.C. and Nash, T.E. (1994) Complementation of an *Escherichia coli* glycolysis mutant by *Giardia lamblia* triosephosphate isomerase. *Exp. Parasitol.* **78**: 85–92.

Müller, M. (1996) Energy metabolism of amitochondriate protists, an evolutionary puzzle, in *Christian Gottfried Ehrenberg-Festschrift anlässlich der 14. Wissenschaftlichen Jahrestagung der Deutschen Gesellschaft fur Protozoologie, 9.–11. Marz (1995 in Delitzsch (Sachsen)* (eds M. Schlegel and K. Hausmann), Leipziger Universitätsverlag, Leipzig, pp. 63–76.

Ouzounis, C. and Sander, C. (1992) TFIIB, an evolutionary link between the transcription machineries of archaebacteria and eukaryotes. *Cell* **71**: 189–90.

Rivera, M. and Lake, J.A. (1992) Evidence that eukaryotes and eocyte prokaryotes are immediate relatives. *Science* **257**: 74–76.

Rowlands, T., Baumann, P. and Jackson, S.P. (1994) The TATA-binding protein: a general transcription factor in eukaryotes and archaebacteria. *Science* **264**: 1326–1329.

Saitou, N. and Nei, M. (1987). The neighbor-joining method: a new method for reconstructing phylogenetic trees. *Mol. Biol. Evol.* **4**: 406–425.

Schmidt, M., Svendsen, I. and Feierabend, J. (1995) Analysis of the primary structure of the chloroplast isozyme of triosephosphate isomerase from rye leaves by protein and cDNA sequencing indicates a eukaryotic origin of its gene. *Biochim. Biophys. Acta* **1261**: 257–264.

Schnarrenberger, C., Jacobshagen, S., Müller, B. and Krüger, I. (1990) Evolution of isozymes of sugar phosphate metabolism in green algae, in *Isozymes: Structure, Function, and Use in Biology and Medicine*, (eds S. Ogit and J.S. Scandalios), Wiley-Liss, New York, pp. 743–746.

Schnarrenberger, C., Pelzer-Reith, B., Yatsuki, H. *et al.* (1994) Expression and sequence of the only detectable aldolase in *Chlamydomonas reinhardtii. Arch. Biochem. Biophys.* **313**: 173–178.

Schnarrenberger, C., Flechner, A. and Martin, W. (1995) Enzymatic evidence indicating a complete oxidative pentose phosphate in the chloroplasts and an incomplete pathway in the cytosol of spinach leaves. *Plant Physiol.* **108**: 609–614.

Schönheit, P. and Schäfer, T. (1995) Metabolism of hyperthermophiles. *World J. Microbiol. Biotechnol.* **11**: 26–57.

Van de Peer, Y., Rensing, S., Maier, U.-G. and De Wachter, R. (1996) Substitution rate calibration of small subunit RNA identifies chlorarachniophyte endosymbionts as remnants of green algae. *Proc. Natl Acad. Sci. USA* **93**: 7732–7736.

Weeden, N.F. (1981) Genetic and biochemical implications of the endosymbiotic origin of the chloroplast. *J. Mol. Evol.* **17**: 133–139.

Wiemer, E.A.C., Hannaert, V., van den IJssel, P.R.L.A. *et al.* (1995) Molecular analysis of glyceraldehyde-3-phosphate dehydrogenase in *Trypanoplasma borelli*. Evolutionary scenario of subcellular compartmentation in kinetoplastida. *J. Mol. Evol.* **40**: 443–454.

Woese, C.R., Kandler, O. and Whellis, M.L. (1990) Towards a natural system of organisms: proposal for the domains Archaea, Bacteria and Eucarya. *Proc. Natl Acad. Sci. USA* **87**: 4576–4579.

Zillig, W., Klenk, H.-P., Palm, P. *et al.* (1989) Did eukaryotes originate by a fusion event? *Endocyt. Cell Res.* **6**: 1–25.

Dating the age of the last common ancestor of all living organisms with a protein clock

29

Ronald M. Adkins and Wen-Hsiung Li

SUMMARY

Using highly conservative proteins which have evolved in an approximately clock-like manner and taking into consideration the variability of rates among sites, we estimated the date of the divergence between eubacteria and eukaryotes/achaebacteria, i.e. the age of the last common ancestor (LCA) of all living organisms, to be about 2.2 billion years ago. This estimate, which is close to that by Doolittle *et al.*, may not necessarily be incompatible with the appearance of microfossils before 3.5 billion years ago because prokaryotes might have existed long before this divergence occurred. The minimum age for the progenitor lineage may be estimated by calculating the timing of ancient gene duplications which preceded the divergence of the three modern domains of life. Using three pairs of duplicate genes for which reasonable alignments could be obtained, we estimated the ages of the duplications to be roughly 1.3 to 1.6 billion years older than the age of the LCA, or roughly 3.5 to 3.8 billion years old. Therefore, the estimate of 2.2 billion years for the age of the LCA does not appear so recent as to be incompatible with the microfossil record.

29.1 INTRODUCTION

The extant forms of life are currently divided into three domains: the Bacteria (eubacteria), the Archaea (archaebacteria), and the Eukarya (eukaryotes). Molecular data strongly indicate that the bacterial and archaebacterial lineages diverged first, and later the eukaryotic lineage branched off from the archaebacterial lineage (e.g. Iwabe *et al.*, 1989; Brown and Doolittle, 1995; Baldauf *et al.*, 1996; Lawson *et al.*, 1996). Therefore, the date of divergence between the bacterial and the archae-bacterial–eukaryotic lineages represents the last time that all living organisms shared a common ancestor – that is, the age of the last common ancestor (LCA) of all extant forms of life, or the age of the cenancestor (Fitch and Upper, 1987). This age is obviously of great interest to evolutionary biologists. Recently, Doolittle *et al.* (1996) used extensive protein sequence data to obtain an estimate of between 2 and 2.2 billion years for this age, and also an estimate of between 1.8 and 1.9 billion years for the Archaea–Eukarya divergence (i.e. the prokaryote–eukaryote divergence). Although these estimates are within the wide range (from 1.3 to 2.6 billion years) of the previous estimates of the prokaryote–eukaryote divergence based on limited molecular data (e.g. Jukes, 1969; McLaughlin and Dayhoff, 1970; Kimura and Ohta, 1973), they strongly contradict the view of some biologists and paleontologists that the prokaryote–eukaryote divergence occurred 3.5 billion years ago (e.g. Knoll, 1992; Martin 1996), the fossil date for the first appearance of cellular life (Schopf, 1993). It is therefore worth obtaining another estimate by restricting the analysis to proteins that are highly conservative and appear to have evolved in a fairly regular (i.e. clock-like) manner, and also by taking into account the rate of variation among amino acid sites, which can have a strong effect on date estimation.

It is worth noting that the estimate that the eukaryote lineage arose only about 2 billion years ago is not necessarily incompatible with the fossil record that prokaryotes emerged before 3.5 billion years ago, because prokaryotes could have existed long before the emergence of eukaryotes (Mooers and Redfield, 1996). Considerable insight into this issue may be obtained by estimating the date of a gene duplication that occurred before the Bacteria Archaea divergence and gave rise to two genes that still exist in both prokaryotes and eukaryotes and can be readily recognized as duplicate genes. Note that for two genes to be readily recognizable as duplicate genes, their structure and function must have been well conserved and should be similar to the structure and function of their common ancestor. In other words, the structure and function of the ancestral gene was already or almost already as sophisticated as the present-day duplicates. Since it would take a long time for a primitive protein to evolve into a sophisticated one, the duplication date might have been much younger than the emergence of prokaryotes. Therefore, if the duplication date can be shown to be considerably older than the age of the LCA, there would be no conflict between the estimate of 2 to 2.2 billion years for the age of the LCA (or the estimate of 2 billion years for the age of the eukaryote lineage) and the fossil record evidence that the prokaryotes emerged before 3.5 billion years ago. We have found three sets of ancient duplications that are suitable for this purpose.

29.2   MATERIALS AND METHODS

Thirteen data sets taken from Doolittle *et al.* (1996) were used for estimating the date of the eubacteria–eukaryote divergence (Table 29.1). These sequences were provided by Russell F. Doolittle. Three protein data sets with ancestral duplicates obtained from representatives of eukaryotes, archaebacteria and eubacteria were also used: leucyl-tRNA, isoleucyl-tRNA and valyl-tRNA synthetases; elongation factors EF-Tu and EF-G; and the internal duplication within the synthetase domain of carbamoylphosphate synthetase. An alignment of the tRNA synthetase sequences (Brown and Doolittle, 1995) was provided by James R. Brown, and an alignment of the carbamoylphosphate synthetase sequences (Lawson *et al.*, 1996) was provided by Fiona S. Lawson. Elongation factor sequences were. taken from GenBank. These sequences and the

Table 29.1 Estimation of the age of the last common ancestor (LCA) of all living organisms

Locus	N[a]	Alpha[b]	Age of LCA (Myr)
Argininosuccinate lyase	7	0.91	1480
1,4 alpha-glucan branching enzyme	8	1.4	3410
Acetyl CoA C-acetyl transferase	8	1.27	1805
Dihydrolipoamide dehydrogenase	9	1.30	1970
Glucose 6-phosphate dehydrogenase	8	1.07	1767
Glutamine fructose 6-phosphate transaminase	4	1.05	2185
Glycine dehydrogenase	4	0.81	1101
Glycine hydroxymethyl transferase	10	0.99	1800
Ribonucleotide reductase (large subunit)	6	1.68	3985
Threonine tRNA ligase	4	1.96	2006
Valine tRNA ligase	5	0.88	1253
Phosphoglycerate kinase	10	1.15	2786
Isoleucine tRNA ligase	5	1.83	2994
Average			2196

[a] Number of sequences used.
[b] Alpha parameter of gamma distribution of amino acid substitution rates among sites.

sequences from Doolittle *et al.* (1996) were aligned using the program CLUSTAL W 1.6 (Thompson *et al.*, 1994). Alignments were modified manually and regions of uncertain alignment were excluded.

Each set of protein sequences was first analyzed by the maximum likelihood method, using the program PROTML in the MOLPHY 2.2 package (Adachi and Hasegawa, 1992). For the elongation factor and carbamoylphosphate synthetase sequences, the phylogenies were rooted using each paralog. For the tRNA synthetase sequences, leucyl-tRNA synthetase was used to root the tree for the valyl- and isoleucyl-tRNA synthetase phylogenies. Trees were constructed by an exhaustive search or by the method of star-decomposition, depending on the number of sequences, and employed the substitution matrix of Jones *et al.* (1992). Maximum likelihood estimates of the alpha parameter of a gamma distribution describing the variability of substitution rates across sites was calculated by using the phylogenies produced by PROTML as input to the program CODEML of the PAML 1.1 package (Yang, 1995).

Gamma-corrected distances were calculated by the method of Nei *et al.* (1976) using the alpha values derived from CODEML, and neighbor-joining trees (Saitou and Nei, 1987) were constructed using the program MEGA (Kumar *et al.*, 1993). Taxa which exhibited obvious differences in substitution rates were then excluded from determinations of evolutionary divergence times. The divergence of animals and fungi at 960 million years ago and animals and plants at 1000 million years ago were used as calibration points; we did not use divergences between animals as calibration points because for highly conservative proteins the divergences between animal sequences are small and so are subject to large sampling errors. These dates and the average pairwise distances between animals/fungi and animals/plants were used to calculate substitution rates in proteins. These rates were then used to convert average eukaryote/archaebacteria and eukaryote/eubacteria pairwise distances to divergence times. The final alignments and sets of taxa used are available from W.-H. Li upon request.

29.3   RESULTS

29.3.1   AGE OF THE LAST COMMON ANCESTOR OF LIVING ORGANISMS

We used 13 protein data sets to estimate the eubacteria–eukaryote divergence or the age of the last common ancestor (LCA) of living organisms. The result of our analysis is shown in Table 29.1. As noted above, the alpha parameter of the gamma distribution indicates the strength of rate heterogeneity in a protein; the values of 0.5, 1 and 2 may be taken as strong, intermediate and weak, respectively. The majority of these proteins have an alpha value not far from one and therefore have intermediate rate heterogeneity. Two of the proteins, threonine tRNA ligase and isoleucine tRNA ligase, have an alpha value close to two and so have only weak rate heterogeneity.

Our estimates of the age of the LCA vary greatly among proteins and there are some extreme values. For example, the estimate from glycine dehydrogenase is only 1101 million years (Myr). This could possibly be due to a rate acceleration after the animal–yeast divergence, leading to an overestimate of the substitution rate, or it could be due to a horizontal gene transfer either from a prokaryote to a eukaryote or the other way around; however, we have no sequence from archaea and so we cannot decide whether any horizontal transfer has occurred. At the other extreme, the estimate from ribonucleotide reductase (large subunit) is 3985 Myr; this could possibly be due to a slowdown in rate after the yeast–animal divergence, leading to an underestimate of the substitution rate. Because of these uncertainties, our estimates should be taken with much caution but, surprisingly, our average estimate of 2196 Myr is almost identical to Doolittle *et al.*'s estimate of 2156 Myr (after correction for rate heterogeneity). We may therefore take 2200 Myr (i.e. 2.2 billion years) as an estimate of the age of LCA. At any rate, our main purpose is to see if this estimate is too young as claimed by Martin (1996) and others.

29.3.2 DATES OF ANCIENT DUPLICATIONS

We have been able to find three pairs of ancient duplicate proteins for each of which a reliable alignment can be obtained for a substantial part of the sequences. The first one is carbamoylphosphate synthetase. Although this protein was not duplicated, it contains an internal duplication that is present in all living organisms; the internal duplicates are called duplicates 1 and 2 and have been used to root the tree of life (Lawson *et al.*, 1996). Figure 29.1 shows a tree constructed from the two duplicates of animal, yeast, archaea and eubacteria carbamoylphosphate synthetase sequences. Although we have eliminated some of the sequences with extreme rates, there are still some sequences that show very different rates; for example, the two eubacterial (*Neisseria* and *Pseudomonas stutzeri*) duplicate 2 sequences show a much slower rate than the other duplicate 2 sequences. As proposed by Li and Tanimura (1987), to estimate divergence dates it is better to use only lineages that have evolved in a relatively regular manner. For example, to estimate the eukaryote–archaea divergence date (T_e) from the duplicate 2 sequences, we will use the yeast lineage; the animal lineages will give similar estimates because their branch lengths are similar to that of the yeast lineage, whereas the *Neisseria* and *P. stutzeri* lineages are short and will give biased estimates. In the yeast lineage the branch length from the

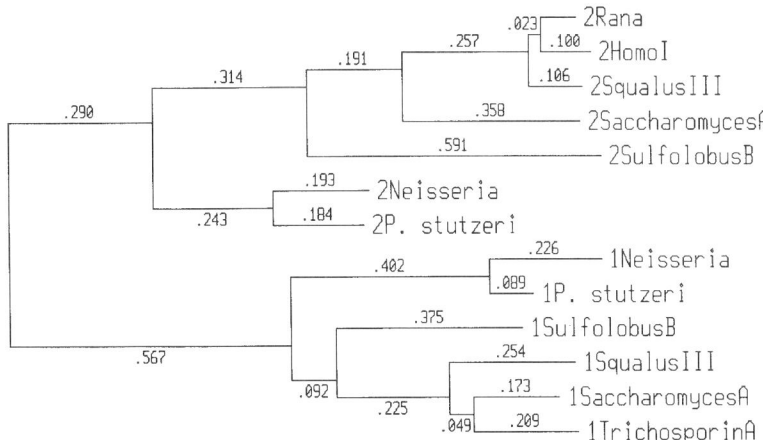

Fig. 29.1 Neighbor-joining tree constructed from γ-corrected distances (α = 1.14) calculated from the two internal duplicates of the synthetase domain of carbamoylphosphate synthetase. Prefixes 1 and 2 refer to the separate duplicates. Suffixes I, III, A and B refer to the locus designations used by Lawson *et al.* (1996). Generic names are given for each taxon except for *Pseudomonas stutzeri* (P. stutzeri). Branch lengths are in terms of the number of amino acid substitutions per site and are indicated above the branches.

eukaryote-eubacterium node to the animal–yeast node is 0.191 and that from the latter node to the present yeast sequence is 0.358. So, T_e can be estimated as [(0.191 + 0.358)/0.358] x 0.96 billion years = 1.47 billion years. Similarly, we estimate the age of the LCA as [(0.314 + 0.191 + 0.358)/(0.191 + 0.358)] x 1.47 = 2.31 billion years. For duplicate 1, if we use the Squalus III lineage, then we obtain 1.81 and 2.16 billion years for the two dates. So, the average estimate for the age of LCA is 2.24 billion years, which is very close to the estimate in Table 29.1. We now estimate the age of the internal duplication. Note that although the root of the tree is clearly in between the two duplicates, we do not know exactly where the root should be placed. However, we can compute the total branch length between the two eukaryote–eubacterium divergence nodes inferred from the two duplicates as 0.290 + 0.567 = 0.857. From this we estimate that the duplication is 1.32 billion years older than the age of the LCA, or the duplication occurred 2.2 + 1.3 = 3.5 billion years ago.

Figure 29.2 shows a tree constructed from elongation factors EF-Tu and EF-G. We used only sequences that appeared to have evolved at similar rates. However, it seems that the rate of amino acid substitution has slowed down in the eukaryotic lineages in recent times because the branch lengths from the animal–plant–fungus node to *Hydra*, *Arabidopsis*

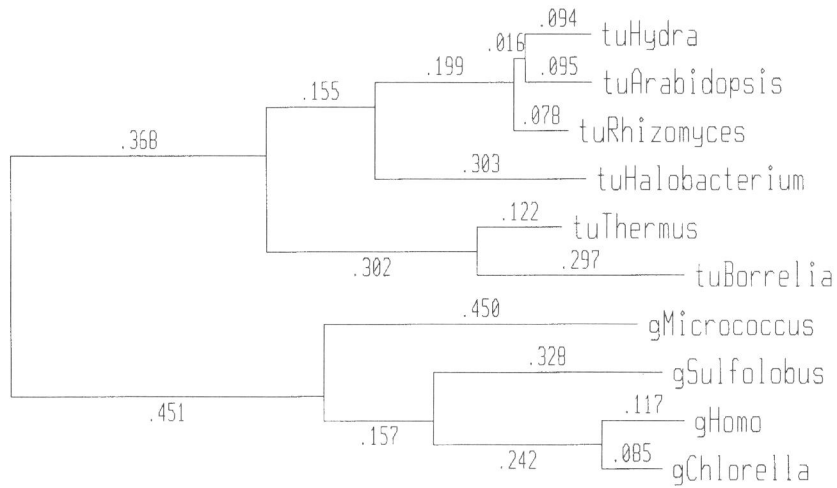

Fig. 29.2 Neighbor-joining tree constructed from γ-corrected distances (α = 1.22) calculated from elongation factor Tu and G sequences. The two loci are indicated by the prefixes tu or g. Branch lengths are shown above the branches.

and *Rhizomyces* are only about one half of the branch length between their common node and the archaea–eukaryote divergence node. For this reason, if we use the animal–fungus (or animal–plant) divergence to calibrate the substitution rate, we will obtain an estimate of about 3 billion years for the archaea–eukaryote divergence, an estimate of about 4.5 billion years for the age of the LCA, and an estimate of > 6 billion years for the gene duplication, which is older than the age of the Earth! In any case, it is clear that the duplication date is considerably older than the age of the LCA. Let us estimate this date by assuming that the age of the LCA is 2.2 billion years. For EF-Tu we shall use the *Borrelia* lineage, which is the longest among the EF-Tu lineages and will give the smallest estimate of the duplication date. The total branch length of this lineage since the LCA is 0.302 + 0.297 = 0.599. For the same reason, for EF-G we shall use the *Homo* lineage; its total length since the LCA is 0.157 + 0.242 + 0.117 = 0.516. The sum of the two preceding lineages is 0.599 + 0.516 = 1.115. On the other hand, the total branch length between the two LCA nodes inferred from EF-Tu and EF-G sequences is 0.368 + 0.451 = 0.819. Therefore, the duplication date is estimated to be (0.819/1.115) x 2.2 = 1.62 billion years older than the LCA, or to be 1.6 + 2.2 = 3.8 billion years old.

We now consider the two duplications that gave rise to the leucyl-tRNA, isoleucyl-tRNA, and valyl-tRNA synthetase genes. In Fig. 29.3, where these three synthetases are denoted l, i and v, respectively, the root was simply chosen to be the mid-point between the two longest lineages, which are the iTetrahymena and lNeurospora lineages. Therefore, it may not represent the true root, but this assumption was intended to be unfavorable for inferring the age of the second duplication. Under this assumption, the first gene duplication gave rise to leucyl-tRNA synthetase and the common ancestor of the isoleucyl- and valyl-tRNA synthetases, and the second duplication gave rise to the isoleucyl- and valyl-tRNA synthetases. Note that both of the branches from the latter duplication to the common node of all the isoleucyl-tRNA sequences and the common node of all the valyl-tRNA synthetase sequences are long (0.649 and 0.584), suggesting that the duplication is much older than the LCA. The isoleucyl-tRNA synthetase appears to have an accelerated rate in the branch between the branching nodes for the eukaryote and the *Saccharomyces* lineages; the branch length is 0.495. So, if we assume that the *Saccharomyces* lineage arose 0.96 billion years ago and use this lineage to calibrate the rate, we will obtain a date of [(0.495 + 0.22)/0.22] x 0.96 = 3.1 billion years for the age of the eukaryote lineage and an even older date for the age of the LCA. To use unfavorable conditions for estimating the date of the second duplication, we shall assume 2.2 billion years for the age of the LCA and use the *Tetrahymena* lineage (the longest lineage); its branch length from the LCA to the present is 0.083 + 0.495 + 0.070 + 0.312 = 0.960. The duplication date is estimated as [(0.649 + 0.960)/0.960] x 2.2 = 3.7 billion years ago. We now use the valyl-tRNA synthetase sequences to obtain another estimate. Note that the *Lactobacillus* and *E. coli* lineages are shorter than the *Neurospora* and *Homo* lineages, suggesting either a rate slowdown in the former lineages or a rate acceleration in the latter. Either case will lead to an underestimate of the age of the LCA. For example, if we use the *Homo* lineage (the longest lineage) and assume that this lineage arose 0.96 billion years ago, then the age of the LCA is estimated as [(0.063 + 0.114 + 0.014 + 0.352)/0.352] x 0.96 = 1.4 billion years old, which appears to be too young. Using the same assumptions, the duplication date is estimated as (0.584/0.352) x 0.96 = 1.6 billion years older than the LCA. If we assume that the age of the LCA is 2.2 billion years old, then the duplication date is 1.6 + 2.2 = 3.8 billion years old, which is close to the first estimate. Note that the first duplication should be older than the second duplication.

29.4   DISCUSSION

Our estimate of 2.2 billion years for the age of the LCA involves some uncertainties. First, the number of proteins used was not large and the individual estimates showed large variation. To reduce the sampling effect, more proteins should be used. Second, it did not consider the possibility of horizontal gene transfer. Martin (1996) has emphasized the

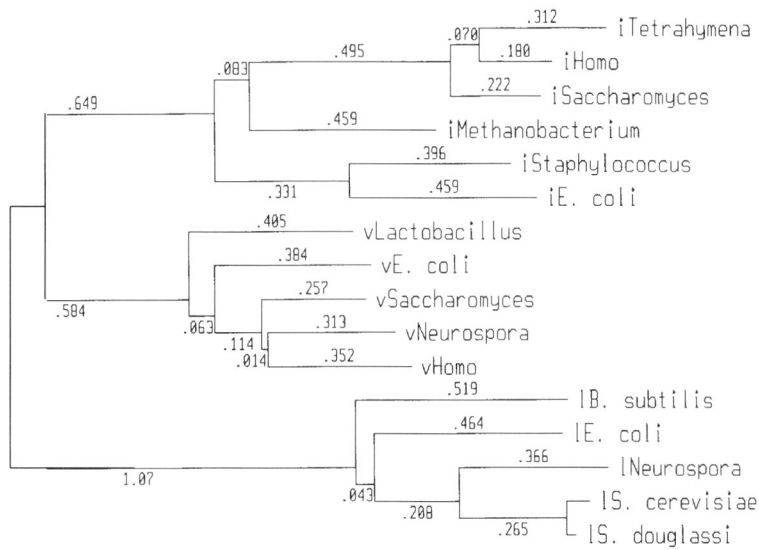

Fig. 29.3 Neighbor-joining tree constructed from γ-corrected distances (α = 1.26) calculated from leucyl-, valyl- and isoleucyl-tRNA synthetases. The three loci are identified by the prefixes i (isoleucine), v (valine) or l (leucine). Branch lengths are shown above the branches. Generic names are given for taxa except for *Escherichia coli* (E. coli), *Bacillus subtilis* (B. subtilis) and *Saccharomyces* (S. cerevisiae and S. douglassi).

importance of endosymbiotic gene transfer because there is increasing evidence that many genes in the nucleus (e.g. GAPDH, PGK, TPI) might have been acquired from mitochondrial or chloroplast genomes. This type of transfer is unidirectional (i.e. from eubacteria to nucleus) and can cause underestimation of the age of the LCA. We are unable to rule out this possibility in many of the proteins we used because not many sequences from eubacteria and archaebacteria are presently available for a thorough phylogenetic analysis. Despite these uncertainties, our estimate is very close to that of Doolittle *et al.* (1996), though we used only a subset of their data and used different methods of analysis.

Is the estimate of 2.2 billion for the age of the LCA really incompatible with the fossil record showing that life emerged more than 3.8 billion years ago (Schidlowski, 1988) and that prokaryotes existed before 3.5 billion years ago? To answer this question we considered the dates of three ancient duplications and obtained estimates of at least 3.5 billion years. In our estimation we assumed a constant rate of amino acid substitution since the duplication. This assumption may not hold well because duplicate genes may evolve faster following the duplication if relaxation in functional requirement or positive selection for functional changes or modifications occurs (Goodman, 1976, 1981; Li, 1985; Ohta, 1994). However, even if we assume that the average substitution rate between the duplication and the LCA is two times that of the average rate after the LCA, our estimates are still 2.9, 3.0 and 3.0; also remember that the first duplication for the aminoacyl tRNA synthetases should be even older. As noted above, at the time of duplication each of these proteins had already or almost already evolved the sophisticated function and protein structure we see today. Since it is likely to take much time for a primitive protein to evolve to a protein with a sophisticated function and structure, there is not much room for us to push back the duplication date. Therefore, these ancient duplication dates, which were estimated on the assumption of 2.2 billion years for the age of the LCA, suggest that there is no severe conflict between the estimated age of the LCA and the fossil record that life started more than 3.8 billion years ago.

Martin (1996) noted that if all present forms of life can be traced to a common ancestor only 2.2 billion years ago, then there would be no members of the cyanobacteria as we know them today that were photosynthesizing 3.5 billion years ago. This difficulty would not exist if all extant forms of life descend via direct filiation from cyanobacteria-like forefathers, but this is highly unlikely. It is also unlikely that the cyanobacteria-like organisms that were apparently very successful and produced the first large amount of oxygen on Earth became extinct, whereas the present-day cyanobacteria emerged later from other eubacteria. However, this problem remains unless we assume the age of the LCA to be nearly as ancient as 3.5 billion years old. In fact, it remains a problem as long as we believe that all extant eubacteria are a monophyletic group and share a common ancestor younger than, say, 3 billion years of age, because we still have to explain what happened to the cyanobacteria-like organisms that existed 3.5 billion years ago.

ACKNOWLEDGMENTS

This study was supported by NIH grants. We thank Russell F. Doolittle, James R. Brown and Fiona S. Lawson for sending us the sequence data in Table 29.1, the alignment of the tRNA synthetases and the alignment of the carbamoylphosphate synthetases, respectively.

REFERENCES

Adachi, J. and Hasegawa, M. (1992) *Computer science monographs, no. 27. MOLPHY: programs for molecular phylogenetics, I. PROTML: maximum likelihood inference of protein phylogeny*, Institute of Statistical Mathematics, Tokyo.

Baldauf, S.L., Palmer, J.D. and Doolittle, W.F. (1996) The root of the universal tree and the origin of eukaryotes based on elongation factor phylogeny. *Proc. Natl Acad. Sci. USA* **93**: 7749–7754.

Brown, J.R. and Doolittle, W.F. (1995) Root of the universal tree of life based on ancient aminoacyl-tRNA synthetase gene duplications. *Proc. Natl Acad. Sci. USA.* 92: 2441–2445.

Doolittle, R.F., Feng, D.-F., Tsang, S. *et al.* (1996) Determining divergence times of the major kingdoms of living organisms with a protein clock. *Science* **271**: 470–477.

Fitch, W.M. and Upper, K. (1987) The phylogeny of tRNA sequences provides evidence for ambiguity reduction in the origin of the genetic code. *Cold Spring Harbor Symp. Quant. Biol.* **52**: 759–767.

Goodman, M. (1976) Protein sequences in phylogeny, in *Molecular Evolution*, (ed. F.J. Ayala), Sinauer, Sunderland, MA, pp. 141–159.

Goodman, M. (1981) Decoding the pattern of protein evolution. *Prog. Biophys. Mol. Biol.* **38**: 105–164.

Iwabe, N., Kuma, K.-I., Hasegawa, M. *et al.* (1989) Evolutionary relationship of archaebacteria, eubacteria, and eukaryotes inferred from phylogenetic trees of duplicated genes. *Proc. Natl Acad. Sci. USA* **86**: 9355–9359.

Jones, D.T., Taylor, W.R. and Thorton, J.M. (1992) The rapid generation of mutation data matrices from protein sequences. *Comp. Appl. Biosci.* **8**: 275–282.

Jukes, T.H. (1969) Recent advances in studies of evolutionary relationships between proteins and nucleic acids. *Space Life Sciences* **1**: 469–490.

Kimura, M. and Ohta, T. (1973) Eukaryotes–prokaryotes divergence estimated by 5S ribosomal RNA sequences. *Nature New Biol.* **243**: 199–200.

Knoll, A.H. (1992) The early evolution of eukaryotes: a geological perspective. *Science* **256**: 622–627.

Kumar, S., Tamura, K. and Nei, M. (1993) *MEGA: Molecular Evolutionary Genetics Analysis, version 1.01*, The Pennsylvania State University, University Park, PA.

Lawson, F.S., Charlebois, R.L. and Dillon, J.-A.R. (1996) Phylogenetic analysis of carbamoylphosphate synthetase genes: complex evolutionary history includes an internal duplication within a gene which can root the tree of life. *Mol. Biol. Evol.* **13**: 970–977.

Li, W.-H. (1985) Accelerated evolution following gene duplication and its implication for the neutralist–selectionist controversy, in *Population Genetics and Molecular Evolution*, (eds T. Ohta and K. Aoki), Japan Scientific Societies Press, Tokyo, pp. 333–352.

Li, W.-H. and Tanimura, M. (1987) The molecular clock runs more slowly in man than in apes and monkeys. *Nature* **326**: 93–96.

Martin, W.F. (1996) Is something wrong with the tree of life? *BioEssays* **18**: 523–527.

McLaughlin, P.J. and Dayhoff, M.D. (1970) Eukaryotes versus prokaryotes: an estimate of evolutionary distance. *Science* **168**: 1469–1471.

Mooers, A.O. and Redfield, R.J. (1996) Digging up the roots of life. *Nature* **379**: 587–588.

Nei, M., Chakraborty, R. and Fuerst, P.A. (1976) Infinite allele model with varying mutation rate. *Proc. Natl Acad. Sci. USA.* **73**: 4164–4168.

Ohta, T. (1994) Further examples of evolution by gene duplication revealed through DNA sequence comparisons. *Genetics* **138**: 1331–1337.

Saitou, N. and Nei, M. (1987) The neighbor-joining method: a new method for reconstructing phylogenetic trees. *Mol. Biol. Evol.* **4**: 406–425.

Schidlowski, M. (1988) A 3,800-million-year isotopic record of life from carbon in sedimentary rocks. *Nature* **333**: 313–318.

Schopf, J.W. (1993) Microfossils of the early Archaen apex chart: New evidence of the antiquity of life. *Science* **260**: 640–646.

Thompson, J.D., Higgins, D.G. and Gibson, T.J. (1994) CLUSTAL W: improving the sensitivity of progressive multiple sequence alignment through sequence weighting, position specific gap penalties and weight matrix choice. *Nucl. Acids Res.* **22**: 4673–4680.

Yang, Z. (1995) *Phylogenetic analysis by maximum likelihood (PAML), version 1.1.* Institute of Molecular Evolutionary Genetics, The Pennsylvania State University, University Park, PA.

The prokaryotic origin of eukaryotes

30

James A. Lake and Maria C. Rivera

SUMMARY

The prokaryotic origins of the eukaryotic cell have been enigmatic. Within the last decade, with the availability of DNA sequences and entire genomes from diverse organisms, it has become possible to investigate the remote origins of eukaryotes. This chapter describes the current understanding of relationships among the major groups of prokaryotes and eukaryotes and emphasizes an ongoing search to determine which prokaryotes contributed genes, and even gene lineages, to eukaryotes. It now appears that nuclear genes have originated from both eubacterial and eocyte (crenarchael) prokaryotes.

30.1 INTRODUCTION

The prokaryotes, unlike their multicellular eukaryotic relatives – the animals, plants, and fungi – have few morphological features that are useful for phylogenetic studies. Within the last decade, with the availability of DNA sequences from phylogenetically diverse organisms, it has become possible to use sequence data to probe the relationships among the most diverse prokaryotic groups and even to investigate the remote prokaryotic origins of eukaryotes.

Understanding the origin of eukaryotes is made more difficult because the nucleus is a chimera of genes from various sources. Many eukaryotic nuclear genes have been imported from eukaryotic organelles – for example, the chloroplast and the mitochondrion (Gray, 1993; Baldauf and Palmer, 1990). Even the mechanism by which the nucleus was formed may have involved an endosymbiosis between two bacterial types (Lake, 1982).

Several theories exist for the origin of the nucleus. Most scientists subscribe to the **karyogenic hypothesis,** shown in Fig. 30.1. In this theory the nucleus and its enclosing membranes were gradually acquired through some (unspecified) segregating process. The competing theory is less well known: the **endokaryotic hypothesis** is based on the observation that the nucleus, like the other eukaryotic organelles enclosed in double membranes (the chloroplast and mitochondrion), has been derived through capture by an engulfing bacterium (Lake, 1982). The latter proposal is simple and parsimoniously explains the origin of all double-membrane organelles through a single mechanism rather than requiring two different mechanisms (one for the mitochondrion and chloroplast and another for the nucleus). Until recently there were few data to support either theory, but Gupta *et. al.* (1994) argued, on the basis of sequences of the 70 kDa heat shock protein (HSP70), that the endokaryotic hypothesis is supported. Clearly deciding between both theories will be important for any complete understanding of the origin of eukaryotes. This review will emphasize the search for the prokaryotic group of organisms that has contributed the **majority** of genes to the eukaryotic nucleus.

KARYOGENIC HYPOTHESIS

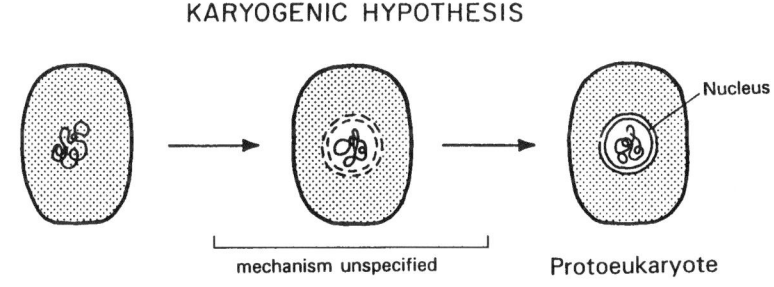

ENDOKARYOTIC HYPOTHESIS

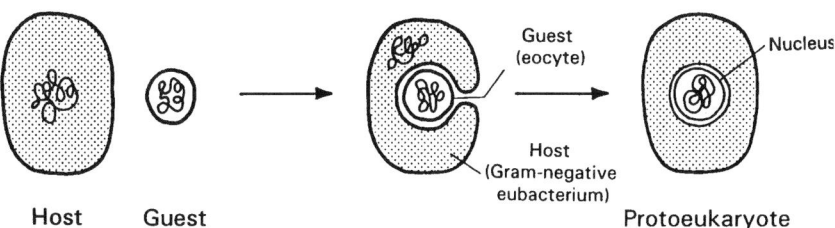

Fig. 30.1 Competing hypotheses for the evolution of the nucleus (Lake and Rivera, 1994).

30.2 DIFFICULTIES WITH THE TRADITIONAL PROKARYOTIC–EUKARYOTIC CLASSIFICATION

Early attempts to understand the diversity of life on Earth emphasized eukaryotic cells, since their large nuclei were readily visible by light microscopy. According to these studies, all life could be divided into either eukaryotes (organisms with a nucleus) or prokaryotes (organisms that lack a nucleus). Although the nucleus is a positive (derived) feature that unites the eukaryotes into a monophyletic group (a group containing the last common ancestor of the group and all of its descendants; de Queiroz and Gauthier, 1990), the lack of a nucleus in prokaryotes is a negative feature and cannot be used to define a group (Eldrige and Cracraft, 1986). Gradually it has become clear that the prokaryotes are not a valid, monophyletic group, because they are grouped by a negative feature. A major challenge in microbiology is to ascertain the phylogenetic relationships among the prokaryotes. The central question which arises – 'Which group of prokaryotes is the closest relative of the eukaryotic nucleus?' – is addressed in this review.

30.3 PROKARYOTIC DIVERSITY

Before attempting to identify the prokaryotic group that is most closely related to the eukaryotic nucleus (sister taxon) it is useful to survey briefly their phenotypic and phylogenetic diversity. A phylogenetic tree representing the major known bacterial groups is shown in Fig. 30.2. The relationships shown are generally accepted by the microbiological community except that the origin of the eukaryotic lineage (the subject of this chapter) has been controversial. There are several groups with prokaryotic organization, any one of which might be related to the eukaryotic nucleus. These include the eubacteria, the halobacteria, the methanogens, relatives of the methanogens, and sulfur-metabolizing, high-temperature organisms known as the eocytes.

The eubacteria are a diverse group that includes all the photosynthetic bacteria (except for the halobacteria) as well as many non-photosynthetic groups. Most eubacteria are mesophiles, but in Fig. 30.2 the eubacteria are represented by the extreme thermophiles, *Thermotoga maritima* and *Aquifex pyrophilus*, which can grow in temperatures up to 90°C and 95°C, respectively, utilizing an unknown fermentative process (Huber and Stetter, 1992). The lipids of eubacteria are primarily of the ester type although *Thermotoga*, *Aquifex* and their relatives also contain branched ether lipids.

The halobacteria are extreme halophiles. They are carbon heterotrophs that can use an unusual photosynthesis system – namely, a light-driven proton pump based on bacteriorhodopsin. Surprisingly, they have many molecular properties in common with the eubacteria that are not found

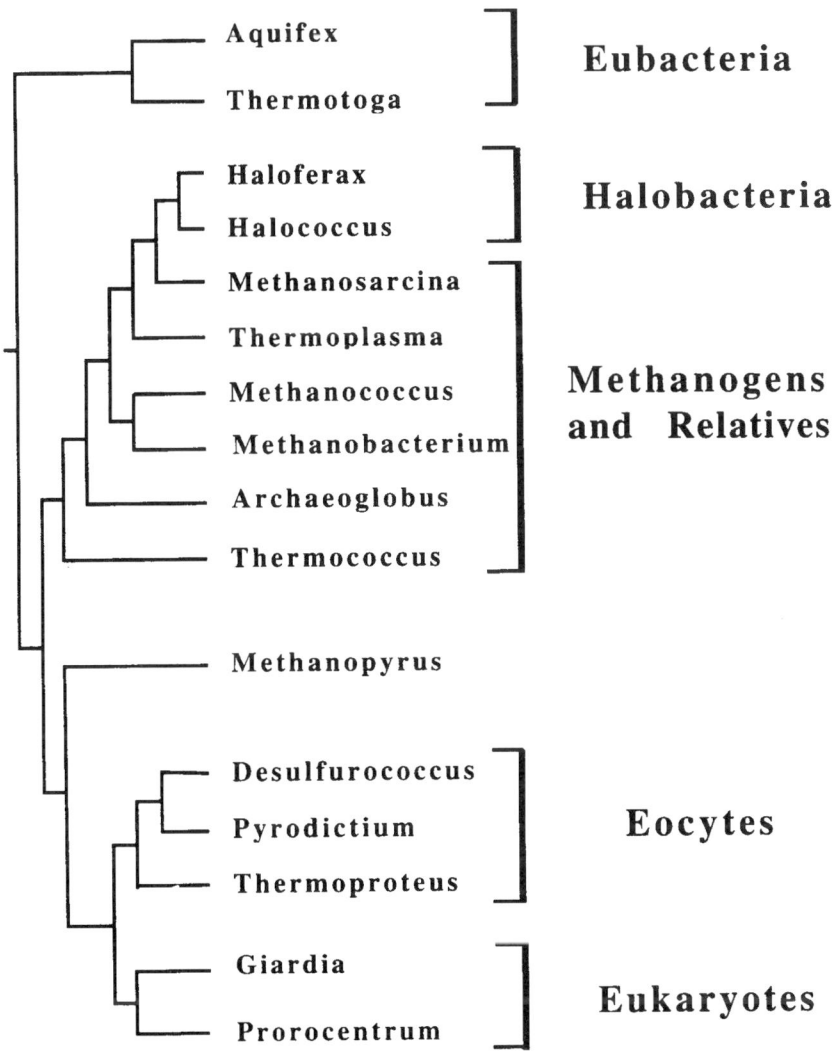

Fig. 30.2 The rooted phylogenetic tree relating prokaryotic and eukaryotic organisms reconstructed from 16S and 18S ribosomal RNA sequences using paralinear distances (Lockhart *et al.*, 1994; Lake, 1994) and the bootstrappers gambit multi-taxon tree reconstruction algorithm (Lake, 1995). The Halobacteria + Methanogens and Relatives clade at the 99% level and the eocytes + eukaryotes clade at the 100% level were supported by 200 bootstrap replicates.

in other bacterial groups. For example, like eubacteria, they contain the biochemical pathways for the synthesis of C40 and C50 carotenoids (Goodwin, 1980).

The methanogens are a phylogenetically diverse group, despite the fact that they share a common phenotype in that they are strict anaerobes with the ability to chemically reduce carbon compounds to methane to provide energy. According to most rRNA phylogenetic studies, some methanogens, such as *Methanospirillum* and *Methanosarcina*, are more closely related to the halobacteria than they are to other methanogens.

Associated with the methanogens is a phenotypically diverse group of organisms represented by such organism as *Thermococcus celer*, *Archaeoglobus fulgidus* and, possibly, *Methanopyrus kandleri* (Rivera and Lake, 1996). *Methanopyrus kandleri*, which is phenotypically a methanogen, grows at temperatures up to 112°C, but genotypically is probably not a methanogen (Burggraf *et al.*, 1991; Rivera and Lake, 1996). In the tree in Fig. 30.2, *Methanopyrus* is not within the same clade as the methanogens.

The final prokaryotic group, the eocytes, consists of thermophilic, mostly sulfur-metabolizing organisms, many of which can grow at temperatures in excess of 100°C. The eocytes include *Sulfolobus*, *Desulforococcus*, *Thermoproteus*, *Pyrodictium*, *Pyrobaculum*, etc. *Sulfolobus sulfataricus* oxidizes sulfur to H_2S. Others, such as *Acidianus infernus*, can oxidize or reduce S^0 to H_2SO_4 or to H_2S, respectively. The organisms with the highest maximum growth temperatures (112°C) are *Pyrodictium occultum* and *Pyrodictium abyssum*. The group is metabolically diverse, uniformly thermophilic, and phylogenetically monophyletic.

30.4   ORIGIN OF THE EUKARYOTES

The ability of eukaryotes and prokaryotes to transfer genes laterally is well known. Numerous genes originally contained in the mitochondrial and chloroplast genomes have been transported and are now encoded in the nucleus. Well-documented examples include the incorporation of diverse mitochondrial and chloroplast genes into the nuclear genome (for a thoughtful review, see Smith *et al.*, 1992). Thus not all nuclear genes are indicative of the ancestry of the bulk of the nuclear genes. As a result, in this review we concentrate on the genes of the translational and transcriptional machinery. Since these genes are so well integrated into the cellular machinery they are unlikely to have been imported from outside sources. They are generally accepted to be good markers of the evolutionary history of the nucleus.

The two most intensively studied prokaryotic genes are the 16/18S ribosomal RNA genes and the genes of protein synthesis factor EF-Tu (EF-1α in eukaryotes). Based on their analyses, two theories have been proposed to explain the origin of the eukaryotic nucleus: the archaebacterial (archaea) theory and the eocyte theory. Trees corresponding to both theories, reconstructed from 16/18S rRNA sequences, are shown in Fig. 30.3. Both are rooted in the branch leading to the eubacteria, as in Fig. 30.2 (Gogarten *et al.* 1989; Iwabe *et al.*, 1989). The fundamental difference between these two theories is that in the eocyte tree (at the top of the figure) the eukaryotic nucleus shares a most recent common ancestor solely with the eocytes, whereas in the archaebacterial theory the eukaryotes are most closely related to an ancestral organism that gave rise to the halobacteria, the methanogens and their relatives and to the crenarchaea (= eocytes). These two theories are based on the topology of the phylogenetic trees. This makes the two theories mutually exclusive and hence eminently testable, because if one theory is correct the other must be incorrect.

Molecular sequences are our most informative source of data for testing these theories, but there are significant artifacts associated with their analysis. When sequences have diverged extensively, three different artifacts can cause long branch attraction: unequal rate effects that result when tree reconstruction algorithms fail to account adequately for multiple substituions; site-to-site variation that results when the variation of rates of evolution within sequences is not properly accounted for; and alignment artifacts. The result is that rapidly evolving taxa (long branches) are placed with other rapidly evolving taxa, whether or not the taxa are phylogenetically related. These three types of artifacts are relevant to the origin of eukaryotes because the archaebacterial tree groups the two longest branches of the tree together. It places the rapidly evolving eubacteria with the rapidly evolving eukaryotes and therefore could be caused by these artifacts. One comprehensive paper, unusual for its attention to detail, indicates that the archaebacterial topology is probably caused by unequal rate effects (Volters and Erdmann, 1988).

30.5   ELONGATION FACTOR GENES

Because of the long branch attraction artifacts, we searched for molecular sequences which contained structural features, such as inserted segments, that would evolve much more slowly than individual nucleotides, be more easily interpreted, and therefore be much less sensitive to long branch artifacts.

The molecule we chose to study was protein synthesis elongation factor EF-Tu (EF-1α in eukaryotes) (Rivera and Lake, 1992). EF-Tu is a ubiquitous protein that transports aminoacyl tRNAs to the ribosome and participates in their selection by the ribosome. The structure of the guanosine diphosphate (GDP)-binding domain of EF-Tu from *Escherichia coli* has been determined by X-ray diffraction. Within this domain, the amino acid sequence, KNMITG$_{94}$, which is strictly conserved in EF-1α and EF-Tu sequences, terminates an α helix and is followed by a β strand

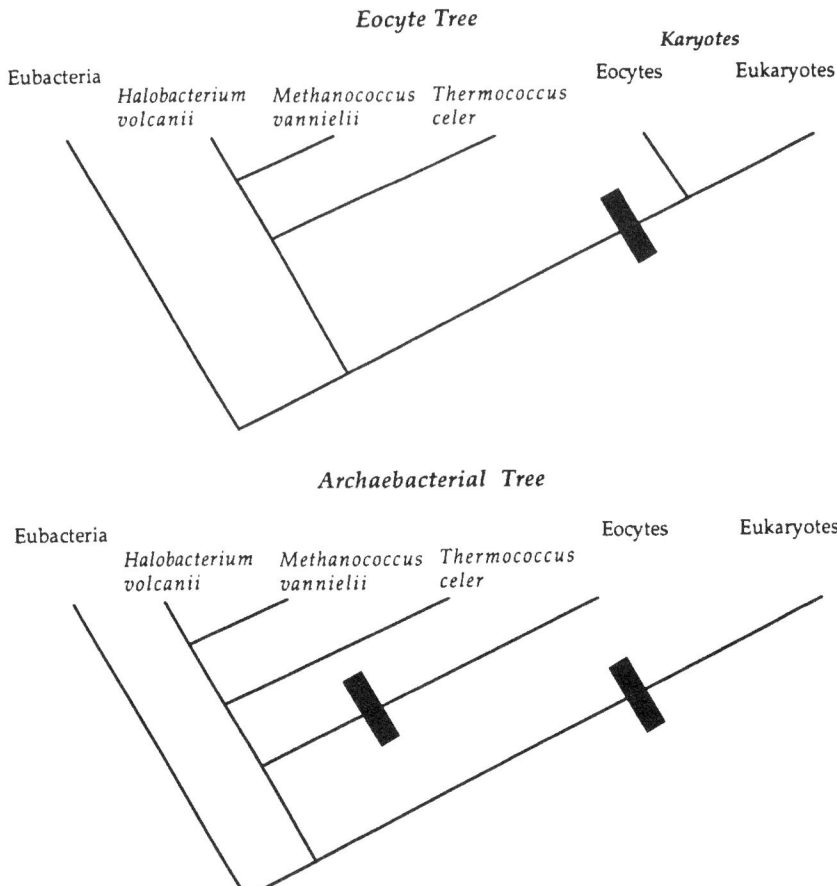

Fig. 30.3 Rooted trees illustrating the two theories proposed to explain the origin of the eukaryotic nucleus. The trees corresponding to both theories are reconstructed from 16S/18S sequences and rooted in the branch leading to the eubacteria. The solid boxes indicate changes from the 4 amino acid segment to the 11 amino acid form. The eocyte tree is favored by parsimony since it requires only a single change to the 11 amino acid segment whereas the archaebacterial tree is opposed since it requires two independent changes (the same distribution could also be explained by one appearance of the 11 amino acid form and reappearance of the 4 amino acid form, but still two changes would be required).

that is terminated by GPMP$_{113}$ at the GDP binding site. The sequence QTREH$_{118}$ then starts a 3$_{10}$ helix. The amino acid motifs of the eukaryotic EF-1α are similar, except that the four-amino acid sequence GPMP$_{113}$ is replaced by the 11-amino acid sequence GEFEAGISKDG and its variants (Table 30.1).

Since the eukaryotic 11-amino acid insert is so well conserved among eukaryotic sequences, we thought that eocyte sequences might also contain the 11-amino acid insert. Using the polymerase chain reaction and DNA primers designed for use with the KNMITG and QTREH sites, we amplified, cloned and sequenced the insert region, with the results shown in Table 30.1. The eocyte amino acid sequences, translated from DNA, shared the eukaryotic motif (11 amino acids) rather than that found in methanogens, halobacteria and eubacteria (four amino acids). The longer 11-amino acid segment, present in eocytes and eukaryotes, shares little obvious similarity with the shorter, four-amino acid segment found in other prokaryotes.

In order to ascertain which form of the segment existed first, we compared the sequences of the related (paralogous) proteins EF-2 (termed EF-G in eubacteria) and IF-2 because both diverged from EF-1α before the last common ancestor of eukaryotes and prokaryotes (Gogarten *et al.*, 1989; Iwabe *et al.*, 1989). Since all sequences from EF-2 and IF-2 contained the four-amino acid segment, this indicated that $GPMP_{113}$ (or one of its variants) is the original (plesiomorphic) form of this segment and that the 11-amino acid segment must be the derived (synapomorphic) form.

Our conclusions are shown in Fig. 30.3. We have mapped the changes onto the trees representative of both theories. Starting from the four-amino insert at the root of the tree, each solid box indicates a change from the four-amino acid segment to the 11-amino acid form. The eocyte tree is favored because it requires only a single change, whereas the archaebacterial tree requires two independent but identical changes. The archaebacterial tree could also be explained by one appearance of the 11-amino acid form and one reappearance of the four-amino acid form, but even so two changes would be required. The EF-1α synapomorphy cannot discriminate among the possible branchings below the eocyte–eukaryote node, so that any tree having eocytes and eukaryotes as sister taxa would be consistent with these data. In Fig. 30.4, these results are compared with the larger set of taxa originally shown in Fig. 30.2. Again, the agreement is complete.

Several lines of reasoning buttress the interpretation that eocytes are the closest relatives of the eukaryotes. First, the 11-amino acid segments present in eocytes and eukaryotes are very likely homologous. Eight of 11 amino acids (seven in *Sulfolobus* and *Acidianus*) are identical to the consensus eukaryotic sequence. Amino acid shuffling of the segments produced random alignments that scored 6–7 standard deviations lower than those found for the eukaryotic–eocyte alignment, thereby implying homology (Waterman and Eggert, 1987). Second, the alignments are well defined. No gaps are needed to align the eukaryotic and eocytic EF-1α sequences, and no gaps are needed to align the eubacteria, methanogen

Table 30.1 Comparison of the *Methanopyrus kandleri* EF-1 sequence to the sequences from methanogens, halobacteria, eubacteria, eocytes and eukaryotes

Taxon	Organism		11 amino acid segment	4 amino acid	
Methanogens and relatives	Mp. kand.	KNMITGASQADAAILVVAADD		---GVMP	qtreh
	T. celer	KNMITGASQADAAVLVVAVTD		---GVMP	QTKEH
	P. woes.	KNMITGASQADAAVLVVAATD		---GVMP	QTKEH
	A. fulg.	knmitgASQADAAVLVMDVVE		---KVQP	qtreh
	Mc. vann.	KNMITGASQADAAILVVNVDD		AKSGIQP	QTREH
	T. acido.	KNMITGTSQADAAILVISARD		-GEGVME	QTREH
Halobacteria	H. maris.	KNMITGASQADNAVLVVAADD		---GVQP	QTQEH
Eubacteria	Th. mar.	KNMITGAAQMDGAILVVAATD		---GPMP	QTREH
	D. sal.	KNMITGAAQMDGAIIVCSAAD		---GPMP	QTREH
	E. coli	KNMITGAAQMDGAILVVAATD		---GPMP	QTREH
Eocytes	Su. acid.	KNMITGASQADAAILVVSAKK	GEYEAGMSAEG		QTREH
	Td. mari.	KNMITGASQADAALLVVSARK	GEFEAGMSAEG		qtreh
	P. occu.	knmitgASQADAAILVVSARK	GEFEAGMSAEG		qtreh
	D. muco.	knmitgASQADAAILVVSARK	GEFEAGMSAEG		qtreh
	A. infe.	knmitgASQADAAIIAVSAKK	GEFEAGMSEEG		qtreh
Eukaryotes	Giardia	KNMITGTSQADVAILVVAAGQ	GEFEAGISKDG		QTREH
	Tetrahy.	KNMITGTSQADVAILMIASPQ	GEFEAGISKDG		QTREH
	Yeast	KNMITGTSQADCAILIIAGGV	GEFEAGISKDG		QTREH
	Tomato	KNMITGTSQADCAVLIIDSTT	GGFEAGISKDG		QTREH

Droso.	KNMITGTSQADCAVQIDAAGT	GEFEAGISKND	QTREH
Rat	KNMITGTSQADCAVLIVAAGV	GEFEAGISKNG	QTREH
Human	KNMITGTSQADCAVLIVAAGV	GEFEAGISKNG	QTREH

Four amino acid and 11 amino acid segments are underlined. Small letters represent sequences from the PCR primers. The sequences compared are the following.

- Methanogens and their relatives: Mp. kan., *Methanopyrus kandleri*; T. celer, *Thermococcus celer*; P. woesei, *Pyrococcus woesei*; A. fulg., *Archaeoglobus fulgidus*; Mc. van., *Methanococcus vannielii*; T. acido., *Thermoplasma acidophilum*.
- Halobacteria: H. maris., *Halobacterium marismortui*.
- Eubacterial sequences: thermophilic eubacteria Tt. mar., *Thermotoga maritima*; halophilic cyanobacteria D. salina, *Dactylococcopsis salina*.
- Eocytes: Td. mar., *Thermodiscus maritimus*; P. occu., *Pyrodictium occultum*; A. infe., *Acidianus infernus*; S. acid., *Sulfolobus acidocaldarius*.
- Eukaryotes: Giardia, *Giardia lamblia* Tetrahy., *Tetrahymena pyriformis*; yeast, *Saccharomyces cerevisiae*; tomato, *Lycopersicon esculentum*; Droso., *Drosophila melanogaster*; rat, *Rattus norvegicus*; human, *Homo sapiens*.

Original sources for the sequences are listed in Rivera and Lake (1996).

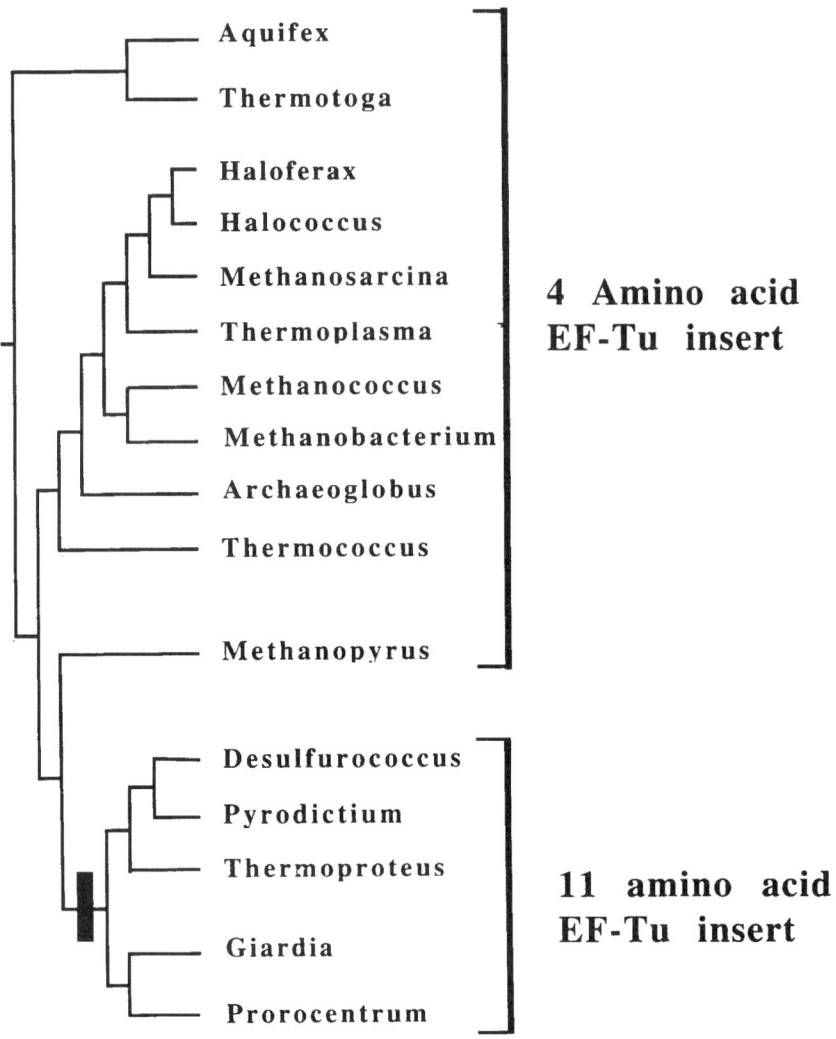

Fig. 30.4 The phylogenetic tree from Fig. 30.2 illustrating the distribution of 4 and 11 amino acid EF-Tu inserts. The dark box indicates the approximate location for the change from the 4 amino acid insert to the 11 amino acid insert.

and halobacterial sequences. Third, the sequences encoding EF-1α are not likely to have been laterally transferred between organisms, since EF-1α is present in all cells and, during protein synthesis, interacts with cellular components encoded by genes dispersed throughout the bacterial genome, including aminoacyl tRNAs, ribosomal proteins, elongation factor EF-Ts and 16S and 18S ribosomal RNAs (Hill *et al.*, 1990). Thus these results lend strong support to the proposal that the eukaryotes and eocytes are sister taxa within the 'tree of life'.

30.6   ORGANIZATION OF RIBOSOMAL OPERONS

A number of fundamental molecular properties have been thought to have an idiosyncratic distribution on the tree of life, principally because they did not fit the archaebacterial tree. Yet these same molecular properties fit the eocyte theory perfectly. For a discussion of transcriptional promoters, for example, see Lake and Rivera (1996). In the following section the organization of ribosomal rRNA operons is used to illustrate their support for the eocyte tree, and their inconsistency with the archaebacterial one.

Because small subunit ribosomal RNA sequences are the standard for defining the phylogenetic positions of organisms, a large data base of ribosomal RNAs exists and one knows far more about the organization of ribosomal operons than about any other operons. Eubacteria, halobacteria, methanogens and eocytes contain three rRNAs – 16S, 23S and 5S – which are homologous to the eukaryotic 18S, 5.8S + 28S, and 5S. (For simplicity we will refer to both the eukaryotic and prokaryotic homologs using the prokaryotic labels.) The number of ribosomal rRNA transcriptional units varies between one and four in the halobacteria and methanogens. Ribosomal operons are arranged in the same general pattern in eubacteria, halobacteria and methanogen, namely 16S-tRNA-23S-5S. Occasionally an additional tRNA gene will be found between the 16S and 23S genes or following the 5S gene (reviewed in Brown *et al.*, 1989). *Thermoplasma*, which is phylogenetically related to the methanogens, is an exception to this general rule and, unlike any other prokaryotes, contains unlinked 16S, 23S and 5S genes (Tu and Zillig, 1982). The pattern in eocytes and eukaryotes is different from the eubacteria, halobacteria, and methanogens. In the eocytes the 16S-23S genes are linked without a tRNA spacer and there is a variable linkage of 5S rRNA encoding genes to the 16S-23S unit. The non-operon-associated 5S rRNA gene of *D. mobilis* forms its own transcriptional unit (Kjems and Garrett, 1988) but those of many other eocytes contain a 16S-23S-5S unit. The eukaryotic pattern is similar, with a 16S-23S (equivalent) transcription unit lacking tRNA spacers and with the 5S either separately transcribed or linked (Gerbi, 1985). A exception to this rule is found among the Cryptomonads where the rRNA genes are unlinked (Gray, 1992).

Although this pattern of rRNA operon organization cannot be easily explained by the archaebacterial theory, it easily fits the eocyte tree. In Fig. 30.5, the tree is labeled with the types of operons. Note that to accommodate this distribution on the eocyte tree, only a single change of operon type is required: the 16S-tRNA-23S-5S pattern found in eubacteria, halobacteria and methanogens is substituted by the derived 16S-23S type at the position on the tree shown by the box. Depending on the operon organization in *Methanopyrus* (presently unknown), the site will be either before or after *Methanopyrus* branches. In either case still only a single change will be required.

30.7   CONCLUSIONS

Of all the sequences obtained to date, the EF-Tu molecule seems to offer the most reliable indication of early divergences. It is one of slowest evolving (if not the slowest) sequences yet found and is unlikely to be laterally transferred between organisms because it is present in all cells and, during protein synthesis, interacts with cellular components that are dispersed throughout the bacterial genome. Furthermore, direct phylogenetic analyses of EF-Tu sequences by almost all authors support the eocyte tree. Major publications on these genes in the last five years and their conclusions are shown in Table 30.2. Significant support for the eocyte tree also comes from the observations that eukaryotic ribosomal operons are organized like those of *Sulfolobus*, *Desulfurococcus* and *Thermoproteus* and not organized like the tRNA containing rRNA operons of halobacteria, methanogens and eubacteria.

We are just starting to fathom the diversity of life on Earth. Although the relationship of the nuclear genes of eukaryotes with those of prokaryotes have been controversial, it now appears that eukaryotes are very likely the sister group of the eocyte prokaryotes. The tremendous progress that is being made toward determining the relationships among bacteria and eukaryotes gives one optimism that at least some of the major events in the tree of life may soon be understood.

Table 30.2 Recent sequence analyses of EF-1α sequences

Supports *Archae*	Supports *Eocyta*	References
	+	Baldauf *et al.*, 1996
	+	Hashimoto and Hasegawa, 1996
	+	Runnegar, 1993
	+	Lake, 1994
	+	Hasegawa *et al.*, 1993
	+	Cousineau *et al.*, 1992
+		Creti *et al.*, 1991

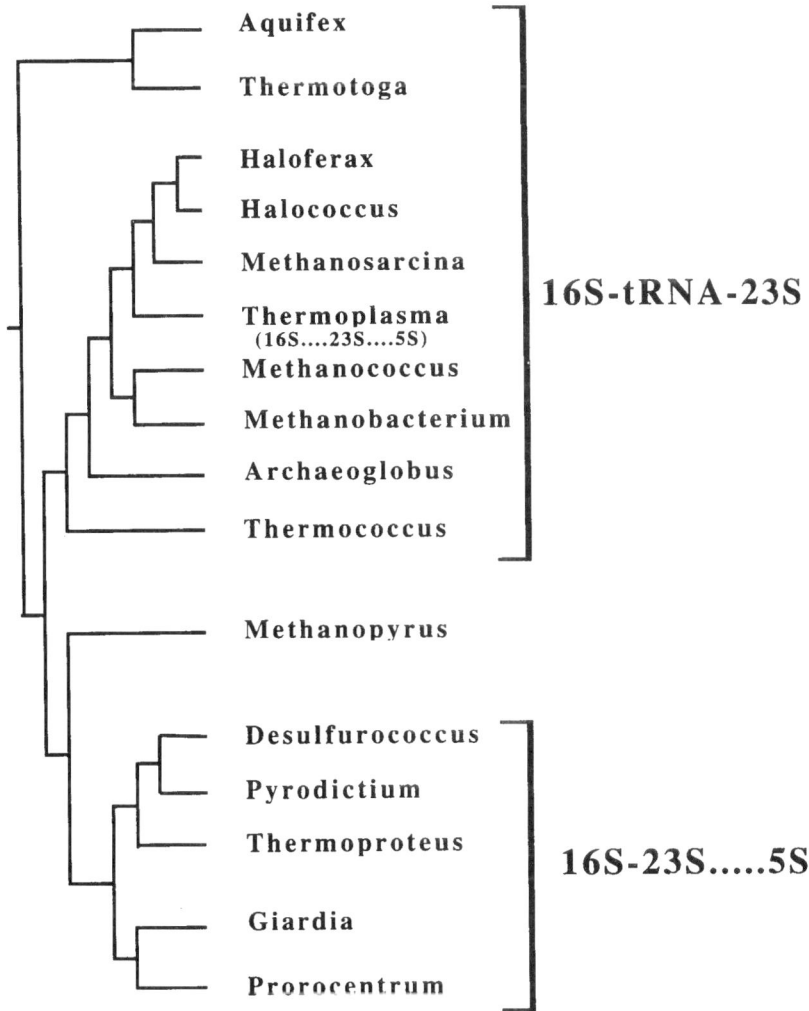

Fig. 30.5 The phylogenetic tree from Fig. 30.2 illustrating the distribution of ribosomal operon types. In the 16S-tRNA-23S type the operon contains 16S rRNA + tRNA (possibly two) + 23S rRNA + 5S rRNA. In the 16S-23....5S type the operon contains the 16S rRNA + 23S rRNA with the 5S rRNA frequently transcribed separately.

NOTE IN PROOF

Recent analyses of complete genomes (Rivera, Jain, Moore and Lake, submitted) indicate that prokaryotes consist of two ancestral lineages which code for broadly defined functional gene classes. The deepest diverging **informational lineage** codes for translation, transcription, and replication genes, and also includes GTPases, tRNA synthetases, and vacuolar ATPase homologs. The more recently diverging, **operational lineage**, codes for amino acid synthesis, cofactor biosynthesis, cell envelope, energy metabolism, intermediary metabolism, fatty acids and phospholipids, nucleotide biosynthesis, and regulatory functions.

REFERENCES

Baldauf, S L. and Palmer, J.D. (1990) Evolutionary transfer of the chloroplast tufA gene to the nucleus. *Nature* **344**: 262–263.

Baldauf, S., Palmer, J. and Doolittle, W.F. (1996) The root of the universal tree and the origin of eukaryotes based on elongation factor phylogeny. *Proc. Natl Acad. Sci. USA* **93**: 7749–7754.

Brown, J.W., Daniels, C.J. and Reeve, J.N. (1989) Gene structure, organization, and expression in archaebacteria. *CRC Critical Reviews in Microbiology* **16**: 287–338.

Burggraf, S., Stetter, K.O., Rouviere, P. and Woese, C.R. (1991) *Methanopyrus kandleri*: an archeal methanogen unrelated to all other known methanogens. *System. Appl. Microbiol.* **14**: 346–351.

Cousineau, B., Cerpa, C., Lefebvre, J. and Cedergren, R. (1992) The sequence of the gene encoding elongation factor Tu from *Chlamydia trachomatis* compared with those of other organisms. *Gene* **120**: 33–41.

Creti, R., Citarell, F., Tiboni, O. *et al.* (1991) Nucleotide sequence of a DNA region comprising the gene for elongation factor 1α from the ultrathermophilic archaeote *Pyrococcus woesei*: phylogenetic implications. *J. Mol. Evol.*, **33**: 332–342.

Eldridge, N. and Cracraft, J. (1986) *Phylogenetic Patterns and the Evolutionary Process*, Columbia University Press, New York.

de Queiroz, K. and Gauthier, J. (1990) Phylogeny as a central principle in taxonomy – phylogenetic definitions of taxon names. *Systematic Zoology* **39**: 307–322.

Gerbi, S.A. (1985) Evolution of ribosomal DNA, in *Molecular Evolutionary Genetics*, (ed. R.J. MacIntyre), Plenum Publishing, pp. 215–223.

Gogarten, J.P., Kibak, H., Dittrich, P. *et al.* (1989) Evolution of the vacuolar H⁺-ATPase: implications for the origin of eukaryotes. *Proc. Natl Acad. Sci. USA* **86**: 6661–6665.

Goodwin, T.W. (1980) *The Biochemistry of the Carotenoids*, Vol. I, Chapman & Hall, London and New York.

Gray, M.W. (1992) The endosymbiont hypothesis revisited. *Int. Rev. Cytol.* **141**: 233–357.

Gray, M.W. (1993) Origin and evolution of organelle genomes. *Curr. Opin. Genet. Dev.* **3**: 884–890.

Gupta, R.S., Aitken, K., Falah, M. and Singh, B. (1994) Cloning of *Giardia lamblia* heat shock protein HSP70 homologs: implications regarding origin of eukaryotic cells and of endoplasmic reticulum. *Proc. Natl Acad. Sci. USA*, **91**: 2895–2899.

Hasegawa, M., Hashimoto, T. and Adachi, J. (1993) Origin and evolution of eukaryotes as inferred from protein sequence data, in *The Origin and*

Evolution of Prokaryotic and Eukaryotic Cells, (eds Hartman and Matsuno), World Sci. Publ., Singapore and New York, pp. 45–64.

Hashimoto, T. and Hasegawa, M. (1996) Origin and early evolution of eukaryotes inferred from the amino acid sequences of translation elongation factors 1α/Tu and 2/G. *Adv. Biophys.* **32**: 73–120.

Hill, W.E., Dahlberg, A., Garrett, R.A. *et al.* (eds) (1990) *The Ribosome: Structure, Function, and Evolution*, American Society of Microbiology Press, Washington, DC.

Huber, R. and Stetter, K.O. (1992) The *Thermotogales*: hyperthermophilic and extremely thermophilic bacteria, in *Thermophilic Bacteria*, (ed. J.K. Kristansson), CRC Press, Boca Raton, Florida, pp. 185–194.

Iwabe N., Kuma, K., Hasegawa, M. *et al.* (1989) Evolutionary relationship of archaebacteria, eubacteria, and eukaryotes inferred from phylogenetic trees of duplicated genes. *Proc. Natl Acad. Sci. USA* **86**: 9355–9359.

Kjems, J. and Garrett, R.A. (1988) Novel expression of the ribosomal RNA genes in the extreme thermophile and archaebacterium, *Desulfurococcus mobilis*. *EMBO J.* **6**: 3521–3527.

Lake, J.A. (1982) Mapping evolution with ribosome structure: intralineage constancy and interlineage variation. *Proc. Natl Acad. Sc. USA* **79**, 5948–5952.

Lake, J.A. (1994) Reconstructing evolutionary trees from DNA and protein sequences: paralinear distances. *Proc. Natl Acad. Sci. USA* **91**: 1455–1459.

Lake, J.A. (1995). Calculating the probability of multitaxon evolutionary trees: Bootstrappers Gambit. *Proc. Natl Acad. Sci. USA* **92**: 9662–9666.

Lake, J.A. and Rivera, M.C. (1994) Was the nucleus the first endosymbiont? *Proc. Natl Acad. Sci. USA* **91**: 2880–2881.

Lake, J.A. and Rivera, M.C. (1996) The prokaryotic ancestry of eukaryotes, in *Evolution of Microbial Life* (eds Roberts, Alberson, Sharp and Collins), Cambridge University Press.

Lockhart P.J., Steel, M.A., Hendy, M.D. and Penny, D. (1994) Recovering evolutionary trees under a realistic model of sequence evolution. *Mol. Biol. Evol.* **11**: 605–612.

Rivera M.C. and Lake, J.A. (1992) Evidence that eukaryotes and eocytes prokaryotes are immediate relatives. *Science* **257**: 74–76.

Rivera, M.C. and Lake, J.A. (1996) The phylogeny of *Methanopyrus kandleri*. *Int. J. Syst. Bacteriol.* **46**: 348–351.

Runnegar, B. (1993) Proterozoic eukaryotes: evidence from biology and geology, in *Early Life on Earth* (ed. S. Bengtson), Cambridge University Press.

Smith, M.W., Feng, D.F. and Doolittle, R.F. (1992) Evolution by acquisition: the case for horizontal gene transfers. *Trends Biochem. Sci.*, **17**: 489–493.

Tu, J. and Zillig, W. (1982) Organization of rRNA structural genes in the archaebacterium *Thermoplasma acidophilum*. *Nucl. Acids Res.* **10**: 7227–7231.

Volters, J. and Erdmann, V.A. (1989) The structure and evolution of Archaebacterial ribosomal RNAs. *Can. J. Microbiol.* **35**: 43–51.

Waterman, M.S. and Eggert, M. (1987) A new algorithm for best subsequence alignments with application to tRNA–rRNA comparisons. *J. Mol. Biol.* **197**: 723–728.

PART FIVE

Macroevolutionary Trends: Does Horizontal Gene Transfer Resolve Homoplasy?

Character parallelism and reticulation in the origin of angiosperms

31

Valentin A. Krassilov

SUMMARY

Angiosperms have appeared as assemblages of different life forms in association with the advanced gnetoid and other proangiosperms. Their origin was not a solitary event but rather a result of parallel evolution. Typical angiosperm characters, such as vessels, areolate venation, enclosed ovules, extraovular pollen germination, double fertilization, etc., appear scattered among different lineages of proangiospermous plants. Their assembly by horizontal gene transfer seems even more plausible due to the recently obtained direct evidence of interaction between proangiospermous plants and pollinivorous insects, with certain pollen characters, such as taeniate exine or columellate infrastructure, spreading across taxonomic boundaries. Insects might facilitate horizontal gene transfer in plants by transferring microorganisms capable of gene transduction. Major events in angiosperm evolution occurred during widespread environmental crises making plant populations more receptive to extraneous genetic material. Evolutionary significance of horizontal gene transfer is discussed.

31.1 INTRODUCTION

Angiosperm origin is still sometimes perceived as a single evolutionary event. However, a wealth of data drawn from different sources seems more reconcilable with the notion of a long and intricate angiospermization process in which a lateral spread of genes between lineages evolving in parallel might play a certain role as a mechanism of sharing a genetic basis for the characters that appeared in a single or a few groups and then became, as it were, a common heritage (Krassilov, 1977; Syvanen, 1994). These might be characters under simple genetic control, as in pollen grains (haploid structures giving wonderful examples of translineage parallelisms) or the more complex characters arising from developmental accelerations or sequential changes caused by insertions of regulatory elements.

Morphological evidence of horizontal gene transfer is, by necessity, indirect. There are some points of interest, however, pertaining to angiosperm prehistory, angiosperm diversity and the ecology of angiosperm appearance, including the role played by insects.

31.2   SEED PLANT PHYLOGENY

Angiosperm evolution starts with protected ovules that appeared in the mid-Devonian time, about 400 million years ago. Currently, seed plants are thought to be rooted in progymnosperms of which both the heterosporous archaeopterids and homosporous aneurophytes are considered as potential ancestors (Rothwell and Erwin, 1987). Moreover, there were herbaceous plants of simpler axial anatomy but with elaborate cupule-like sporangial clusters, as in *Lenlogia* (Krassilov and Zakharova, 1995). Early seed plant diversity is insufficiently known for establishing progenitorial relations otherwise than in a very general form. The possibility of gymnosperm anatomy and seeds originating in different lineages and then being combined by horizontal gene transfer cannot be excluded.

Insofar as phylogeny primarily conveys our understanding of homology, it cannot be more objective than the latter. The objective elements in it are chronological relationships as well as morphological continuities furnished by intermediate forms that show character states midway in the morphocline, and by mosaic forms that combine typical characters of different taxonomic units, thus serving as phylogenetic links or stepping stones. Contemporaneous plant groups are likely to have independent origins or, if connected by intermediate forms, fraternal rather than progenitorial relationships. On the other hand, members of successive age groups, if connected by intermediate or mosaic forms, are likely to have progenitorial relationships. A phylogenetic tree based on the combined chronological information and intermediate/mosaic morphologies of the linking forms (Fig. 31.1) shows the following general tendencies.

1. The increase of the higher rank diversity that, for the non-angiosperm seed plants, peaked in the early Mesozoic, at the time of proangiosperm appearances. The subsequent decrease, at the expense of proangiospermous orders, might reflect the latter evolving into conventional angiosperms.

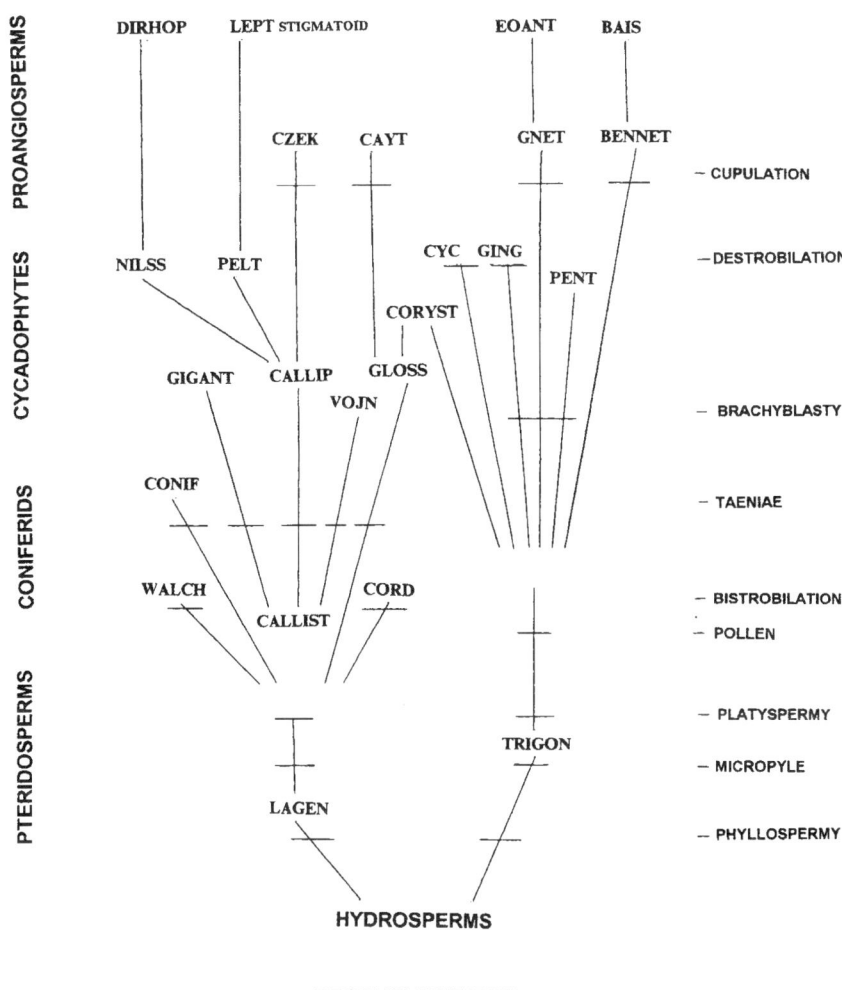

Fig. 31.1 Grades and phylogeny of gymnosperm and proangiosperm higher taxa based on chronological relations and intermediate or mosaic forms. Horizontal dashes show parallel appearances of characters indicated at the right margin.

2. The genetic continuity of seed plant lineages, few of which disappeared without leaving descendants that in turn contributed genetic material to the next evolutionary stage up to the proangiosperm level. Angiosperms thus inherited most of the genetic information accumulated during their prehistory.

3. New groups of gymnosperms and proangiosperms appearing to occur in clusters rather than by a succession of dichotomies.

4. A widespread homoplasy of fundamental derived characters simultaneously appearing in fraternal lineages (a notable feature of the nodal arrangement). For example: phyllospermy in lagenostomaleans and trigonocarps; true pollen and platyspermic ovules replacing prepollen and radiospermic ovules in trigonocarps, callistophytes and cordaites; compound strobili in cordaites, walchians and conifers; radial synangiate and ovuliferous structures in pteridosperms, glossopterids, vojnovskyaleans and bennettites; taeniate pollen grains in glossopterids, peltasperms and conifers and their simultaneous transition to asaccate morphotypes; specialized laminar ovuliphores (seed-scales) in peltasperms, nilssonialeans, conifers, etc.; their fusion to subtending bracts assisting in seed dispersal in glossopterids, conifers and gnetaleans; peltate imbricate cone scales in peltasperms, cycads and several groups of conifers; destrobilation in *Ginkgo, Cycas,* Taxaceae and other conifers as well as Mesozoic gnetophytes; secondary cupules in caytonialeans, czekanowskias and other proangiosperms; asaccate anasulcate pollen morphologies in Mesozoic nilssonialeans, cycads, ginkgoaleans, etc.

5. The cyclicity of morphological evolution; in particular, the appearance of cupulate ovules in the Late Devonian–Early Carboniferous as the first round of angiospermization followed by gymnospermization in the late Paleozoic, in turn followed by a second round of angiospermization in the late Mesozoic. A morphological distance between cupulate structures of extinct seed plants and angiosperm gynoecia appears much shorter than between the latter and the scaly ovuliphores of extant gymnosperms.

31.3   PROANGIOSPERMS AS A MORPHOLOGICAL POOL

Seed plant evolution has proceeded through a series of successional hydrosperm, pteridosperm, cycadophyte and proangiosperm grades (Fig. 31.1) before reaching angiosperm level. The proangiosperm grade became apparent due to a series of paleobotanical findings (Krassilov, 1975, 1977, 1982, 1984, 1986; Krassilov and Bugdaeva, 1982) that brought to light seed plants chronologically preceding or contemporaneous with the earliest angiosperm records and showing some critical characters but lacking in other critical characters on which conventional recognition of angiosperms is based. Thus vessels appeared in gnetophytes and bennettites; paracytic stomata in bennettites; graminoid leaves in gnetophytes; compound-palmate to lobatopalmate leaves with reticulate venation in the *Sagenopteris–Scoresbya* group (Caytoniales and allies); 'dendroid' androclades anticipating fasciculate androecia in *Caytonanthus*; pollen grains with protocolumellar infrastructure and zonal, as well as porous, protoapertures in *Classopollis* (Hirmerelllaceae); cupular gynoecia in *Dirhopalostachys* of cycadophyte descent (with a solitary anatropous ovule) and *Basia* of bennettitalean descent (with a solitary orthotropous ovule); four-membered cupulate gynoecia with bracteate perianth in *Eoantha* (gnetophytes); ascidiform cupule with many ovules in *Caytonia*; syncupulate capsules with stigmatic crests in *Leptostrobus* (Czekanowskiales), etc. Thus proangiosperm groups are complementary in forming a morphololological pool containing an almost complete set of typical angiosperm characters.

In angiosperms we find derived states of the characters that appeared in their prototypal states in proangiosperms, but no single proangiosperm lineage could conceivably have given rise to the basic angiosperm diversity evidenced by the mid-Cretaceous fossil record. Rather the diversity of angiosperm morphologies could have arisen by recombination of the diversity of proangiosperm morphologies. Angiosperms as a whole are 'pachyphyletic' in the sense that they descended from proangiosperms as a whole. At the same time, the basic dichotomy of seed plant morphoclines with receptacular orthotropous ovules versus appendicular anatropous ovules is traceable through the gymnosperm and proangiosperm grades to the early angiosperms, giving a clue to their progenitorial relations; for example, linking the gnetophyte–bennettite group of proangiosperms to graminoid monocots and the caytonialean–czekanowskialean proangiosperms to ranunculids and related dicot and monocot orders.

31.4   ANGIOSPERM DIVERSITY

Traditional taxonomy represents angiosperms as a diverse but fairly integral group bound up by such critical characters as areolate leaf venation, vessels, flower, carpel, double fertilization, etc., the structures and the process lacking (at least nominally) in other seed plants, thus constituting a unique diagnostic complex that could arise only once in the history of the plant kingdom. The morphological diversity of angiosperms is then reducible to a single ancestral prototype.

The comparative morphology of angiosperms is governed by this creed. A drive for unity makes a taxonomist feel unsatisfied until all the items are brought to one end. In effect, monophyletic systems are much more popular than polyphyletic systems. Cladistics provide a seemingly objective approach to the problem by numerical estimates of intergroup versus outgroup similarities. In the case of angiosperms, the objectivity of results (invariably indicating monophily) are doubtful for several reasons, notably the semantic. Strict morphological definitions are wanting for most of the characters used. Therefore their presence or absence

Table 31.1 Example of comparison of groups

	Gnetophytes	Monocotyledons	Dicotyledons
Flower	-	+	+
Stamen	-	+	+
Carpel	-	+	+
Cupule	+	-	-

depends on what we mean by them. Consider the example shown in Table 31.1 above.

This table is pointless unless we know that 'flower', 'carpel', etc. mean the same thing for all the compared groups. Incidentally, flowers, familiar by association with weddings and funerals, have escaped precise morphological definition. Historically 'flower' was scarcely considered as a discriminative feature: Linnaeus said that all plant species had flowers and fruits, even if concealed from our eyes (Linnaeus, 1751). While 'flower' is used discriminately at present as a special feature of angiosperms separating them from other seed plants, it seems logical to define 'flower' in relation to reproductive structures of non-angiospermous seed plants. In the majority of angiosperms, flowers consist of sporangiophores and/or ovuliphores formed of floral meristem that is similar to apical meristem of vegetative shoots but is mitotically more active and of a less distinct zonal structure. Gymnosperms also have specialized reproductive shoots, but their apices are not fully fertile, bearing sterile scales (in bennettites) or, in *Ginkgo*, fully developed leaves intermingled with ovuliphores. Although differing from typical flowers, these structures correspond to some anomalous 'flowers' with bracts in the gynoecial zone, e.g. in *Eupomatia* where they form the 'inner corolla' between carpels and stamens. Such floral structures occur in a number of angiosperm families and are actually pre-flowers rather than typical flowers.

The situation with other typical structures is similarly biased semantically. Characters assigned to a certain morphological type are not necessarily homological and, in fact, are rarely so. For example, follicles, often thought of as basic leaf-like gynoecial structures, are either monomerous ascidiform (Rohweder, 1967) or pseudomonomerous (Vink, 1978). The unitegmic condition can result from fusion or reduction or integumental shifting (Bouman and Callis, 1977) while the bitegmic condition can result from splitting or modification of peripheral nucellar tissues with respect to the pollen-conducting function of the inner integument (Heslop-Harrison *et al.*, 1985). Even some characteristic biochemical compounds are end-products of dissimilar biosynthetic pathways (Kubitzki, 1973). Pseudohomology is a prolific source of phylogenetic misconceptions, such as morphological integrity of angiosperms.

Comparative morphological analysis shows that none of the critical characters is shared by all the species currently assigned to angiosperms: there are forms lacking distinct stratification of apical meristem, vessels, typical sieve element companion cells, with atypical double fertilization, as in Onagraceae, with embryogenesis nuclear up to 64–128 or even 256-nucleate stage, as in *Paeonia*, etc. On the other hand, such characters are not exclusively confined to angiosperms but occur, though less consistently, in seed plants that are not formally recognized as angiospermous; this includes not only pre-flowers and carpel-like cupules, but also double fertilization, as in *Ephedra*. The morphological boundary between angiosperms and gymnosperms is thus not absolute.

Notably the anomalous character states occur in angiosperm taxa that are generally considered primitive. An assembly of all such characters may be closer to an ancestral form than the paradygmatic angiosperm. This form may not be conventionally classified as angiosperm, thus making angiosperms cladistically paraphyletic with an implication that the typical angiosperm characters, not yet occurring in the common ancestor, appeared in parallel in the descendant lines and, therefore, are not uniquely derived.

Actually shared by all angiosperms is developmental acceleration resulting in highly condensed and/or chimeric structures of great morphological plasticity, but this is an evolutionary trend developing simultaneously in different seed plant groups.

31.5 COLLECTIVE BREAKTHROUGH

The angiosperm origin is here perceived as a process that does not have strictly definable starting points and deadlines. A semitectate pollen grain represents a step in morphological evolution towards angiospermy rather than the existence of angiosperms. Such records go back to the Triassic (Cornet, 1979) and even further. Assembled records of several angiosperm traits accumulate, after a period of single-trait appearances, in the Barremian/early Aptian, about 115 –120 million years ago. Close to this date are the associated records of authentic angiosperms and various proangiosperms, notably the advanced gnetaleans, in central Asia, the Middle East, the Atlantic coasts and Australia. They not only testify to early angiosperm appearances in these areas, but also give evidence of the ongoing process of angiospermization.

Remarkably, the most important localities have yielded not only the first angiosperms and angiosperm-like fossils, but also the remains of advanced proangiosperms. In the Barremian Baisa locality in the upper reaches of the Vitim River, Transbaikalia, angiosperm leaves from *Dicotylophyllum pusillum* and pollen grains from *Asteropollis* and *Tricolpites* (Vakhrameev and Kotova, 1977) are accompanied by the abundant achene-like disseminules of *Baisia*, a one-seeded cupule on persistent bristled receptacles (Krassilov and Bugdaeva, 1982). Recently, intact inflorescences of this plant were studied by Krassilov and Bugdaeva (in press), who confirmed the previously suspected bennettitalean derivation of this proangiospermous plant. From the same plant-bed came *Eoantha*, a bracteate preflower with a four-lobed gynoecium and with *Ephedripites* pollen grains in the pollen chambers of orthotropous ovules (Krassilov, 1986). A recently found attached flower (Fig. 31.2) enabled us to assign the associated graminoid leaves to this plant. New finds have added inflorescences with staminate pre-flowers of the gnetalean type (Fig. 31.2).

Even more diverse are angiospermoid and proangiosperm fossils in the roughly contemporaneous localities of Manlay, Gurvan-Eren and Bon-Tsagan in the western Gobi, Mongolia (Krassilov, 1982). Angiospermoid fruits from *Gurvanella* and *Erenia* are accompanied there by the monocot-like *Cyperacites, Graminophyllum* and unassigned *Sparganium*-like and *Potomageton*-like fruiting axes, as well as a pappose reed-mace-like *Typhaera*. Their preservation is unfavorable for detailed morphological studies, thereby making their angiospermous or proangiospermous status uncertain. Similar associations were recently found in northeastern China (Sun Ge and Dilcher, 1996).

Such situations are described in other parts of the world. In the Potomac flora of the Atlantic coast of North America, early angiosperms (Doyle and Hickey, 1972) appeared in association with *Drewria*, a herbaceous gnetalean plant (Crane and Upchurch, 1987). In the English Wealden, the entry of angiosperms in the pollen record was paralleled by the rise of gnetoid pollen (Hughes and McDougal, 1987). In Koonwarra, southern Australia, ceratophyllean fruits (Dilcher *et al.*, 1996) and racemose inflorescences (Taylor and Hickey, 1990) are joined by ephedroids (Douglas, 1969; Krassilov *et al.*, 1996) and perhaps some other gnetaleans represented by ovulate bracts and bracteate pollen cones (Drinnan and Chambers, 1986). In a small collection from the Aptian 'amphibian bed' of Makhtesh-Ramon, Israel, angiospermoid fruits are found together with *Sagenopteris*-type leaves of caytonialean proangiosperms (Krassilov and Dobruskina, 1995).

Thus the appearance of angiosperms was not a lonely breakthrough against a static background, pushing other plants aside, but, rather, was a collective breakthrough involving a number of parallel lineages that grew side by side as members of breakthrough plant communities.

31.6   INSECT ROLE

The communal association of proangiosperms makes horizontal gene transfer between them at least plausible. Sporangia of microscopic

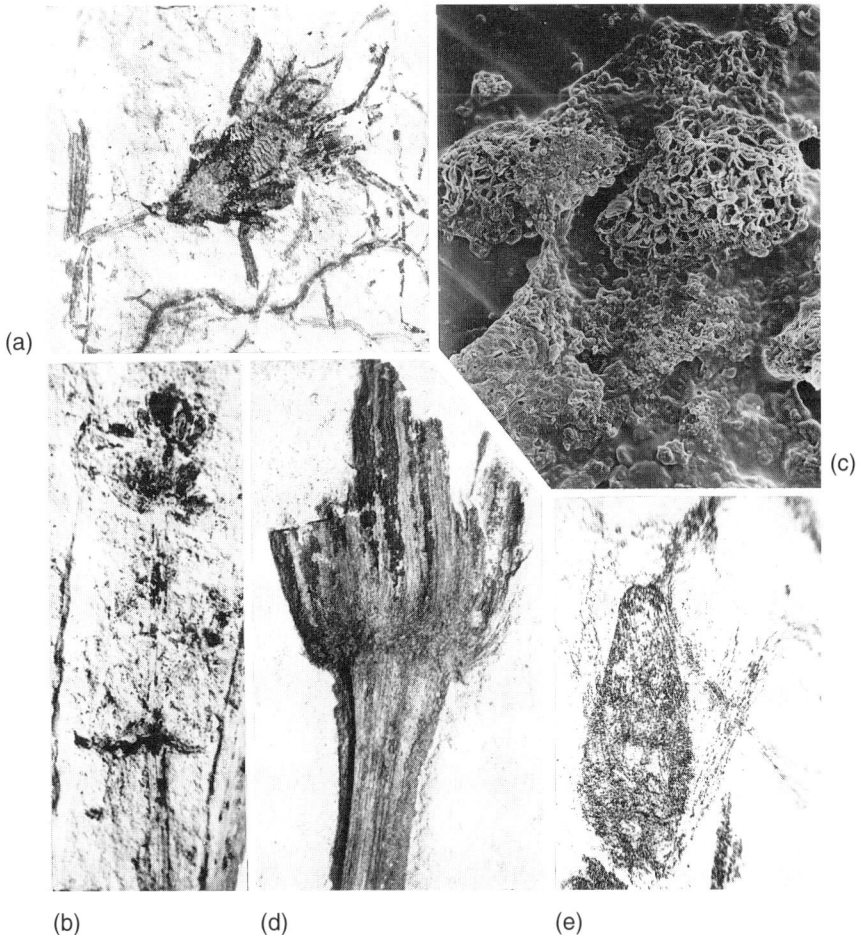

Fig. 31.2 Proangiosperms of the Baisian assemblage, Early Cretaceous, Transbaikalia. (a) *Eoantha*, attached ovulate preflower, × 8; (b) gnetalean inflorescence with pollen preflowers, × 8; (c) single ovulifore with several sporangia from the same specimen, SEM, × 170; (d) graminoid leaf of *Eoantha* plant, × 8; (e) *Basia*, a cupule on persistent receptacle with bristles, × 12.

endoparasitic fungi frequently found in fossil pollen grains (e.g. Krassilov, 1987) may indicate a potential transducing agent. Interactions with insects and other animal components of breakthrough communities might in turn mediate the transfer of such agents.

Direct evidence of plant–insect interaction in biotic communities of the geological past is provided by pollen preserved in the gut compressions of fossil insects first obtained from the Early Cretaceous Xyelidae

(Krassilov and Rasnitsyn, 1982) from the Baisa proangiosperm/early angiosperm locality (see above). Insects are known to transfer various gene-transducing microorganisms that can confer parallel genetic changes in the target plants. This mechanism might have already been in action in the Permian, as evidenced by the recently described striate pollen in the gut compressions of hypoperlid and grylloblattid insects of this age (Rasnitsyn and Krassilov, 1996).

Striate (taeniate) exinal structure gives one of the most spectacular examples of a single feature morphological parallelism appearing simultaneously in major gymnosperm groups, glossopterids, peltasperms and conifers, dominating late Paleozoic plant communities of Eurasia and Gondwanaland (Clement-Westerhof, 1974; Zavada, 1985). Our most recent finding of widespread angiospermoid *Classopollis* pollen grains in the intestines of Jurassic grasshopper-like katydid insects (Krassilov *et al.*, 1997) (Fig. 31.3) suggests a similar role of Mesozoic insects in the innovation of reproductive morphology occurring in proangiosperm lineages that evolved in parallel in the direction of angiospermy.

The katydid example shows that proangiosperms and, by implication, their succeeding early angiosperms with their small gregarious flowers

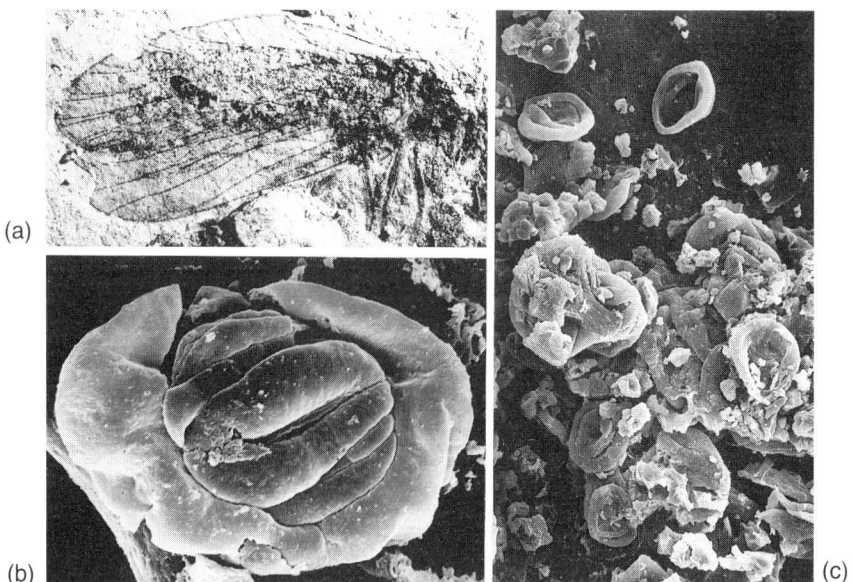

(a)

(b)

(c)

Fig. 31.3 Pollen in the gut compressions of fossil insects. (a) *Idelopsocus*, a Permian hypoperlid insect, × 9; (b) taeniate pollen *Lunatisporites* from the intestine of the same specimen, SEM, × 1700; (c) *Classopollis* from the gut of Jurassic katydid insect, SEM, × 700.

could use unconventional pollinators, while beetles and other then existing anthophilous insects were engaged with more conspicuous bennettite pre-flowers (for the insects, they were flowers, irrespective of what plant morphologists might think of them). With the decline of bennettites at the end of Early Cretaceous, their insect retinue passed to the angiosperms, perhaps mediating horizontal gene transfer between these groups. In fact, solitary bennettite-like flowers of many parts (Dilcher and Crane, 1984) then first appeared in angiosperms.

31.7   ECOLOGICAL CRISES AS A GENE TRANSFER SITUATION

It follows from the above discussion that angiosperm origin was a communal event. The above-mentioned localities of early angiosperms and their accompanying proangiosperms reflect xeromorphic brachyphyllous communities widespread in the ecotonal zones of temperate summer-green and subtropical evergreen to winter-green vegetation, at about 50°N and 40°S. Angiospermization might conceivably have been going on throughout the extent of these zones but most of the actual records are confined to downfaulted grabens and semigrabens of the Early Cretaceous rift systems. Thus, the Transbaikalian and Mongolian basins are linear depressions of the extensive rift system striking northeast from Mongolia to the Sea of Okhotsk. The lacustrine facies of the rift zone are typical of stratified lakes, with thick, finely laminated black shale sequences interleaved with psammitic and carbonate interbeds, the latter abounding in fish and aquatic insect larvae remains. Large dragonfly, mayfly and beetle (coptoclavid) larvae pile up on the bedding planes, suggesting mass mortality, perhaps related to abrupt pH fluctuations caused by volcanogenic acid rains. The taphonomic data suggest stressful environments as a factor impelling the developmental acceleration and condensation of morphological structures characteristic of both proangiosperms and early angiosperms.

Major evolutionary novelties appear after major environmental crises for at least two reasons. The first is abbreviation of seral sequences (Krassilov, 1992). Successional species are, as a rule, more 'fine-grained' than climax species. Their relatively broad ecological niches are potentially splittable into narrower ecological niches. Elimination of climax phase thus leaves a community more open to entries by new species, either by invasion or by speciation. In addition, in perturbed communities, a decrease of stabilizing selection pressure provides opportunities for evolutionary experimentation. While species occupying well-defined ecological niches have to insulate their finely adapted genetic system from invading genes that are likely to decrease their fitness, the opposite is true for post-crisis species, which tend to be highly polymorphic. In the pioneer stage of ecological expansion, introgression of genetic material is likely to be advantageous as a source of additional genetic variability. Therefore gene transfer, both vertical (by hybridization) and horizontal (by microorganisms), is promoted by ecological crises. Actually, new groups appearing after ecological crises show not only elaboration, but also recombination of characters occurring in their preceding groups. Thus recombination of proangiosperm characters in angiosperms can be taken as indirect evidence of interspecies gene transfer.

31.8   CONCLUDING REMARKS

The horizontal gene transfer concept is potentially of great importance for leading evolutionary thinking beyond the ossified tenets of 'synthetic' theory, first of all by introducing a long-sought mechanism – vigorously denied by traditionalists – by which macromutations can spread. The ubiquitious evolutionary parallels receive a new explanation and intra-communal interactions between coevolving organisms appear in a new light as not only competitive, but also cooperative, including at least episodic sharing of a communal gene pool, thus enforcing the idea of community as an evolutionary unit.

Although, at the first glance, horizontal gene transfer may seem accidental, the suggested association with ecological crises means that there could be method in its accidentality: it is effective when actually required as a mechanism of genetic enrichment promoting adaptive innovation and ecological expansion. At least, interpretation of the fossil record seems easier with horizontal gene transfer than without it, which can be taken as indirect evidence in favor of the mechanism.

REFERENCES

Bouman, F. and Calis, J.L.M. (1977) Integumentary shifting – a third way to unitegmy. *Ber. Dt. bot. Ges.* **90**: 15–28.

Clement-Westerhof, J.A. (1974) *In situ* pollen from gymnospermous cone of the Upper Permian of the Italian Alps – a preliminary account.*Rev. Palaeobot. and Palynol.* **17**: 65–73.

Cornet, B. (1979) Angiosperm-like pollen with tectate columellate wall structure from the per Triassic (and Jurassic) of the Newark Supergroup, USA. *Palynology* **3**: 281–282.

Crane, P.R. and Upchurch, R. Jr (1987) *Drewia potomacensis* gen. et sp. nov., an early Cretaceous member of Gnetales from the Potomac Group of Virginia. *Amer. J. Bot.* **74**: 1722–1736.

Dilcher, D.L. and Crane, P.R. (1984) *Archaeanthus*: an early angiosperm from the Cenomanian of the western interior of North America. *Ann. Mo. Bot. Gard.*, **71**: 351–383.

Dilcher, D.L., Krassilov, V.A. and Douglas, J.G. (1996) Angiosperm evolution: fruits with affinities to Ceratophyllales from the Lower Cretaceous. In: *Abstr. Fifth Conf. Int. Org. Paleobot., Santa Barbara, California*, p.23.

Douglas, J.G. (1969) The Mesozoic floras of Victoria, 1-2. *Mem. Geol. Surv. Victoria* **28**: 1–310.

Doyle, J.A. and Hickey, L.J. (1972) Coordinated evolution in Potomac Group angiosperm pollen and leaves. *Amer. J. Bot.* **59**: 660.

Drinnan, A.N. and Chambers, T.C. (1986) Flora of the Lower Cretaceous Koonwarra fossil bed (Korumburra Group) South Gippsland, Victoria. *Mem. Ass. Austr. Palaeontol.* **3**: 1–77.

Heslop-Harrison, Y., Heslop-Harrison, J. and Reger, B.J. (1985) The pollen–stigma interaction in the grasses. 7. Pollen-tube guidance and the regulation of tube number in *Zea mays* L. *Acta bot. neer.* **34**: 193–211.

Hughes, N.F. and McDougall, A.B. (1987) Record of angiospermid pollen entry into the English Early Cretaceous succession. *Rev. Palaeobot. and Palynol.* **50**: 255–272.

Krassilov, V.A. (1975) Dirhopalostachyaceae – a new family of proangiosperms and its bearing on the problem of angiosperm ancestry. *Palaeontographica* **153B**: 100–110.

Krassilov, V.A. (1977) The origin of angiosperms. *Bot. Rev.* **43**: 143–176.

Krassilov, V.A. (1982) Early Cretaceous flora of Mongolia. *Palaeontographica*, **181B**: 1–43.

Krassilov, V.A. (1984) New paleobotanical data on origin and early evolution of angiospermy. *Ann. Mo. Bot. Gard.* **71**: 577–592.

Krassilov, V.A. (1986) New floral structure from the Lower Cretaceous of Lake Baikal area. *Rev. Palaeobot. and Palynol.* **47**: 9–16.

Krassilov, V.A. (1987) Fungi sporangia in the pollen of the Early Cretaceous conifers. In *Palynology of the Soviet Far East*, (ed. V.S. markevich), Far East Science Center, Vladivostok, pp. 6–8.

Krassilov, V.A. (1992) Ecosystem theory of evolution and social ethics. *Rivista Biol.* **87**: 87–104.

Krassilov, V.A. (1995) *Scytophyllum* and the origin of angiospermous leaf characters. *Paleont. Jour.* **29**: 110–115.

Krassilov, V.A. and Bugdaeva, E.V. (1982) Achene-like fossils from the Lower Cretaceous of the Lake Baikal area. *Rev. Palaeobot. and Palynol.* **36**: 279–295.

Krassilov, V.A. and Dobruskina, I.A. (1995) Angiosperm fruit from the Lower Cretaceous of Israel and origins in rift valleys. *Paleont. J.* **29**: 63–74.

Krassilov, V.A. and Rasnitsyn, A.P. (1982) Unique finding: pollen in the guts of the Early Cretaceous xyelotomid insects. *Palaeontol. Zh.* **4**: 83–96.

Krassilov, V.A. and Zakharova, T.V. (1995) *Moresnetia*-like plants from the Upper Devonian of Minusinsk Basin, Siberia. *Paleont. J.* **29**: 35–43.

Krassilov, V.A., Shilin, P.V. and Vachrameev, V.A. (1983) Cretaceous flowers from Kazakhstan. *Rev. Palaeobot. and Palynol.* **40**: 91–113.

Krassilov, V.A., Dilcher, D.L. and Douglas, J.G. (1996) Ephedroid plant from the Lower Cretaceous of Koonwarra, Australia. In *Abstr. Fifth Conf. Int. Org. Paleobot., Santa Barbara, California*, p. 54.

Krassilov, V.A., Zherikhin, V.V. and Rasnitsyn, A.P. (1997) *Classopollis* in the gut of Jurassic insects. *Palaeontology* **4**: 1095–1101.

Kubitzki, K. (1973) Probleme der Grosssystematik der Blütenpflanzen. *Ber. Dt. bot. Ges.* **85**: 259–277.

Linnaeus, C. (1751) *Philosophia Botanica*, Stockholm.

Rasnitsyn, A.P. and Krassilov, V.A. (1996) Pollen in the gut of Early Permian insects: first direct evidence of pollinivory in the Paleozoic. *Paleoht. Zh.* **3**: 1–6.

Rohweder, O. (1967) Karpellbau und Synkarpie bei Ranunculaceen. *Ber. Schweiz. bot. Ges.* **77**: 376–425.

Rothwell, G.W. and Erwin, D.M. (1987) Origin of seed plants: an aneurophyte/seed fern link elaborated. *Amer. J. Bot.* **74**: 970–973.

Sun Ge and Dilcher, D.L. (1996) Evolution of angiosperm flower and leaf flora from Northeast China. In *Abstr. 5th Conf. Int. Org. Paleobot., Santa Barbara, California*, p. 98.

Syvanen, M. (1994) Horizontal gene transfer: evidence and possble consequences. *Annu. Rev, Genet.* **28**: 237–261.

Taylor, D.W. and Hickey, L.J. (1990) An Aptian plant with attached leaves and flowers: implications for angiosperm origin. *Science*, **247**: 702–704.

Vakhrameev, V.A. and Kotova, I.Z. (1977) Early angiosperms and their accompanying plants from the Lower Cretaceous of Transbaikalia. *Paleont. Zh.* **4**: 101–109.

Vink, W. (1978) The Winteraceae of the Old World. 3. Notes on the ovary of *Takhtajania. Blumea* **24**: 521–525.

Zavada, M.S. (1991) The Ultrastructure of Pollen Found in the Dispersed Sporangia of *Arberiella* (Glossopteridaceae). *Bot. Gaz.* **152**: 248–255.

Graptolite parallel evolution and lateral gene transfer

32

William B.N. Berry and Hyman Hartman

SUMMARY

The horizontal transfer of genes by infectious agents such as viruses or viroids is suggested to have been a mechanism for the parallel evolution of morphological characters in the paleontological record. In particular, the case of stunting of graptolite colonies which took place in a number of genera after a near-extinction of graptolites due to oxygenic stress about 437 million years ago is considered. An infectious viroid model for the development of stunting is proposed as a possible horizontal transfer of a control gene.

32.1   INTRODUCTION

A symposium held at the University of Chicago in 1959 celebrated the centenary of publication of Darwin's *Origin of Species*. The three volumes of papers that resulted from that symposium provide ample evidence of the vitality of Darwinian natural selection. The major focus of the symposium was the synthesis of Mendelian genetics with Darwinian theory of natural selection. The so-called synthetic theory of evolution, as discussed at the Chicago symposium, was applied to several fields of biology, such as taxonomy and paleontology. The centerpiece of the synthetic theory was the problem of the formation of species, which was defined as a population of sexually reproducing organisms that shared a common gene pool. Speciation would result when this population was split into two or more separate gene pools by geographic isolation for an extended period of time. No genes could flow from the gene pool of one species to that of other species. A species was an isolated gene pool.

In the field of paleontology, however, the fossil record posed certain problems to the synthetic theory. Among those discussed at the symposium was the observation of parallel evolution in the paleontological record. Parallel evolution was defined at this symposium by Olson (1960) as:

> . . . parallelism in major morphologic structures, and especially in suites of major structures, in evolving lines of populations related only at rather high categorical levels, and with remote common ancestors in which common structures did not exist.

This is a phenomenon observed frequently in the fossil record.

It should be noted that parallel evolution as defined by Olson poses certain difficulties for the synthetic theory. For example, there is no common ancestor detectable in the fossil record, and also the complex morphological character appears in populations which in taxonomic terms are above the level of species and hence are not capable of exchanging genes. The only possible explanation for parallel evolution, using this sense of species and speciation, is similarity in selective forces operating on the genetically isolated populations.

In recent years, due to advances in molecular biology, possibilities for transfer of genes between species by means of viral and other agents have been demonstrated (Syvanen, 1994). Therefore, when one observes parallel evolution, there is also the possibility of explaining the observations by means of horizontal transfer of genes across species or higher taxonomic levels. This chapter examines a case of parallel evolution in graptolites, an extinct fossil group.

32.2   GRAPTOLITES

Graptolites, because they have an extensive paleontological record resulting from a great many collections of large numbers of individuals from closely spaced rock layer intervals, are a favorable group in which to seek evolutionary patterns. Graptolite morphologies, systematics and habitats have been discussed by Berry (1987).

The most common graptolites found in the fossil record are the remains of colonial marine plankton. They lived from early in the Ordovician into the Early Devonian (a time span of about 500 to 390 million years ago). Graptolites are commonly found as silhouettes on rock surfaces. Rare finds of little chemically altered, uncrushed specimens that may be freed from the rock matrix using acids provide most of the known morphological information concerning graptolites. That information is used to interpret what is seen in silhouettes.

The majority of planktonic graptolites lived in tropical oceans, where they inhabited waters over the outer parts of continental shelves and upper parts of adjacent continental slopes of the time. The greatest number of planktonic graptolite colonies lived within or close to sites of oceanic upwelling. Some graptolites inhabited surface waters, but most of them appear to have lived at some depth below the surface. Many graptolites seem to have lived on the margins of and within hypoxic waters, where they encountered high concentrations of denitrifying bacteria that utilized nitrous oxides (Berry *et al.*, 1987).

Based on a few impressions and on diameters of zooidal tubes that comprise the graptolite colony shell, or test, zooids ranged in size from about 0.05 to 0.8 mm in diameter. The zooids seem to have possessed a lophophore (a tentacle-bearing arm-like structure) and they appear to have budded serially from the preceding zooid. The budding pattern is considered to have been similar to that of modern rhabdopleurans (Berry, 1987; Rigby, 1994; Urbanek, 1994). Study of the growth of colonies of the pterobranch hemichordate *Cephalodiscus* led Dilly (1993) to suggest that living cephalodiscids and graptolites are so similar that cephalodiscids are living graptolites. Urbanek (1994), however, concluded from an analysis of budding patterns seen in certain graptolites and in the living pterobranch *Rhabdopleura* that graptolites were more closely similar to rhabdopleurans and that *Rhabdopleura* is a living graptolite and, therefore, a 'living fossil'.

The first-formed part of the shell, or test, of the colony (the sicula) housed a presumably sexually produced individual. That part of the test differs in morphological details from the test of the remainder of the colony. Test material appears to have been proteinaceous in composition for collagen-like fibers have been identified in it.

Development of the graptolite colony records changes in zooid size and shape during passage from the immature to mature state (Fig. 32.1). In many colonies, zooidal tubes in the immature part of the colony are small-diameter, slender cones. Later in colony development, as the colony became more mature, zooidal tubes achieved greater diameters. Such changes from slender to more robust zooidal tubes gave the whole colony test a tapered appearance. That appearance reflects the change from small to relatively larger zooids produced as the colony became more mature. The example in Fig. 32.1 is a monograptid (it has uniserial colony form) and it is used to illustrate how changes in zooidal cup size in the development of the colony result in distinctive colony shape. That shape is described hereafter as tapering.

32.3   CONDITIONS AT A NEAR-EXTINCTION AMONG GRAPTOLITES

Graptolites nearly became extinct and were restricted to a number of relatively small habitats in tropical oceans during a 2-million-year interval of

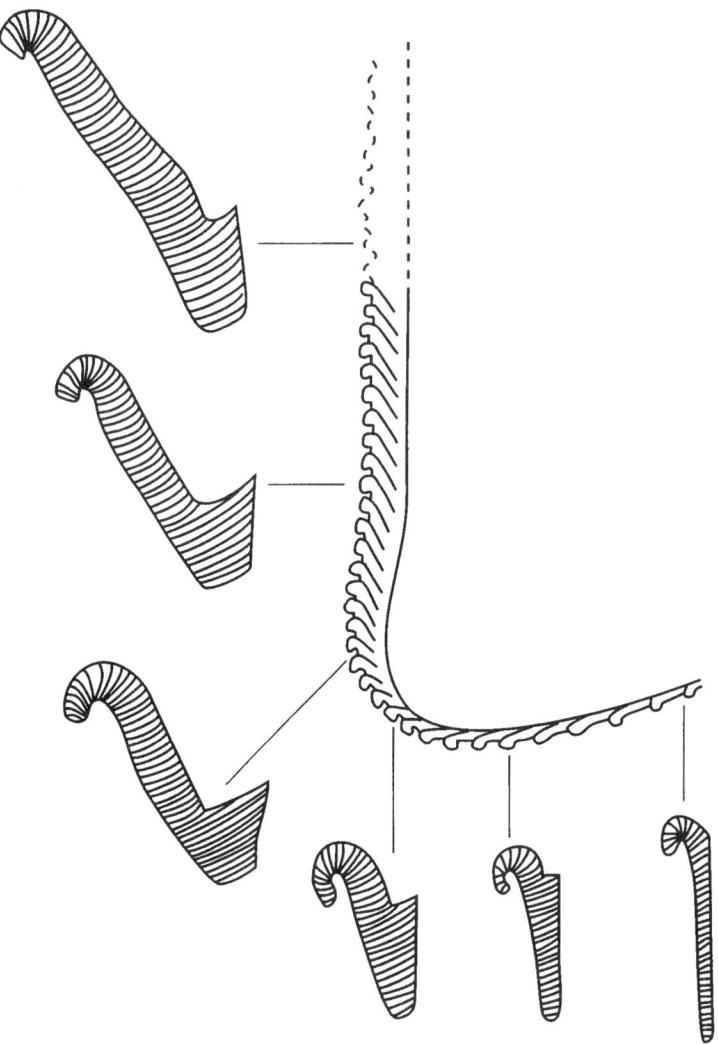

Fig. 32.1 Monograptid graptolite showing change in size and shape of zooidal cups from immature to mature part of the colony test (adapted from Bulman, 1970, Fig. 43).

massive continental glaciation in the Late Ordovician (approximately 435–437 million years ago) (Berry *et al.*, 1995). Continental ice covered much of the land mass of a large southern hemisphere continent, Gondwanaland, at that time. The record of Late Ordovician graptolites and the geochemistry of the rocks bearing them demonstrate that the oxygen-poor, nitrous oxide-rich waters inhabited by most graptolites diminished greatly during the interval of extensive continental glaciation (Berry *et al.*, 1995). A consequence of the extensive continental ice was marked reductions of continental shelf seas. Habitats preferred by most graptolites were reduced during glaciation, both by falls in glacio-eustatic sea-levels and by ventilation of the shelf seas more fully than they had been for millions of years prior to glaciation. The microaerophilic graptolites were being stressed by oxygenic waters during glaciation. The rock record bears the evidence that graptolites nearly became extinct as habitats shrank and colonies became oxygenated (Rickards *et al.*, 1977; Melchin and Mitchell, 1991; Berry *et al.*, 1995).

Stunted or dwarfed forms typify the majority of the colonies that survived the near-extinction. At certain sites, large numbers of such colonies occur crowded together on rock surfaces that accumulated during and immediately after the near-extinction. Stunting is expressed both in slender, markedly tapered colonies and in small colonies composed of a few small zooidal tubes. Colonies before and after near-extinction are illustrated in Fig. 32.2.

32.4   RE-RADIATION AFTER NEAR-EXTINCTION

When climates warmed and glaciers melted, sea-levels rose worldwide and ocean surface waters warmed. The hypoxic, nitrous oxides-rich habitats re-developed and expanded. As these habitats that graptolites preferred expanded, many new lineages developed (Rickards *et al.*, 1977; Melchin and Mitchell, 1991). Many of the initial representatives in these new lineages occurring at the genus or family level have markedly tapered colonies. Rickards (1988) stated specifically in this regard that colonies of species within the new re-radiation genera are characterized by a 'drawn-out, thorn-like' first-formed part of the colony and that the initial zooidal cups in these colonies are long slender cones that are significantly thinner and more tapering than zooidal cups in the remainder of the colony. Rickards *et al.* (1977) discussed the re-radiation lineages, indicating that colonies in the genera *Akidograptus*, *Atavograptus*, *Dimorphograptus*, *Glyptograptus*, *Paraorthograptus* and *Raphidograptus* are typified by markedly tapered colonies. Examples of two of these new genera are illustrated in Figs 32.1 and 32.2b and the pattern of appearances of the new, re-radiation genera is shown in Fig. 32.3. Tapered colony morphologies include those in which the zooidal tubes are arranged biserially and colonies in which zooidal tubes have uniserial arrangement in the immature part of the colony but become biserial in the mature part. The tapered colony morphology characterizes most monograptid graptolites (Fig. 32.1) – those graptolites in which zooidal cups are arranged uniserially throughout the colony. The size of

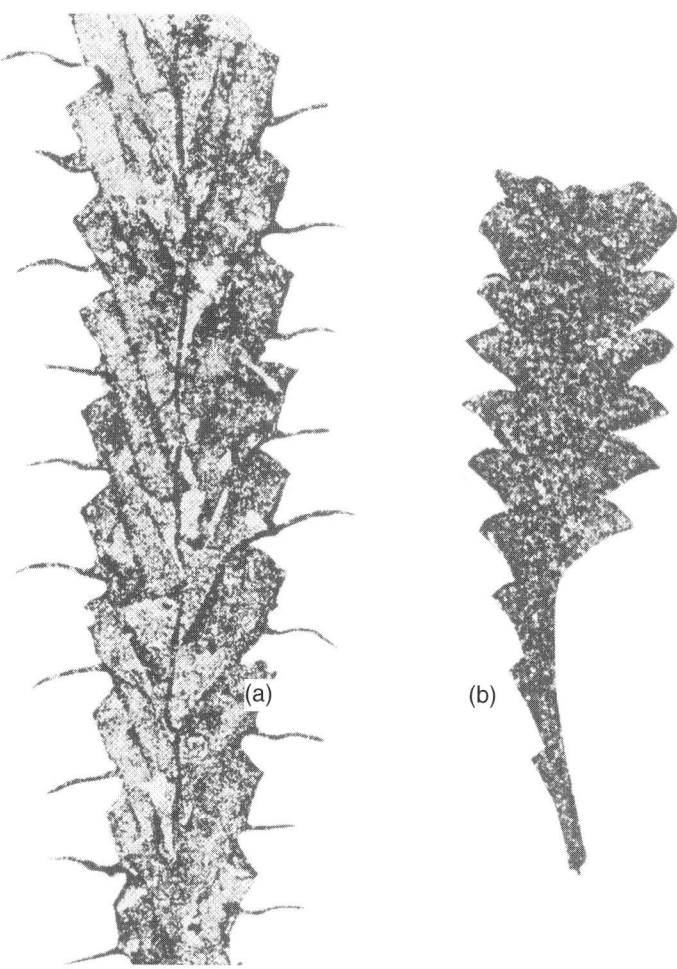

Fig. 32.2 Pre- and post-Late Ordovician near-extinction graptolites. (a) *Paraorthograptus pacificus*, a pre-near extinction species showing relatively 'square' immature part of colony test. This species disappeared at the Extinction Event. (b) *Dimorphograptus conferus*, a post-near extinction species showing tapered aspect of initial part of colony test. Appearance of this genus is indicated in Fig. 32.3.

the zooidal tubes changes from the immature to mature portion of the colony (Fig. 32.1). The uniserial morphologies of monograptids parallel the uniserial immature parts of those colonies that become biserial in the mature parts of the colony, such as the colony shown in Fig. 32.2b. Slender immature region zooidal tubes (Figs 32.1 and 32.2b) characterize most re-radiation taxa and post-extinction lineages.

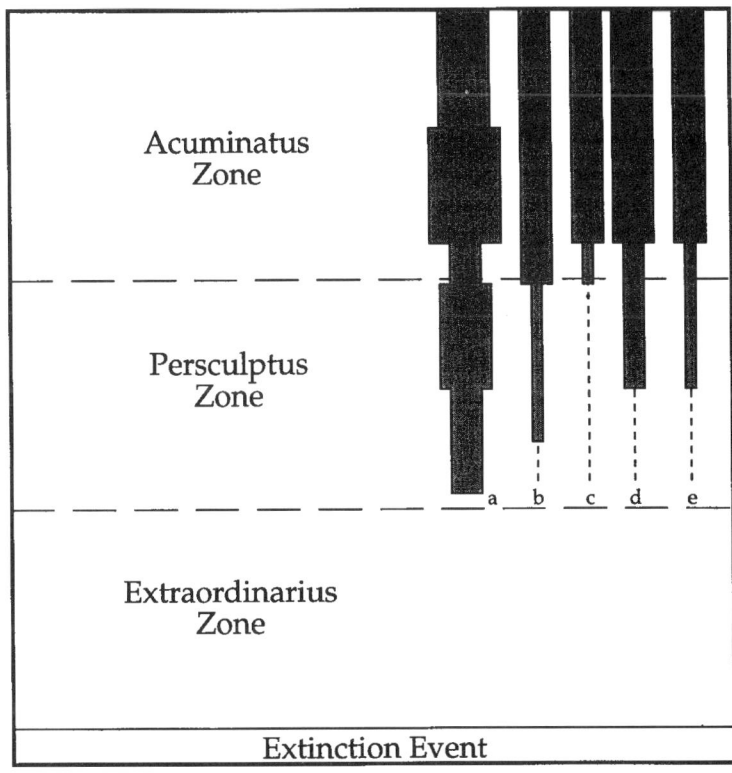

Acuminatus
Zone

Persculptus
Zone

a   b   c   d   e

Extraordinarius
Zone

Extinction Event

Fig. 32.3 Chart illustrating parallelism among new, post-near extinction, re-radiation genera (adapted from Melchin and Mitchell, 1991, Fig. 5). Appearances where known in the fossil record are indicated by shading. Widths of the shaded areas are indicative of the relative number of species. The narrowest shaded area is one species. The time interval indicated from the Extinction Event through the Acuminatus Zone is somewhat more than a million years. The widths indicated for the zones do not reflect relative time durations: the Extraordinarius Zone may have been only a few hundreds of thousands of years in duration and fossils indicative of this zone are rarely preserved. New appearances seen in the rock record of the Persculptus Zone are controlled by preservation of grap- tolite-bearing rocks. (a) *Pseudorthograptus*; (b) *Akidograptus*; (c) *Parakidograptus*; (d) *Dimorphograptus*; (e) *Atavograptus*.

The common explanation of the relatively sudden appearance of the new genera is that they developed from undetected ancestral popula- tions that were small and isolated. Despite carefully conducted searches for potential ancestors of the new, post near-extinction genera (Rickards *et al.*, 1977; Melchin and Mitchell, 1991), none has been identified.

32.5   GENE EXCHANGES

Eukaryotic cells have the ability to form endosymbioses with bacteria. This ability resulted in the formation of mitochondria and chloroplasts. This, of course, is a way to transmit genes above the species level.

Molecular biologists have also discovered an extensive system of gene exchanges mediated by plasmids, episomes and viruses among bacteria, especially *E. coli*. These observations led Anderson (1970) to suggest that viral mechanisms existing in sexual populations may transport genes across species barriers. In particular, Anderson (1970) indicated that parallel evolution is due to continuous gene flow among populations at higher levels than that of the species. Went (1971) independently noted that parallel evolution is the rule rather than the exception in plant evolution, and he suggested that a viral mechanism may transfer genes between plant populations representative of different families. Reanny (1974) reviewed ideas on the potential significance of viral mechanisms for lateral gene transfer.

Syvanen (1985, 1994) discussed gene transfer by viruses, bacteria and eukaryotes. He interpreted certain instances of parallel evolution among fossil invertebrates to be the result of cross-species gene transfer (Syvanen, 1985) and included an analysis of parallel evolution among graptolites, which had been discussed originally by Bulman (1933). In another analysis of the fossil record, Jeppsson (1986) suggested cross-species gene transfer by viruses as a mechanism for convergence in five lineages of the extinct fossils, conodonts. Erwin and Valentine (1984) commented on the possibility of linking horizontal transfer of genetic information by RNA-based viruses with the Early Cambrian radiation among metazoans. The fossil record appears to include evidence suggesting that lateral gene transfer has been a significant element in certain radiations. Among extant organisms, Benneviste (1985) proposed retroviruses as a mechanism for gene transfer among mammalian species. His study indicates that lateral gene transfer occurs in multicellular, eukaryotic, sexually reproducing organisms.

32.6   VIROIDS

An interesting discovery in recent years suggests that viroids may be a mechanism for lateral gene transfer. Viroids are infective agents of plants. They are single-stranded RNAs of low molecular weight. The number of nucleotides in viroids ranges from 245 to 370. They do not code for any protein and they are replicated by enzymes found in the plant host. Semancik and Conjero-Tomas (1987) reviewed the properties of viroids and suggested that viroids could be escaped regulatory agents. Sequence homology between small nuclear RNAs and viroids has been observed (Diener, 1979). Furthermore, numerous sequence homologies between Group I introns and viroids have been demonstrated (Dinter-Gottlieb, 1987). Diener (1979) indicated that viroids are recent in origin, but the assumption is made that a viroid originated in the wild and spread from 'a wild carrier species to a susceptible cultivated plant species'. The major observable effect of viroid infection is stunting in the infected host plant.

The feature of stunting of infected hosts may explain the described example of parallel evolution observed in the extinct fossils, graptolites. Parallel evolution appears to be prominent in post near-extinction re-radiation among Late Ordovician graptolites. Stunting is one characteristic of certain survivor and re-radiation taxa.

32.7   DISCUSSION

It is clear that a number of possible scenarios may explain the slender tapered colony morphologies observed among the new post near-extinction, re-radiation graptolite colonies that represent new genera. The explanation of the observed phenomenon as having been the result of lateral gene transfer by viruses or viroids has the merit of directing the experimentalist to search for such agents in present-day evolution. Viroid infection by plants suggests a mechanism for what could have taken place among graptolites in the Late Ordovician near-extinction and re-radiation. Potentially, a viroid-like RNA passed among environmentally stressed graptolites, causing stunting and delayed budding (the long slender zooidal tubes in immature parts of colonies reveal that budding must have been delayed when compared with budding seen among pre-near-extinction colonies) among infected colonies. The RNA was incorporated into the host genome, leading to permanent delayed budding from the initial asexually-produced zooids in infected colonies. In many re-radiation colonies, frequency of budding did increase eventually. In certain of them, the monograptids, frequency of budding remained delayed throughout maturation of the colony. The origins of monograptids have been an unresolvable issue, thus the possibility of monograptid origination as a consequence of lateral gene transfer is plausible and reasonable. The restricted, nutrient-poor environments in which survivor graptolites lived during the interval of mass mortalities would have facilitated transfer of a viroid-like RNA. Based upon fossil occurrences in the near-extinction rock record, individual colonies appear to have been crowded together in small areas. Crowding of colonies would have created opportunities for passage of infecting RNA from colony to colony. The signal for the horizontal transfer of viroids was the oxygenic stress which the graptolites experienced at that time.

Horizontal transfer of genes may be an adaptive response set off by a stressed population of organisms.

To date, much of the discussion of lateral gene transfer has centered on structural genes. The latest Ordovician graptolite near-extinction and subsequent re-radiation may be an example of lateral transfer of genes that control development. The evidence from the latest Ordovician graptolites suggests that lateral transfer of such genes did take place.

In a discussion of the origin of the eukaryotic cell, it was postulated that small RNAs were used to control complex patterns of differentiation in multicellular eukaryotes (Hartman, 1986). These RNAs could possibly be remnants of an ancient RNA-based cell. It is ironic that the study of parallel evolution in the paleontological record may lead to the search for molecular mechanisms of development in multicellular organisms: ontogeny may capitulate to phylogeny (or paleontology).

REFERENCES

Anderson, N.S. (1970) Evolutionary significance of virus infection. *Nature* **227**: 1346–1347.

Beneviste, R.E. (1985) The contribution of retroviruses to the study of mammalian evolution, in *Molecular Evolutionary Genetics*, (ed. R.J. MacIntyre), Plenum Press, New York, pp. 359–385.

Berry, W.B.N. (1987) Phylum Hemichordata (including Graptolithina), in *Fossil Invertebrates*, (eds R.S. Boardman, A.H. Cheetham and A.J. Rowell), Blackwell Scientific Publications., Palo Alto, California, pp. 612–635.

Berry, W.B.N., Wilde, P. and Quinby-Hunt, M.S. (1987) The oceanic non-sulfidic oxygen minimum zone: a habitat for graptolites?. *Geol. Soc. Denmark Bull.* **35**: 103–114.

Berry, W.B.N., Quinby-Hunt, M.S. and Wilde, P. (1995) Impact of Late Ordovician glaciation–deglaciation on marine life, in *Effects of Past Global Change on Life*, Studies in Geophysics series, NRC Board on Earth Sciences and Resources, National Academy Press, Washington, DC, pp. 34–46.

Bulman, O.M.B. (1933) Programme-evolution in the graptolites. *Biol. Revs Cambridge Philos. Soc.* **8**: 311–334.

Bilman, O.M.B. (1970) *Graptolithina with Sections on Enteropneusta and Pterobranchia. Treatise on Invertebrate Paleontology, Part V*, The Geological Society of America, Boulder, Colorado.

Diener, T.O. (1979) *Viroids and Viroid Diseases*, Plenum Press, New York.

Dilly, P.N. (1993) *Cephalodiscus graptoloides* sp.nov., a probably extant graptolite. *J. Zool.* **229**: 69–78.

Dinter-Gottlieb, G. (1987) Possible viroid origin: viroids, virusoids and Group I introns, in *The Viroids*, (ed. T.O. Diener), Plenum Press, New York, pp. 189–204.

Erwin, D.H. and Valentine, J.W. (1984) 'Hopeful Monsters', transponsons, and Metazoan radiation. *Proc. Natl Acad. Sci. USA* **81**: 5482–5483.

Hartman, H. (1986) The origin of the eukaryotic cell. *Speculations in Science and Technology* **7**: 77–81.

Jeppsson, L. (1986) A possible mechanism in convergent evolution. *Paleobiology* **12**: 80–88.

Melchin, M.J. and Mitchell, C.E. (1991) Late Ordovician extinction in the graptoloidea, in *Advances in Ordovician Geology* (eds C.R. Barnes and S.H. Williams), Geological Society of Canada Paper 90-9, pp. 143–156.

Olson, E.C. (1960) Morphology, paleontology, and evolution, in *Evolution after Darwin*, (ed. Sol Tax), University of Chicago Press, Chicago, Illinois, pp. 523–545.

Reanney, D. (1974) Viruses and evolution. *Int. Rev. Cytology* **37**: 21–52.

Rickards, R.B. (1988) Graptolite faunas at the base of the Silurian. *Br. Mus. Nat. Hist. (Geol.) Bull.* **43**: 345–349.

Rickards, R.B., Hutt, J.E. and Berry, W.B.N. (1977) Evolution of Silurian and Devonian Graptoloids. *Br. Mus. Nat. Hist. (Geol.) Bull.* **28**: 1–120.

Rigby, S. (1994) Hemichordate skeletal growth: shared patterns in *Rhabdopleura* and Graptoloids. *Lethaia* **27**: 317–324.

Semancik, D.S. and Conjero-Tomas, V. (1987) Viroid pathogenesis and expression of biological activity, in *Viroids and Viroid-like Pathogens*, (ed. J.S. Semancik), CRC Press, Boca Raton, Florida, pp. 71–126.

Syvanen, M. (1985) Cross-species gene transfer: implications for a new theory of evolution. *J. Theoret. Biol.* **112**: 333–343.

Syvanen, M. (1994) Horizontal gene transfer: evidence and possible consequences. *Ann. Rev. Genet.* **28**: 237–261.

Urbanek, A. (1994) When is a pterobranch a graptolite? *Lethaia* **27**: 324.

Went, F.W. (1971) Parallel evolution. *Taxon* **20**: 197–226.

Larval transfer in evolution

33

Donald I. Williamson

SUMMARY

Larval transfer is the hypothesis that, during the evolution of some animals, genes coded for a body form in one lineage have occasionally been transferred to another to introduce a larval form. Hybridization provides the probable mechanism. It is now suggested that all transferred larvae originated as non-larval forms.

Larval transfers have probably taken place in many phyla. Sponge-crabs (Dromiidae) and echinoderms are used here as examples. The bilateral larva and the radial juvenile of the same individual of the starfish *Luidia sarsi* can coexist for several months after metamorphosis. Such a life history suggests fusion of two lineages in the evolution of this species. Other echinoderms supply rather less spectacular examples, as do all animals with trochophore larvae and urochordates with tadpole larvae. Experimental hybrids between ascidian eggs and sea-urchin sperm are described. More information will come from more hybridizations, including those between less distantly related species. Classifications and cladograms of the animal kingdom that ignore larval transfer are unsound.

33.1 INTRODUCTION

Larval transfer is the hypothesis that, during the evolution of some animals, genes coded for a body form in one lineage have occasionally been transferred to another to introduce a larval form. Hybridization is the probable method of transfer. To Darwin (1859), evolution was an essentially gradual process which occurred within separate lines of descent, but larval transfer is a comparatively rapid form of evolution involving fusion of lineages. To this extent it is, like endosymbiosis

(Margulis, 1981), non-Darwinian, but both these processes are complementary to Darwinian evolution, not substitutes. Lineages have occasionally fused, producing saltations in the otherwise gradual process of lineal evolution. Lineal and synlineal evolution are discussed by Williamson (1996). The book, *Larvae and Evolution* (Williamson, 1992), presents evidence for larval transfer in representatives of eight animal phyla. The present chapter illustrates the principles of larval transfer and updates and expands parts of *Larvae and Evolution*, to which readers are referred for additional references.

Larval transfer has resulted in the spreading of larval forms between lineages, but in some cases the new larva in the second lineage was an animal that matured without metamorphosis in the first. In other words, non-larval forms can also be transferred. This chapter introduces the concept that all transferred body forms can be traced back to non-larval forms.

The larval transfer hypothesis presupposes that some or all of the genes that prescribe embryological and larval features act largely independently of those that prescribe juvenile and adult features. The suggestion that transferred larval forms can all be traced back to non-larval forms goes some way to explaining the postulated independence of the developmental phases. The hypothesis also assumes that, while hybrids between closely related species may show a simultaneous mixture of characters of both parents, those between distantly related species seldom do. The result may be a sequential chimera, in which the larval form of one of the parents may be followed by the juvenile phase of the other, with no mixing of characters within each phase (Williamson, 1991).

33.2   LARVAE

All the larvae considered in this chapter are marine and planktonic, and convergent evolution of larvae has been suggested as the explanation for several of the anomalies discussed below. Convergent evolution results from organisms that are not necessarily closely related adopting similar shapes as an adaptation to life in the same environment or to similar patterns of behavior. There are undeniable cases of convergent evolution in many environments, and the streamlined shapes of many aquatic, fast-moving animals, both living and extinct, provide well-known examples. Streamlining is an adaptation to swimming quickly, but there is no one adaptation to swimming slowly or drifting. The vast variety of shapes found in marine plankton implies that this environment imposes little constraint on form. There is not one planktonic shape but thousands, and, judging by their success, they are all adapted to the environment. In consequence, while it would be rash to say that there are no cases of convergent evolution in marine larvae, such cases must be rare, and the concept of convergent evolution does not explain away all unexpected larval similarities, particularly when the similarities go far beyond external shape. Also, adaptation to the environment should not be used to explain away all unexpected dissimilarities between marine larvae, particularly when the forms occur together, eat a similar range of foods and exhibit similar behavior.

It is generally accepted that larvae and adults have both evolved, but some larvae appear to suggest different classifications, and hence different phylogenies, from the corresponding adults. Under conventional evolutionary theory, which assumes lineal evolution in which all parts of a life history evolved together, such incongruities are attributed to misclassification of the larvae or the adults, or both. Where it can be shown that there is no misclassification, conventional theory must be questioned.

In most cases, the larvae of decapod crustaceans can be classified into the same groups as the adults. This is what we should expect if the larval and adult forms had evolved within one lineage, irrespective of the origin of the larval form. The great majority of brachyuran crabs have very distinctive larvae, quite unlike those of hermit-crabs and other anomurans. Adult sponge-crabs and related families (superfamily Dromioidea) appear to be rather primitive brachyurans, but while some have no zoea larvae, others have larvae resembling those of hermit-crabs. Those seeking to resolve this paradox along conventional lines have suggested:

(a) that the adults are only superficially like brachyuran crabs but are really crab-like hermit-crabs; or
(b) that the larvae are only superficially like those of hermit-crabs, some suggesting that their anomuran features arose by convergent evolution in response to life in the plankton; or
(c) that brachyuran crabs evolved from hermit-crabs and that the dromioids have retained the ancestral larval condition; or
(d) that brachyurans, anomurans and dromioids all evolved from former dromioids with anomuran-type larvae.

Thirty years of study convinced me that the brachyuran characters of adult dromioids and the anomuran characters of the larvae are not superficial, that no other evidence supports the evolution of brachyurans from hermit-crabs, and that there are great difficulties in deriving adult anomurans and larval brachyurans from dromioid ancestors.

Some of the evidence which seems at variance with suggestions (c) and (d) comes from homoloid larvae. The Homoloidea and the Dromioidea are both widely regarded as primitive crabs and are often grouped together in the Dromiacea. Some skeletal characters of adult homoloids appear to be more primitive than those of dromioids, and the oldest homoloids predate those of the oldest dromioids in the fossil record. Homoloid larvae, however, are not anomuran or pre-anomuran, but they show many features from which brachyuran larvae may be derived. Larval and adult characters are thus consistent with the derivation of brachyurans from homoloids, but the larval characters of dromioids seem to preclude their close relationship to either the homoloids or the brachyurans (for references, see Williamson, 1988a, b). The dromioid problem seemed to defy explanation in terms of accepted evolutionary theory, but it could be solved if one accepted the hypothesis that an animal could acquire the larval form of another and that the larval form would be inherited. In this case it is postulated that a dromioid with no larva acquired the larval form of a hermit-crab. Serious consideration of the larval transfer hypothesis began as an attempt to explain the unexpected larvae of dromioids, although it gradually emerged that many comparable or more bizarre developmental anomalies in many phyla may also lend themselves to similar solutions. Now there is a molecular study relevant to the phylogeny of the dromioids and to the hypothesis of larval transfer.

Spears *et al.* (1992) published cladograms based on 18S ribosomal RNA showing the relationships of species representing several families of crabs, two species of dromiids, a hermit-crab, and a caridean shrimp. The study was later extended to cover several more species of Dromiidae (L.G. Abele, personal communication). The 18S rRNA sequences show a considerable gap between the brachyurans (*sensu stricto*) and the only anomuran investigated, the diogenid hermit-crab *Clibanarius vittatus*. Most of the dromiids showed sufficient affinity to the brachyurans to merit inclusion in the Brachyura, *sensu lato*, but the two species of *Hypoconcha* (including one investigated after the published report) showed remarkable similarity to *Clibanarius*. No one had previously questioned the classification of *Hypoconcha* as a dromiid, and it certainly bears little morphological resemblance to a hermit-crab. The 18S gene, of course, tells us nothing of morphology. Relationships inferred by parsimony analysis (Spears *et al.*, 1992) do not entirely preclude the suggestion that the last common ancestor of the Brachyura and the Anomura was a dromioid, but such a conclusion does not explain the brachyuran affinity of most dromiids and the diogenid affinity of *Hypoconcha*. The ancestral form in question could equally well have been an early brachyuran, an early anomuran, or (as I favor) a more generalized reptant decapod.

The results of the 18S rRNA investigation might have been expected to show whether the Dromiidae were brachyurans (*sensu lato*), as their adult characters suggest, or anomurans, as their larval characters suggest, or a mixture, consistent with the suggestion that modern dromiids are descended from a hybrid. In the event, they showed that the Dromiidae as a whole show both brachyuran and anomuran genetic affinities, but no one species showed a mixture. The only conventional explanation that has been offered implies that *Hypoconcha* is a diogenid in disguise, but no one who has examined adult specimens accepts this. A solution in terms of larval transfer has been proposed by Williamson and Rice (1996), who suggested that the dromiids acquired their larvae by two or more separate hybridizations. It was further suggested that the cross which produced the first *Hypoconcha* with larvae might have been the reciprocal of that which gave rise to other dromiids with larvae. Such an explanation presupposes that some genes of hybrids between more or less distantly related species resemble those of one parent only. The evidence, which is limited to one such hybrid (section 33.4), suggests that this might be so.

The hypothesis that the Dromiidae acquired their larvae by hybridization between members of different infraorders, originally based purely on morphology, is certainly strengthened by the molecular evidence linking different members of this supposed family to the same two infraorders. Obviously more investigations are needed into the phylogeny of these and other crab-like animals, including molecular studies looking at more genes and more species. It is hoped that these studies will include members of the Homolidae and *Dorhynchus thomsoni*. The zoeal carapace of this latter species has 14 spines arranged in a pattern described previously only in homolid larvae, but the adult characters and most larval characters, apart from the carapace, place it in the Majidae, subfamily Inachinae. A hybrid origin for this species was proposed by Williamson (1992).

The crustacean examples attributed to larval transfer by Williamson (1992) are now seen as comparatively recent examples of this process, which, it is postulated, has occurred at intervals throughout the evolution of the group. Williamson and Rice (1996) propose that most crustacean larvae, other than megalopas and postlarvae, evolved from transferred larvae. The fact that larval classifications are usually compatible with adult classifications of the Crustacea suggests that the larvae and adults evolved as part of the same lineage for much of the history of the group, but it tells us little about the origins of the larval forms. In the more recent examples ascribed to larval transfer it is possible to identify the probable sources of the transferred larvae, but this is rarely so for older transfers. Crustacean nauplii show features that are quite atypical of most crustaceans, and it is suggested that this larval form was originally transferred from a non-crustacean arthropod. Cambrian arthropods resembling nauplii and zoeas are described, and it is postulated that such animals provided the original sources of modern crustacean larvae.

Echinoderm larvae present an obvious incongruity in having an entirely different form of symmetry from the corresponding adults. The conventional explanation, repeated in numerous textbooks, is that the original echinoderms were bilaterally symmetrical throughout life. The larvae have remained so, but the adults adopted radial symmetry as an adaptation to a sedentary life. This solution seems commendably simple, but it is entirely lacking in supporting evidence. The oldest known echinoderms were not bilaterally symmetrical, and they were not all sessile (Paul, 1979). Proponents of the conventional explanation usually put forward the homolozoans (including carpoids and cinctans) as bilaterally symmetrical Cambrian echinoderms. Paleontologists debate whether the homolozoans were bilaterally symmetrical or asymmetrical, and there is no general agreement on the relationship of this group to the echinoderms, but there seems to be no doubt that there were radially symmetrical echinoderms before, during and after the appearance of the homolozoans. The ancestral bilateral echinoderm remains purely hypothetical.

The assumption that a group of sedentary, bilaterally symmetrical animals could have evolved radial symmetry should also not go unchallenged. There is no precedent for this in any group. All known members of the Bryozoa and the Brachyopoda are sessile and bilateral, and the fossil record suggests that they have always been so.

I postulate that adult echinoderms have always been radially symmetrical. Some early echinoderms might have had ciliated gastrulae, but otherwise they had no larva until one acquired a bilateral larva from a hemichordate by hybridization. Further hybridizations within the echinoderms, together with continuing gradual evolution of both adults and larvae, led to the existing situation.

All echinoderm larvae are enterocoelous deuterostomes (the coelom develops from offshoots of the enteron, and the blastopore does not become the mouth) and this includes those with abbreviated larval development, often with much modified larvae. Other echinoderms have secondarily adopted direct development, with vestigial larvae persisting as embryos, and these also develop as enterocoelous deuterostomes. The phylum, however, also includes a number of brittle-stars and the sea-daisies (Concentricyclomorpha) (Rowe et al., 1988) that show no trace of larvae and are radially symmetrical throughout life. The brittle-stars in question are schizocoelous protostomes (the coelom develops from splits in the mesenchyme, and the blastopore becomes the mouth) and the sea-daisies probably develop in a similar way, though some features are undescribed. For over a century the two types of coelom and mouth formation have been widely regarded as marking a fundamental cleavage in metazoan evolution. If this is so, either the enterocoelous, deuterostomatous echinoderms evolved from quite different ancestors from the schizocoelous, protostomatous forms, or (as I believe) the larvae were added later. This postulated later addition of bilateral larvae could have resulted from the fertilization of eggs of a Carboniferous holothurian with sperm from a hemichordate. Resemblances between the holothurian auricularia larva and the hemichordate tornaria are widely accepted.

The Hemichordata includes the Enteropneusta, some of which have tornaria larvae, *Planctosphaera*, a pelagic animal which resembles a giant tornaria, and the Pterobranchiata, some of which have larvae resembling trochophores. Trochophore larvae are discussed below. *Planctosphaera* is inadequately known, but since it can grow to at least 22 mm (Barnes *et al.*, 1988), it may well mature without metamorphosing. If this is confirmed, perhaps one of its ancestors provided the original source of the larvae of both enteropneusts and echinoderms. Enteropneusts with tornaria larvae probably include descendants of the original hybrid between an enteropneust and a planctosphaeroid and possibly descendants of subsequent hybridizations between enteropneusts with and without larvae. The lack of larvae in some enteropneusts and the occurrence of pterobranch larvae which resemble trochophores suggest that the trochophore is not a typical hemichordate larva, and these observations are consistent with the suggestion that other enteropneusts acquired their larvae by transfer from a planctosphaeroid. A hybrid between an echinoderm and either a planctosphaeroid or an enteropneust with tornaria larvae produced the first echinoderm with bilateral larvae, and subsequent hybridizations within the echinoderms led to the spread of such larvae throughout most of the phylum. The sea-daisies and some brittle-stars, however, have retained the ancestral method of direct development of echinoderms. A trimerous coelom occurs in *Planctosphaera*, tornaria larvae and echinoderm larvae, and there is insufficient evidence to indicate whether the original echinoderm larvae were acquired directly or indirectly from a planctosphaeroid.

33.3   METAMORPHOSIS

If all stages in development have evolved, methods of metamorphosis must also have evolved. The metamorphoses of many animals are difficult or impossible to explain in terms of lineal evolution but suggest fusion of lineages.

Figure 33.1 shows the swimming, bilaterally symmetrical larva and the crawling, radially symmetrical juvenile of the starfish *Luidia sarsi* shortly after their separation (Tattersall and Sheppard, 1934). These two body forms developed from the same egg, and, in this instance, they continued their independent existence for a further three months before the larva eventually died. Clearly the larval *Luidia* does not 'develop into' the juvenile, but how did this situation evolve? What was the evolutionary

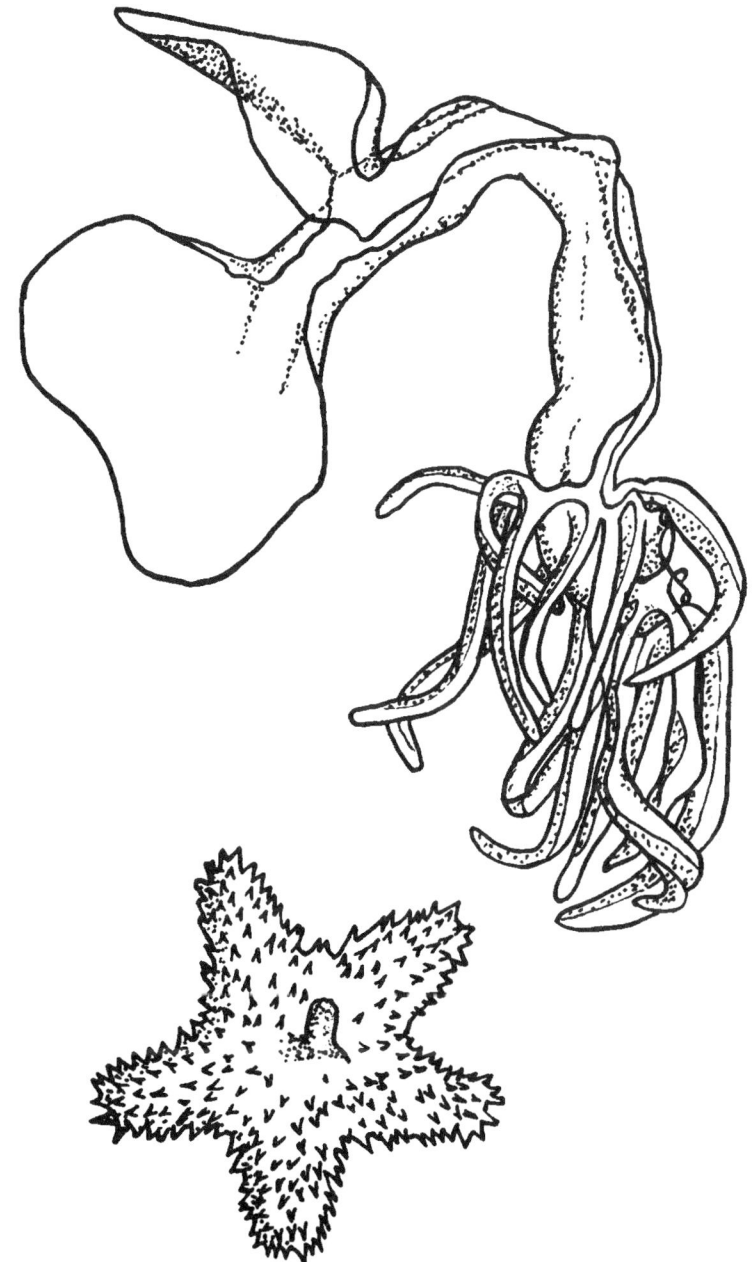

Fig. 33.1 Larva and juvenile of the starfish *Luidia sarsi* shortly after their separation. Larva *c.* 16 mm; juvenile *c.* 3.5 mm. (Redrawn from Tattersall and Sheppard, 1934.)

process that produced an egg with the genetic recipes for two different body forms, each with its own nervous system? The explanation in terms of larval transfer is that *Luidia* is descended from a hybrid between an echinoderm and a hemichordate. It is postulated that the genes specifying the bilateral hemichordate larva and the radial echinoderm juvenile have continued to function independently ever since the hybridization, although both developmental phases must have continued to evolve gradually.

I have yet to see an explanation of the evolution of *Luidia* in terms of natural selection or any other form of lineal evolution. The radial rudiment of the juvenile first appears in a group of undifferentiated cells lining the left mesocoel of the bilateral larva. It incorporates some larval ectoderm as it grows like a parasite within the larva, eventually migrating to the outside before separating. If the original echinoderms were bilateral, natural selection would have acted against the development of a radial quasiparasite. Nature can select only functioning animals and functioning organs, and, as Darwin insisted, natural selection is gradual. It cannot look ahead in evolutionary time and foster a gradually evolving second body form which is not yet capable of independent existence.

The development of *Luidia sarsi* is a particularly clear case of a very widespread phenomenon. The unusual feature is the length of overlap between the larval and juvenile phases, but metamorphosis in echinoderms always involves some overlap. No echinoderm larva 'develops into' the juvenile, but, unlike *Luidia*, the larva usually dies when the juvenile separates. The amount of larval tissue assimilated by the juvenile varies between different groups of echinoderms, and it is greatest in sea-cucumbers, sea-lilies and feather-stars, all of which go through a doliolaria larval stage. The juvenile grows within the barrel-shaped doliolaria and eventually incorporates virtually all of it. In these cases there is no larva left to die, but the radial juvenile nevertheless originated within the bilateral larva. Even when the juvenile uses larval nervous tissue, it is employed to make an entirely different nervous system from that of the larva, and the two systems function independently. Any animal with two distinct body plans, overlapping in time and space and developed from the same egg, seems quite inexplicable in terms of evolution within one lineage. Such a situation, however, is consistent with the suggestion that early echinoderms developed radially from the undifferentiated cells of the gastrula, just as the sea-daisies and some brittle-stars do today. Most modern echinoderms, however, have bilateral larvae, acquired by larval transfer. In such cases, the radial juvenile develops from cells lining part of the larval coelom, in much the same way that ancestral echinoderms developed from cells lining the blastocoel.

Echinoderms are the only animals to change from bilateral to radial symmetry during development, but they are far from the only animals in which the juvenile nervous system is quite distinct from that of the larva. Some annelids, echiurans, sipunculans and molluscs hatch as trochophore larvae, which are planktonic protostomes with a characteristic pattern of cilia. The pilidium larva of nemertines is similar. The wriggling, segmented juvenile annelid can, in some cases, be seen protruding from the late trochophore, and the juvenile nemertine worm can be seen wriggling within the late pilidium. In these cases, the independence of the larval and juvenile body forms is self-evident. Even when the distinction between the larva and the juvenile is less obvious, the nervous systems of the trochophore and the subsequent phase are quite distinct and overlap in time. Ascidians and doliolids are chordates with tadpole larvae, and here again the larval and juvenile nervous systems are quite distinct and overlap. In the development of *Doliolum*, the complete juvenile zooid can be seen at the anterior end of the late tadpole larva. In all these groups, the egg gives rise to two distinct and overlapping nervous systems, a condition which seems quite inexplicable in terms of evolution within one lineage.

Bryozoans cover a considerable range in larval form. Some have only a short planktonic life, and metamorphosis to the adult is by the gradual modification of the larva. Others spend weeks in the plankton either as a trochophore or as a shelled larva called a cyphonautes, and in these cases the subsequent metamorphosis is drastic. The larva, whether trochophore or cyphonautes, has well-developed nervous, locomotory and digestive systems, but it eventually settles and undergoes histolysis to produce two undifferentiated vesicles. No larval organs are preserved, and the juvenile grows from these vesicles. Here again, the juvenile does not 'develop from' the larva. Such a life history seems to contradict most of von Baer's laws of development, particularly those which state that, during development, an animal progresses from general to specialized characters, and departs more and more from the form of other animals. Also, the evolution of animals that have two very different body forms, connected only by undifferentiated cells, is almost as difficult to explain in conventional terms as that of animals with two overlapping body forms.

A larval polyclad turbellarian is known as Müller's larva. It resembles a trochophore, but its metamorphosis is quite different. As it grows, it gradually changes shape, retaining and modifying the larval organs. Garstang (1966), in a poem published posthumously but written before 1922, said:

Johannes Müller's larva is the primal Trochophore
That shows how early worms grew up from fry in days of yore:
No drastic metamorphosis! – each youngster keeps her skin:
Her larval frills are not thrown off, but eaten from within.

Until recently, I agreed with Garstang that Müller's larva was 'the primal trochophore', which, I postulated, had been transferred to other groups by hybridization. But if this larval form had evolved in ancestors of the Turbellaria, why is it today restricted to polyclads, and why do none of the other trochophorate groups show any non-larval polyclad features? I now propose that the primal trochophore was not Müller's larva but a trochophore-like rotifer. *Trochosphaera* and other rotifers show a marked resemblance to trochophores, and this led Hatschek to suggest that rotifers evolved from pedomorphic trochophores (Hyman, 1951). If rotifers antedated trochophores, a polyclad flatworm could have acquired a trochophore larva by hybridizing with a rotifer, and there is 'no drastic metamorphosis' because none is required to transform a rotifer into a flatworm. Other groups acquired trochophores either from rotifers or from groups that had acquired their larvae from rotifers. The transient segmentation in the development of echiurans suggests that members of this group acquired their trochophores from polychaets (Williamson, 1992).

This section has postulated that tornaria larvae and echinoderm larvae were derived from planctosphaeroids, urochordate tadpole larvae from appendicularians, and trochophore larvae fom rotifers. In each case, the original source of the larval form was an animal that matured without metamorphosis, and I propose that all transferred larval forms originated as non-larvae.

33.4   HYBRIDIZATION

Larvae are, by definition, different from adults, and, whatever their origins, there must be mechanisms to prevent the larval and adult forms mixing and to regulate the order and timing of development. In postulating that larval transfer can result from hybridization, it is assumed that these same mechanisms will continue to operate in hybrids.

The urochordate *Ascidia mentula* normally hatches as a tadpole larva, and the sea-urchin *Echinus esculentus* normally hatches as a blastula which develops into a pluteus, with arms supported by calcareous rods. These species are not only very different as adults and larvae but their eggs are also very different. The *Ascidia* egg has a firm but nodulose outer membrane, and the inner egg is surrounded by follicle cells. The *Echinus* egg is surrounded by a transparent, sticky jelly layer, and there are no follicle cells. These differences are very obvious even at low magnifications.

Williamson (1992) described laboratory experiments, carried out in 1988/90, in which eggs of *A. mentula* were cross-fertilized with sperm of *E. esculentus*. Specimens of *A. mentula* were obtained from an old, covered, seawater storage tank at the Marine Laboratory, Port Erin, Isle of Man, and specimens of *E. esculentus* were collected from the open sea nearby. Ripe eggs were gently squeezed from large ascidians, and sperm was obtained from *Echinus* inverted over beakers for up to two hours. Similar procedures were followed in both experiments that produced many hybrid larvae (see below). *Echinus* sperm was obtained one morning and one drop used to fertilize *Echinus* eggs. This culture of developing *Echinus* eggs and all unused eggs were removed from the laboratory. *Ascidia* eggs were obtained the following morning, washed on a mesh in filtered seawater, then counted. Ascidians are hermaphrodites, and eggs obtained by squeezing would sometimes be self-fertilized. Only batches that remained undivided for at least three hours were used further. In these cases, the water was filtered from the eggs, very concentrated *Echinus* sperm was added to the eggs on the mesh and left for 20–30 minutes, and this was followed by repeated washing to remove excess sperm. The eggs were then washed off the mesh, counted again, and kept under observation for cleavage.

In most experiments, the *Ascidia* eggs did not divide. In four cases, however, they did, and hatched next day as ciliated blastulae. These developed into plutei, indistinguishable from those of *Echinus*. The first three experiments yielded one pluteus, over 200 plutei and one pluteus, respectively, but none metamorphosed. In an experiment in 1990, over 3000 hybrid pluteus larvae were obtained and some of these did metamorphose. Sea-urchin rudiments developed in the wall of the left larval mesocoel in over 70 cases. Some died at various stages during the growth of the rudiment but, 37–50 days after fertilization, 20 free-living sea-urchins were counted. Of the four that survived for a year, the two largest were pentaradial, like normal *Echinus*, and the other two were tetraradial. The smallest reached a diameter of 9 mm by the end of the second year, and, without growing further, died at 3 years 3 months. The other three all produced eggs at 3 years 6 months and at intervals over the next 9 months. Some eggs from each were fertilized with wild *Echinus* sperm and produced healthy pluteus larvae. These three hybrid urchins died at the age of 4 years 3 months when the seawater circulation failed. Their respective diameters were 65, 52 and 43 mm.

Of the 3000 hybrid pluteus larvae, those that developed *Echinus* rudiments were only a minority. Many larvae died, but several hundred, after attaining full development as plutei and a length of about 1.0 mm, gradually resorbed their arms and rods to become ciliated spheroids, each with a small protuberance at one end. Each spheroid was about 0.25 mm in diameter and could attach itself by the protuberance, release, and re-attach. This stage was reached in 57–77 days from fertilization, and all died without developing further. We can only speculate whether, under different conditions, they would have been capable of metamorphosing and to what. Several older publications describe the development of *Echinus* and other sea-urchins in detail, but they do not mention spheroids developing from plutei. These can tentatively be regarded as the type of larvae that ascidians had before they acquired tadpoles, but this must remain conjecture until more specimens are obtained.

Hart (1996) investigated nucleotide sequences for the COI mitochondrial gene and the 28S ribosomal gene extracted from tube-feet of three hybrid urchins, and in each case found near identity with wild *Echinus esculentus*, with no ascidian components. Unfortunately this work was carried out some three years after the death of the spheroidal larvae. Hart concluded that, because the COI and 28S sequences agreed with those of *Echinus*, the putative hybrid urchins could not have hatched from ascidian eggs, and he suggested that 'a hermaphrodite used in the cross-fertilization experiments to provide sperm may have provided eggs as well'. *Echinus* eggs are very difficult to filter because they stick to any mesh, but the hermaphrodite hypothesis implies that the eggs I recovered from the mesh were all *Echinus* , that their number tallied with my earlier count of *Ascidia* eggs (> 200 in 1989; > 3000 in 1990), and that they were still undivided, in spite of having been in concentrated *Echinus* sperm for more than 24 hours. In practice, *Echinus* eggs exposed to this concentration of sperm for more than a minute do not hatch, probably as a result of polyspermy. Obviously I reject the hermaphrodite supposition and also Hart's alternative of contaminated cultures. Contamination could not explain the spheroids into which the majority of the hybrid plutei developed. The nucleotide sequences are unexplained, but there is no doubt that the urchins in question came from *Ascidia* eggs.

Clearly the nucleotide sequences of more hybrids should be investigated, but Hart's findings on my hybrid urchins are relevant to the anomalies of the dromiids (mentioned earlier). To summarize: 18S rRNA suggested a wide divergence between brachyuran crabs and hermit-crabs, and while the affinities of dromiids of several genera were with the brachyurans, those of members of one genus were with a hermit-crab (Spears *et al.*, 1992). Incongruities between adult and larval morphologies in the dromiidae had earlier prompted the suggestion that this family is descended from one or more hybrids between dromiids and hermit-crabs (Williamson, 1988a,b, 1992). Unaccountable facts require unconventional solutions, and, to repeat, some genes of hybrids between species that are not closely related may resemble those of one parent rather than both. Such a suggestion seems to fit the dromiids and my *Ascidia* × *Echinus* crosses.

Hatching occurred in about one in four attempts to fertilize eggs of *Ascidia mentula* with sperm of *Echinus esculentus* in 1988/90. In two experiments all the eggs hatched as tadpoles; in another two, some hatched as blastulae that quickly died and others as tadpoles. The *A. mentula* were all from the old storage tank, but this was leaking so badly that it had to be drained in 1990. In over 100 subsequent attempts to repeat this hybridization with *A. mentula* from other sources, no eggs hatched as blastulae. In some experiments some eggs hatched as tadpoles, and in others some eggs divided once or twice, only to re-fuse. Occasional experiments with other species of ascidian gave similar results. In one experiment with *Ciona intestinalis*, two eggs were rapidly changing shape and repeatedly dividing and re-fusing 24 hours after exposure to *Echinus* sperm.

These experiments show that, in spite of many failures, species in different phyla may occasionally hybridize in the laboratory. The eggs may hatch as maternal or paternal larvae, and there is no mixing of maternal and paternal features. Generalizations based on a few attempted hybridizations can clearly be misleading. The experiments also imply that the source of specimens may be important. The only paternal hybrid larvae were from eggs of specimens of *Ascidia mentula* from the old storage tank. This population was growing on a substrate of old, crumbling concrete, and the specimens may well have been genetically similar, originating from one or two tadpole larvae, but the relative importance of these factors is unknown.

An ascidian and an echinoderm were chosen for these experiments because their larvae are very different and could not be confused, but there is no suggestion that representatives of these groups have hybridized in nature. The hypothesis of larval transfer postulates only a few interphylar hybrids and rather more hybridizations between species within the same phylum or the same class. There is evidence that cross-fertilizations between less remotely related species are easier to achieve in the laboratory, and Giudice (1973) lists many within the echinoderms. The regular sea-urchin *Strogylocentrotus purpuratus* and the sand dollar *Dendraster excentricus* have been hybridized on three occasions (Flickinger, 1957; Moore, 1957; Brookbank, 1970). Unfortunately none of these hybrid larvae were kept long enough to see whether they would metamorphose, and all the experiments were conducted before the age of gene sequence analysis. Now is the time to repeat them. Experimental cross-fertilizations between species at all levels of relationship are needed before we can generalize on the morphology and genetics of the larval and adult hybrids.

33.5   CLASSIFICATION

Today many zoologists are trying to reconcile molecular phylogenies of the animal kingdom with classifications based on morphology (for references, see Raff, 1996). Most of these morphological classifications rely heavily on the characters of embryos and larvae, and they assume that types of development have always evolved with adult morphologies. If the forms of embryos and larvae have occasionally been transferred between phyla, the whole basis of these classifications must be questioned. For example, all echinoderm larvae are enterocoelous deuterostomes, and this has been regarded as justification for claiming affinity between echinoderms and chordates, which are also enterocoelous deuterostomes. If it is accepted that the bilateral larval form of echinoderms was a later addition to the radial form of the adults, no such affinity with the chordates can be claimed. Even if this is rejected, the observations that some echinoderms develop directly as schizocoelous protostomes cannot be ignored. Comparable incongruities occur in other phyla, including the gastropod mollusc that develops directly as a deuterostome, while mollusc larvae are protostomes (Williamson, 1992).

It has been claimed (e.g. by Raff, 1996) that molecular systematics, based on 18S rRNA, upholds the grouping of echinoderms and chordates in the deuterostome superphylum, with the associated characters of radial cleavage and enterocoely. I contend that all echinoderms with larvae are descended from a hybrid between a protostome and a deuterostome but some develop directly as schizocoelous protostomes. Brittle-stars of the families Ophiomyxidae and Gorgonocephalidae come in this latter category, and their molecular affinities should be investigated. The molecular findings on the Dromiidae and on my hybrid urchins are not inconsistent with the possibility that some genes of hybrids may resemble those of one parent only, and, at present, we cannot predict which. If this is so, it will greatly complicate molecular phylogeny. This is not to advocate abandonment of this branch of science, but classifications based on one or two genes should be treated with caution and taxonomists should at least consider the possibilty that there may be hybrids in the ancestry of many phyla.

Raff (1996) admits that some groups pose particular problems for molecular phylogeny. One such group is the chaetognaths, which have no trace of a larva, and so presumably their history was not affected by larval transfer. They develop by radial cleavage, deuterostomy and enterocoely, and they are therefore usually classified in the chordate superphylum. Their 18S rRNA, however, apparently excludes them from the deuterostomes (Telford and Holland, cited in Raff, 1996). I suggest that they are true deuterostomes and that the phylogeny of most other so-called deuterostomes is complicated by an ancestral hybrid in each lineage. The molecular phylogeny of groups traditionally regarded as chordates or near-chordates is also unexpected. Turbeville *et al.* (cited in Raff, 1996) found that maximum parsimony analysis linked urochordates (ascidians) with hemichordates rather than with chordates. Raff considered that 'the morphological features ... that link ascidians with chordates are so persuasive that the 18S rRNA-based inference is unacceptable without strong indepenent support'. The main links between chordates and urochordates are through the tadpole larvae of

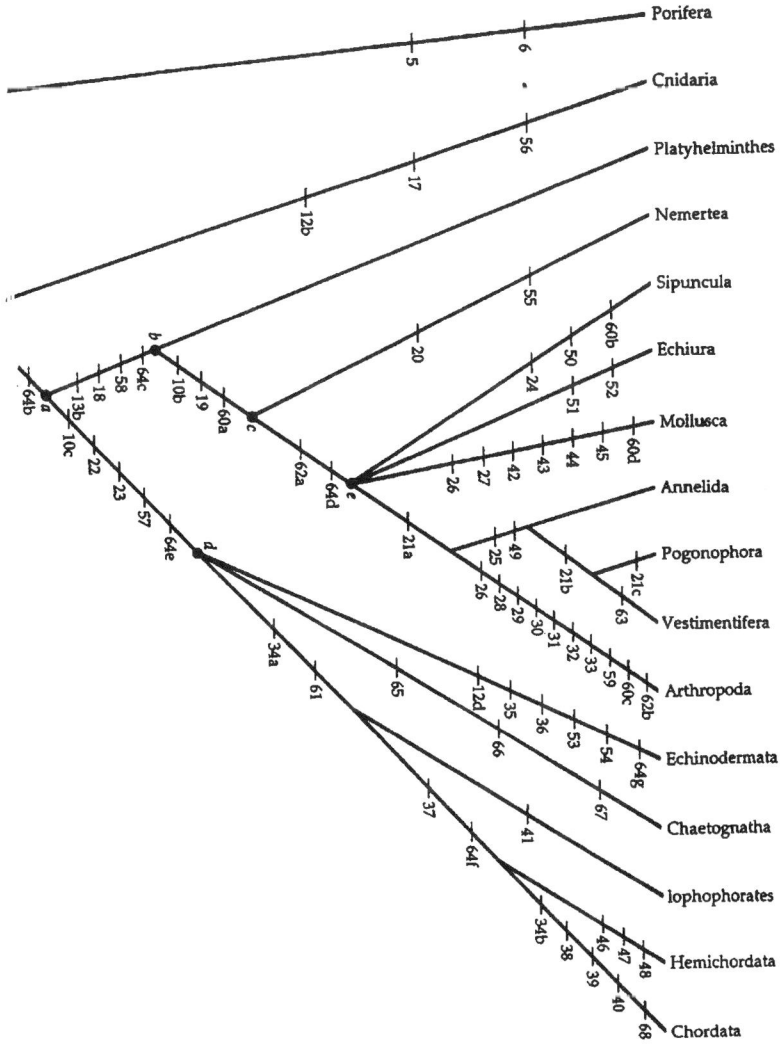

Fig. 33.2 A cladogram depicting the relationships of the major animal phyla. (After Brusca and Brusca, 1990.)

the latter group, and overlapping of nervous systems of tadpoles and ascidians points to this larval form having been transferred from another group. I have also suggested that the tornaria larva of hemichordates was acquired by larval transfer. The 18S rRNA of appendicularians and *Plactosphaera* should be investigated. An appendicularian was the suggested source of the ascidian tadpole, and a planctosphaeroid was the suggested source of the hemichordate tornaria.

The phylogenetic tree of the animal kingdom in Barnes (1980) was reproduced in Williamson (1992) and criticized as a typical example of a dendrogram based largely on larval features without considering the origins of the larvae. Two commentators on Williamson (1992) considered Barnes's tree outdated, and suggested that a computer-generated cladogram based on Hennigian systematics would have been beyond reproach. Figure 33.2 gives an example of just such a cladogram from Brusca and Brusca (1990). It uses the same larval criteria as Barnes (1980) to separate the phyla and arrives at much the same groupings, except that Brusca and Brusca place the lophophorates in the deuterostome superphylum. Such a grouping apparently ignores the trochophore larvae of some bryozoans. Trochophores are protostomes and bryozoans undoubtedly have lophophores, but the metamorphosis from a bryozoan trochophore to the juvenile (described earlier) is incompatible with evolution within one lineage. Hennig accepted Darwin's assumption that all organisms have evolved by lineal descent, but I do not. I agree with both Barnes (1980) and Brusca and Brusca (1990) that methods of mouth formation and coelom formation mark a fundamental cleavage in animal evolution, but the main divisions in the tree and the cladogram reflect differences in larval development. I am convinced that most animal phyla contain larvae which originated in other groups. Such larvae tell us nothing about the evolution of the adults in the group to which they were transferred.

The phylogenetic origins of animal phyla should be re-investigated.

REFERENCES

Barnes, R.D. (1980) *Invertebrate Zoology*, 4th edn, Saunders, Philadelphia, 632 pp.

Barnes, R.S.K., Callow, P. and Olive, P.J.W. (1988) *The Invertebrates: a New Synthesis*, Blackwell Scientific Publications, Oxford, 582 pp.

Brookbank, J.W. (1970) DNA synthesis and development in reciprocal interordinal hybrids of a sea urchin and a sand dollar. *Dev. Biol.* **21**: 29–47.

Brusca, R.C. and Brusca, C.G. (1990) *Invertebrates*, Sinaer Associates, Sunderland, Massachusetts, 922 pp.

Darwin, C. (1859) *The Origin of Species by Means of Natural Selection or the Preservation of Favoured Races in the Struggle for Life*, John Murray, London, 477 pp.

Flickinger, R.A. (1957) Evidence from sea urchin sand dollar hybrid embryos for a nuclear control of alkaline phosphatase activity. *Biol. Bull.* **112**: 21–27.

Garstang, W. (1966) *Larval Forms and other Zoological Verses*, Blackwell, Oxford, 76 pp.

Giudice, G. (1973) *Experimental Biolology of the Sea Urchin Embryo*, Academic Press, New York, 469 pp.

Hart, M.W. (1996) Testing cold fusion of phyla: maternity in a tunicate × sea urchin hybrid determined fron DNA comparisons. *Evolution* **50**(4): 1713–1718.

Hyman, L.H. (1951) *The Invertebrates: Acanthocephala, Aschelminthes, and Entoprocta. The Pseudocoelomate Bilateria*, Vol. III, McGraw-Hill, New York, Toronto and London, 572 pp.

Margulis, L. (1981) *Symbiosis in Cell Evolution*, W.H.Freeman, San Francisco, 419 pp.

Moore, A.R. (1957) Biparental inheritance in the interordinal cross of sea urchin and sand dollar. *J. Experim. Zool.* **135**: 75–83.

Paul, C.R.C. (1979) Early echinoderm radiation, in *The Origin of Major Invertebrate Groups*, (ed. M.R. House), Systematics Association Special Vol. No. 12, Academic Press, London, pp. 443–481.

Raff, R.A. (1996) *The Shape of Life. Genes, Development, and the Evolution of Animal Form*, University of Chicago Press, Chicago and London, 520 pp.

Rowe, F.W.E., Baker, A.N. and Clark, H.E.S. (1988) The morphology, development and taxonomic status of *Xyloplax* Baker, Rowe and Clark ((1986) (Echinodermata, Concentricycloidea), with description of a new species. *Proc. Royal Soc. Lond.* **233**: 431–459.

Spears, T., Abele, L.G. and Kim, W. (1992) The monophyly of brachyuran crabs: a phylogenetic study based on 18S rRNA. *System. Biol.* **41**: 446–461.

Tattersall, W.M. and Sheppard, E.M. (1934) Observations on the bipinnaria of the asteroid genus *Luidia*, in *James Johnstone Memorial Volume*, Liverpool University Press, Liverpool, pp. 25–61.

Williamson, D.I. (1988a) Incongruous larvae and the origin of some invertebrate life-histories. *Prog. Oceanog.* **19**: 87–116.

Williamson, D.I. (1988b) Evolutionary trends in larval form, in *Aspects of Decapod Crustacean Biology*, (eds A.A. Fincham and P.S. Rainbow), Symposia of the Zoological Society of London No. 59, pp. 11–15.

Williamson, D.I. (1991) Sequential chimeras, in *Organism and the Origin of Self*, (ed. A.I. Tauber), Kluwer, Dordrecht, Netherlands, pp. 299–336.

Williamson, D.I. (1992) *Larvae and Evolution: Toward a New Zoology*, Chapman & Hall, New York and London, 223 pp.

Williamson, D.I. (1996) Types of evolution. *J. Nat. Hist.* **30**: 1111–1112.

Williamson, D.I. and Rice, A.L. (1996) Larval evolution in the Crustacea. *Crustaceana* **69**(3): 267–287.

Analysis of animal and plant cytochrome-*c* sequences by probability of character compatibility

34

Christopher A. Meacham and Hyman Hartman

SUMMARY

This chapter presents a technique for judging the amount of homoplasy in a data set that does not require a full parsimony analysis and that does not depend on the fit of data to a particular tree. The technique, Probability Analysis of Character Compatibility, consists of comparing the compatibilities observed for a particular position with the compatibilities expected at random for that position. We compare published data on cytochrome-*c* sequences for plants and animals and show, without deriving a phylogenetic tree, that the plant sequences have a higher level of homoplasy. This result agrees with previously published analyses.

34.1   INTRODUCTION

Phylogenetic reconstruction from molecular sequence data has been attempted with varying degrees of success for different groups of organisms. Under current practice, the suitability of a particular molecular data set for phylogenetic reconstruction is evaluated by first attempting a phylogenetic analysis using the criterion of maximum parsimony. This procedure results in an evaluation of the data set as a whole, and, because each position in the sequence is evaluated in terms of a fit to a particular tree, this procedure does not result in a tree-independent evaluation of the level of homoplasy at each position. Here we briefly present an approach that allows us to evaluate molecular sequence data before attempting a phylogenetic reconstruction. In this chapter we examine two cytochrome-*c* data sets: one from animals and another from plants. We have chosen these two data sets to illustrate this technique because the one from animals has been shown to produce much better trees when analyzed by maximum parsimony whereas the plant data set produces distinctly poorer results (Syvanen *et al.*, 1989). The procedure described below has been applied to problems in the early evolution of angiosperms by Meacham (1994), who showed that parsimony analyses confirmed the ability of this procedure to sort characters into subsets with high and low levels of homoplasy, as judged by the consistency index.

In standard terminology, the amino acid positions in a protein are called **characters**, and the different amino acids found in different taxa at an aligned position define different **character states**. In nearly every real data set examined, all the character states cannot be fitted to any single phylogenetic tree without parallel origins of the same state, reversals of advanced states to more primitive states, or by tree reticulation. This condition of character state parallelism, reversal or reticulation is called **homoplasy**. Homoplasy presents a problem for phylogenetic reconstruction in that the number of optimal trees tends to increase with increased homoplasy and the optimality of these trees tends to fall as measured by the consistency index.

Although early theoretical phylogenetic literature mentions only parallelism or reversal as significant sources of homoplasy, horizontal gene transfer may play a much more important role in the origin of homoplasy than previously suspected. We can view an instance of homoplasy as an occurrence that can be explained, historically, by some combination of parallelism, reversal or horizontal transfer. It is of primary importance to identify clearly and evaluate the instances of homoplasy, and the identification and evaluation of instances of homoplasy are the subjects of this chapter. The review by Syvanen (1994) discusses the theoretical implications of horizontal gene transfer and the problems for phylogeny reconstruction that may arise from it. Katz (1996) presents the evidence for horizontal transfer of the phosphoglucose isomerase gene between different kingdoms of organisms. If horizontal gene transfer has occurred between kingdoms, we can suspect that it may be even more common between more closely related organisms and, at the same time, much more difficult to identify as instances of horizontal transfer. Once occurrences of homoplasy have been discovered and evaluated, we can go on to examine alternative hypotheses of the causes of homoplasy by other procedures in the manner of Katz (1996), Syvanen (1994, and Chapter 25) and Doolittle (Chapter 23).

34.2 COMPATIBILITY ANALYSIS

Compatibility analysis is a technique that focuses on identifying conflict between characters. Compatibility analysis compares individual characters to identify situations where absence of homoplasy in one character requires homoplasy to have occurred in the other. Conflicts between characters, in this sense, provide us with evidence of homoplasy on the actual tree. When we discover that two characters are incompatible, then we know that at least one of the two characters involved a parallelism or reversal during the evolution of the taxa in the data set (because there is no possible tree on which both characters can be fitted without homoplasy). In this way, compatibility may provide a means to infer which characters are more likely to be homoplastic. Meacham (1980, 1981, 1984, 1994) has explained the underlying theory in greater detail. Meacham and Estabrook (1985) have reviewed the earlier literature on character compatibility analysis.

Here we use a probability model for character compatibility to evaluate the individual characters (positions) in protein sequence data. The technique, called PACC (Probability Analysis of Character Compatibility), depends on the idea that a random model of character state assignment across taxa provides a reference point for evaluating compatibilities. Given a particular character, one can count the total number of compatibilities with all the other characters in the data set. This number is called the **compatibility count** for the given character. This is a number that we can calculate for every character (position) in a data set of sequences. A low compatibility count at a particular character indicates many conflicts with other characters in the data set, an indication of homoplasy. To evaluate how probable a particular count is, we compare the count that we have observed with the count that we would obtain at random.

The random compatibility count for a particular character can be estimated by a simple procedure. We start with a data matrix, where the rows correspond to taxa and the columns to characters, which in this study are the positions on the polypeptide chain. For example, given a data matrix for 12 taxa, if at the first position a protein sequence shows five taxa with GLY and seven taxa with LYS at the same position, then a random assignment of GLY to five taxa and LYS to seven taxa at the first position is made. All character state assignments for other characters are kept constant. The compatibility count is calculated for this random character with the other unchanged characters. This randomization is performed many times. We can estimate how likely we are to obtain the observed compatibility count by calculating the frequency with which the compatibility count for the random character matches or exceeds the compatibility count actually observed for the original, unrandomized character. This frequency is referred to as the Frequency of Compatibility Attainment (C_f).

The C_f for a particular character is a measure of the extent to which the character states at that position are distributed independently of the character states of the other characters in the data set. A character whose character states are distributed independently of other characters can be expected to have a C_f of about 0.5. A character whose character state distribution is highly clumped with respect to the character state distributions of other characters will have a higher observed number of compatibilites. Because of the higher observed number, it will be less probable for the randomized character to attain the observed number. So, a character with a clumped character state distribution will have a C_f below 0.5. Characters like this will tend to fit with relatively lower levels of homoplasy to a maximum parsimony tree. A character whose character state distribution is widely scattered (hyperdispersed) with respect to character state distributions of other characters can nearly always achieve the observed number of compatibilities when randomized. A character with a hyperdispersed character state distribution will have fewer observed compatibilities, but will achieve more compatibilities when randomized. A hyperdispersed character will thus have a C_f above 0.5. Characters like this will tend to show higher amounts of homoplasy when fitted to a maximum parsimony tree. In summary, characters whose states are clumped when compared with the states of other characters will tend to have a C_f below 0.5, while characters whose states are hyperdispersed when compared with the states of other characters will tend to have a C_f above 0.5.

Previous studies of the properties of character compatibility analysis have exclusively focused on clique analysis. It is important to emphasize that the method used here is not based on compatible cliques, so that criticisms of clique methods do not apply to this method. There is no requirement that characters that do well by this criterion be members of a clique. This method makes no assumption about the evolutionary processes and is not based on comparison with any postulated evolutionary tree. This method only compares observed patterns of character state distribution with those expected at random, which would occur if character state evolution were unrelated to evolutionary history.

34.3   RESULTS

The sequences analyzed here are those used by Syvanen *et al.* (1989), who showed that although the cytochrome-*c* from animals seems to yield good phylogenetic trees with a high consistency index (0.78) and low levels of homoplasy, the cytochrome-*c* from plants yields trees that show obvious problems in terms of reconstructed phylogenetic relationships as reflected by the lower consistency index (0.59) and much greater homoplasy.

To investigate the difference between the plant and animal data sets using this technique, which does not depend on consistency indices calculated on the basis of fit to a particular tree, we have analyzed the cytochrome-c sequences from animals and from plants as described above, calculating C_f on the basis of 1000 trials for each position. Tables 34.1 and 34.2 give these results. The expected number of compatibilities at random was estimated by averaging the number of compatibilities obtained over the 1000 trials. These tables show that for the cytochrome-c

Table 34.1 The 26 informative animal cytochrome-c amino acid positions ranked by frequency of compatibility attainment, C_f, the frequency with which a randomized position, in 1000 trials, met or exceeded the number of compatibilities actually observed for that position

C_f	Exp.	Obs.	Position number
0.000	6.195	16	44
0.001	6.828	16	104
0.002	13.369	24	13
0.002	3.690	12	103
0.003	11.285	21	3
0.004	13.532	23	11
0.004	14.391	24	60
0.005	7.170	16	15
0.005	13.143	23	12
0.006	13.754	23	83
0.010	17.527	25	88
0.012	6.571	14	62
0.025	3.253	9	89
0.043	17.206	24	46
0.125	5.515	9	100
0.159	12.342	16	50
0.185	8.740	12	92
0.339	17.375	19	70
0.358	16.507	18	9
0.424	12.985	14	33
0.451	17.108	18	54
0.470	17.208	18	36
0.557	10.170	10	58
0.578	17.199	17	28
0.765	13.895	12	47
0.777	17.098	15	35

Exp., expected number of compatibilities at random, estimated by averaging the number of compatibilities obtained at random over the 1000 trials. Obs., observed number of compatibilities for the actual position.

Table 34.2 The 28 informative plant cytochrome-c amino acid positions ranked by frequency of compatibility attainment, C_f

C_f	Exp.	Obs.	Position number
0.000	4.807	10	4
0.008	3.369	8	11
0.016	18.928	25	36
0.053	9.256	13	71
0.097	6.271	10	5
0.190	10.692	14	8
0.235	11.098	14	52
0.240	4.320	6	64
0.306	5.418	7	69
0.371	4.019	5	110
0.518	13.402	14	15
0.642	10.698	10	24
0.654	18.558	18	51
0.665	11.071	10	98
0.674	9.975	9	112
0.787	9.470	8	10
0.791	18.782	17	30
0.824	1.667	1	67
0.846	14.026	11	75
0.846	14.932	12	109
0.854	7.662	5	13
0.856	9.533	7	12
0.866	11.023	9	99
0.908	18.519	15	105
0.924	11.041	7	2
0.931	4.072	2	29
0.974	13.984	8	58
0.989	13.901	7	41

Exp., expected number of compatibilities at random, estimated by averaging the number of compatibilities obtained at random over the 1000 trials. Obs., observed number of compatibilities for the actual position.

sequences of plants examined, only 10 out of 28 informative positions have a C_f of less than 0.5, while for the animals examined 22 out of 26 informative positions have a C_f of less than 0.5.

The C_f histograms shown in Fig. 34.1 reveal substantial differences between the animal and plant data sets. This result suggests that homoplasy is much greater in the plant data set than in the animal data set in accordance with the findings of Syvanen *et al.* (1989), discussed further in Syvanen (1994). The lower consistency indices of trees generated from plant protein sequences was pointed out by Bremer (1988). Also, it has been pointed out by Archie (1989) that the consistency index tends to

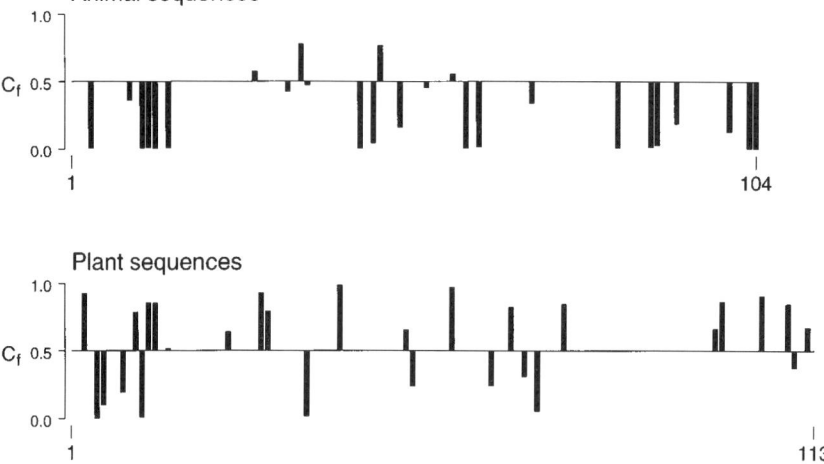

Fig. 34.1 Histograms of C_f values for positions in the animal and plant cytochrome c sequences. A random character would be expected to have a C_f of 0.5. Positions with a C_f of less than 0.5 have more compatibilities with other positions than expected at random; positions with a C_f of more than 0.5 have fewer compatibilities with other positions than expected at random. Only 'informative' positions are plotted.

drop as the number of taxa increases. He pointed out the need for a methodology detecting homoplasy that was independent of the number of taxa in the data set. His method, the Homoplasy Excess Ratio, evaluates the entire data set rather than the homoplasy of individual characters as is done by the procedure we present here.

34.4   DISCUSSION

The PACC technique allows one to judge, by a preliminary analysis that is not influenced by the number of taxa or characters, whether a data set can yield a tree with a high consistency index when subsequently analyzed by maximum parsimony or other methods. PACC allows one to evaluate the level of homoplasy in individual characters before attempting to reconstruct a phylogeny. For some data sets, this method will show that any results obtained should be considered suspect and that other methods explicitly recognizing the possibility of reticulation or rampant homoplasy should be investigated. The ability of this technique to evaluate characters without a full parsimony analysis becomes more significant as larger molecular sequence data sets are gathered. The vast increase in the number of taxa and positions treated in phylogenetic analyses makes it essential to develop techniques, such as PACC, that can be used to focus phylogenetic techniques on positions that are most likely to hold phylogenetic information.

REFERENCES

Archie, J.W. (1989) A randomization test for phylogenetic information in systematic data. *Syst. Zool.* **38**: 239–252.
Bremer, K. (1988) The limits of amino acid sequence data in angiosperm phylogenetic reconstruction. *Evolution* **42**: 795–803.
Darwin, C. (1859) *On the Origin of Species by Means of Natural Selection*, Murray, London.
Estabrook, G.F. (1972) Cladistic methodology: a discussion of the theoretical basis for the induction of evolutionary history. *Ann. Rev. Ecol. Syst.* **3**: 427–456.
Estabrook, G.F. and Landrum, L.R. (1975) A simple test for the possible simultaneous evolutionary divergence of two amino acid positions. *Taxon* **24**: 609–613.
Fitch, W.M. (1975) Toward finding the tree of maximum parsimony, in *Proceedings of the Eighth International Conference on Numerical Taxonomy*, (ed. G.F. Estabrook), W.H. Freeman, San Francisco, pp. 189–230.
Fitch, W.M. (1977) On the problem of discovering the most parsimonious tree. *Am. Nat.* **111**: 223–257.
Fitch, W.M. and Margoliash, E. (1970) The usefulness of amino acid and nucleotide sequences in evolutionary studies. *Evol. Biol.* **4**: 67–109.
Katz, L.A. (1996) Transkingdom transfer of the phosphoglucose isomerase gene. *J. Mol. Evol.* **43**: 453–459.
Le Quesne, W.J. (1969) A method of selection of characters in numerical taxonomy. *Syst. Zool.* **18**: 201–205.
Meacham, C.A. (1980) Phylogeny of the Berberidaceae with an evaluation of classifications. *Syst. Bot.* **5**: 149–172.
Meacham, C.A. (1981) A manual method for character compatibility analysis. *Taxon* **30**: 591–600.
Meacham, C.A. (1984) The role of hypothesized direction of characters in the estimation of evolutionary history. *Taxon* **33**: 26–38.
Meacham, C.A. (1994) Phylogenetic relationships at the basal radiation of angiosperms: further study by character compatibility analysis. *Syst. Bot.* **19**: 506–522.
Meacham, C.A. and Estabrook, G.F. (1985) Compatibility methods in systematics. *Ann. Rev. Ecol. Syst.* **16**: 431–446.
Swofford, D.L. (1990) *PAUP: Phylogenetic Analysis Using Parsimony*, version 3.0h, Illinois Natural History Survey, Champaign.
Syvanen, M. (1994) Horizontal gene transfer: evidence and possible consequences. *Ann. Rev. Genet.* **28**: 237–261.
Syvanen, M., Hartman, H. and Stevens, P.F. (1989) Classical plant ambiguities extend to the molecular level. *J. Mol. Evol.* **28**: 536–544.

Index